A Systemic Perspective on Cognition and Mathematics

Communications in Cybernetics, Systems Science and Engineering

ISSN: 2164-9693

Book Series Editor:

Jeffrey Yi-Lin Forrest
International Institute for General Systems Studies, Grove City, USA
Slippery Rock University, Slippery Rock, USA

Volume 1

A Systemic Perspective on Cognition and Mathematics

Jeffrey Yi-Lin Forrest
College of Economics and Management, Nanjing University of Aeronautics and Astronautics, Nanjing, China; and Department of Mathematics, Slippery Rock University, Slippery Rock, PA, USA

CRC Press
Taylor & Francis Group
Boca Raton London New York

CRC Press is an imprint of the
Taylor & Francis Group, an **informa** business
A BALKEMA BOOK

CRC Press
Taylor & Francis Group
6000 Broken Sound Parkway NW, Suite 300
Boca Raton, FL 33487-2742

First issued in paperback 2018

CRC Press/Balkema is an imprint of the Taylor & Francis Group, an informa business

ISBN-13: 978-1-138-00016-2 (hbk)
ISBN-13: 978-1-138-37262-7 (pbk)

Typeset by MPS Limited, Chennai, India

Library of Congress Cataloging-in-Publication Data

Applied for

Published by: CRC Press/Balkema
P.O. Box 11320, 2301 EH, Leiden, The Netherlands
e-mail: Pub.NL@taylorandfrancis.com
www.crcpress.com – www.taylorandfrancis.com

Visit the Taylor & Francis Web site at
http://www.taylorandfrancis.com

and the CRC Press Web site at
http://www.crcpress.com

To my parents – without their patient and comprehensive home teaching I would not have become who I am today

To my wife and children – Kimberly, Dillon, Alyssa, and Bailey for their love, support, and understanding

Table of contents

Editorial board

About the author

Dr. Yi Lin, also known as Jeffrey Yi-Lin Forrest, holds all his educational degrees (BS, MS, and PhD) in pure mathematics from Northwestern University (China) and Auburn University (USA) and had one year of postdoctoral experience in statistics at Carnegie Mellon University (USA). Currently, he is a guest or specially appointed professor in economics, finance, systems science, and mathematics at several major universities in China, including Huazhong University of Science and Technology, National University of Defense Technology, Nanjing University of Aeronautics and Astronautics, and a tenured professor of mathematics at the Pennsylvania State System of Higher Education (Slippery Rock campus). Since 1993, he has been serving as the president of the International Institute for General Systems Studies, Inc. Along with various professional endeavors he organized, Dr. Lin has had the honor to mobilize scholars from over 80 countries representing more than 50 different scientific disciplines.

Over the years, he has and had served on the editorial boards of 11 professional journals, including Kybernetes: The International Journal of Systems, Cybernetics and Management Science, Journal of Systems Science and Complexity, International Journal of General Systems, and Advances in Systems Science and Applications. And, he is the editor of the book series entitled "Systems Evaluation, Prediction and Decision-Making", and the editor of the book series "Communications in Cybernetics, Systems Science and Engineering", both published by Taylor and Francis with the former since 2008 and later since 2011.

Some of Dr. Lin's research was funded by the United Nations, the State of Pennsylvania, the National Science Foundation of China, and the German National Research Center for Information Architecture and Software Technology.

Professor Yi Lin's professional career started in 1984 when his first paper was published. His research interests are mainly in the area of systems research and applications in a wide-ranging number of disciplines of the traditional science, such as mathematical modeling, foundations of mathematics, data analysis, theory and methods of predictions of disastrous natural events, economics and finance, management science, philosophy of science, etc. As of the end of 2011, he had published nearly 300 research papers and nearly 40 monographs and edited special topic volumes by such prestigious publishers as Springer, Wiley, World Scientific, Kluwer Academic (now part of Springer), Academic Press (now part of Springer), and others. Throughout his career, Dr. Yi Lin's scientific achievements have been recognized by various professional organizations and academic publishers. In 2001, he was inducted into the honorary fellowship of the World Organization of Systems and Cybernetics.

Acknowledgements

This book contains many research results previously published in various sources, and I am grateful to the copyright owners for permitting me to use the material. They include the International Association for Cybernetics (Namur, Belgium), Gordon and Breach Science Publishers (Yverdon, Switzerland, and New York), Hemisphere (New York), International Federation for Systems Research (Vienna, Austria), International Institute for General Systems Studies, Inc. (Grove City, Pennsylvania), Kluwer Academic and Plenum Publishers (Dordrecht, Netherlands, and New York), MCB University Press (Bradford, UK), Pergamon Journals, Ltd. (Oxford), Springer, Taylor and Francis, Ltd. (London), World Scientific Press (Singapore and New Jersey), and Wroclaw Technical University Press (Wroclaw, Poland).

I also like to use this opportunity to express my sincere appreciation to many individuals who have helped to shape my life, career, and profession. Because there are so many of these wonderful people from all over the world, I will just mention a few. Even though Dr. Ben Fitzpatrick, my PhD degree supervisor, has left this material world, he will forever live in my professional works. His teaching and academic influence will continue to guide me for the rest of my professional life. My heartfelt thanks go to Shutang Wang, my MS degree supervisor. Because of him, I always feel obligated to push myself further and work harder to climb high up the mountain of knowledge and to swim far into the ocean of learning. To George Klir – from him I acquired my initial sense of academic inspiration and found the direction in my career. To Mihajlo D. Mesarovic and Yasuhiko Takaraha – from them I was affirmed my chosen endeavor in my academic career. To Lotfi A. Zadeh – with personal encouragements and appraisal words I was further inspired to achieve high scholastically. To Shoucheng OuYang and colleagues in our research group, named Blown-Up Studies, based on their joint works, Yong Wu and I came up with the systemic yoyo model, which eventually led to the completion of the earlier book *Systemic Yoyos: Some Impacts of the Second Dimension* (published by Auerbach Publications, an imprint of Taylor and Francis in 2008) and *Systemic Structure behind Human Organizations: From Civilizations to individuals* (Springer in 2011). To Zhenqiu Ren—with him I established the law of conservation of informational infrastructure. To Gary Becker, a Nobel laureate in economics – his rotten kid theorem has brought me deeply into economics, finance, and corporate governance.

Preface

In 1987, at the invitation of Ronald Mickens I had the honor to join some very well-known scholars from around the globe, such as Wendell Holladay (Vanderbilt University, USA), Saunders Mac Lane (University of Chicago, USA), John Polkinghorne (Cambridge, UK), Robert Rosen (Dalhousie University, Canada), and others, to express my opinion from the angle of systems research on Nobel laureate Eugene P. Wigner's assertion about "the unreasonable effectiveness of mathematics":

> The miracle of the appropriateness of the language of mathematics for the formulation of the laws of physics is a wonderful gift which we neither understand nor deserve. We should be grateful for it and hope that it will remain valid in future research and that it will extend, for better or for worse, to our pleasure even though perhaps also to our bafflement, to wide branches of learning. (Wigner, 1960)

My paper, entitled "A Few Systems-Colored Views of the World" was eventually published in (Mickens, 1990). Continuing this train of thought, based on the observation that when faced with practical problems, new abstract mathematical theories are developed to investigate these problems, predictions made based on the newly developed theories are very "accurate," in 1997, with Hu and Li, I posited the following question:

> Does the human way of thinking have the same structure as that of the material world?

The thinking behind this question is that the accuracy in our predictions, derived from the newly developed theories, is an indication that the human way of thinking is roughly the same as how the material world is constructed. In other words, the human way of thinking and the material world have the same structure. But how can this end be proven? The difficulty here is:

> What is meant by *structure*? And how could I start to uncover this structure?

With these questions in mind, starting in 1995, I, together with colleagues, began my search in two directions, one in theory and the other in applications. Along the line of theory, my colleagues and I analyzed what makes some theories in history successful and long lasting. Evidently, mathematics, as a system of theories, has been representative of such a theory. Its appearance can be traced back to the very beginning of human existence. And as a form of life itself, mathematics still goes on very strongly. Along the line of applications, I notice that some human endeavors have been underway since the dawn of recorded history. So, the following questions arise naturally:

What is the connection here?
What keeps mathematics young and thriving?
And what underlies the human motivation to pursue those difficult and long-lasting endeavors?

Over the years since 1995, I, together with colleagues, have published in theory, in applications, and about interactions between theories and applications. Finally, in 2002, we discovered a structure that underlies all physical and imaginary systems (Wu and Lin, 2002). This structure is given the name of systemic (Chinese) yoyo model in (Lin, 2007). In that publication, the yoyo model is applied to generalize Newton's laws of motion, and to shed new insights on Kepler's laws on planetary motion, Newton's law of universal gravitation, and the study of the three-body problem. And in a number of papers (Lin and Forrest, 2008b, c), this yoyo model is successfully employed to study the rotten kid theorem, initially established by Nobel laureate Gary Becker (1974), in household economics, child labor in labor economics, interindustrial wage differentials in the economics of wage structure, and the interactions between the CEO and the board of directors of corporate governance. By making use of this yoyo model, many facts taken for granted in sociology are shown to either hold true or fail to hold (Lin and Forrest, 2011).

On the other hand, the development of the yoyo model is also motivated by the greater landscape of modern science, where at the same time when disciplines are further and further refined and narrowed, interdisciplinary studies appear in abundance. As science further develops and human understanding of nature deepens, it is discovered that many systems interact nonlinearly with each other and do not satisfy the property of additivity as in $1+1=2$. Their emergent irreversibility and sensitivity cannot be analyzed and understood by using the methodology of the traditional reductionism and quantitative analysis. Facing this challenge, systems research appeared in response of time. The most fundamental characteristic of this new science is the concept of "emergence": The whole that consists of a large number of individuals that interact with each other possesses some unprecedented, complicated properties. That is, the whole is greater than the sum of its parts ($1+1>2$). The basic tasks of systems science consist of the exploration of complexity and the discovery of elementary laws that govern complex systems of different kinds; so by making use of the principles of systems science, one can explain complicated and numerous matters and events of the kaleidoscopic world and provide and take different control measures.

However, even with such a magnificent promise, since the time when the concept of general systems was initially hinted at by von Bertalanffy in the 1920s, the systems movement has experienced over eighty years of ups and downs. After the highs of the movement in the 1960s and 1970s, the heat wave started to cool gradually in the past twenty, thirty years. And in the last few years, what has been witnessed is the fast disappearance of major systems science programs from around the USA (in the rest of the world, the opposite has been true). By critically analyzing the recent and drastic cooling of the systems movement in the USA, one can see the following problems facing systems research:

(1) With over 80 years of development, systems research has mainly stayed at the abstract level of philosophy without developing its own clearly visible and tangible methodology and consequent convincingly successful applications in the traditional disciplines.

This end is extremely important, because systems scientists are humans too, they also need a means to make a living. And, only when some of the unsolvable problems in the traditional science, together with the capability of solving new problems that arise along with the appearance of systems science, can be resolved using systems methods, will systems research be embraced by the world of learning as a legitimate branch of knowledge.

(2) Those specific theories, claimed to be parts of systems science by systems enthusiasts, can be and have been righteously seen as marbles developed within their individual conventional areas of knowledge without any need to mention systems science.

This fact partially explains why programs in systems science across the USA have been disappearing in recent years. In other words, most of the systems science programs that disappeared in recent years from around the USA have been tightly associated with applied fields, such as engineering, management, etc. So, when the funding agencies cut their financial resource allocations in the area of systems science, the corresponding programs in systems science lost their need for existence.

(3) Systems engineering considers various practical systems that do not really have much in common even at the level of abstract thinking.

This end constitutes a real challenge to the systems movement. Specifically, when a systems project or idea needs public support, such as locating reputable reviewers for a research grant application, the principal investigator or the funding agency in general has a hard time to crystallize the base of supporters other than a few possible personal contacts. For this reason, it generally takes a long time, if it ever happens, for a new rising star in systems research to be recognized even within the areas of systems science. Without a steady supply of new blood, the effort of systems research will surely stay on the sideline and remain secondary to the traditional sciences instead of being complementary to the traditional sciences.

Comparing this state of systems science to that of, for instance, calculus, one can see the clear contrast. In particular, systems science does not have a tightly developed system of theory and methods, which newcomers can firstly feel excited about and consequently identify strongly with, and scientific practitioners can simply follow established procedures to produce their needed results. On the other hand, calculus gives one the feeling of a holistic body of thought where each concept is developed on the previous ones in a well accepted playground, the Cartesian coordinate system. Beyond that, calculus possesses a high level of theoretical beauty and contains a large reservoir of procedures scientific practitioners can follow to obtain their desired consequences. In other words, by either further developing calculus and related theories or using these theories, thousands of people from around the world from the generations both before us and after us in the foreseeable future have made and will continue to make a satisfactory living. That in turn feeds back into the livelihood of calculus.

It is also on the basis of this understanding of the history of the systems movement that the yoyo model is introduced (Lin, 2008) to play several crucial roles:

1) It will be the intuition and playground for all systemic thinking in a similar fashion as that of Cartesian coordinate systems in modern science. To this end, of course each of the characteristics of the model needs to be specified in each individual scenario of study.
2) It will provide a means and ground of logic thinking for one to see how to establish models in the traditional fields, the first dimension of science, in an unconventional sense in order to resolve the problem at hand.
3) It will help to produce understandings of nature from an angle that is not achievable from the traditional science alone.

Comparing to the items 1) and 2) above, it is line 3) that will provide the needed strength of life for systems science to survive the current slowdown of development in the USA and to truly become an established second dimension of science, as claimed by George Klir (1985).

Now, in this book, as suggested by the title, I devote my effort to the study of human thought, its systemic structure, and the historical development of mathematics both as a product of thought and as a fascinating case analysis. After demonstrating that systems research constitutes the second dimension of modern science, as suggested by George Klir, I employ the yoyo model, a recent ground-breaking development of the systems research, which has brought forward revolutionary applications of systems research in various areas of the traditional disciplines, the first dimension of science, to establish the basic composites of the mind by carefully analyzing why self-awareness, on which imagination, conscience, and free will emerge naturally, not as previously believed, innately exists. After the systemic structure of thought is factually revealed, I analyze mathematics, as a product of thought, by using the age-old concepts of actual and potential infinities.

As soon as actual and potential infinities are shown to be different by using both counterexample and formal logic reasoning, the existence of infinite sets becomes a problem, and Berkeley's paradox from three hundred years ago reappears. In other words, not only were the past three crises in the foundations of mathematics not resolved as believed in history, but also a new crisis appears. In particular, the well-applied calculus once again becomes a scientific theory without any rigorous foundation. To salvage calculus because of its magnificent successes in all areas of science, I outline an approach of re-establishing calculus without using the concept of limits.

In an attempt to rebuild the system of mathematics, I first provide a new look at some of the most important paradoxes, each of which had played a crucial role in the development of mathematics, by proving what these paradoxes really entail. I then turn my attention to constructing the logical foundation of two different systems of mathematics, one assuming that actual infinity is different than potential infinity, and the other that these infinities are the same.

At this juncture, I hope you will enjoy reading this book. As a conclusion of systems research, we know that as a team, consisting at least of you and me, each member does more than when he is alone. So should you have any comment or suggestion, please contact me at Jeffrey.forrest@sru.edu or Jeffrey.forrest@yahoo.com.

Yi Lin (also known as Jeffrey Yi-Lin Forrest)

Part I

Basics

Chapter 1

Where everything starts

The breadth and diversity the vast amount of the literature on systems science and engineering covers point to a golden opportunity for the next development stage of the systems movement. By comparing the origin where numbers are from and the places systems are seen, this chapter establishes a reason for the 2-dimensional landscape of knowledge. By pointing to the advantage of the additional dimension, it is shown that because of the maturing development of systems research, some age-old problems that have challenged mankind for thousands of years can now hopefully be resolved. After introducing the systemic yoyo model as the common intuition and playground for general systems thinking and reasoning, this chapter looks at how even at the ground root level, traditional science has purposefully ignored the internal structures of objects it studies. That is, to successfully employ numbers and quantities, systemhood has to be ignored. However, the natural world consists mainly of structures and organizations, that is, systems, so that systems researchers should start working on improving almost all, if not all, the basic concepts and elementary procedures of traditional science so that problems of systemhood can be effectively addressed.

By taking the responsibility of reshaping the traditional science, from its very bottom up, as part of the future works of systems research, one will be able to truly make systems science the second dimension of knowledge. In terms of the future development of systems research, calculus should be employed as a reference. And when systems science is further developed, along all the fundamentals operational procedures should also be established in order for practitioners to produce some more or less definite outcomes. That is actually why calculus has been used by generations upon generations. In other words, since its inception over three hundred years ago, that is how calculus has provided a livelihood for millions of people, and in turn these people had helped to create the prosperity of calculus.

This chapter is organized as follows. Section 1.1 emphasizes why systems research constitutes the second dimension of knowledge by looking at the difference between numbers and systems and by introducing the systemic yoyo model, as the intuition and playground of systems science. To make this volume self-contained, Section 1.2 outlines the justifications for why such an intuitive yoyo model holds true for the general system from three different angles: theoretical, empirical, and social. Section 1.3 looks at the problem of what systems thinking and methodology can address satisfactorily and successfully, while pointing to the appearance of new frontiers of knowledge. To help the reader with what will follow in the rest of this book, Section 1.4 lists some of the recent achievements accomplished by using the systemic yoyo model, where an example is given to illustrate how a workplace could be modeled as a spinning field, and why the concept of fluids has become very useful. Section 1.5 concludes this chapter by outlining the organization and highlights of this book.

1.1 SYSTEMS RESEARCH: THE SECOND DIMENSION OF KNOWLEDGE

This section establishes the importance of systems science as the second dimension of knowledge. By revealing the purposeful ignorance of internal structures of objects investigated by traditional science, what follows in the following paragraphs shows the wide range of applicability of systems research not only in new territories created by the research but also in the studies of all the age-old quest of mankind.

1.1.1 A brief history

Since 1924 when von Bertalanffy pointed out that the fundamental character of living things is their organization, and that the customary investigation of individual parts and processes cannot provide a complete explanation of the phenomenon of life, this holistic view of nature and social events has spread over all corners of science and technology (Lin and Forrest, 2011).

Accompanying this realization of the holistic nature, in the past 80 some years, studies in systems science and systems thinking have brought forward brand new understandings and discoveries to some of the major unsettled problems in conventional science (Lin, 1999; Klir, 1985). Due to these studies of wholes, parts, and their relationships, a forest of interdisciplinary explorations has appeared, revealing the overall development trend in modern science and technology of synthesizing all areas of knowledge into a few major blocks, and the boundaries of conventional disciplines have become blurred ("Mathematical Sciences," 1985).

By subscribing to this trend, we can see the united effort of studying similar problems in different scientific fields on the basis of wholeness and parts, and of understanding the world in which we live by employing the point of view of interconnectedness. As tested in the past 80 plus years, the concept of systems has been widely accepted by the entire spectrum of science and technology (Blauberg, et al., 1977; Klir, 2001).

1.1.2 Numbers and systems, what is the difference?

Similar to how numbers are theoretically abstracted, systems can also be proposed out of any and every object, event, and process. For instance, behind collections of objects, say, apples, there is a set of numbers such as 0, 1, 2, 3, ...; and behind each organization there is an abstract system within which the relevant whole, component parts, and the related interconnectedness are emphasized. In other words, when internal structures can be ignored, numbers can be and have been very useful. When internal structures and organizational wholes are attended to, the world consists predominantly of systems.

Historically speaking, the development of traditional science is founded on numbers and quantities and along with systemhood comes systems science. That very fact gives rise to a 2-dimensional spectrum of knowledge, where classical science, which is classified by the thinghood it studies, constitutes the first dimension, and systems science, which investigates structures and organization, forms the genuine second dimension (Klir, 2001).

That is, systems thinking focuses on those properties of systems and associated problems that emanate from the general notion of structures; the division of classical science has been done largely on properties of particular objects. Therefore, systems research naturally transcends all the disciplines of classical science and becomes a force making the existing disciplinary boundaries totally irrelevant and superficial.

The importance of this supplementary second dimension of knowledge cannot be in any way over-emphasized. For example, when studying dynamics in a n-dimensional space, there are difficulties that cannot be resolved within the given space without getting help from a higher-dimensional space. In particular, when a one-dimensional flow is stopped by a blockage located over a fixed interval, the movement of the flow has to cease. However, if the flow is located in a two-dimensional space, instead of being completely stopped, the 1-dimensional blockage would only create a local (minor) irregularity in the otherwise linear movement of the flow (that is how nonlinearity appears (Lin, 2008)). Additionally, if one desires to peek into the internal structure of the 1-dimensional blockage, one can simply take advantage of the second dimension by looking into the blockage from either above or below the blockage. That is, when an extra dimension is available, science will gain additional strength in terms of solving more problems that have been challenging the survival of mankind.

1.1.3 Challenges systems science faces

Even though systems research holds such a strong promise, the systems movement has suffered a great deal in the past 80 some years of development due to the reason that this new science does not have its own speaking language, and does not have its own thinking logic, either. Conclusions of systems research, produced in this period of time, draw either on ordinary language discussions or utilize the conventional mathematical methods, making many believe naturally that systems-thinking is nothing but a clever way of rearranging conventional ideas.

Due to the lack of an adequate tool for reasoning and an adequate language for speaking, systems research has been treated with less significance than they were thought initially since the 1970s when several publications criticized how systems enthusiasts derived their results without sufficient rigorous means (Berlinski, 1976; Lilienfeld, 1978), even though most of the results turned out to be correct if seen with 20-20 hindsight.

Considering the importance of the Cartesian coordinate system in modern science (Kline, 1972), (Wu and Lin, 2002) realizes that the concepts of (sizeless and volumeless) points and numbers are bridged beautifully within the Cartesian coordinate system so that this system plays the role of intuition and playground for modern science to evolve; and within this system, important concepts and results of modern mathematics and science are established.

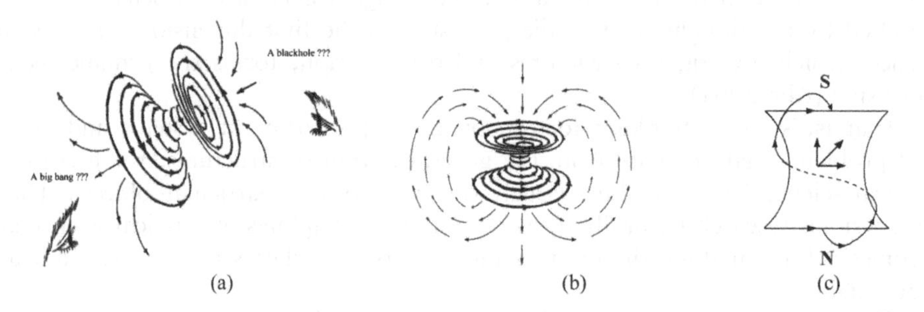

Figure 1.1 The eddy motion model of the general system.

1.1.4 Intuition and playground of systems research

Recognizing the lack of such an intuition and playground for systems science, on the basis of the blown-up theory (Wu and Lin, 2002), the yoyo model in Figure 1.1 is formally introduced by (Lin, 2007) in order to establish the badly needed intuition and playground for systems science.

In particular, on the basis of the blown-up theory and the discussion on whether or not the world can be seen from the viewpoint of systems (Lin, 1988; Lin, Ma and Port, 1990), the concepts of black holes, big bangs, and converging and diverging eddy motions are coined together in the model shown in Figure 1.1.

This model was established in (Wu and Lin, 2002) for each object and every system imaginable. In other words, each system or object considered in a study is a multi-dimensional entity that spins about its either visible or invisible axis. If we fathom such a spinning entity in our 3-dimensional space, we will have a structure as shown in Figure 1.1(a). The side of the black hole sucks in all things, such as materials, information, and energy. After funneling through the short narrow neck, all things are spit out in the form of a big bang. Some of the materials, spit out from the end of big bang, never return to the other side and some will (Figure 1.1(b)). For the sake of convenience of communication, such a structure as shown in Figure 1.1(a), is called a (Chinese) yoyo due to its general shape.

More specifically, what this model says is that each physical entity in the universe, be it a tangible or intangible object, a living being, an organization, a culture, a civilization, etc., can be seen as a kind of realization of a certain multi-dimensional spinning yoyo with either an invisible or visible spin field around it. It stays in a constant spinning motion as depicted in Figure 1.1(a). If it does stop its spinning, it will no longer exist as an identifiable system. What Figure 1.1(c) shows is that due to the interactions between the eddy field, which spins perpendicularly to the axis of spin, of the model, and the meridian field, which rotates parallel to the axis of spin, all the materials returning to the black-hole side travel along a spiral trajectory.

1.2 THE BACKGROUND TO THE SYSTEMIC YOYO MODEL

To make this presentation plausible for the reader, let us briefly look at why such a yoyo model of systems holds true in general. For a more in depth explanation on the theoretical and empirical justifications of this model, please consult the appendices at the end of this book.

1.2.1 The theoretical foundation

In theory, the justification for such a model of general systems is the blown-up theory (Wu and Lin, 2002). In particular, it is shown that each nonlinearity represents a movement in a curvature space. Each acting force, as described by Newton's second law of motion, generally results in spinning movement. What is very important is that forces exist naturally both within and between objects. That is totally different from what Newton's mechanics claim: The acting force is from outside the object that is being acted on. In other words, as long as the density of an object or the organizational structure of a social event is uneven or non-homogenous, the object and the event will possess a fundamental characteristic of spin.

1.2.2 The empirical foundations

At the same time, the systemic yoyo model can also be seen as a practical background for the law of conservation of informational infrastructures. More specifically, based on empirical data from various and diversely different disciplines, the following law of conservation is proposed (Ren, Lin and OuYang, 1998): For each given system, there must be a positive number a such that

$$AT \times BS \times CM \times DE = a \tag{1.1}$$

where A, B, C, and D are some constants determined by the structure and attributes of the system concerned, and T stands for the time as measured in the system, S the space occupied by the system, M and E the total mass and energy contained in the system.

Because M (mass) and E (energy) can exchange to each other and the total of them is conserved, if the system is a closed one, equ. (1.1) implies that when time T evolves to a certain (large) value, space S has to be very small. That is, in a limited space, the density of mass and energy becomes extremely high. So, an explosion (a big bang) is expected. Following the explosion, space S starts to expand. That is, time T starts to travel backward or to shrink. This end gives rise to the well-known model for the universe as derived from Einstein's relativity theory (Einstein, 1983; Zhu, 1985).

In terms of systems, what this law of conservation implies is: Each system goes through such cycles as: ... → expanding → shrinking → expanding → shrinking → ... Now, the geometry of this model of the universe established from Einstein's relativity theory is given in Figure 1.1.

1.2.3 The social foundations

Socially, the multi-dimensional yoyo model in Figure 1.1 is manifested in different areas of life. For example, each human being, as we now see it, is a 3-dimensional realization of such a spinning yoyo structure of a higher dimension.

To this end, consider two simple and easy-to-repeat experiences. For the first one, imagine we go to a swim meet. As we enter the pool area, we immediately fall into a frenzied pot of screaming and jumping spectators, cheering for their favorite swimmers competing in the pool.

Now, let us pick a person standing or walking on the pool deck for whatever reason, either for her beauty or for his strange look or body posture. Magically enough, before long, the person from quite a good distance will feel our stare and she/he will be able to locate us in a very brief moment out of the reasonably sized and frenzied audience.

The reason for the existence of such a miracle and silent communication is because each side is a high dimensional spinning yoyo. Even though we are separated by space and possibly by informational noise, the stare from one side to the other has directed that side's spin field of the yoyo structure into the spin field of the yoyo structure of the other side. That is the underlying mechanism for the silent communication to be established.

For the second example, let us look at the situation of a human relationship. When an individual A has a good impression of another individual B, magically, individual B also has a similar and almost identical impression of A. When A does not like B and describes B as a dishonest person with various undesirable traits, it has been clinically proven in psychology that what A describes about B is exactly who A is himself (Hendrix, 2001). Once again, the underlying mechanism for such a quiet and unspoken evaluation of each other is because each human being stands for a spinning yoyo and its rotational field. Our feelings about other people are formed through the interactions of our invisible yoyo structures and their spin fields.

1.2.4 Unevenness implies spinning

To intuitively see why uneven density and non-homogeneous structure leads to spin, let us imagine that we look at two points $P_1 = (x_1, y_1, z_1)$ and $P_2 = (x_2, y_2, z_2)$ within an entity of density (or structural homogeneity) $\rho = \rho(x, y, z)$, as depicted in Figure 1.2, that are a certain distance away from each other. Then each of these points experiences the act of a gradient force

$$\vec{P}_l \left(\frac{\partial \rho(x_i, y_i, z_i)}{\partial x}, \frac{\partial \rho(x_i, y_i, z_i)}{\partial y}, \frac{\partial \rho(x_i, y_i, z_i)}{\partial z} \right) \tag{1.2}$$

where $i = 1$, 2. Therefore, as long as the density (or structural homogeneity) $\rho = \rho(x, y, z)$ is not constant, a twisting force between the points $P_1 = (x_1, y_1, z_1)$ and $P_2 = (x_2, y_2, z_2)$ is created, leading to a spinning motion. That also explains why solids do not exhibit any clear feature of spin, because the internal density of a solid is constant so that at each chosen point within the solid, the gradient force is $\vec{0}$.

1.3 PROBLEMS ADDRESSABLE BY USING SYSTEMS THINKING AND METHODOLOGY

Because of the availability of the yoyo model, we now have an intuition and playground that is commonly available for systems theorists and practitioners to house

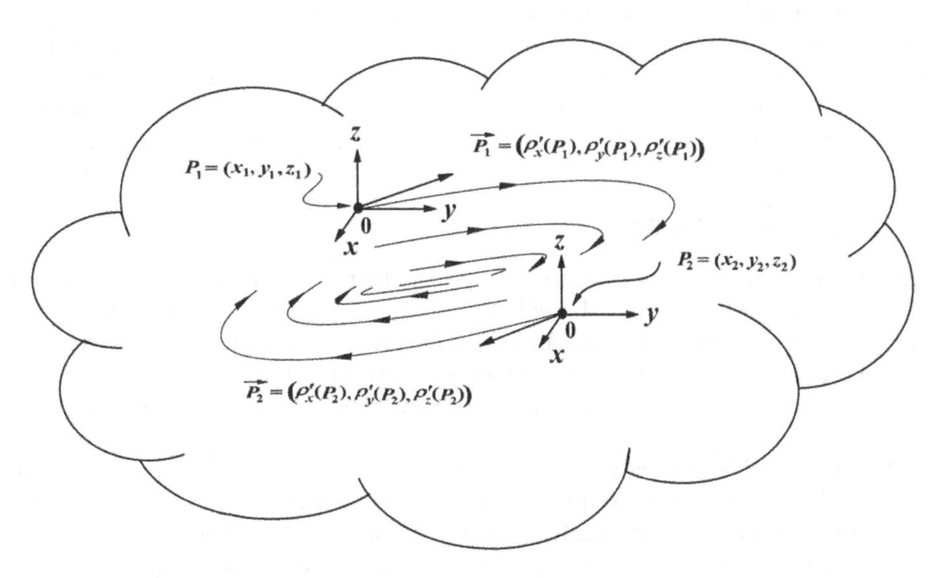

Figure 1.2 How forces naturally exist along with uneven structures.

their abstract reasoning and thinking. That is exactly like what people are accustomed to do with the Cartesian coordinate system when they think about how to resolve a problem in classical science. Now, one of the most important aspects of systems research is what kinds of problems systems science could and should address and attempt to resolve.

Historically speaking, the importance of this question is that only if systems science can provide new and powerful means to resolve at least some of the age-old whys which have challenged mankind since the dawn of recorded history, systems science will have a chance to become firmly recognized as a legitimate branch and the second dimension of knowledge. The never-fading effort of those invested in studying these forever important whys fundamentally signals the relevance of these endeavours to the very survival of the human race.

1.3.1 Kinds of problems systems researchers address

There has been a common belief in the community of systems researchers about what kinds of problems systems science could and should attempt to address (Armson, 2011):

> Systems science resolves problems that are related to systemhood instead of thinghood. In other words, systems science is good at addressing such a problem that when one tries to start looking at one aspect, he realizes that several other aspects of the issue should be first addressed. That is, the issue seems to be messy with neither any beginning nor an ending and is surely not a linear causality. Many factors influence the outcome, while the outcome simultaneously affects the influencing factors.

Although such a belief is in line with how systems science is perceived, it has been quite misleading. In fact, based on this belief, it should be readily recognized that the limitation of modern science is that it ignores all structure-related aspects of issues it addresses. For instance, for the very basic arithmetic fact that $1+1=2$ to hold true, modern science has purposefully given away many structural characteristics of the issue of concern.

1.3.2 Is 1 + 1 really 2?

To answer this question, we have to first specify the relevant meanings. For instance, if $1+1=2$ stands for the fact that when one object is placed together with another object, there will be two objects totally, then the internal structures of the objects have been ignored. It is because when the internal structures of the objects are considered, the togetherness of the objects can be zero, or one, or two, or any other possible natural number.

To illustrate this fact, let us assume that $1+1=2$ represents placing two systems together. That is, the objects we put together now have their individual internal structures. Does that mean consequently we will have two systems totally? The answer is: Not necessarily. In particular, each system now is a spin field, as we have seen in the previous section. When we place two spin fields together, what do we have? Do we really have two spinning fields?

To answer this question, let us consider all possibilities when two spin fields *N* and *M* are placed alongside each other, Figure 1.3. Considering their directions of spin and the divergence and convergence of the fields *N* and *M*, Figure 1.3(a) will produce the outcome of two systems. The spin fields *N* and *M* in Figure 1.3(b) will remain separate while creating a joint rotational field. Due to their convergence, the fields *N* and *M* in Figure 1.3(c) will merge together to become one bigger field. If we look back at Figure 1.3(b), then they are simply the "big-bang" sides of the convergent fields in Figure 1.3(c). So, if the fields *N* and *M* do combine with each other, then there will be only one rotational field resulting. If these fields do not combine into a greater field, then while they stay separate from each other, they also create many smaller fields in the areas between themselves. Similar to the situation in Figure 1.3(a), the fields *N* and *M* in Figure 1.3(d) will stay separate without creating any small field.

Similar analysis indicates that the fields *N* and *M* in Figure 1.3(e, i) will either destroy each other so that no more rotational field results, or if they stay separate, they will create many smaller fields in the zones between *N* and *M*. For the fields *N* and *M* in Figure 1.3(f, h), they either destroy each other or simply stay separate.

In short, $1+1=2$ is a very special case among all the eight possibilities as depicted in Figure 1.3. As a matter of fact, we can conclude the same fact that $1+1=2$ is a very special case as follows:

- When one positron is placed together with an electron, the outcome is nothing; that is $1+1=0$.
- When a woman and man combine into a family, the outcome can be (in theory) any number of people.

In comparison, of course, the previous spin-fields analysis is systemically more scientifically significant than this short version of modelling.

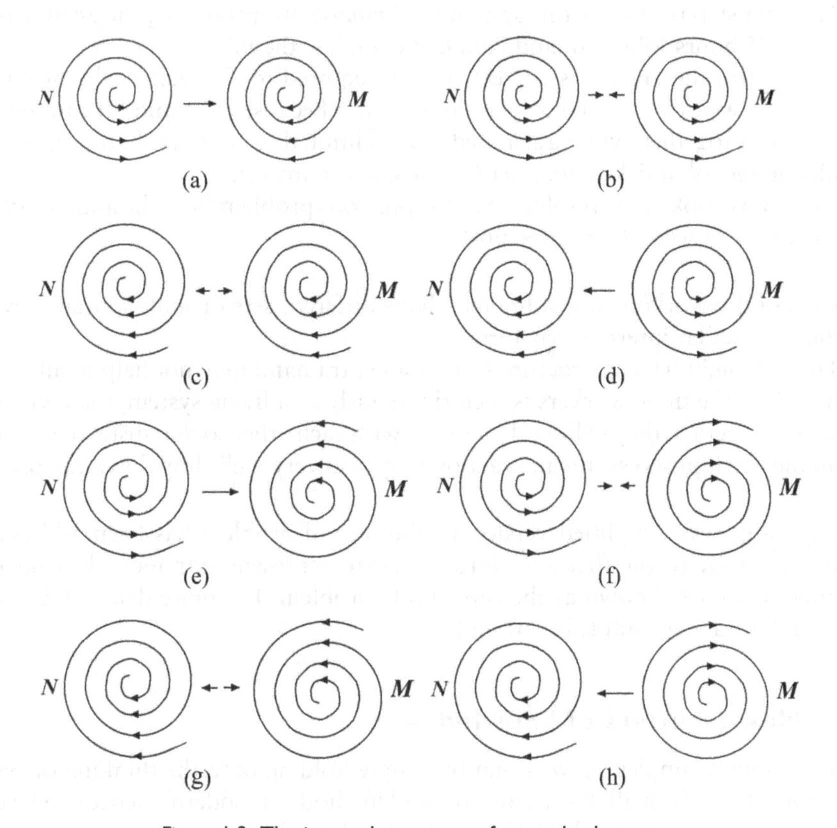

Figure 1.3 The internal structure of a two-body system.

1.3.3 An example of how modern science resolves problems

For our purpose here, let us analyse the following basic algebraic scenario: Suppose that John and Ed together can finish a job in 2 hours, John and Paul together take do the same job in 3 hours, while Ed and Paul together take 4 hours. The question that needs to be resolved is: If John, Ed, and Paul work together, how long do they need to finish the job?

The standard method of solving this algebraic problem at the middle school stage and early years of college level is first assume that John can do the job alone in x hours, Ed alone in y hours, and Paul alone in z hours. Then, the following system is established to describe the relationship among the quantities x, y, and z.

$$\begin{cases} \dfrac{1}{x} + \dfrac{1}{y} = \dfrac{1}{2} \\ \dfrac{1}{x} + \dfrac{1}{z} = \dfrac{1}{3} \\ \dfrac{1}{y} + \dfrac{1}{z} = \dfrac{1}{4} \end{cases} \tag{1.3}$$

The next step is to solve this system mathematically, producing the answer that in $24/13 \approx 1.85$ hours John, Ed, and Paul can complete the job.

To confirm the answer is correct, one is required to go back to the problem to check the answer. Here is one way to do just that: Because John and Ed can finish the job in 2 hours together, with Paul added, the additional manpower should surely make completing the job quicker. So, 24/13 is the correct answer.

Now, if we look at the resolution of the previous problem from the angle of systems thinking, we can see at least three problems:

- Each of the workers in real life may have different sets of skills so that they have their particular internal structures.
- The job might have a structure so that an extra hand may not help at all.
- If each of the three workers is seen righteously as a living system, then when they are put together, they will surely interact with each other so that instead of speeding up the work progress, the interaction may also very well slow down the progress.

Although a greatly simplified version of the second problem has been addressed in operations research, the first and third problems represent extremely difficult issues for modern science, known as the three-body problem. For more detailed discussion to this end, please consult (Lin, 2008b).

1.3.4 New frontiers of knowledge

What previous examples show is that by simply holding onto the thinking of systemhood, almost all, if not all, basic concepts and methods of modern science can be either generalized or improved or both. For example, the following are some of the success stories among many others of thinking and doing along these lines:

1. By applying the yoyo model to the forecasting of near-zero probability disastrous weather conditions, the current prediction accuracy that is commercially available has been greatly improved (Lin, 2008b; Lin and OuYang, 2010).
2. By employing the rotation structure of the yoyo model and the observational facts of the dishpan experiment, we are led to the discovery of the fourth crisis in the foundations of mathematics (Lin (guest editor), 2008a).
3. By relying on the yoyo model as a road map, a sufficient and necessary condition is established for when Becker's rotten kid theorem can hold true, where the theorem is widely used in the research of economics.

In short, our discussion in this section implies that by taking the responsibility of reshaping modern science, from its very bottom up, as part of the future works of systems research, we will be able to truly make systems science the second dimension of knowledge. To individual scholars, what this end means is a widely opening virgin land of knowledge that is awaiting each of us to enter and to pick up those previously unseen "beautiful sea shells" as Newton did over three hundred years ago when modern science was initiated.

What a golden moment of history you and I live in!

1.4 SOME SUCCESSFUL APPLICATIONS OF SYSTEMIC YOYOS MODEL

To show this yoyo model can indeed, as expected, play the role of intuition and playground for systems research, let us list a few past achievements.

1.4.1 Some recent achievements

Most of the recent achievements that are related to the systemic yoyo model and applications are summarized in the following publications:

- *Mystery of Nonlinearity and Lorenz's Chaos* (Lin (guest editor), 1998),
- *Beyond Nonstructural Quantitative Analysis: Blown-Ups, Spinning Currents and Modern Science* (Wu and Lin, 2002),
- *Systemic Yoyos: Some Impacts of the Second Dimension* (Lin, 2008b),
- *Systematic Studies: The Infinity Problem in Modern Mathematics* (Lin (guest editor), 2008a),
- *Digitization of Information and Prediction* (OuYang el al., 2009),
- *Irregularities and Prediction of Major Disasters* (Lin and OuYang, 2010),
- *Research Studies: Systemic Yoyos and Their Applications* (Lin (guest editor), 2010), and
- *Systemic Structure behind Human Organizations: From Civilizations to Individuals* (Lin and Forrest, 2011).

What is most noteworthy of mentioning includes specifically, among others:

- A new figurative analysis method, composed of spin fields, is introduced. After establishing its theoretical and empirical foundations, this method is used to generalize Newton's laws of motion by addressing several unsettled problems in the history.
- On the basis of the characteristics of whole evolutions of converging and diverging fluid motions, the concept of time is revisited using this yoyo model.
- A creative explanation is developed for why planets travel along elliptical orbits, why no external forces are needed for systems to revolve about one another without colliding into each other as described by the law of universal gravitation, and why binary star systems, tri-nary star systems, and even n-nary star systems can exist, for any natural number $n \geq 2$.
- By checking the current state of research of the three-body problem, a brand new method is provided to analyze the movement of three stars, be they visible or invisible.
- By treating materials and objects as non-particle spinning yoyos, the concept of stirring energy is introduced. Then, possible conservations of stirring energies and some fundamental laws of evolution science are established. As a direct application, the long-term technology of urban disaster reduction and prevention of floods, caused by suddenly appearing torrential rains, is established.

- It is shown that in a market of free competition, a concept as fundamental as demand and supply is about mutual reactions and mutual restrictions of different forces under equal quantitative effects. Hence, each economic entity can be naturally modeled and simulated as an economic yoyo or a flow of such yoyos.
- Due to the wide applicable scope and practical significance of the Rotten Kid Theorem (Becker 1974) and the excellent analysis of Bergstrom (1989) concerning the fact that this theorem is not generally true, after developing a yoyo model for various economic entities, a sufficient and necessary condition under which the Rotten Kid Theorem holds true in general in light of whole systems evolution is established.
- As a consequence, an astonishing corollary, which is named the Theorem of Never-Perfect Value Systems, is derived. This theorem states that no matter how a value system is introduced and reinforced, the system will never be perfect.
- By using the systemic yoyo model as the foundation to establish economic yoyos and their spin fields, one can create a qualitative means (also called intuition) to foretell what could happen, how one should construct traditional models to verify the predictions, and why some of the economic observations must be universally true.
- When looking at how people think, one can show the existence of the systemic yoyo structure in human thoughts. So, the human way of thinking is proven to have the same structure as that of the material world.
- After highlighting all the relevant ideas and concepts, which are behind each and every crisis in the foundations of mathematics, it becomes clear that some difficulties in human understanding of nature originated from confusing actual infinities with potential infinities, and vice versa. By pointing out the similarities and differences between these two kinds of infinities, one can then handily pick out some hidden contradictions existing in the system of modern mathematics. Then, theoretically, using the yoyo model, it is predicted that the fourth crisis in the foundations of mathematics has appeared. The value and originality of this application are that it shows the first time in history that human thought, the material world, and each economic entity, share a common structure – the systemic yoyo structure.
- The yoyo model has been successfully employed to improve the forecasting accuracies of such (nearly) zero-probability disastrous weathers as suddenly appearing severe convective weathers, small regional, short-lived fogs and thunderstorms, windstorms and sandstorms, and abnormally high temperature weather conditions. Most of these disastrous conditions have been extremely difficult for the current meteorologists to predict.

1.4.2 How a workplace is seen as a spinning field

At this juncture let us look at how a workplace can be investigated theoretically as such a spinning structure. In fact, each social entity is an objectively existing system that is made up of objects, such as people and other physical elements, and some specific relations between the objects. It is these relations that make the objects emerge as an organic whole and a social system. For example, let us look at a university of higher education. Without the specific setup of the organizational whole (relationships), the people, the buildings, the equipment, etc., will not emerge as a university (system). Now, what the yoyo model says is that each imaginable system, which is defined as the totality of some objects and some relationships between the objects (Lin, 1999), possesses the yoyo structure so that each chosen social system, as a specific system involving people, has its own specific multi-dimensional yoyo structure with a rotational field.

To this end, there are many different ways for us to see why each social entity spins about an invisible axis. In particular, let us imagine an organization, say a business entity. As is well known in management science, each firm has its own particular organizational culture. Differences in organizational cultures lead to varied levels of productivity. Now, the basic components of an organizational culture change over time. These changes constitute the evolution of the firm and are caused by inventing and importing ideas from other organizations and consequently modifying or eliminating some of the existing ones. The concept of spin beneath the systemic yoyo structure of the firm comes from what ideas to invent, which external ideas to import, and which existing ones to eliminate. If idea A will likely make the firm more prosperous with higher level of productivity, while idea B will likely make the firm stay as it has been, then these ideas will form a spin in the organizational culture. Specifically, some members of the firm might like additional productivity so that their personal goals can be materialized in the process of creating the extra productivity, while some other members might like to keep things as they have been so that what they possess, such as income, prestige, social status, etc., will not be adversely affected. These two groups will fight against each other to push for their agendas so that theoretically, ideas A and B actually "spin" around each other. At one moment, A is ahead; at the next moment B is leading. And at yet another moment no side is ahead when the power struggle might very well return to the initial state of the affair. In this particular incidence, the abstract axis of spin is invisible, because no one is willing to openly admit his underlying purpose for pushing for a specific idea (either A or B or other ones).

As for the concept of black hole in a social organization, it can be seen relatively clearly, because each social organization is an input-output system, no matter whether the organization is seen materially, holistically, or spiritually. The input mechanism will be naturally the "black hole," while outputs of the organization the "big bang". Again, when the organization is seen from different angles, the meanings of "black hole" and "big bang" are different. But, together these different "black holes" and "big bangs" make the organization alive. Without the totality of "black holes" and that of "big bangs", no organization can remain physically upright. Other than intuition, to this end the existing literature on civilizations, business entities, and individual humans readily does testify.

1.4.3 Fluids in yoyo fields?

From the previous example, a careful reader might have sensed the fact that in this yoyo model, we look at each system, be it a human organization, a physical entity, or an abstract intellectual being, as a whole that is made up of the 'physical' body, its internal structure, and its interactions with the environment. This whole, according to the systemic yoyo model, is a high dimensional spin field. Considering the fact that the body is the carrier of all other (such as cultural, philosophical, spiritual, psychological, etc.) aspects of the system, in theory the body of the system is a pool of fluid realized through the researcher's sensory organs in the three-dimensional space. The word "fluid" here is an abstract term totalling the flows of energy, information, materials, etc., circulating within going into, and emanating from the body. And in all published references we have searched these flows are studied widely in natural and social science using continuous functions, which in physics and mathematics mean flows of fluids, widely known as flow functions. On the other hand, it has been shown and concluded in (Lin, 2008b; Lin and Forrest, 2011) that the universe is a huge ocean of eddies, which changes and evolves constantly. That is, the totality of the physically existing world can be legitimately studied as fluids.

1.5 ORGANIZATION OF THIS BOOK

This book consists of five parts. Part 1 contains two chapters, where other than this first chapter the second chapter presents the elementary properties of the systemic yoyo model, such as the quark structure of the yoyo fields, formation and classification of these fields, movements of yoyo dipoles, and the laws of state of motion.

Part 2 presents topics of the mind. It contains Chapters 3, 4, and 5. The first chapter shows how and why the human body can be and should be treated as systems by introducing some of the fundamental concepts of systems science, while visiting the underlying principles of Chinese traditional medicine and the ancient classic, Tao Te Ching. Chapter 4 develops the systemic yoyo explanation for the four endowments of man: self-awareness, imagination, conscience, and free will. Different from the existing literature, it is shown that other than self-awareness, all other three endowments are simply consequences of self-awareness. After providing systemic explanation for character, Chapter 5 investigates thought, desire, and enthusiasm. In terms of thought, the chapter considers the formation of thought, how thoughts and desirable outcomes are related, how the mind could be controlled, guided, and directed, the relationship between mental creation and physical materialization, and mental inertia. In terms of desire, the chapter looks at the systemic mechanism of desire, how desire and extraordinary capability are related, and how desire could be artificially installed. Within the topic of enthusiasm, leadership, self-suggestion, and self-control are also considered.

As a case study of the theory contained in Part 2, mathematics, a product of thought, is seen as a systemic flow in Part 3. This part contains Chapters 6–9. In particular, Chapter 6 provides a brief history of mathematics by visiting all those major crises in the foundations of mathematics and how each of them was resolved. Chapter 7 studies the concepts of actual and potential infinite to see how historically they have been seen as different with one representing completed perfect tense and the other presently progressive tense. What is different from the existing literature is that this chapter clearly points out the fact that analytically these two concepts of infinite are also different, leading to contradictory outcomes. Chapter 8 continues what is started in Chapter 7 by addressing particularly whether or not actual and potential infinite are the same. What is discovered includes the following astonishing answers:

1 Yes, actual and potential infinite are the same!
2 No, actual and potential infinite are different!

whereby the first answer is mainly a convention of mathematics in order to obtain certain desirable results, while the second answer is derived using several particular examples. After these contradictory answers are derived, the chapter presents two classes of contradictions that are implicitly hidden in the system of modern mathematics. To this end, in order not to create panic, the chapter is concluded with a revisit of the historical role that paradoxes played in the development of mathematics.

Based on what is presented earlier, Chapter 9 investigates the problem of existence of infinite sets. As it turns out, it is found that such sets cannot exist within the ZFC system as expected by some of the masters of the past. Additionally, the Berkeley paradox, which caused the second crisis in the foundations of mathematics, has found its way back, rendering the established theory of limits useless and inadequate.

Considering the importance of calculus in particular and mathematics in general, Part 4 of this book is devoted to re-establish the system of modern mathematics. Specifically, this part contains three chapters. Chapter 10 presents a recent approach on how to develop the same calculus without using the concept of limits. As indicated by the literature, this attempt can be and has been successful. Chapter 11 provides a symbolic system that is beneficially useful in terms of classifying some of the historically most important paradoxes, such as Plato-Socrates' paradox, the liar's paradox, the barber's paradox, Russell's paradox, the catalogue paradox, and Richard's paradox, into the category of contradicting fallacies. Chapter 12 presents a working plan for splitting the system of modern mathematics into two systems: Mathematics of potential infinities and mathematics of actual infinities.

A section, entitled Afterword, follows right after Part 4. It provides a systemic yoyo model prediction about the future state of mathematics.

This book concludes with Part 5, where two appendices show both theoretically and empirically why the systemic yoyo model holds true for each and every system. In particular, Appendix A provides the theoretical backing of the model on the basis of the blown-up theory by visiting the concept of blown-ups, properties of transitional changes, a new understanding of the quantitative infinity, and the so-called equal quantitative movements and effects. Appendix B lists several empirical evidences for the existence of the systemic yoyo model, such as circulations that exist in fluids, the law of conservation of informational infrastructure, and silent human evaluations that are done daily by all people around the world.

Chapter 2

Elementary properties of systemic yoyos

To prepare the theoretical foundation for the rest of this book, in this chapter, we will study based on the basic attributes of the yoyo model the structure of meridian fields that helps to hold the dynamic spin field of the yoyo model together and some of the elementary properties of this model. For the completeness of the presentation in this book, brief theoretical and empirical justifications on why each and every system can be seen as a rotational yoyo field are provided in the appendices at the end of this book.

This chapter is organized as follows. Section 2.1, which is based on (Lin, 2010, pp. 204–217), investigates the detailed structures of the spin field of systemic yoyos, the important quark structure of the general systemic yoyo model, and the field structures of electrons and positrons. Section 2.2, which is based on (Lin, 2010, pp. 218–232), looks at the two components of yoyo fields, the eddy and meridian fields, how like-poles of spinning yoyos attract each other and opposite poles repel, how systemic yoyos tend to align so that their axes of rotation could be parallel to each other, the interactions between particle yoyos, the intensities of the eddy fields of general particle yoyos, and the movement patterns of yoyo dipoles in yoyo fields. Section 2.3 presents four laws on the states of motion, where the first two laws generalize the first two Newton's laws of motion, while the third and the fourth laws generalize Newton's third law of motion. At the end of this section, we look at why figurative analysis can be valid in scientific reasoning.

2.1 QUARK STRUCTURE OF SYSTEMIC YOYO FIELDS

Each systemic yoyo possesses two aspects: The whole of the underlying yoyo structure can be seen as a particle with internal structure; and the structure of the yoyo is a spinning field extending indefinitely outward with a weakening field intensity as it is further away from the axis of spinning through the narrow neck. That is, each systemic yoyo shows the duality of a particle (as an entity) and a field, which is made up of two directional rotations that reinforce on each other: The (horizontal) eddy field and the (vertical) meridian field. In this section, we look at the phenomenon of yoyo particles.

More specifically, we will list more evidence, such as the Stern-Gerlach experiment, from various areas of scientific research to analyze the spin and the spin field of the general systemic yoyo model. Then, this model is further investigated by using known laboratory observations or theoretical studies in quantum mechanics and particle physics as our supporting evidence. Accordingly, the concepts of mass and charge of a systemic yoyo are introduced. By using the three-jets events as our supporting evidence, we study the quark structures of general systems, showing that each electron and/or positron is a two-quark yoyo with one pole ill formed. After constructing the 3-quark spinning yoyo structures for protons and neutrons, we provide a brand new explanation for the process of β^+ and β^- decays of radioactive elements. Similar to the argument of why n-nary star systems might theoretically exist (Lin, 2007), for any natural number n, we provide an explanation for why m-quark yoyos might exist theoretically, for any natural number m.

2.1.1 The spin of systemic yoyos

When a system is identifiable, it possesses a high-dimensional (either visible or invisible) systemic yoyo structure. Only when the yoyo structure spins about its axis, does the system physically or epistemologically exist. Otherwise, all parts of the system would be captured by other viable systems.

In (Lin, 2007), it is stated, and in (Lin, 2010, pp. 190–203; Lin and Forrest, to appear), it is proved that each physical entity in the universe, be it tangible or intangible object, a living being, an organization, a culture, a civilization, etc., can be seen as a kind of realization of a certain multi-dimensional spinning yoyo with either a visible or invisible spin field around it. Here, by using the word "spin," we try to capture the concept of angular momentum or the presence of angular momentum intrinsic to a body that is either a physical entity or a cognitive and abstract system, as opposed to orbital angular momentum, which is the movement of the object about an external point. For example, the spin of the earth stands for the earth's daily rotation about its polar axis. The orbital angular momentum of the earth is about the earth's annual movement around the sun. In general, a two-dimensional object spins around a center (or a point), while a three-dimensional object rotates around a line called an axis. Here, the center and the axis must be within the body of the object.

Mathematically, the spins of rigid bodies have been understood quite well. If a spin of a rigid body around a point or axis is followed by a second spin around the same point (respectively axis), a third spin results. The inverse of a spin is also a spin. Thus, all possible spins around a point (respectively axis) form a group of mathematics. However, a spin around a point or axis and a spin around a different point (respectively axis) may result in something other than a rotation, such as a translation. Spins around the x-, y-, and z-axes in the 3-dimensional Euclidean space are called principal spins. Spin around any axis can be performed by taking a spin around the x-axis, followed by a spin around the y-axis, and then followed by a spin around the z-axis. That is, any 3-dimensional spin can be broken down into a combination of principal spins.

In astronomy, spin (or rotation) is a commonly observed phenomenon. Stars, planets, and galaxies all spin around on their axes. The speeds of spin of planets in the solar system were first measured by tracking visible features. This spin induces a centrifugal acceleration, which slightly counteracts the effect of gravity and the phenomenon of precession, a slight "wobble" in the movement of the axis of a planet.

In social science spin appears in the study of many topics. That is why we have the old saying: Things always "go around and come around". As an example, the theory and practice of public relations heavily involve the concept of spin, where a person,

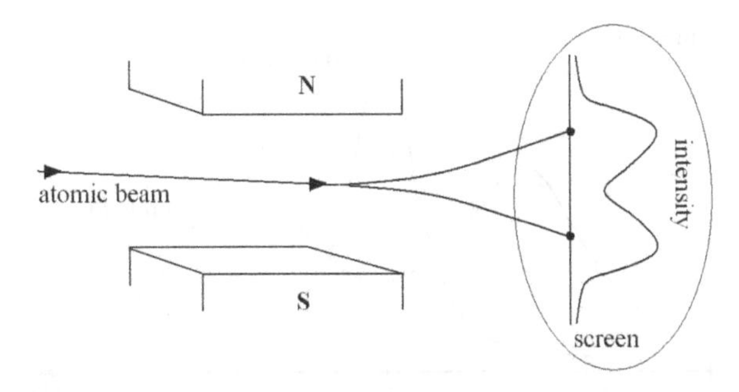

Figure 2.1 The principle underlying the Stern-Gerlach experiment.

such as a politician, or an organization, such as a publicly traded company, signifies his often biased favor of an event or situation. While traditional public relations may rely on creative presentation of the underlying facts, "spin" tends to imply disingenuous, deceptive and/or highly manipulative tactics used to influence public attitudes and opinions (Stoykov and Pacheva, 2005; Bernays, 1945).

In quantum mechanics, spin is particularly important for systems at atomic length scales, such as individual atoms, protons, or electrons. Such particles and the spin of quantum mechanical systems possess unusual or non-classical features. For such systems, spin angular momentum cannot be associated with the concept of rotation precisely, but instead, refers to the presence of angular momentum (Griffiths, 1998).

The electron was the first elementary particle whose spin was detected (Greiner, 1993). The Stern-Gerlach experiment, which was initially designed in 1992, first revealed the spin of electrons. Figure 2.1 shows the principle of this experiment. More specifically, a beam of hydrogen atoms (in the original experiment, silver atoms) is sent through an inhomogeneous magnetic field. The atoms are in the ground state, meaning that the electrons are in the 1s state so that they have no orbital angular momentum, and consequently, should not have any magnetic moment. However, a split of the beam into two components is observed. The distribution of the atoms after passing through the magnetic field is measured. Classically, one would expect a broadening of the beam due to the varying strength of the magnetic field. However, what is observed is that the beam is split into two distinct partial beams. The intensity distribution on the screen is shown qualitatively in Figure 2.2. This double peaked distribution means that the magnetic field moment of the hydrogen atoms cannot orient itself arbitrarily with respect to the magnetic field. Instead, only two opposing orientations of the magnetic moment in the field are possible. This split has its origin in a force

$$\vec{F} = -\nabla(-\vec{M} \cdot \vec{B}) = \nabla(\vec{M} \cdot \vec{B}) = (\vec{M} \cdot \nabla)\vec{B}, \tag{2.1}$$

which acts on the magnetic moment $\vec{M}$ in the inhomogeneous magnetic field $\vec{B}$. This split gives rise to the discovery that each electron has an intrinsic magnetic moment. Because the beam is split into components of equal intensity, it follows that all electrons have a magnetic moment with the same absolute value. They also have two possible orientations: either parallel or antiparallel to the magnetic field. (Note: A

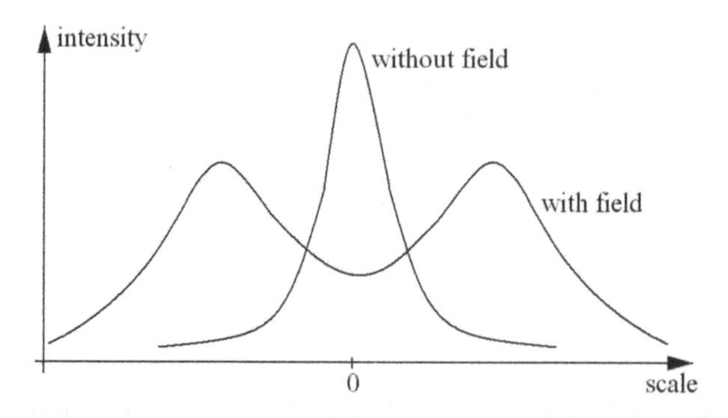

Figure 2.2 Intensity distribution of hydrogen atoms after their transition through an inhomogeneous magnetic field.

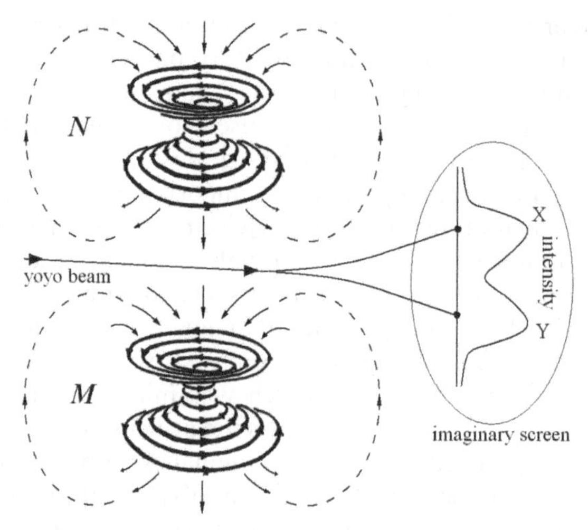

Figure 2.3 A splitting yoyo beam.

careful analysis shows that it is impossible for the magnetic moment to be originated in the nucleus of hydrogen atoms). Further proofs of the existence of electron spin are provided by the multiplet structure of atomic spectra and Einstein and de Hass experiment that was initially designed in 1915. All details are omitted here. Interested readers should refer to (Griffiths, 1998; Greiner, 1993).

Even though the Stern-Gerlach experiment was employed to uncover the intrinsic physical property – spin – of electrons, in terms of our yoyo structure this experiment in fact shows the following systemic property: Let N and M be two yoyo structures of roughly the same scale with N placed on top of M so that there is a gap in between in such a way that N's big bang side is right above the black hole side of M. If the directions of the eddy fields of N and M are the same, then when a beam of tiny yoyos, where compared to the scale of N and M they are tiny, is sent through the meridian field in between the gap, and if these tiny yoyos' eddy spins are in the same direction as that of N and M, then this beam would split into two components as in the Stern-Gerlach experiment. See Figure 2.3 for details.

The basic idea for why the split of the yoyo beam occurs is that due to the harmonic spin directions of all the yoyos involved, each tiny yoyo m, sent through the space between N and M, has a tendency to combine with either N or M. Because of the force pushing this tiny yoyo m through the space between N and M, yoyo m would be influenced by both the upward moving meridian field of N and the downward moving meridian fields of N and M heading into the black hole of M.

For the existence of non-zero spins possessed by elementary particles, which are particles that cannot be divided into any smaller units, such as the photon, the electron, and the various quarks, theoretical and experimental studies have shown that the spin possessed by these particles cannot be explained by postulating that they are made up of even smaller particles rotating about a common center of mass. The spin of composite particles, such as protons, neutrons, atomic nuclei, and atoms, is made up of the spins of the constituent particles and their total angular momentum is the sum of their spin and the orbital angular momentum of their motions around one another. Composite particles are often referred to as having a definite spin, just like elementary particles. This is understood to refer to the spin of the lowest-energy internal state of the composite particles, i.e., a given spin and orbit configuration of the constituents. In short, there has been plenty of evidence in terms of observations, laboratory experiments, and theories to support our systemic model of spinning yoyos.

Also, in theory, we can think of the totality of all materials that can be physical, tangible, intangible, or epistemological and that are contained in a systemic yoyo, if this yoyo is situated in isolation from other yoyo structures. So, the concept of mass for a systemic yoyo can be defined as in conventional physics. Since systems are of various kinds and scales, the universe can be seen as an ocean of eddy pools of different sizes, where each pool spins about its visible or invisible center or axis. At this juncture, one good example in our 3-dimensional physical space is the spinning field of air in a tornado. In the solenoidal structure, at the same time when the air within the tornado spins about the eye in the center, the systemic yoyo structure continuously sucks in and spits out air. In the spinning solenoidal field, the tornado takes in air and other materials, such as water or water vapor on the bottom, lifts up everything it took in into the sky, and then it continuously sprays out the air and water from the top of the spinning field. At the same time, the tornado also breathes in and out with air in all horizontal directions and elevations. If the amounts of air and water taken in by the tornado are greater than those expelled, then the tornado will grow larger with increasing effect on everything along its path. That is the initial stage of formation of the tornado. If the opposite holds true, then the tornado is in the process of dying out. If the amounts of air and water that are taken in and expelled reach an equilibrium, then the tornado can last for at least a while. In general, each tornado (or a systemic yoyo) experiences a period of stable existence after its initial formation and before its disappearance.

Similarly, for the general systemic yoyo model, it also constantly takes in and spits out materials. For the convenience of our discussion in the rest of this chapter and most of this book, we assume that the spinning of the yoyo structures follows the left hand rule 1 below:

Left Hand Rule 1: When holding our left hand, the four fingers represents the spinning direction of the eddy plane and the thumb points to the direction along which the yoyo structure sucks in and spits out materials at its center (the narrow neck).

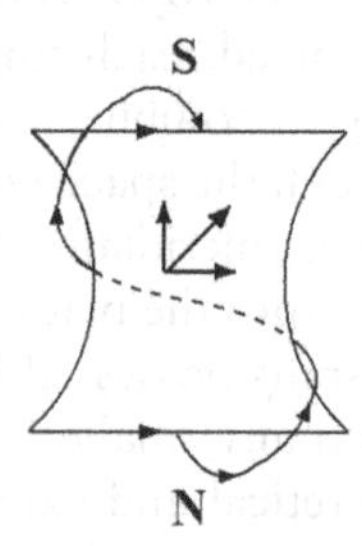

Figure 2.4 Slanted meridians of a yoyo structure.

Note: It can be seen that in the physical world, systemic yoyos do not have to comply with this left hand rule. This fact explains why many scientific theories have their limitations of validity, because they are established for situations satisfying this left hand rule.

As influenced by the eddy (horizontal) spin, the meridian direction movement of materials in the yoyo structure is actually slanted instead of being perfectly vertical (Figure 2.4). In Figure 2.4, the horizontal vector stands for the direction of spin on the yoyo surface toward the reader and the vertical vector the direction of the meridian field, which is opposite of that in which the yoyo structure sucks in and spits out materials. Other than breathing in and out materials from the black hole (we will call it the south pole of the yoyo) and big bang (it will accordingly be named the north pole of the yoyo) sides, the yoyo structure also takes in and expels materials in all horizontal directions and elevations, just as in the case of tornadoes discussed earlier.

In the process of taking in and expelling materials, an outside surface of materials is formed that is mostly imaginary of our human mind; this surface holds most of the contents of the spinning yoyo. The density of materials of this surface decreases as one moves away from the yoyo structure. The maximum density is reached at the center of the eddy field. As the spin field, which is the field of the combined eddy and meridian fields, constantly takes in and expels materials, no clear boundary exists between the yoyo structure and its environment, which is analogous to the circumstance of a tornado that does not have a clear-cut separation between the tornado and its surroundings.

2.1.2 The quark structure of systemic yoyos

To make our presentation complete, let us first look at a well-known laboratory observation from particle physics. The so-called three-jet event is an event with many particles in a final state that appear to be clustered in three jets, each of which consists of particles that travel in roughly the same direction. One can draw three cones from the interaction point, corresponding to the jets, Figure 2.5, and most particles created in the reaction appear to belong to one of these cones. These three-jet events are currently the most direct available evidence for the existence of gluons, the elementary particles that cause quarks to interact and are indirectly responsible for the

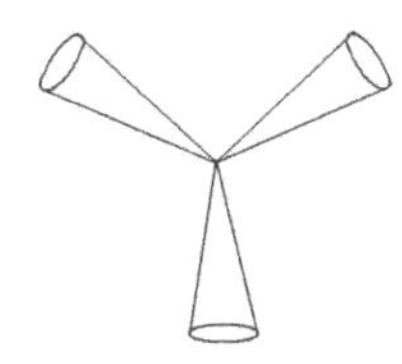

Figure 2.5 A "snapshot" in time and two spatial dimensions of a three-jet event.

binding of protons and neutrons together in atomic nuclei (Brandelik, et al., 1979). For our new model on how protons and neutrons combine, see (Lin, 2010, pp. 218–232; pp. 251–262). Because jets are ordinarily produced when quarks hadronize, the process of the formation of hadrons out of quarks and gluons, and quarks are produced only in pairs, an additional particle is required to explain such events as the three-jets that contain an odd number of jets. Quantum chromodynamics indicates that this needed particle of the three-jet events is a particularly energetic gluon, radiated by one of the quarks, which hadronizes much as a quark does. What is particularly interesting about these events is their consistency with the Lund string model. And, what is predicted out of this model is precisely what is observed.

Now, let us make use of this laboratory observation (the three-jet events) to study the structure of systemic yoyos. To this end, let us borrow the term of quark structure from (Chen, 2007), where it is argued that each microscopic particle is a whirltron, a similar concept as that of systemic yoyos.

Out of the several hundreds of different microscopic particles, other than protons, neutrons, electrons, and several others, most only exist momentarily. That is, it is a common phenomenon for general systemic yoyos to be created and to disappear constantly in the physical microscopic world. According to (Chen, 2007, p. 41) all microscopic systemic yoyos can be classified on the basis of laboratory experiments into two classes using the number of quarks involved. One class contains 2-quark yoyos (or whirltrons), such as electrons, π-, κ-, η-mesons, and others; and the other class 3-quark yoyos (whirltrons), including protons, neutrons, Λ-, Σ-, Ω-, Ξ (Xi) baryons, etc. Here, electrons are commonly seen as whirltrons without any quark. However, Chen (2007) showed that yes, they are also 2-quark whirltrons. Currently, no laboratory experiment has produced 0-quark or n-quark whirltrons, for natural number $n \geq 4$. For the completeness of this paper, let us rewrite Chen's (2007) argument for why electrons have two quarks here using our yoyo structures.

Following the notations of (Chen, 2007), each spinning yoyo, as shown in Figure 2.4, is seen as a 2-quark structure, where we imagine the yoyo is cut through its waist horizontally in the middle, then the top half is defined as an absorbing quark and the bottom half a spurting quark. Now, let us study 3-quark yoyos by looking at a proton P and a neutron N. At this juncture, the three jet events are employed as the evidence for the structure of 3-quark yoyos, where there are two absorbing quarks and one spurting quark in the eddy field. The proton P has two absorbing u-quarks and one spurting d-quark (Figure 2.6), while the neutron N has two spurting d-quarks and one absorbing u-quark (Figure 2.7). In these figures, the graphs (b) are the simplified flow charts with the line segments indicating the imaginary axes of rotation of each

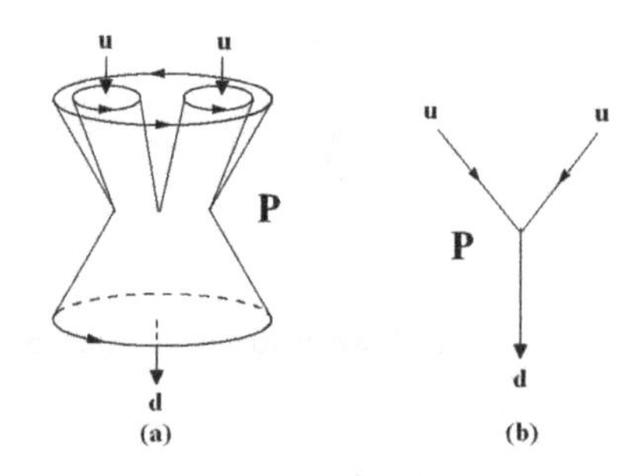

Figure 2.6 The quark structure of a proton P.

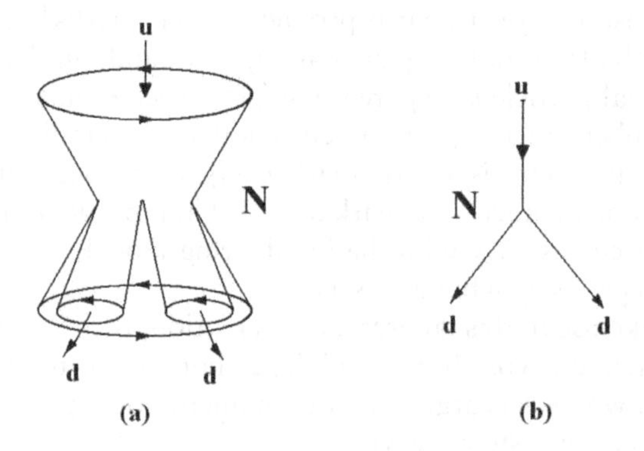

Figure 2.7 The quark structure of a neutron N.

local spinning column. Here, in Figure 2.6, the absorbing u-quarks stand for local spinning pools while together they also travel along in the larger eddy field of which they are a part. Similarly in Figure 2.7, the spurting d-quarks are regional spinning pools. At the same time when they spin individually, they also travel along in the large yoyo structure of the neutron N. In all these cases, the spinning directions of these u- and d-quarks are the same except that each u-quark spins convergently (inwardly) and each d-quark divergently (outwardly).

Different yoyo structures have different numbers of absorbing u-quarks and d-quarks. And, the u-quarks and d-quarks in different yoyos are different due to variations in their mass, size, spinning speed and direction, and the speed of absorbing and spurting materials. This end is well supported by the discovery of quarks of various flavors, two spin states (up and down), positive and negative charges, and colors. That is, the existence of a great variety of quarks has been firmly established.

Now, if we fit Fultz's dishpan experiment to our discussion above by imagining both the top and the bottom of each yoyo as a spinning dish of fluids, then the patterns as observed in the dishpan experiment suggest that in theory, there could exist such a yoyo structure that it has n u-quarks and m d-quarks, where $n \geq 1$ and $m \geq 1$ are arbitrary natural numbers, where each of these quarks spins individually and along with each other in the overall spinning pool of the yoyo structure.

From discussions in (Lin, 2007; Lin, 2010, pp. 190–203), it can be seen that due to uneven distribution of forces, either internal or external to the yoyo structure, the quark structure of the spinning yoyo changes, leading to different states in the development of the yoyo. This end can be well seen theoretically and has been well supported by laboratory experiments, where, for example, protons and neutrons can be transformed into each other. When a yoyo undergoes changes and becomes a new yoyo, the attributes of the original yoyo in general will be altered. For example, when a two-quark yoyo is split into two new yoyos under an external force, the total mass of the new yoyos might be greater or smaller than that of the original yoyo. And, in one spinning yoyo, no matter how many u-quarks (or d-quarks) exist, although these quarks spin individually, they also spin in the same direction and at the same angular speed. Here, the angular speeds of u-quarks and d-quarks do not have to be the same, which is different to what is observed in the dishpan experiment, because in this experiment everything is arranged with perfect symmetry, such as the flat bottom of the dish and perfectly round periphery.

2.1.3 The field structure of electrons

It is common knowledge that when an electric current goes through a wire, a magnetic field is created around the wire where the direction of the current and the direction of the field satisfy the following left hand rule 2:

Left Hand Rule 2: When holding our left hand with the thumb pointing in the direction of the electric current, the four fingers point to the south direction of the magnetic field.

In comparison with the well-established convention, let us cite the conventional right hand rule as follows:

Right Hand Rule 1: When holding our right hand with the thumb pointing in the direction of the electric current, the four fingers point to the north direction of the magnetic field. In the rest of this chapter, when no specific statement is given, we use the Left Hand Rule 2.

With our systemic yoyo model in place, the previous common knowledge in fact implies that the yoyo structure of each electron spins clockwisely and the electron possesses both positive and negative electric fields (the naturally existing meridian field, one side of which goes into the axis of spin of the electron yoyo and the other goes out) with the former being latent (invisible) and the latter explicit (visible). In other words, the big bang side (the north pole) of the electron yoyo is visible, while the black hole side (the south pole) is invisible.

Now, let us look at the process of decay of a radioactive element, where a neutron in the nucleus is converted to a proton with an electron released. In the process when the neutron N (Figure 2.8) is (β^-) decayed into a proton P and an electron E, one of the d-quarks in N is split off. Without loss of generality, let us assume that d_2-quark is split off from N. When d_2-quark leaves N, it takes a piece of the u-quark (the black hole side) with it. Right before the piece of the u-quark leaves the neutron N (Figure 2.9(a)), the original even flow of materials in the black hole side (the south pole) of N is greatly affected. In particular, in Figure 2.9(a) no more material is supplied into area A so

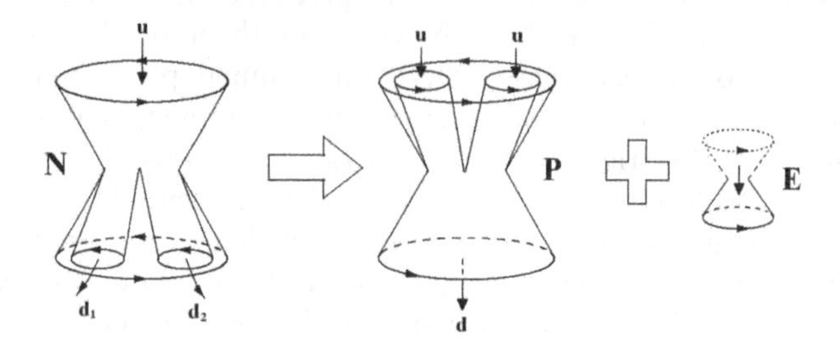

Figure 2.8 A neutron N is decayed into a proton P and an electron E.

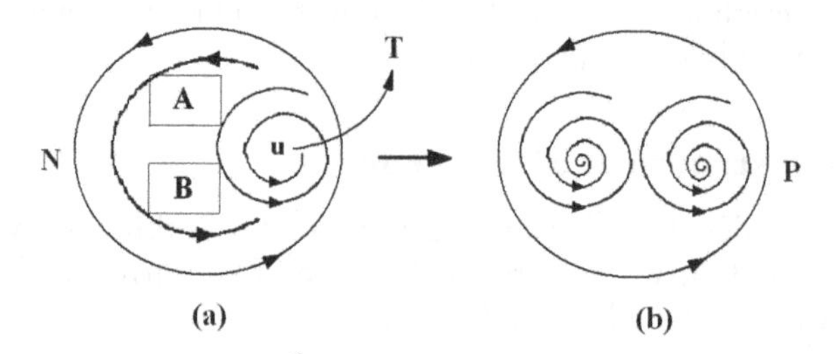

Figure 2.9 The mechanism for the formation of two converging eddy leaves.

that a relative vacuum is created, while due to conflicts in spinning directions, area B is jammed with extra materials. At the same time when the original spinning flows plus the accumulated strength of pushing in area B throws the newly formed regional u-quark out of the eddy field of N along the direction of the arrow, the congestion in area B and the vacuum in area A establish a new u-quark on the left-hand side to these areas. So, the flow pattern in Figure 2.9(b), as seen from above the black hole side (the south pole) of the original yoyo N, is formed, where the local eddy motion on the right-hand side is the residual pool left behind by the departed u-quark in Figure 2.9(a). This illustration also provides an explanation for why the top part of E yoyo (Figure 2.8) is invisible, because in comparison to the bottom part the top part of the structure is less organized.

Similarly, when a radioactive element is (β^+) decayed, the proton P in the nucleus is converted into a neutron N and a positron E^+, Figure 2.10a, where one of the u-quarks, say u_1-quark on the right in P, separates itself from the black hole side (the south pole) of the proton P. When u_1-quark leaves P, it takes a piece of the d-quark of the big-bang side (the north pole) with it. Right before the local eddy field in Figure 2.10b(a) leaves the big bang side of P, the original even, diverging flow of materials in the big bang side (the north pole) of P is greatly affected. In particular, in area A in Figure 2.10b(a), the normal supply of materials is cut off by the formation of the local eddy d, while area B experiences a jam with extra materials accumulated due to conflicts in spinning directions between coming into this region from the left side as the original flow continues and the appearance of the local eddy field d. At the same time when the original spinning flows together with the strengthening pushing effect in area B throw the weakly formed d-quark out of the diverging eddy field of P along the

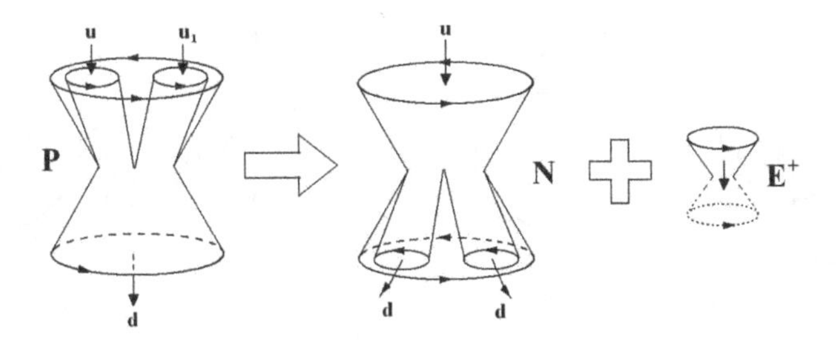

Figure 2.10a A proton P is decayed into a neutron N and a positron E^{+}.

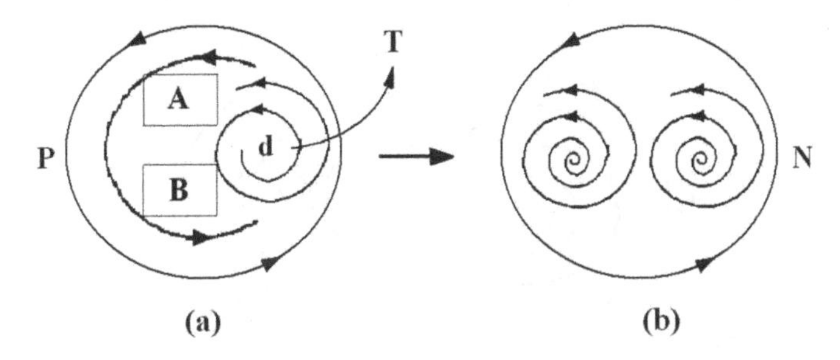

Figure 2.10b The mechanism for the formation of two diverging eddy leaves.

direction of the arrow T, the congestion in area B and the vacuum in area A establish a new d-quark in the region left to these areas. So, the flow pattern in Figure 2.10b(b), as seen from above the black hole side (the south pole) of the original yoyo P, is formed, where the local eddy field on the right-hand side is what is left behind by the d-quark in Figure 2.10b(a). Since the bottom half of the positron E^{+} in Figure 2.10a is poorly formed when compared to the top half, this yoyo E^{+} only shows visibly the top half with the bottom hidden.

What is analyzed above also provides an explanation for why in both β^{-} and β^{+} decays of a radioactive element, the newly formed proton P and the newly created neutron N repel their accompanying electron E and positron E^{+} like a canon shooting a cannonball. In particular, the force that propels the u-quark in Figure 2.9(a) and the d-quark in Figure 2.10b(a) is at least as powerful as

$$F = 2g\frac{C_M c_m}{R^2}, \tag{2.2}$$

where R stands for the distance between the center of the u-quark in Figure 2.9(a) or the d-quark in Figure 2.10b(a) and area B, C_M the mass of the materials in area B, c_m the mass of the u-quark or the d-quark, and $g \geq 7.55 \times 10^{28}$ the action constant of the nuclear field (Chen, 2007, p. 60). In other words, the u-quark and the d-quark leave their mother fields at a speed more than twice that of the spinning speed of the mother fields. Because the force that bonds the protons and neutrons in a nucleus works within the distance of about 10^{-15} m (Chen, 2007, p. 60), from equ. (2.2) we can establish

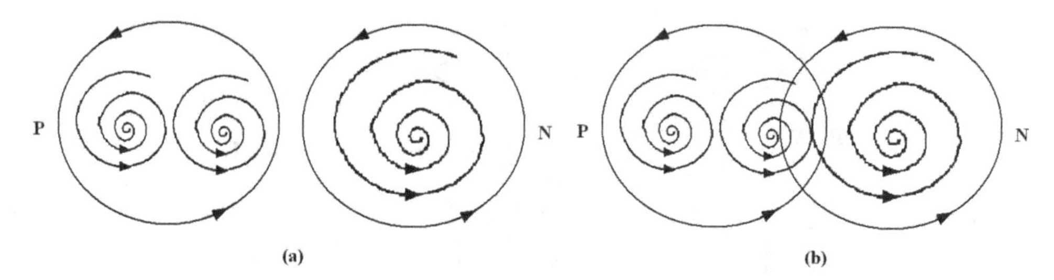

Figure 2.11 Interactions between the converging eddy fields of P and N.

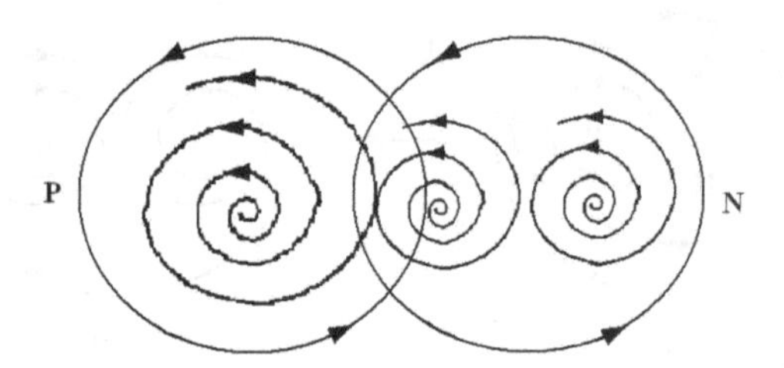

Figure 2.12 Interaction between the diverging eddy fields of P and N.

the magnitude of the force that shoots the electron E or the positron E^+ out of their mother eddy fields.

Similarly, we are able to estimate a lower bound for the pressure needed to make an atomic nucleus go through decay, using the model in Figures 2.11 and 2.12, where this model is developed to explain why one proton and one neutron can combine to form nuclei. For more details on this discussion, please go to (Lin, 2010, pp. 218–232).

Laboratory experiments show that within an atomic nucleus, no electron or positron exists. Our systemic yoyo model analysis above provides an explanation for this observational fact. Because if either an electron or a positron does get into the spin field and the meridian field of a nucleus, which we assume to have only one proton and one neutron (Figures 2.11 and 2.12), the electron or the positron will be readily and instantly absorbed by the yoyo structure of the nucleus due to the fact that the yoyo fields of the electron and the positron are not well established.

From the well-known phenomenon that electricity can be created by rubbing actions (for more details see the experiment on electric charges in (Lin, 2010, pp. 242–250)), it follows that the rubbed plastic and glass rods become electricity carriers. Each electricity carrier contains a certain number of charged particles, which are manifested as spinning yoyos. Charged particles are classified into positively charged and negatively charged, which are respectively known in physics as protons and electrons. The masses of a proton and an electron are respectively 1.67×10^{-27} kg and 9.11×10^{-31} kg.

As what is discussed earlier, when the big bang side (the north pole) of a yoyo is visible while the black hole side (the south pole) is ill formed, one has a negatively charged yoyo, to which an electron corresponds. When the south pole of the yoyo is visible while the other pole is ill formed, the yoyo is positively charged, to which a positron corresponds. For each neutral object, meaning that the object does not carry any electric charge, be it negative or positive, its number of positively charged component yoyos is equal to that of the negatively charged component yoyos, where the interactions between one positive and one negative particle yoyos satisfy the laws on state of motion (Lin, 2007). That is, the concept of yoyo charges can be established here. Just as electric charge is a characteristic of some subatomic particles, for each spinning yoyo, the yoyo charge of this systemic yoyo represents the elementary attribute of the yoyo that characterizes its strength of rotation, absorption, and ejection of materials in and out of the spinning field. As for an object consisting of component particle yoyos, its yoyo charge stands for the elementary attribute of the object that characterizes the strength of its abstract, high-dimensional rotation, absorption, and ejection of materials.

When two objects touch each other, it is possible for some negatively or positively charged particle yoyos to move from one object to another so that both of these objects have the potential to be either positively or negatively charged. Let q be the amount of yoyo charge of an object. Then, we have

$$q = N_p e - N_n e = (N_p - N_n)e, \tag{2.3}$$

where N_p and N_n are respectively the numbers of positively and negatively charged component yoyos in the object, and e the unit yoyo charge. When $N_p > N_n$, the object carries a positive yoyo charge; when $N_p < N_n$, the object carries a negative yoyo charge.

2.2 SYSTEMIC YOYO FIELDS AND THEIR STRUCTURES

In this section, we look into more detailed structures of eddy and meridian fields, be they visible or invisible, of the general systemic yoyos, and how these fields work together to exchange materials with the environments. It is shown that just like magnetic fields, for yoyo fields, they also exhibit the properties that like poles repel and opposite poles attract.

By making use of some of the characteristics of spinning pools, we provide a new explanation for what a nuclear field is and why it is powerful but short ranged. After laying down the relevant foundations, we generalize Coulomb's law, which was initially established for computing the acting forces on electric charges, for particle yoyo charges and establish the principle of superposition of eddy field intensities.

After analyzing the potential formation of yoyo dipoles, we investigate the movement patterns of yoyo dipoles in uniform and uneven yoyo fields. In terms of yoyo fluxes, we generalize Gauss's Theorem, which was formally developed for magnetic fluxes, to the case of yoyo fluxes.

2.2.1 The formation of yoyo fields

It has been well documented that there exist various types of fields in the natural world (Landau and Lifshitz, 1971). They not only behave like a kind of force, but can also interact with each other over distance. For instance, as long as there is an electric charge, there exists around the charge an electric field, which is a special kind of material. One of the basic characteristics of this electric field is that it exerts an acting force on any static or moving electric charge; this force acts as follows: The direction of the force acting on a positive charge is the same as that of the field intensity, while the direction of the force on a negative charge is opposite of that of the field intensity. When an observer is at relative rest with regard to an electric charge, the field of the charge he observes is referred to as an electrostatic field. If the charge moves with respect to the observer, then other than the static electric field, there also appears a magnetic field. Electric fields can be created by electric charges as well as variable magnetic fields. These electric fields, caused by magnetic fields, are referred to as eddy electric fields or induced electric fields.

Magnetic fields are also special materials and exist widely in the natural world. There are magnetic fields around our earth, each star, every galaxy, each of the planets, any satellite, and in the spaces between galaxies (Zeilik, 2002). In modern technology and our daily lives, magnetic fields can be seen everywhere. Electricity generators, electrical motors, transformers, telegraphs, telephones, televisions, accelerators, devices of thermonuclear fusions, instrumentations of electromagnets, etc., are all related to the phenomenon of magnetism (Mohd, 2005). Even within of human bodies, along with the phenomenon of life, some parts and organs also create magnetic fields of slight intensity (Tierra, 1997).

Similar to electric fields, each magnetic field is also a vector field that is distributed continuously within a certain spatial region. Each electromagnetic field is the medium for electric charge and magnetism to work on each other and is a whole. Electric fields and magnetic fields are closely related to each other and dependent on each other.

As we have seen in (Lin, 2010, pp. 204–217), each electric charge is a systemic yoyo. In particular, each positive charge stands for a spinning yoyo with its big bang side (the north pole) less formed; and each negative charge for a spinning yoyo with a less formed black hole side (the south pole). In fact, each magnetic body, as shown in (Lin, 2010, pp. 242–250), is also a spinning yoyo. So, as long as there is a yoyo, there will be around this systemic yoyo a special kind of entity or material, referred to as a yoyo field. Hence, electric and magnetic fields are special cases of the general yoyo fields. For instance, if the eddy field of a system behaves like an electric field, then the meridian field will be the corresponding magnetic field, or vice versa. These fields exist simultaneously along side each other, while they might both be invisible or one of them is visible or both are visible. They are inseparable aspects of any system.

In a systemic yoyo, there are two kinds of field the eddy field (which is vertical to the invisible central axis of spin) and the meridian field (which travels from the big bang side to the black hole side parallel to the center of the spin of the yoyo). However, under the influence of the eddy field, each actual meridian field line travels along a spiral trajectory (Figure 2.13)). That is, each systemic yoyo possesses the duality of a particle (with internal structure) and a field that can be broken down into the eddy field and the meridian field. Several yoyos can interact with each other through their ever expanding fields and by colliding into each other when influenced by external and more powerful yoyos. Due to differences in the contents of materials, strengths of movement, and directions of movements, different yoyos have different fields, which can be either visible or invisible to our human eyes and scientific equipments. As indicated by the law

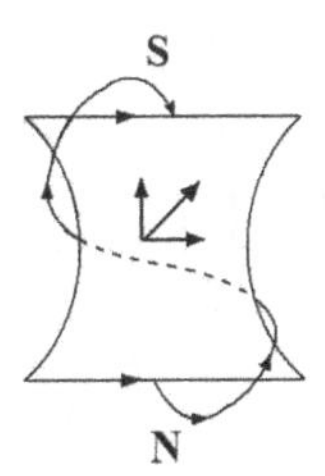

Figure 2.13 Slanted meridians of a yoyo structure.

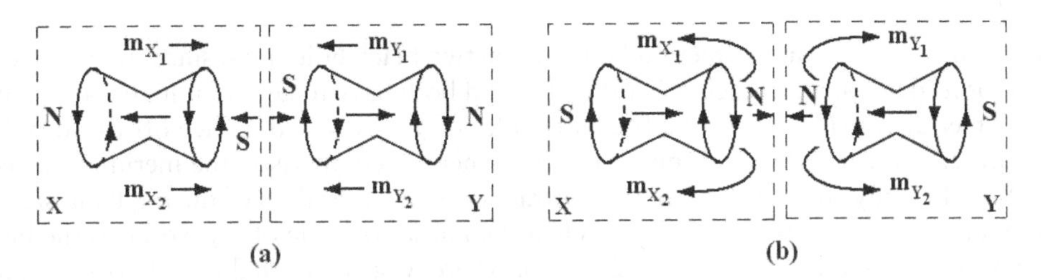

Figure 2.14 Repulsion of like yoyo fields.

of conservation of informational infrastructure (Ren, Lin and OuYang, 1998), each specific yoyo field has its own particular parameters. The content of a yoyo moves in three different forms:

1 All materials in the yoyo spin about their center both horizontally and vertically;
2 The materials travel in an orderly fashion by either being sucked in or spurted out of the yoyo along the center axis of spin; and
3 Various materials are constantly breathed in and out of the side of the yoyo in all horizontal directions and elevations.

Here, the yoyo exchanges its content with the environment in two different ways: one is its sucking in and spurting out along its axis of spin and the other breathing in or out from all round its side. Through these means the yoyo constantly exchanges with the environment and renews itself.

2.2.2 Classification of yoyo fields

Just as electric or magnetic fields, where like fields, such a positive (or negative) electric fields and the S (or N) poles of magnets, repel each other and opposite fields attract, for yoyo fields, the same principle holds true, where the like-kind ends (the black hole sides or the big bang sides) repel and the opposite attract. In particular, when the black hole sides (the south pole) of two yoyos face each other, Figure 2.14(a), where S stands for the black hole side and N the big bang side, although the two black hole sides have the tendency to attract each other, the meridian fields X and Y actually repel each other where m_{X_1} repels against m_{Y_1} and m_{X_2} against m_{Y_2}. So, in this case,

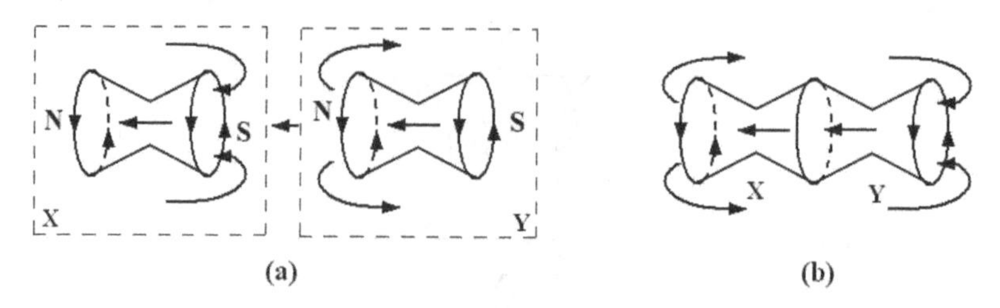

Figure 2.15 How two yoyos can potentially become one yoyo.

there does not exist any attraction between the two black hole sides, since these S sides cannot feed anything to each other directly, and both have to get input materials from their meridian fields. When the big bang sides of yoyos X and Y face off (head on), Figure 2.14(b), the yoyos are pushed apart, where in comparison the meridian fields of X and Y only come into place with attraction when the edges of the big bang sides of X and Y directly face each other. When the black hole side of a yoyo faces the big bang side of another yoyo (Figure 2.15(a)), where what is spurted out of yoyo Y can go and does go directly into yoyo X. At the same time, the meridian fields of both yoyos X and Y have the tendency to combine into one field too (Figure 2.15(b)).

The (horizontal) spin field of a yoyo forms a field of eddy currents also called an eddy field; the yoyo's (vertical) absorbing/spurting and (all directional) taking in/giving out create an active field that can be visibly seen as a field with source (or active field). In the physical world, gravitational fields, magnetic fields, electric fields, and nuclear fields are some of the examples of both active fields and eddy fields. In particular, in a given yoyo, if its eddy field is identified as its magnetic field, then the active (the meridian) field will be corresponding to the electric field. In this case, the so-called N and S poles of the magnetic field correspond respectively to the diverging and the converging sides of the yoyo structure, and the positive and negative electric fields to the converging and diverging eddy fields. Similarly, the eddy field can be identified as an electric field and so correspondingly the meridian field as the magnetic field. Now, if the eddy field is seen as a gravitational field, then the meridian field will be the corresponding magnetic field. In this case, we can see that the gravitational field can be both attraction and repulsion that exist side by side, the former corresponds to the converging side of the eddy field, and the latter to the diverging side. From this discussion, it can be seen that all forms of fields must have two opposite effects, such as the N and S poles of a magnetic field, where the opposite effects have to coexist at all times and no one side effect can exist without the other opposite effect, even though the opposite effects do not have to be visible at the same time or both of them do not have to be visible at all.

At this junction, a natural question arises: What is a nuclear field? According to laboratory experiments, it is discovered that the reason why protons and neutrons can combine to form nuclei is because there is a certain kind of field between them; this field possesses extremely strong acting power with extremely limited acting distance. Because this kind of field has been mainly associated with the bonding of protons and neutrons within nuclei, it has been called a nuclear field. It is believed that each proton

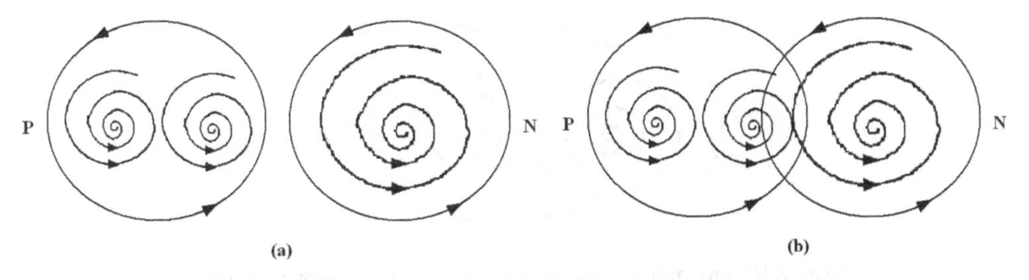

Figure 2.16 Interactions between the converging eddy fields of P and N.

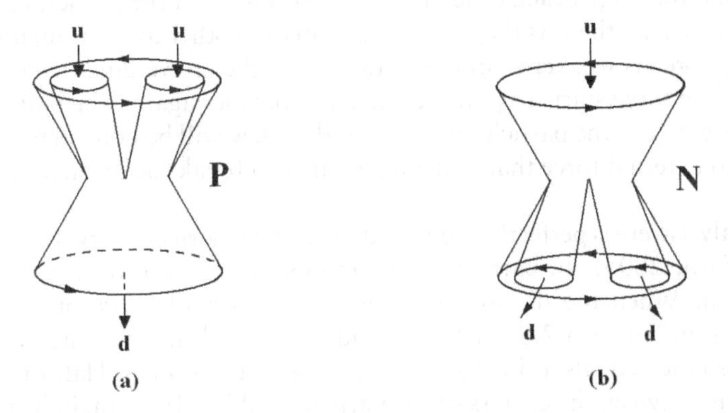

Figure 2.17 Quark structures of proton and neutron.

or neutron naturally has its own nuclear field, which is why they can combine and form nuclei (Mannel, 2005).

Since we have our yoyo model in place, we can now provide a new explanation for the formation of nuclei similarly to how we explained the existence of n-ary star systems (Lin, 2007), for any natural number n. In particular, from the quark structures (Lin, 2010, pp. 204–217) of a proton P and a neutron N as shown in Figure 2.17, when these particles P and N are near each other within their fields' acting distance, they would interact with each other as follows. First, let us look at the converging (the black hole) sides of the yoyos (Figure 2.16(a)). Due to their harmonic spinning directions, meaning that they spin in the same direction and the minor pulling effects of their meridian fields, which flow out of the page toward the reader and then turn to their respective spinning centers that is when the minor attracting (pulling) effects are created, the yoyo structures of the proton P and neutron N are pulled together, as shown in Figure 2.16(b). However, the closeness of the particles P and N does not go much into the spinning field of the right quark in P due to a conflict of spinning. At this moment of stagnation of movement, the diverging eddy fields of the proton P and neutron N on the other side takes over (Figure 2.18). Because of the divergence of all the eddy fields in P and N and the two quarks in N, their conflicting spinning directions, and their repelling meridian fields, which come out from the centers of spinning toward

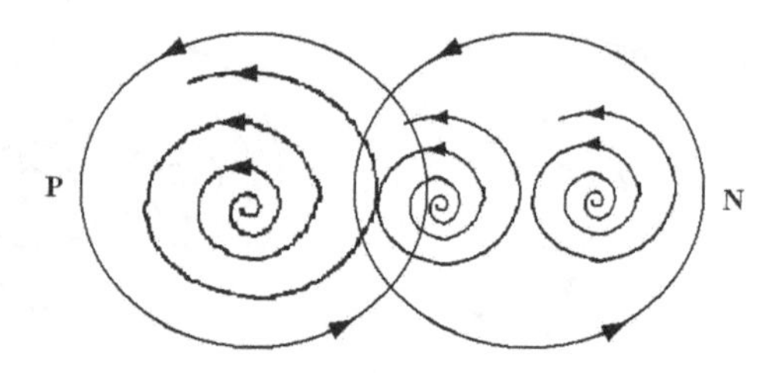

Figure 2.18 Interaction between the diverging eddy fields of P and N.

the reader and then repel each other, the yoyo structures of the particles P and N start to depart from each other. As they are apart from each other to a certain distance when the repelling force is weaker than the attraction of the converging fields on the other side, the pulling force starts to pull the particles together again. The pull and push act alternatively between the particles P and N so that they will be bonded together as long as there is no external force that is strong enough to break the attraction between the particles.

Our analysis here is perfectly supported by the laboratory observations (Basdevant, Rich and Spiro, 2005). In particular, the radius of a proton or a neutron is about 0.8×10^{-15} m. When the distance between the centers of a proton and a neutron is somewhere in between 2.8×10^{-15} m and 1.4×10^{-15} m, they attract each other. When the distance equals 1.4×10^{-15} m, they repel each other. Half of this distance 1.4×10^{-15} m between the centers of the particles is 0.7×10^{-15} m, indicating that the proton and neutron have cut into each other by as much as 0.1×10^{-15} m.

As for why the field actions between protons and neutrons are extremely powerful and short ranged, it is because their yoyo structures spin at a very high speed without much mass involved.

To make this model complete, one additional question is: Why do yoyo structures have the tendency to line up side by side with their axes of spin parallel to each other? To address this question, let us look at Figure 2.19. If two yoyos X and Y are positioned as in Figure 2.19(a), then the meridian field A of X fights against C of Y so that both X and Y have the tendency to realign themselves in order to reduce the conflicts along the meridian directions. Similarly in Figure 2.19(b), the meridian field A_1 of yoyo X fights against B_1 of Y. So, the yoyos X and Y also have the tendency to realign themselves as in the previous case.

Our discussion so far in this section indicates that

1 Gravitation can be both attraction and repulsion;
2 The so-called universal fields are indeed universal and exist along with each and every system, be it tangible or not, be it physical or epistemological;
3 The so-called nuclear fields are simply the yoyo field naturally existing with each extremely fast spinning system that has little mass;
4 Although it has been experimentally shown that under ordinary pressure, the greatest nucleus can only contain 270 protons and neutrons (Martin and Shaw, 2008), our analysis above indicates that in nature, where various pressures can exist, it

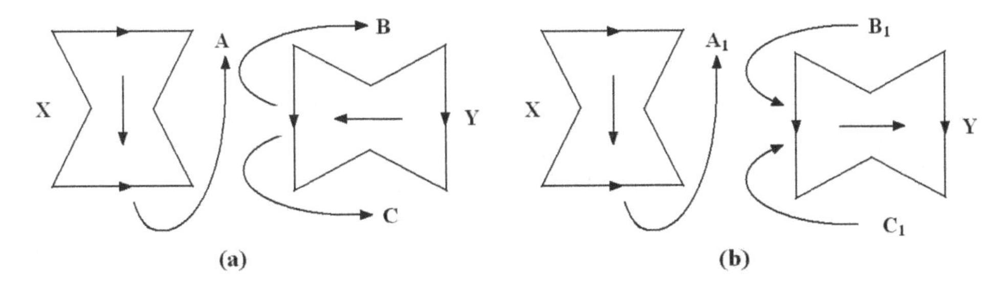

Figure 2.19 The tendency for yoyos to line up.

is possible to construct a nucleus with n protons and neutrons, jus as the possible existence of n-nary star systems, for any natural number n, if the axes of the eddy fields of all the particle yoyos are parallel to each other. (Lin, 2010, pp. 242–250) will show that it is also possible for n spinning yoyos to form relatively stable structures by connecting their big bangs sides with their black holes sides, for any natural number $n \geq 3$.

2.2.3 The Coulomb's law for particle yoyos

Assume that the effective ranges of two yoyos are very limited compared to the distance between them, we can estimate the acting force between these yoyos by using the Coulomb's law, developed in the 1780s by French physicist Charles Augustin de Coulomb (Griffiths, 1998) for electric charges. In scalar form this generalized Coulomb's law can be stated as follows: The magnitude of the yoyo-static force between two point yoyo charges is directly proportional to the product of the magnitudes of each charge and inversely proportional to the square of the distance between the charges.

If we do not require the specific direction of the acting force between the charges, then the simplified version of Coulomb's law will suffice. In particular, the law can be stated as that the magnitude of the acting force on a charge q_1 due to the presence of a second charge q_2 is given by the magnitude of

$$F = k\frac{q_1 q_2}{r^2}, \tag{2.4}$$

where r is the separation of the charges and k the constant of proportionality. A positive force implies a repulsive interaction, while a negative force implies an attractive interaction.

In order to obtain both the magnitude and direction of the acting force on a yoyo charge q_1 at position $\vec{r}_1$, experiencing a spinning field due to the presence of another yoyo charge q_2 at position $\vec{r}_2$, the full vector form of our generalized Coulomb's law can be written as follows:

$$\vec{F} = k\frac{q_1 q_2 (\vec{r}_1 - \vec{r}_2)}{|\vec{r}_1 - \vec{r}_2|^3} = k\frac{q_1 q_2}{r^2}\hat{r}_{21} \tag{2.5}$$

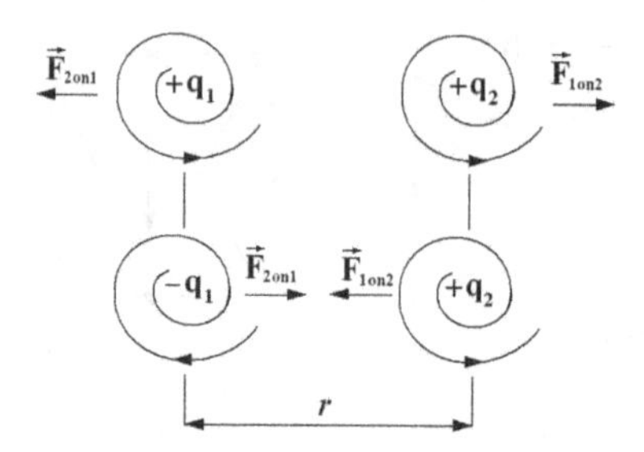

Figure 2.20 Coulomb's interaction between spinning yoyo charges.

where r is the separation of the two yoyo charges, and $\hat{r}_{21}$ the unit vector that is parallel to the line from the yoyo charge q_2 to q_1. If both yoyo charges have the same sign (like yoyo charges) then the product q_1q_2 is positive and the direction of the force on q_1 is given by $\hat{r}_{21}$; the yoyo charges repel each other. If the charges have opposite signs then the product q_1q_2 is negative and the direction of the force on q_1 is given by $-\hat{r}_{21}$. In this case, the yoyo charges attract each other (Figure 2.20).

If we investigate a system of N yoyo charges $q_1, q_2, \ldots, q_N$, then the principle of linear superposition can be used to calculate the combined force acting on a small test yoyo charge q due to this system of N discrete charges as follows:

$$\vec{F}(\vec{r}) = kq\sum_{i=1}^{N} \frac{q_i(\vec{r}-\vec{r}_i)}{|\vec{r}-\vec{r}_i|^3} = kq\sum_{i=1}^{N} \frac{q_i}{R_i^2}\hat{R}_i, \tag{2.6}$$

where $\vec{r}_i$ is the position of the ith charge, $\hat{R}_i$ is a unit vector in the direction of $\vec{R}_i = \vec{r}-\vec{r}_i$ (a vector pointing from charge q_i to charge q), and R_i is the magnitude of $\vec{R}_i$ (the separation between charges q_i and q).

For a regional distribution of great many yoyo charges, we can theoretically think of the distribution as of infinitely many yoyo charges or a single charge of the region. So, an integral over the region containing the charges is equivalent to an infinite summation, treating each infinitesimal element of the region as a point charge dq. For a linear charge distribution, which has been employed as a good approximation for electric charge in a wire, let $\lambda(\vec{r}')$ be the charge per unit length at position $\vec{r}'$, and $d\ell'$ an infinitesimal element of length, then we have

$$dq = \lambda(\vec{r}')\, d\ell'. \tag{2.7}$$

For a surface charge distribution, which has employed as a good approximation for electric charge on a plate in a parallel plate capacitor, then $\sigma(\vec{r}')$ gives the charge per unit area at position $\vec{r}'$, and dA' is an infinitesimal element of area, satisfying

$$dq = \sigma(\vec{r}')dA'. \tag{2.8}$$

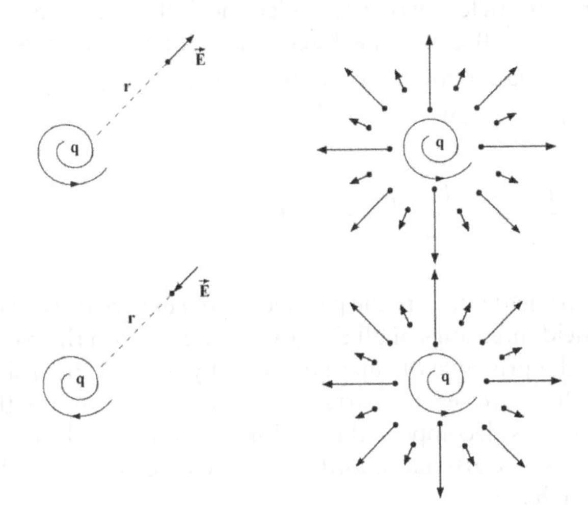

Figure 2.21 The general structure of a yoyo's eddy field.

For a volume charge distribution, which has been investigated as an electric charge within a bulk metal, $\rho(\vec{r}')$ gives the charge per unit volume at position $\vec{r}'$, and dV' is an infinitesimal element of volume, satisfying

$$dq = \rho(\vec{r}')dV'. \tag{2.9}$$

The force on a small test yoyo charge q' at position $\vec{r}$ is given by

$$\vec{F} = q' \int dq \frac{\vec{r} - \vec{r}'}{|\vec{r} - \vec{r}'|^3}. \tag{2.10}$$

2.2.4 The eddy fields of yoyos

The intensity of the eddy field of a general systemic yoyo q at a specific location in the eddy field is defined as

$$\vec{E} = k\frac{q}{r^2}\hat{r}, \tag{2.11}$$

where the narrow neck of the yoyo is idealized into a point so that the originally layered eddy field of the yoyo is abstracted into a planar whirlpool motion, the specific location is at a distance r away from the spinning center of the yoyo, $\hat{r}$ the unit vector pointing from the yoyo center to the specific location, and k the intensity constant of the given yoyo. For an electric yoyo charge, $k = 1/4\pi\varepsilon_0$, where $\varepsilon_0 = 8.85 \times 10^{-12}$ C^2/Nm^2.

Graphically, the magnitudes and directions of the yoyo eddy field intensities $\vec{E}$ at various locations are shown in Figure 2.21, where the length of a vector stands for the relative magnitude at a specific point in the field.

Assume that a particle yoyo q is under the influence of a collection of spinning yoyos q_1, q_2, ... Because the force $\vec{F}_{on\,q}$ that is collectively exerted on the particle yoyo q is related to the yoyo field intensity $\vec{E}_{net}$ by $\vec{E}_{net} = \vec{F}_{on\,q}/q$, and $\vec{F}_{on\,q} = \vec{F}_{1\,on\,q} + \vec{F}_{2\,on\,q} + \cdots$, we have

$$\vec{E}_{net} = \frac{\vec{F}_{on\,q}}{q} = \frac{\vec{F}_{1\,on\,q}}{q} + \frac{\vec{F}_{2\,on\,q}}{q} + \cdots = \vec{E}_1 + \vec{E}_2 + \cdots. \tag{2.12}$$

That is, the net yoyo field intensity the particle yoyo q experiences is equal to the vector sum of the yoyo field intensities of all the yoyos $q_1, q_2, \ldots$ on the particle yoyo q. This property is called the principle of superposition of yoyo field intensities.

Assume that there are several particle yoyos $q_1, q_2, \ldots$ in the three-dimensional Euclidean space. Let us decompose the field intensities of each of these yoyos to the three axes of a chosen Cartesian coordinate system of the space. So, based on the formulas above, we have

$$\begin{aligned}
(E_{net})_x &= (E_1)_x + (E_2)_x + \cdots = \sum_i (E_i)_x, \\
(E_{net})_y &= (E_1)_y + (E_2)_y + \cdots = \sum_i (E_i)_y, \\
(E_{net})_z &= (E_1)_z + (E_2)_z + \cdots = \sum_i (E_i)_z.
\end{aligned} \tag{2.13}$$

So, the following holds true:

$$\vec{E}_{net} = (E_{net})_x\hat{i} + (E_{net})_y\hat{j} + (E_{net})_z\hat{k}. \tag{2.14}$$

When an object contains a great number of systemic yoyos of the same scale, we think of this object as a continuous distribution of yoyos. Let us look at a one-dimensional object of length L with an even distribution of smaller yoyos, such as a plastic rod or a metal wire. The line density (of yoyos) of this object is defined as follows:

$$\lambda = \frac{Q}{L}, \tag{2.15}$$

where Q stands for the total number of same-scale yoyos contained in this object.

For a two-dimensional object with even and continuous distribution of yoyos of some specific scale, its surface density (of yoyos) is defined by:

$$\eta = \frac{Q}{A}, \tag{2.16}$$

where Q represents the total number of yoyos of the specific scale and A the surface area of the object.

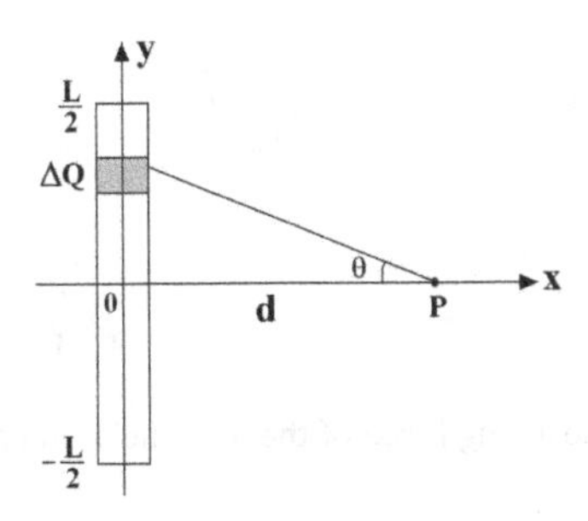

Figure 2.22 The yoyo field intensity of the rod at point P.

For an arbitrary object of uneven distributions of yoyos of a certain scale, based on the following tools:

1 The formula for a particle yoyo of even intensity; and
2 The principle of superposition,

we can calculate the object's yoyo field intensity by following the steps below:

1 Divide the total number Q of yoyos of a certain scale in the object into many small yoyo counts ΔQ of particle yoyos;
2 Find the yoyo field intensity corresponding to each small yoyo count ΔQ; and
3 Compute the sum of the field intensities of these small-yoyo counts ΔQ.

That is, computing the net yoyo field intensity $\vec{E}_{net}$ is equivalent to calculating an integral.

Example 2.1 Find the yoyo field intensity at point P of a rod of length L, Figure 2.22. Assume that the rod's line density of yoyos is $\lambda = \lambda(y)$.

In this case, a small yoyo count ΔQ is given by $\Delta Q = \lambda(y)\Delta y$. So, we have

$$\begin{aligned}(\Delta E)_x &= \Delta E \cos\theta = k\frac{\Delta Q}{r^2}\cos\theta \\ &= k\frac{\lambda(y)\Delta y}{y^2+d^2}\frac{d}{\sqrt{y^2+d^2}} \\ &= kd\frac{\lambda(y)}{(y^2+d^2)^{3/2}}\Delta y.\end{aligned}$$

So, we obtain

$$E_x = kd\int_{-\frac{L}{2}}^{\frac{L}{2}} \frac{\lambda(y)}{(y^2+d^2)^{3/2}}dy.$$

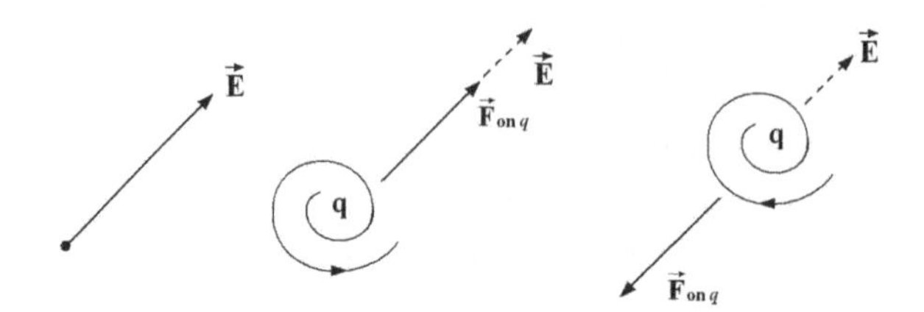

Figure 2.23 The acting force of the yoyo field on particle yoyo q.

Similarly, we have

$$(\Delta E)_y = \Delta E \sin\theta = k\frac{\Delta Q}{r^2}\sin\theta = k\frac{\lambda(y)\Delta y}{y^2+d^2}\frac{y}{y^2+d^2},$$

and so,

$$E_y = k\int_{-\frac{L}{2}}^{\frac{L}{2}} \frac{y\lambda(y)}{(y^2+d^2)^{3/2}}dy.$$

In particular, if in this rod, the distribution of yoyos is even, that is, $\lambda = Q/L$, then we have

$$E_x = k\frac{Q}{d\sqrt{d^2+(\frac{L}{2})^2}}, \quad E_y = 0.$$

Therefore, we obtain

$$E_{rod} = k\frac{|Q|}{d\sqrt{d^2+(\frac{L}{2})^2}}.$$

If the scale of the small yoyos is at the same level of the electrons, then the constant k is equal to $1/4\pi\varepsilon_0$. QED.

Now, let us look at the movement of a particle yoyo in a yoyo field. If we assume that this particle yoyo charge is q with mass m and is placed at a point inside the eddy field of another systemic yoyo, then the force of the eddy field exerted on q is given by

$$\vec{F}_{on\,q} = q\vec{E}, \tag{2.17}$$

where $\vec{E}$ stands for the yoyo field intensity at that specific point, Figure 2.23.

The acceleration created by the force $\vec{F}_{on\,q}$ is given by

$$\vec{a} = \frac{\vec{F}_{on\,q}}{m} = \frac{q}{m}\vec{E}. \tag{2.18}$$

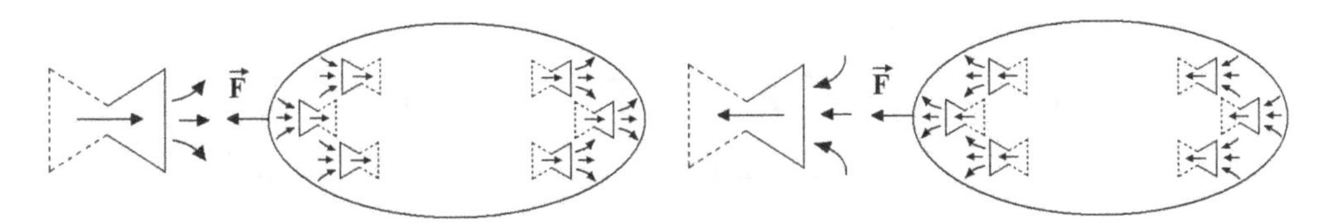

Figure 2.24 Formation of yoyo dipoles.

1 If the eddy field is stable, the movement of the particle yoyo is given by

$$a = \frac{q}{m}E = \text{constant}. \tag{2.19}$$

2 If the yoyo field is unstable, the movement of the particle yoyo is extremely complicated. However, for the special case that the particle yoyo circles around a sphere, we have the following analysis about the movement of the particle yoyo. From Newton's second law,

$$F_{net} = \frac{mv^2}{r}, \tag{2.20}$$

it follows that when

$$|q|E = \frac{mv^2}{r}, \tag{2.21}$$

the particle yoyo circles around a sphere.

2.2.5 Movement of yoyo dipoles in yoyo fields

Electrically charged plastic or glass rods can attract certain non-metal insulators, such as scraps of paper. Does this phenomenon indicate certain properties of yoyo charges? As a matter of fact, when a yoyo-charge neutral object is affected by an external yoyo charge, the object can experience the phenomenon of polarization, just as the scraps of paper being affected by an electric charge. In particular, when a positive yoyo charge is placed near the object, the negative component yoyos inside the object have the potential to move to the side closer to the positive charge, while positive component yoyos move to the side that is furthest away from the external charge. Similarly, if the external charge is negative, the charged component yoyos inside the object will move in directions opposite to those described above. That is, under the influence of an external powerful yoyo charge, a yoyo dipole can be formed, Figure 2.24. Because opposite yoyo poles attract, the external yoyo charge attracts the created yoyo dipole.

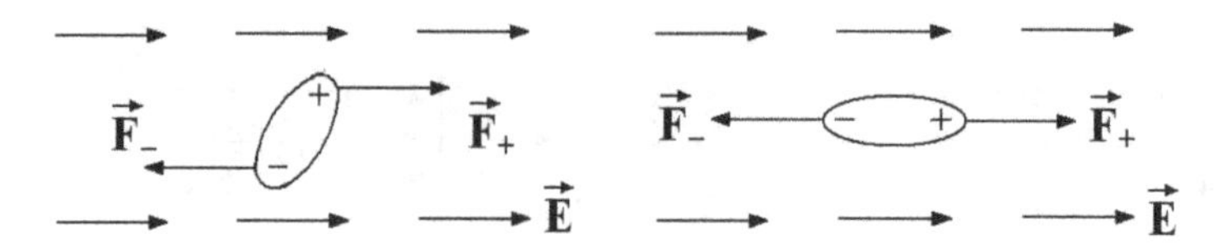

Figure 2.25 The yoyo dipole experiences a torque.

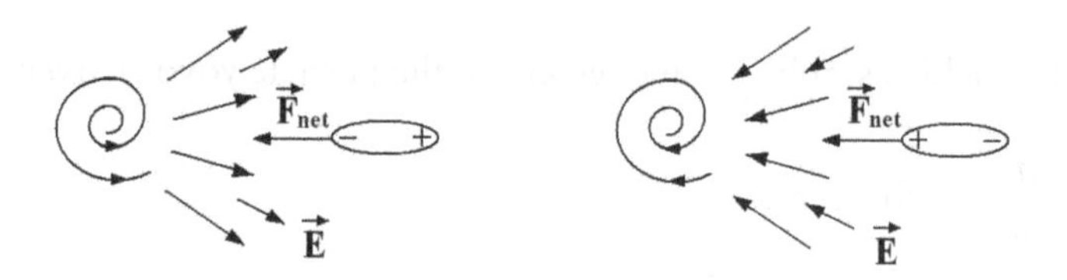

Figure 2.26 A yoyo dipole in an uneven yoyo field.

2.2.5.1 *Movement of yoyo dipoles in uniform yoyo fields*

When a yoyo dipole is situated in an uniform yoyo field $\vec{E}$, its two poles of the dipole will be acted upon by the yoyo field. The acting and reacting forces are of the same magnitudes with opposite directions:

$$\vec{F}_{+} = +q\vec{E} \quad \text{and} \quad \vec{F}_{-} = -q\vec{E}. \tag{2.22}$$

So, the yoyo dipole experiences a torque (Figure 2.25) whose net force is

$$\vec{F}_{net} = \vec{F}_{+} + \vec{F}_{-} = \vec{0}. \tag{2.23}$$

2.2.5.2 *Movement of yoyo dipoles in uneven yoyo fields*

Assume that we place a yoyo dipole inside an uneven yoyo field $\vec{E}$, where in different locations the field intensities and directions are different. For instance, in the field of a particle yoyo, the yoyo dipole will spin until it is parallel to the direction of the field. If the particle yoyo carries a positive charge, then the positive direction of the yoyo dipole agrees with that of the field. Because the yoyo field intensity is different from one location to another, there is a difference between the forces experienced by the two poles of the yoyo dipole. The end closer to the yoyo charge experiences a pull greater than the push at the end pointing to the positive direction. Therefore, the net force acting on the yoyo dipole is not zero and points to the yoyo charge. If the yoyo charge is negative, similar results hold true (Figure 2.26). That is, we have the following result:

Fact 2.1 The force experienced by a yoyo dipole placed in an uneven yoyo field always points to the direction of greater field intensity.

2.2.5.3 *The concept of yoyo flux*

Any closed surface through which a yoyo field penetrates is called a Gauss surface. Studies in electromagnetic theory tell us that if a closed surface encloses a positive yoyo charge, then there is a yoyo field "flowing" out from inside the surface. Conversely, if a closed surface encloses a negative yoyo charge, there is a yoyo field "flowing" into the interior of the surface. If there does not exist any yoyo charge inside a closed surface, then a yoyo field can only pass through the surface with net "flow" zero, because the inflow into the interior of the surface cancels the outflow. This kind of "flow" is referred to a yoyo flux. So, we have the following conclusions:

1. For any closed surface that contains a positive yoyo charge in its interior, there is a yoyo flux flowing outward.
2. For any closed surface that contains a negative yoyo charge in its interior, there is a yoyo flux flowing inward.
3. If a closed surface does not contain any yoyo charge within the inside of itself, then the yoyo flux of the surface is zero.

By making use of the computation of yoyo flow to one side of a surface, we can define the magnitude Φ_e of the yoyo flux as follows:

$$\Phi_e = EA\cos\theta = \vec{E}\cdot\vec{A}, \tag{2.24}$$

where E stands for the intensity of the horizontal yoyo field, A the area of the surface through which the yoyo field passes, θ the angle between $\vec{E}$ and the unit normal vector $\vec{n}$ of the surface, and $\vec{A} = A\vec{n}$ is defined to be the area vector, its direction is perpendicular to the surface and its magnitude is equal to the area of the surface.

If the yoyo field is uneven and the spatial surface is S, then the yoyo flux can be calculated by using the following surface integral:

$$\Phi_e = \int_S \vec{E}\cdot d\vec{A}. \tag{2.25}$$

It can be calculated that the yoyo flux going out of or into a Gauss surface that contains a yoyo charge q is

$$\Phi_e = \frac{q}{4\pi k}, \tag{2.26}$$

whose value evidently has nothing to do with the shape and size of the surface.

If the yoyo charge q is located outside the Gauss surface, then the net yoyo flux passing through the surface is zero. Therefore, we generalize the well-known Gauss's

Theorem that was initially established for magnetic fluxes (Stratton, 2007) to our concept of yoyo fluxes as follows:

Theorem 2.1: (Gauss's Theorem). If there are several yoyo charges located respectively inside and outside a Gauss surface, where the inside charges are $q_1, q_2, \ldots, q_n$ that have the same intensity constant k, denote

$$Q_{in} = q_1 + q_2 + \cdots + q_n, \tag{2.27}$$

then the yoyo flux passing through the Gauss surface is

$$\Phi_e = \frac{Q_{in}}{4\pi k}. \tag{2.28}$$

2.3 STATES OF MOTION

Based on the discovery (Wu and Lin, 2002) that spins are the fundamental evolutionary feature and characteristic of materials, in this section, we will study the figurative analysis method of the systemic yoyo model and how to apply it to establish laws on state of motion by generalizing Newton's laws of motion. More specifically, after introducing the new figurative analysis method, we will have a chance to generalize all the three laws of motion so that external forces are no longer required for these laws to work. As what's known, these laws are one of the reasons why physics is an "exact" science. And, in the rest of this book, we will show that these generalized forms of the original laws of mechanics will be equally applicable to social sciences and humanity areas as their classic forms in natural science. The presentation in this section is based on (Lin, 2007).

2.3.1 The first law on state of motion

Newton's first law says: An object will continue in its state of motion unless compelled to change by a force impressed upon it. This property of objects, their natural resistance to changes in their state of motion, is called inertia. Based on the theory of blown-ups, one has to address two questions not settled by Newton in his first law: If a force truly impresses on the object, the force must be from the outside of the object. Then, where can such a force be from? How can such natural resistance of objects to changes be considered natural?

It is because uneven densities of materials create twisting forces that fields of spinning currents are naturally formed. This end provides an answer to the first question. Based on the yoyo model, the said external force comes from the spin field of the yoyo structure of another object, which is another level higher than the object of our concern. These forces from this new spin field push the object of concern away from its original spin field into a new spin field. Because if there is not such a forced traveling, the said object will continue its original movement in its original spin field. That is why Newton called its tendency to stay in its course of movement as its resistance to changes in its state of motion and as natural. Based on this discussion and the yoyo model developed for each and every object and system in the universe, Newton's first law of mechanics can be rewritten in a general term as follows:

The First Law on State of Motion: *Each imaginable and existing entity in the universe is a spinning yoyo of a certain dimension. Located on the outskirts of the yoyo is a spin field. Without being affected by another yoyo structure, each particle in the said entity's yoyo structure continues its movement in its orbital state of motion.*

Because for Newton's first law to hold true, one needs an external force, when people asked Newton where such an initial force could be from, he answered (jokingly?): "It was from God. He waved a bat and provided the very first blow to all things he created (Kline, 1972)." If such an initial blow is called the first push, then the yoyo model and the stirring forces naturally existing in each "yoyo" created by uneven densities of materials' structures will be called the second stir.

2.3.2 The second law on state of motion

Newton's second law of motion says that when a force does act on an object, the object's velocity will change and the object will accelerate. More precisely, what is claimed is that its acceleration $\vec{a}$ will be directly proportional to the magnitude of the total (or net) force $\vec{F}_{net}$ and inversely proportional to the object's mass m. In symbols, the second law is written:

$$\vec{F}_{net} = m\vec{a} = m\frac{d\vec{v}}{dt} \tag{2.29}$$

Even though equ. (2.29) has been one of the most important equations in mechanics, when one ponders over this equation long enough, one has to ask the following questions: What is a force? Where are forces from and how do forces act on other objects?

To answer these questions, let us apply Einstein's concept of "uneven time and space" of materials' evolution (Einstein, 1997). So, we can assume

$$\vec{F} = -\nabla S(t, x, y, z), \tag{2.30}$$

where $S = S(t, x, y, z)$ stands for the time-space distribution of the external acting object (a yoyo structure). Let $\rho = \rho(t, x, y, z)$ be the density of the object being acted upon. Then, equ. (2.29) can be rewritten as follows for a unit mass of the object being acted upon:

$$\frac{d\vec{v}}{dt} = -\frac{1}{\rho(t, x, y, z)}\nabla S(t, x, y, z). \tag{2.31}$$

If $S(t, x, y, z)$ is not a constant, or if the structure of the acting object is not even, equ. (2.31) can be rewritten as

$$\frac{d(\nabla x \times \vec{v})}{dt} = -\nabla x \times \left[\frac{1}{\rho}\nabla S\right] \neq 0 \tag{2.32}$$

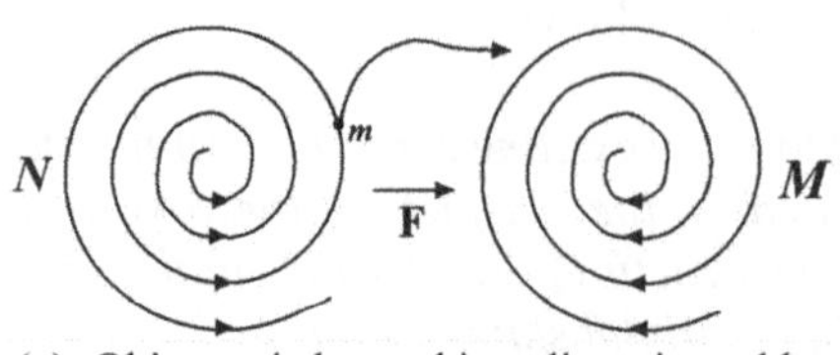

(a) Object m is located in a diverging eddy and pulled by a converging eddy M

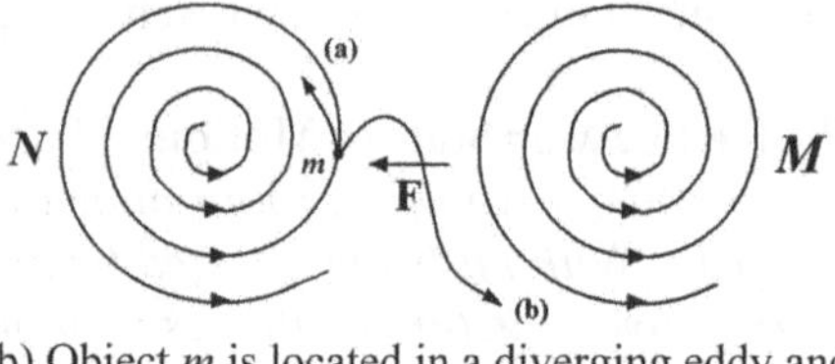

(b) Object m is located in a diverging eddy and pulled or pushed by a diverging eddy M

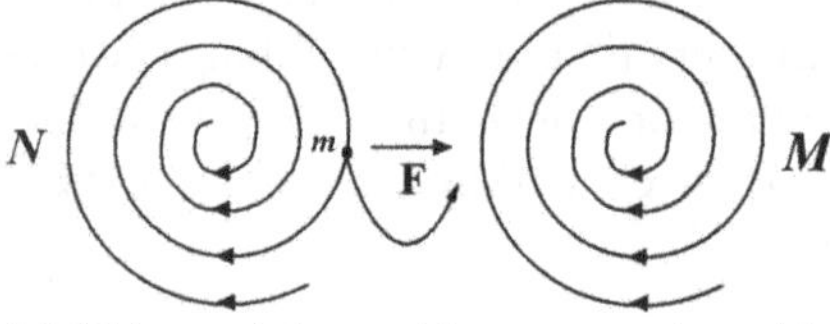

(c) Object m is located in a converging eddy and pulled by a converging eddy M

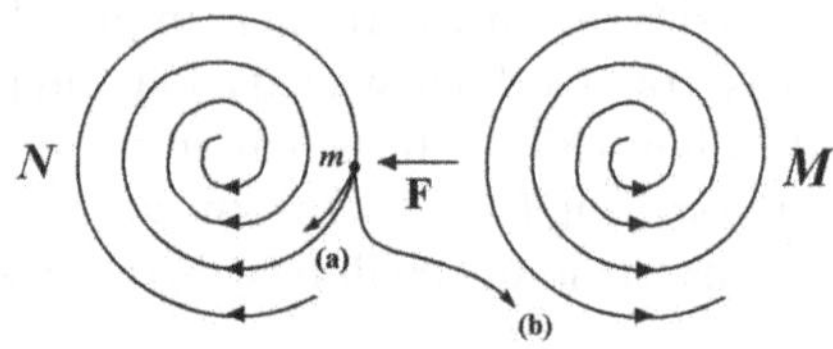

(d) Object m is located in a converging eddy and pulled or pushed by a diverging eddy M

Figure 2.27 Acting and reacting models with yoyo structures of harmonic spinning patterns.

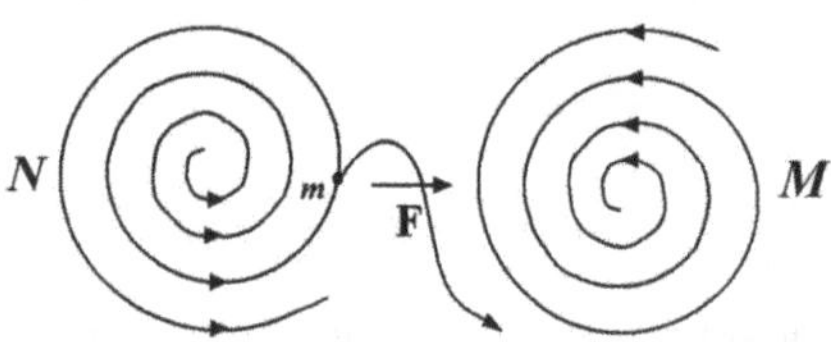

(a) Object m is located in a diverging eddy and pulled by a converging eddy M

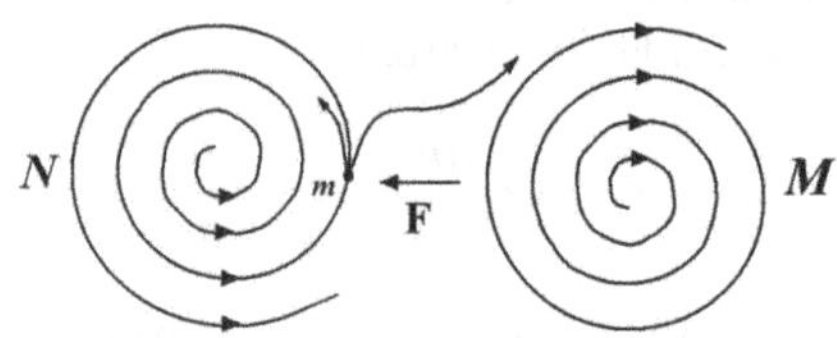

(b) Object m is located in a diverging eddy and pushed or pulled by a diverging eddy M

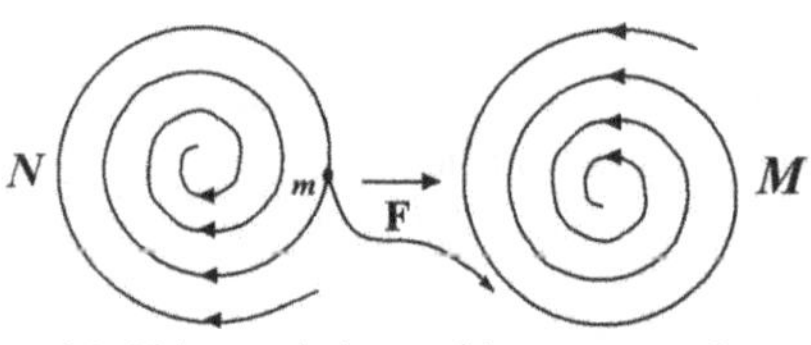

(c) Object m is located in a converging eddy and pulled by a converging eddy M

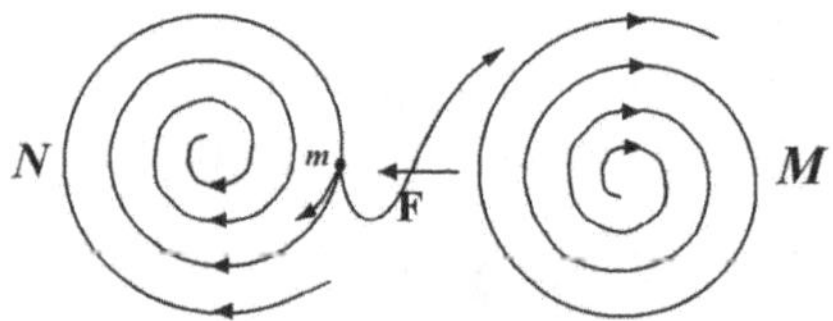

(d) Object m is located in a converging eddy and pushed or pulled by a diverging eddy M

Figure 2.28 Acting and reacting models with yoyo structures of inharmonic spinning patterns.

and it represents an eddy motion due to the nonlinearity involved. That is, when the concept of uneven structures is employed, Newton's second law actually indicates that a force, acting on an object, is in fact the attraction or gravitation from the acting object. It is created within the acting object by the unevenness of its internal structure.

By combining this new understanding of Newton's second law with the yoyo model, we get the models on how an object m is acted upon by another object M (see Figures 2.27 and 2.28).

Now, by summarizing what can be observed in these figures, Newton's Second law can be generalized as follows.

The Second Law on State of Motion: *When a constantly spinning yoyo structure M does affect an object m, which is located in the spin field of another object N, the velocity of the object m will change and the object will accelerate. More specifically, the object m experiences an acceleration $\vec{a}$ toward the center of M such that the magnitude of $\vec{a}$ is given by*

$$a = \frac{v^2}{r} \tag{2.33}$$

where r is the distance between the object m and the center of M and v the speed of any object in the spin field of M about distance r away from the center of M. And, the magnitude of the net pulling force $\vec{F}_{net}$ that M exerts on m is given by

$$F_{net} = ma = m\frac{v^2}{r} \tag{2.34}$$

2.3.3 The third law on state of motion

Newton's third law is commonly known as that to every action, there is an equal, but opposite, reaction. More precisely, if object A exerts a force on object B, then object B exerts a force back on object A, equal in strength but in the opposite direction. These two forces, $\vec{F}_{A-on-B}$ and $\vec{F}_{B-on-A}$, are called an action/reaction pair.

Similar to what has been done earlier, if we analyze the situation in two different angles: two eddy motions act and react to each other's spin field, and two, one spinning yoyo is acted upon by an eddy flow of a higher level and scale, then for the first situation, where two eddy motions act and react to each other's spin field, we have the diagrams in Figure 2.29.

For objects N and M with harmonic spin fields, we have the diagrams in Figure 2.30.

Based on the analysis of these graphs, Newton's third law can be generalized for the case of two eddy motions acting and reacting to each other's spin fields as follows:

The Third Law on State of Motion: *When the spin fields of two yoyo structures N and M act and react on each other, their interaction falls in one of the six scenarios as shown in Figure 2.29(a)–(c) and Figure 2.30(a)–(c). And, the following is true:*

1 *For the cases in (a) of Figures 2.29–2.30, if both N and M are relatively stable temporarily, then their action and reaction are roughly equal but in opposite directions during the temporary stability. In terms of the whole evolution involved, the divergent spin field (N) exerts more action on the convergent field (M) than M's reaction peacefully in the case of Figure 2.29(a) and violently in the case of Figure 2.30(a).*

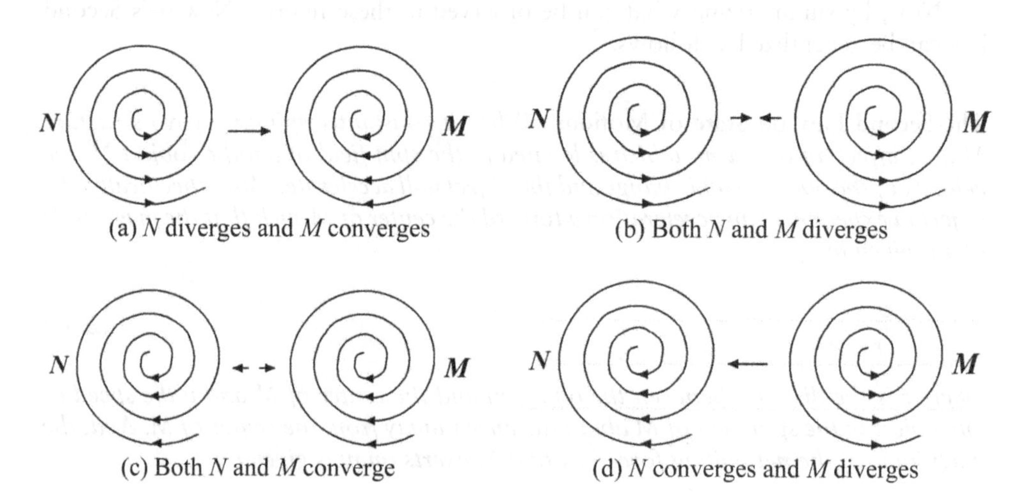

Figure 2.29 Same scale acting and reacting spinning yoyos of the harmonic pattern.

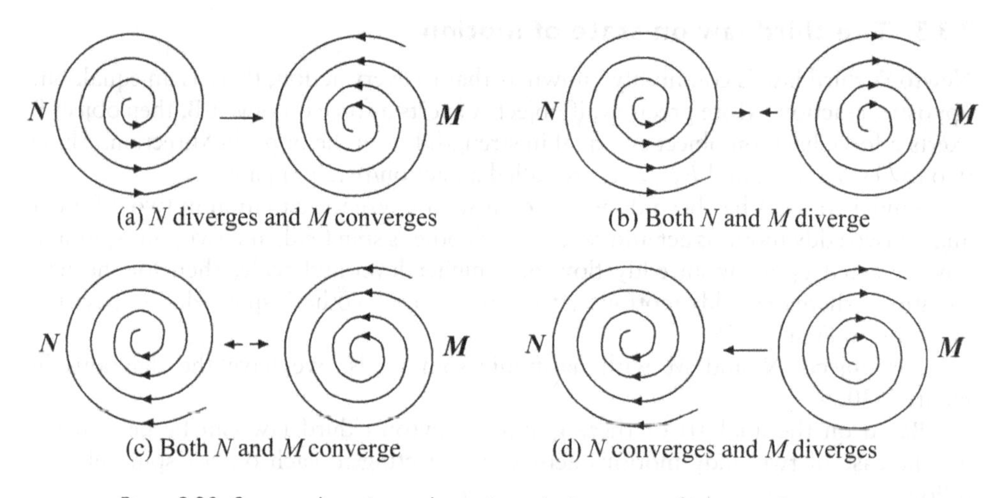

Figure 2.30 Same scale acting and reacting spinning yoyos of inharmonic patterns.

2 *For the cases (b) in Figures 2.29–2.30, there are permanent equal, but opposite, actions and reactions with the interaction more violent in the case of Figure 2.29(b) than in the case of Figure 2.30(b).*
3 *For the cases in (c) of Figures 2.29–2.30, there is a permanent mutual attraction. However, for the former case, the violent attraction may pull the two spin fields together and have the tendency to become one spin field. For the latter case, the peaceful attraction is balanced off by their opposite spinning directions. And, the spin fields will coexist permanently.*

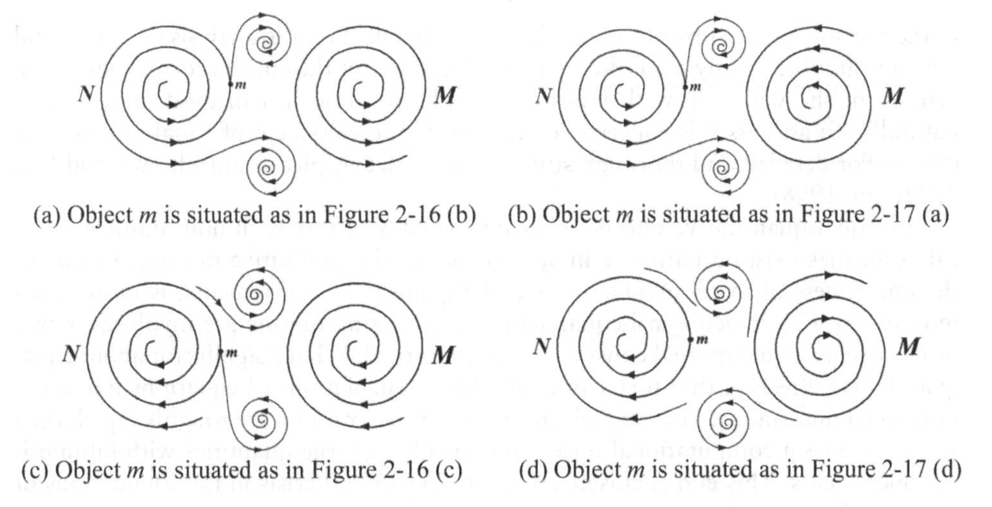

N
m
M
(a) Object m is situated as in Figure 2-16 (b)
(b) Object m is situated as in Figure 2-17 (a)
(c) Object m is situated as in Figure 2-16 (c)
(d) Object m is situated as in Figure 2-17 (d)

Figure 2.31 Object *m* might be thrown into a sub-eddy created by the spin fields of *N* and *M* jointly.

That is to say, Newton's third law holds true temporarily for cases (a), permanently for cases (b) and partially for cases (c) in Figures 2.29–2.30.

2.3.4 The fourth law on state of motion

If we look at Newton's third law from the second angle: One spinning yoyo *m* is acted upon by an eddy flow *M* of a higher level and scale. If we assume *m* is a particle in a higher-level eddy flow *N* before it is acted upon on by *M*, then we are looking at situations as depicted in Figures 2.27 and 2.28. Jointly, we have what is shown in Figure 2.31, where the sub-eddies created in Figure 2.31(a) are both converging, since the spin fields of *N* and *M* are suppliers for them and sources of forces for their spins. Sub-eddies in Figure 2.31(b) are only spinning currents. They serve as middle stop before supplying to the spin field of *M*. Sub-eddies in Figure 2.31(c) are diverging. And, sub-eddies in Figure 2.31(d) are only spinning currents similar to those in Figure 2.31(b).

That is, based on our analysis on the scenario that one object *m*, situated in a spin field *N*, is acted upon by an eddy flow *M* of a higher level and scale, we can generalize Newton's third law to the following form:

The Fourth Law on State of Motion: *When the spin field M acts on an object m, rotating in the spin field N, the object m experiences equal, but opposite, action and reaction, if it is either thrown out of the spin field N and not accepted by that of M (Figure 2.27(a), (d), Figure 2.28(b) and (c)) or trapped in a sub-eddy motion created jointly by the spin fields of N and M (Figure 2.27(b), (c), Figure 2.28(a) and (d)). In all other possibilities, the object m does not experience equal and opposite action and reaction from N and M.*

2.3.5 Validity of figurative analysis

In the previous three subsections, we have heavily relied on the analysis of shapes and dynamic graphs. To any scientific mind produced out of the current formal education system, he/she will very well question the validity of such a method of reasoning naturally. To address this concern, let us start with the concept of equal quantitative effects. For detailed and thorough study of this concept, please consult (Wu and Lin, 2002; Lin, 1998).

By equal quantitative effects, is meant the eddy effects with non-uniform vortical vectorities existing naturally in systems of equal quantitative movements due to the unevenness of materials. Here, by equal quantitative movements, is meant such movements with which quasi-equal acting and reacting objects are involved or two or more quasi-equal mutual constraints are concerned. What's significant about equal quantitative effects is that they can easily throw calculations of equations into computational uncertainty. For example, if two quantities x and y are roughly equal, then $x - y$ becomes a computational uncertainty involving large quantities with infinitesimal increments. This end is closely related to the second crisis in the foundations of mathematics.

Based on recent studies in chaos (Lorenz, 1993), it is known that for nonlinear equation systems, which always represent equal quantitative movements (Wu and Lin, 2002), minor changes in their initial values lead to dramatic changes in their solutions. Such extreme volatility existing in the solutions can be easily caused by changes of a digit many places after the decimal point. Such a digit place far away from the decimal point in general is no longer practically meaningful. That is, when equal quantitative effects are involved, we face either the situation where no equation can be reasonably established or the situation that the established equation cannot be solved with valid and meaningful solution.

That is, the concept of equal quantitative effects has computationally declared that equations are not eternal and that there does not exist any equation under equal quantitative effects. That is why OuYang (Lin, 1998) introduced the methodological method of "abstracting numbers (quantities) back into shapes (figurative structures)". Of course, the idea of abstracting numbers back to shapes is mainly about how to describe and make use of the formality of eddy irregularities. These irregularities are very different of all the regularized mathematical quantifications of structures.

Because the currently available variable mathematics is entirely about regularized computational schemes, there must be the problem of disagreement between the variable mathematics and irregularities of objective materials' evolutions and the problem that distinct physical properties are quantified and abstracted into indistinguishable numbers. Such incapability of modern mathematics has been shown time and time again in areas of practical forecastings and predictions. For example, since theoretical studies cannot yield any meaningful and effective method to foretell drastic weather changes, especially about small or zero probability disastrous weather systems, (in fact, the study of chaos theory indicates that weather patterns are chaotic and unpredictable. A little butterfly fluttering its tiny wings in Australia can drastically change the weather patterns in North America (Gleick, 1987)), the technique of live report has been widely employed. However, in the area of financial market predictions, it has not been so lucky that the technique of live report can possibly applied as effectively.

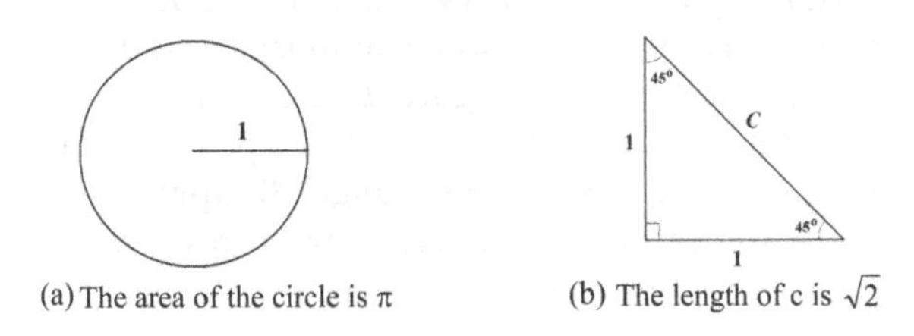

(a) The area of the circle is π (b) The length of c is $\sqrt{2}$

Figure 2.32 Representing π and $\sqrt{2}$ figuratively and precisely.

Due to equal quantitative effects, the movements of prices in the financial market place have been truly chaotic when viewed from the contemporary scientific point of view.

That is, the introduction of the concept of equal quantitative effects has made the epistemology of natural sciences gone from solids to fluids and completed the unification of natural and social sciences. More specifically, after we have generalized Newton's laws of motion, which have been the foundations on which physics is made into an "exact" science, in the previous three subsections, these new laws can be readily employed to study social systems, such as military conflicts, political struggles, economic competitions, etc.

Since we have briefly discussed about the concept of equal quantitative effects and inevitable failures of current variable mathematics under the influence of such effects, then how about figurative analysis?

As for the usage of graphs in our daily lives, it goes back as far as our recorded history can go. For example, any written language consists of an array of graphic figures. In terms of figurative analysis, one early work is the book called *Yi Ching* (or the *Book of Changes*, as known in English (Wilhalm and Baynes, 1967)). For now, no one knows exactly when this book was written and who wrote it. All that is known is that the book has been around since about three thousand years ago. In that book, the concept of Yin and Yang was first introduced and graphic figures are used to describe supposedly all matters and events in the world. When Leibniz (a founder of calculus) got his hands on that book, he introduced the binary number system and base p number system in modern mathematics (Kline, 1972). Later on, Bool furthered this work and laid down the foundation for the modern computer technology to appear.

In our modern days, figures and figurative analysis are readily seen in many aspects of our lives. One good example is the number π. Since we cannot write out this number in the traditional fashion (in either the decimal form or the fraction form), we simply use a figure π to indicate it. The same idea is employed to write all irrational numbers. In the area of weather forecasting, figurative analysis is used each and every day in terms of weather maps. In terms of studies of financial markets, a big part of the technical analysis is developed on graphs. So, this part of technical analysis can also be seen as an example of figurative analysis.

From the recognition of equal quantitative effects and the realization of the importance of figurative analysis, OuYang invented and materialized a practical way to "abstract numbers back into shapes" so that the forecasting of many disastrous small or zero probability weather systems becomes possible. For detailed discussion about this end, please consult Appendix D in (Wu and Lin, 2002) and (Lin, 1998). To simplify the matter, let's see how to abstract numbers π and $\sqrt{2}$ back into shapes with their inherent structures kept. In Figure 2.32(a), the exactly value π is represented using the area of a circle of radius 1. And, the precise value of $\sqrt{2}$ is given in Figure 2.32(b) by employing the special right triangle. By applying these simply graphs, the meaning, the precise values and their inherent structures of π and $\sqrt{2}$ are presented once for all.

Part 2

The mind

Chapter 3

Human body as a system

This chapter shows that each human body is a system; and the nature is also a system, consisting of other systems, including humans. To this end, we make use of the traditional Chinese medicine, the over two thousand years old classic, named Tao Te Ching, and the systemic yoyo model as the foundation of reasoning. The content of this chapter is mainly from (Lin and Forrest, 2010a).

More specifically, this chapter is organized as follows. In Section 3.1, we study the concept of systems and their fundamental properties, such as components, structures, subsystems and levels. In Section 3.2, we show by visiting some of the basic concepts and theory of the traditional Chinese medicine how human body has been studied and treated as an organic whole since the very start of recorded history over 5,000 years ago. In Section 3.3, we look at systems and their environments. Here we consider open and closed systems and interaction of systems. In Section 3.4, we glance over the classic Tao Te Ching, which is the most translated book ever written by man, to see how since antiquity man and nature have been investigated holistically as systems that are related and connected to each other.

3.1 SYSTEMS AND FUNDAMENTAL PROPERTIES

The purpose of this section is to introduce the concept of systems and to study some of the fundamental properties of systems. What is investigated in detail includes subsystems and the level structure of systems.

3.1.1 Systems: What are they?

The English word "system" is originally adopted from the ancient Greek "σύστημα". It means a whole that is made up of parts. *The Great World System*, by the ancient Greek philosopher Democritus, is the earliest book in which this word is employed.

Due to specific and varied focuses of study, the concept of systems is often defined differently from one discipline to another. Systems science investigates those concepts, properties, and characteristics of systems that are not discipline specific. As for the concept of systems well studied in systems science, it has two widely accepted forms.

Definition 3.1 (Qian, 1983b). Each system stands for a whole that is made up of mutually restraining parts and that possesses certain functionality none of the parts possesses.

Systems introduced in such a way are technology based and emphasizes on the attributes, properties, and functionalities of the systems not shared by their parts. Speaking in terms of technology, the purpose of various designs and organizations of management systems is to materialize and to actualize some pre-determined functionalities of the specific systems. Therefore, the essential characteristics that separate one system from another are the systems' specified functionalities.

Definition 3.2 A system is an organically combined entity of some elements along with their connections and mutual interactions. In particular, if in the set of elements are there at least two distinguishable objects and all these elements are associated together through a recognizable expression, then this set is seen as a system.

The concept of systems defined in Definition 3.2 is introduced on the level of fundamental science, emphasizing on the mutual interactions of the elements and the overall effect of the system on each of the elements.

Based on what is presented above, a more comprehensive definition for the concept of systems can be established as follows:

Definition 3.3 Each system is an organic whole, which is made up of some elements and their mutual interactions, with certain structure and functionality.

In order to develop both mathematically and scientifically sound results for the concept of systems of systems science, let us see how systems can be defined and studied using set theory.

A set is defined as a collection of objects, also known as elements. Let X be a given set. This set is ordered, if a binary relation, called an order relation and denoted $\leq$, exists with X as its field such that

1 For all $x \in X, x = x$. That is the relation $\leq$ is reflexive.
2 For $x, y \in X$, if $x \leq y$ and $y \leq x$, then $x = y$; i.e., the relation $\leq$ is antisymmetric.
3 For $x, y, z \in X$, if $x \leq y$ and $y \leq z$, then $x \leq z$. That is, the relation $\leq$ is transitive.

The ordered set X will be denoted as the ordered pair $(X, \leq)$. The notation $x > y$ means the same as $y < x$ and is read either "y precedes x" or "x succeeds y." The field of the order relation $\leq$ is often said to be ordered without explicitly mentioning $\leq$. It should be noted that a set may be ordered by many different order relations. An order relation $\leq$ defined on the set X is linear if for any $x, y \in X$, one of the following conditions holds true: $x \leq y$, $x = y$, or $y \leq x$. If X is not linearly ordered, it is known as partially ordered.

Two ordered sets $(X, \leq_X)$ and $(Y, \leq_Y)$ are similar, if there is a bijection h: $X \rightarrow Y$ satisfying for any $x, y \in X$, $x \leq_X y$ if and only if $h(x) \leq_Y h(y)$. This similarity relation divides the class of all ordered sets into pairwisely disjoint subclasses, each of which contains all sets that are similar to each other. Two ordered sets $(X, \leq_X)$ and $(Y, \leq_Y)$

are said to be the same order type if they are similar. Each order type will be denoted by a lower-case Greek letter.

An ordered set $(X, \leq_X)$ is well-ordered if it is linearly ordered and every nonempty subset of X contains a first element with respect to the relation $\leq_X$. The order type of this well ordered set X is known as an ordinal number. This concept of ordinal numbers is a generalization to that of any (either finite or infinite) sets of the finite ordinal numbers of 1st, 2nd, 3th, ...

With the necessary background of set theory in place, let us see how the concept of systems has been carefully studied in the language of set theory since the late 1980s.

Definition 3.4 (Lin, 1987). A system S is an ordered pair of sets, $S=(M,R)$, such that M is the set of all objects of S, and R a set of some relations defined on M. The sets M and R are respectively called the object set and the relation set of the system S. Here, for any relation $r \in R$, r is defined as follows: There exists an ordinal number $n=n(r)$, which is a function of r, called the length of the relation r, such that $r \subseteq M^n$, where

$$M^n = \underbrace{M \times M \times \cdots \times M}_{n \text{ times}} = \{f : n \to M \text{ is a mapping}\} \tag{3.1}$$

is the Cartesian product of n copies of M.

Remarks: Based on the conventions of set theory, what this definition says includes the following:

1 When a specific study is given, the set of all the elements involved in the study are definitely determined. Although a specific element itself can be a system at a deeper level, in the said study it is simply an element. For instance, in a research of a social organization, the focus is the people involved in the organization so that these particular people will be considered as the elements of the system investigated, although each of these human beings can be further researched as a system of cells.
2 The relation set R contains all the existing descriptions regarding the connections and mutual interactions between the elements in M. In particular, the relations in R together as a totality describe the internal structure of the system S. In the physical world, it is the internal structure that represents the visible attributes and characteristics of the system S.
3 From the discussions that will follow in the rest of this chapter, it can be seen that each system as defined in this definition is static. It represents the state of affairs of the system at a specific moment of time. That is, if the evolution of the system is concerned, one has to connect a sequence of such defined systems together along the time axis so that the concept of time systems will be needed, see below for more details.

Given two systems $S_1=(M_1,R_1)$ and $S_2=(M_2,R_2)$, S_2 is said to be subsystem of S_1 if $M_2 \subseteq M_1$ and for each relation $r_2 \in R_2$ there exists a relation $r_1 \in R_1$ such that $r_2 \subseteq r_1|M_2 = r_1 \cap M_2^{n(r_1)}$. That is, system S_2 is a part of S_1. It is neither a general part of S_1 nor a simple component of the overall system. Instead, it itself also forms a system.

At the same time, as expected, the overall system S_1 also contains other subsystems. When compared to the overall system S_1, each particular subsystem focuses more on a specific region or part of the overall system. Different subsystems represent their distinct spaces, structures, and internal connections.

An entity N is considered a non-system or a trivial system, if it satisfies one of the following conditions:

1 The objects involved with N are not clearly distinguishable; and/or
2 The objects of N are not associated with each other in any way.

Using the language of set theory, a non-system N can still be written as an ordered pair of sets $N=(M, R)$ so that either $M=R=\emptyset$ (the empty set) or even though $M\neq\emptyset$, either $R=\emptyset$ or $R=\{\emptyset\}$ holds true. Similar to the concept of zero in mathematics, in this case, when N satisfies the former condition, N is referred to as a trivial system; when N satisfies the latter condition, N is referred to as a discrete system.

Theorem 3.1: (Lin and Ma, 1993). There does not exist such a system that its object set contains all systems.

This is a restatement of the Zermelo–Russell paradox: By definition of sets, consider the set X of all those sets x that do not contain themselves as elements, that is, $X=\{x: x\notin x\}$. Now, a natural question is whether $X\notin X$ or $X\in X$? If the former $X\notin X$ holds true, then the definition of the set X implies that $X\in X$, a contradiction. If the latter $X\in X$ holds true, then the definition of X implies that $X\notin X$, also a contradiction. These contradictions constitute the well-known Zermelo–Russell paradox.

Proof: Consider the class V of all sets. For each $x\in V$ define a system by $(x, \emptyset)$. Then a 1-1 correspondence h is defined such that $h(x)=(x, \emptyset)$. If there is a system $S=(M, R)$ whose object set contains all systems, then M is a set. According to the Zermelo–Russell paradox, this end is impossible. QED.

3.1.2 Structures and subsystems

Components (and parts) and structures stand for two closely related but very different concepts. The former represents systems' basic or important building blocks (or hard factors) or constructive factors (soft factors). Parts do not involve relationships. However, structures and only structures deal with the relationships between parts.

The concept of structure represents the totality of all connections and interactions between parts. Its focus is how parts are combined together to form the whole system. All physically existing associations and relative movements of the parts belong to the structure of the underlying system. However, in practical investigations it is both impossible and unnecessary to uncover all possible associations and relative movements of the parts. In such a case, the word "structure" stands for the totality of only those associations and movements that play the dominating role in the development of the system with most of the other minor relationships ignored.

The concepts of structures and parts, as just described, are related but different. Structures cannot exist independently from parts and their appearances can only be carried by parts. When parts are considered without the context of structures, they are no longer parts. Without a comprehension of structures, there will be no way to determine parts, without an understanding of parts there will be no basis to talk about structures. That explains why in the concept of systems $S = (M, R)$, as defined using the language of set theory, the object set M and relation set R cannot be talked about separately without considering the other.

As for the structure of a system, it generally contains intensions of many different aspects. So, it needs to be studied from various angles. For instance, let us look at a nation as a system. Then this system needs to be analyzed from such diverse angles as the structure of peoples, that of classes, that of administrative organizations.

When studying systems' structures, one needs to pay attention to generative and non-generative relations. In a school, seen as a system, the relationships between the teachers and students, those between students, those between teachers, and so on, are generative relations, while such relations as those of coming from the same hometown are non-generative. For a biological body, when seen as a system, generative relations might be the biological associations and functions between cells and between organs, while all connections and functionalities in the sense of physics are non-generative. Speaking generally, each decisive connection stands for a generative relation. Of course, some non-generative relations may also produce not ignorable effects on the system of concern. Hence, when analyzing systems, one should also pay attention to non-generative relations. For instance, within a large social organization there are often unorganized groups, which are created by some non-generative relations of the social organization. Theories of organizations often treat these unorganized groups as an important research topic.

Generally systems are structurally classified according to the following two aspects:

1 Frameworks and Movements

Frameworks: the fixed ways of connection between components. They stand for the fundamental ways of association between various components when the system is situated in its inertial state and not affected by any external force.

Movements: The relationships of components when the system is in motion due to the effects of external forces. They represent the mutual reliance, mutual supports, and mutual constraints of the components when the system is in motion.

For example, in an automobile, when seen as a system, the relative locations, connections, spatial distributions, etc., of its body, engine, steering wheel, and other parts are the framework (structure) of the system. The way of how all the components coordinate with each other when the automobile is moving stands for a movement (structure).

2 Spatial and Time Structures

Each physical system and its components exist and evolve within certain space and time period. Any association of the components can only be shown in the form of space and/or time. Hence there are the concepts of spatial and time structures.

Spatial structures: the arrangement and distribution of components in space. They stand for the way of how components are spatially distributed and the consequent relations of mutual supports and mutual constraints between the components.

Time structures: the way of association of the components in the flow of time. They represent the development of the system, as treated as a process, in the time dimension. They can be associations, connections, or transitions of the components from one time period to another

For instance, the structure of a house is a spatial structure. The transitions between childhood, youth, middle age, and old age are time structure. There are of course situations where time and spatial structures are mixed. For instance, the growth rings of trees are mixed time and spatial structures.

Although there are many different kinds of associations between system's components, the associations can be roughly classified into two classes. One is explicit; it can be easily felt, described, and controlled directly. This class is referred to as hard associations. The other kind of associations is implicit, difficult to feel, described, and control. Each such association is known as soft. The totality of the former associations is known as the hard structure of the system, while the latter the soft structure. In principle, each system has both a hard and soft structure. The framework of the system is a hard structure, while the movement reflects a degree of its soft structure. For example, the physical connections of the hardware of a computer represent the system's hard structure, while the associations of the software of the computer stand for the soft structure. For mechanical systems in general, there is no need to consider their soft structures. However, for organic systems, their soft structures generally have to be considered. In particular, the soft structures of humanistic social systems are extremely important. For instance, the success or failure of any management system is very often determined by the system's soft structure.

Typical structures of systems include: chains, rings, nests, pyramids, trees, networks, etc.

If one uses mathematical symbols, the concept of structures has been well studied in universal algebra (Gratzer, 1978). At this juncture, let us take a quick look at this concept.

Let A be a set and n a nonnegative integer. An n-ary operation on A is a mapping f from A^n into A. An n-ary relation r on the set A is a subset of A^n. A type τ of structures is an ordered pair

$$\left((n_0, \ldots, n_\nu, \ldots)_{\nu < O_0(\tau)}, (m_0, \ldots, m_\nu, \ldots)_{\nu < O_1(\tau)}\right) \tag{3.2}$$

where $O_0(\tau)$ and $O_1(\tau)$ are fixed ordinal numbers and n_ν and m_ν nonnegative integers. For every $\nu < O_0(\tau)$ there exists a symbol f_ν of an n_ν-ary operation, and for every $\nu < O_1(\tau)$ we realize r_ν as an m_ν-ary relation.

A structure U is a triplet (A, F, R), where A is a nonempty set. For every $\nu < O_0(\tau)$ we realize f_ν as an n_ν-ary operation $(f_\nu)_U$ on A; and for every $\nu < O_1(\tau)$ we realize r_ν as an m_ν-ary relation $(r_\nu)_U$ on A, and

$$F = \{(f_0)_U, \ldots, (f_\nu)_U, \ldots\}, \quad \nu < O_0(\tau) \tag{3.3a}$$

$$R = \{(r_0)_U, \ldots, (r_\nu)_U, \ldots\}, \quad \nu < O_1(\tau). \tag{3.3b}$$

If $O_1(\tau) = 0$, U is called an algebra and if $O_0(\tau) = 0$, U is called a relational system.

For the rich variety of results on mathematical structures, as defined above, please consult (Gratzer, 1978).

3.1.3 Levels

Comparing to the concept of subsystems, the concept of levels is more important in comprehending systems. This concept constitutes an important content of the part-whole relationship of systems. At the theoretical level, there are several ways to study this concept. For example, each system has at least two levels, one at the overall system level and the other at parts' level. And physically existing systems, almost without exception, possess multi-levels. Here, the classification of levels can be, but does not have to be, closely related to that of subsystems.

Because parts themselves have different senses of levels, it makes the concept of levels extremely difficult to define. As of this writing, there is still no complete level theory of systems. To this end, in this subsection, we will look at one specific concept of level systems, where each system contains only two levels, one at the overall system level and the other at the element level, where some elements are also systems.

Symbolically, a system $S=(M, R)$ is said to have n levels, where n is a fixed natural number, if

1 Each object $S_1=(M_1, R_1)$ in M is a system, called the first-level object system; and
2 If $S_{n-1}=(M_{n-1}, R_{n-1})$ is an $(n-1)$th-level object system, then each object $S_n=(M_n, R_n) \notin M_{n-1}$ is also a system, called the nth-level object system of S.

Definition 3.5 (Lin, 1999, pp. 99). A system $S=(M, R)$ is centralized if each object in S is a system and there exists a nontrivial system $C=(M_C, R_C)$ such that for any distinct elements x and $y \notin M$, say $x=(M_x, R_x)$ and $y=(M_y, R_y)$, then $M_C=M_x \cap M_y$ and $R_C \subseteq R_x|M_C \cap R_y|M_C$. The system C is called a center of S.

The concept of centralized systems captures the realistic phenomena that in a physical system, it is very likely that the system has a central part such that when this part changes slightly, major effects are felt throughout the entire system.

Theorem 3.2: [ZFC (Lin and Ma, 1993)]. Let κ be a natural number and $\theta > \kappa$ the cardinality of the set of all real numbers. Assume that $S=(M, R)$ is a system satisfying

1 $|M| \geq \theta$; and
2 Each object $m \in M$ is a system with $m=(M_m, R_m)$ and $|M_m| < \kappa$.

If there exists an object contained in at least θ objects in M, there then exists a subsystem system $S'=(M', R')$ of S such that S' forms a centralized system and $|M'| \geq \theta$.

The abbreviation ZFC in this theorem and some of the following theorems means that this result holds true in the Zermelo-Fraenkel axiomatic set theory with the axiom of choice.

The proof of this theorem is quite technique and is omitted. For those interested readers, please consult (Lin, 1999).

This theorem is employed in the study of civilizations by (Lin and Forrest, 2010a). These authors provide an insightful theoretical explanation for why any closed society has to go through periodic turmoil in order to adjust itself in its evolutionary development.

Theorem 3.3: [ZFC] (Lin, 1999, pp. 192). Let $S = (M, R)$ be a system and $S_n = (M_n, R_n)$ an nth-level object system of S. Then S cannot be a subsystem of S_n for each natural number n.

Proof: The theorem will be proved by contradiction. Suppose that for a certain natural number n, the system S_0 is a subsystem of the nth-level object system S_n.

Let $S_i = (M_j, R_i)$ be the systems, for $i = 1, 2, \ldots, n-1$, such that $S_i \in M_{i-1}$, $i = 1, 2, \ldots, n$. Define a set X by

$$X = \{M_i : 0 \leq i \leq n\} \cup \{S_i : 0 \leq i \leq n\} \cup \{\{M_i\} : 0 \leq i \leq n\} \tag{3.4}$$

From the axiom of regularity of axiomatic set theory, which states that every non-empty set A contains an element B which is disjoint from A, it follows that there exists a set $Y \in X$ such that $Y \cap X = \emptyset$. There now exist three possibilities:

(i) $Y = M_i$, for some i,
(ii) $Y = S_i$, for some i, and
(iii) $Y = \{M_i\}$, for some i.

If possibility (i) holds, $S_{i+1} \in Y \cap X$ for $i \leq n-1$; and $S_1 \in Y \cap X$ if $Y = M_n$. Therefore, $Y \cap X \neq \emptyset$, contradiction. If (ii) holds, $Y = S_i = (M_i, R_i) = \{\{M_i\}, \{M_i, R_i\}\}$ and $\{M_i\} \in Y \cap X \neq \emptyset$, contradiction. If (iii) holds, $M_i \in Y \cap X \neq \emptyset$, a contradiction. These contradictions show that S_0 cannot be a subsystem of S_n. QED.

In this following section, let us see how human body can be and has been seen as an organic system since the very beginning of the recorded history.

3.2 LOOKING AT HUMAN BODY MORE TRADITIONALLY

For our purpose of showing how human body has been studied and treated as an organic whole since the very start of recorded history over 5,000 years ago, let us look at the Chinese traditional medicine in some detail.

3.2.1 Basic elements of Chinese traditional medicine

The theory of Chinese traditional medicine is based on four primitive terms: Yin, Yang, Qi, and Xue. Here, Xue might be identified with blood in the Western medicine, which by the way no one knows the correctness of this identification for sure. Qi is some kind of vapor, which together with Xue is supposed to carry oxygen and various nutrients throughout the human body, and which guarantees and nurtures the balance of Yin and Yang.

As for Yin and Yang, they also have various meanings. For example, running a fever is Yang, and feeling cold is Yin. Each organ of the human body or the entire body can be controlled by either Yin or Yang.

In general, Chinese traditional medicine is not concerned with microorganisms or details of the body's organs and tissues. The strength of this classical medical discourse lay rather in its sophisticated analysis of how functions were related on many levels, from the vital processes of the body to the emotions to the natural and social environment of the patient, always with therapy in mind. Chinese medicine is best evaluated in the light of this strength rather than according to criteria that could not have been applied until half a century or so ago. For more comments along this line, see (Sivin, 1990).

Some modern physicians who have not troubled themselves to study classical medical doctrines dismiss them as futilities of the feudal past. Others have portrayed classical medicine as a remarkable corpus of theory – based on adaptations of the Yin-Yang and concepts of five-phases – that succeeded in understanding the body as a many-leveled system and treated its ills holistically. To be more specific, let us concentrate on a small branch of Chinese traditional medicine: acupuncture.

3.2.2 Brief history of Chinese traditional medicine

As a means of curing diseases, acupuncture is effected by needling certain related points on the human body to invigorate the meridian system and to harmonize Qi energy with blood circulation. According to the literature, unearthed cultural relics, and the related research on the laws of development of societies, the art of acupuncture had germinated way before the written language was created (Qiu and Zhang, 1985). The study of the cultural relics unearthed in 1963 shows that the use of stone needles can be dated back to more than 14,000 years ago. Accompanying the invention and development of metallurgical technology, by the time the earliest medical book "Canon of Medicine" (500–300 B.C.) was written, ancient stone needles, bone needles, and bamboo needles, has been gradually replaced with bronze needles, iron needles, gold needles, and silver needles. And in our modern time, stainless steel needles have been in use.

As the history indicates, many had considered that China's Yellow Emperor of 2600 B.C. be the founder of acupuncture, and the number of acupuncture points known to the physicians had been gradually increased through generations of practice. A total of 295 acupuncture points, including 25 single points and 135 double points were described in the earliest medical book "Canon of Medicine" (500–300 B.C.).

In the "Jia Yi Classics of Acupuncture," written by Huanfu Mi in 260 A.D., some more points were added, making a total of 649 acupuncture points, including 49 single points and 300 double points. In 1027 A.D., Wei-yi Wang did an investigation along this line and supplemented with more points, reaching a total of 51 single points and 303 double points. Wang also casted the first bronze human figure to show the distribution of the meridian channels and the locations of the points.

In the year of 1601, "Illustrated Bronze-Figure Ming Tang," a chart illustrating acupuncture points and meridian channels, was drawn by Wen-bing Zhao. The illustration was made on the skeleton shown from four different angles: the front, the posterior, the left, and the right posterior. The "Compendium of Acupuncture and Moxibustion" edited by Ji-zhou Yang located 667 acupuncture points, including 51 single points and 308 double points.

According to the "Law for Medical Practice," published in 1742, the number of acupuncture points was further augmented to 670, including 52 single points and 309 double points. Since 1949, the technique of acupuncture treatment has been widely used in medical fields. The successful results of acupuncture anesthesia were published in 1971.

As of today, acupuncture treatment has various effects on 300 some diseases, around 100 of which acupuncture treatment has shown very satisfactory results. The effectiveness of the treatments of cardiovascular and cerebral diseases, gallstone, bacterial dysentery, etc., had not only been confirmed with modern scientific methods, but their working principles had also been giving in terms of physiology, microbiology, and immunology.

3.2.3 Modern interest in acupuncture

The research and clinical practice of acupuncture are not just limited to China. In fact, by around 600 A.D., the theory and clinical practice had been spread to Korea, by around 562 A.D., to Japan; and by the end of 1700 AD, to Europe. Currently, the research and its clinic practice of acupuncture can be found in more than one hundred different countries in the world; and more and more people turn to it for "magic" cure.

For example, in 1994 alone, Americans made some 9 to 12 million visits to acupuncturists for ailments as diverse as arthritis, bladder infections, back pain, and morning sickness. And, internationally, well-known schools, such as Oxford University and University of Maryland, are involved in the fundamental research of the mechanism on why the needles really do what they were said to do. For details, see (Weiss, 1995).

The theory of meridians is about the structure and distribution, physiological functions, pathological changes and its relations with the organs of the human meridian system. It is a major part of the theory of Chinese traditional medicine.

According to this theory, the meridian system consists of two parts. One is called "Jing-Mai," which means pathways and goes up and down and connects human body's interior and exterior. This part dominates the entire meridian system. The other part is called "Luo-Mai," which means network, consists of small branches out of the Jing-Mai, and distributed crisscross all other the body.

The meridian system internally belongs to the classification of organs, and externally networks along with the four limbs. It communicates between the organs and the body surface so that the human body becomes an organic whole in which Qi and Xue are so moved, and Yin and Yang are so nurtured with nutrients that the functions of each part of the body are kept harmonized and balanced. Each clinical application of acupuncture is based on this theory.

3.2.4 The meridian system

The theory of meridians had been formed more than two thousand years ago based on

1 Clinical observations.
2 The logic reasoning through pathological changes on the skin.
3 Inspiration from the knowledge of anatomy and pathology.

Along with clinical observations, the points with similar curing powers are always nicely located on lines. By classifying the acupuncture point with similar functionalities together, the channels of meridians were gradually formed. In various clinical practices, it had been noticed that when a certain organ was not functioning normally, there would be pain under pressure, rashes, color changes, etc., at certain parts of the body. The observation and analysis of these surface pathological phenomena also helped the establishment of the theory of meridians. Through autopsies and vivisection, ancient physicians learned the locations, shapes, and some pathological functions of the organs. And the observation of the distribution of many tubal and stringy structures and their connections with various parts in the human body has inspired the understanding of meridians.

Let us now take a look at the pathological functions of the meridian system. Meridians possess the capability of connecting internal organs and four limbs. Even though the organizational organs, including the vital organs of human body, all the limbs and bones, the five sense organs, the nine apertures, skin, flesh, muscles, etc., have their own different pathological functions, they function collectively in such a way that the interior and the exterior, the upper body and the lower body are harmonized and unified into an organic whole. All these mutual connections and collaborations are realized through the meridian system.

Meridians possess the ability to transport Qi, Xue, and nutrients throughout the body, to nurture Yin and Yang, and to protect the body from various outside attacks and invasions of diseases. All human organs need to be nurtured with Qi and Xue in order to function normally. Qi and Xue are the material foundation for all activities of living human bodies, which must be transported crisscross all over the body through the meridians.

Clinical applications of the theory of meridians consist of three parts:

1 Explain pathological changes of diseases;
2 Guide physicians to make correct diagnoses; and
3 Help physicians to make efficient treatment plans.

Let us now look at each of these three parts with more details.

For the first part, when a person's immune system is low, his meridians become the paths for pathologic bacteria to enter his body. When the body surface is attacked by pathologic bacteria, the ailment can be transported from the exterior of the body to the interior, from slightly discomfort to extremely serious and complicated illness.

For example, when pathologic bacteria attack one's surface, he would experience some of the following symptoms:

- Fever;
- Chill;
- Headache;
- Body-ache, etc.

Since the lungs are closely connected with the skin and hair by meridians, the pathologic bacteria will enter the dwell in the lungs through the meridians. That in turn will cause symptoms of lung diseases, including coughing, short breath, chest pain, and so on. Since the meridians establish the communications between the vital organs, and between the internal and the surface organs, they also transport illness of one part of the body to another. For instance,

- The heart sends heat to small intestine;
- Liver diseases affect the stomach;
- Stomach abnormalities interfere with the spleen.

Abnormalities of the internal organs can be reflected through the meridian system on the external organizational organs. For instance,

- Liver diseases can cause pain in the upper part of the side of the human body;
- Kidney diseases lumbago; stomach heat can cause the appearance of sores on the tongue;
- Heat of the large intestine or stomach can cause gum infection.

For the second part about how the theory of meridians guides physicians to make correct diagnoses, because meridians represent related organs, they can be used as indicators of the conditions of the relevant organs. So, in clinical practice, diagnoses can be made based on symptoms together with abnormalities of the meridians. For example, headache can be analyzed based on the distribution of meridians on the head as follows:

- If the pain is located at the forehead, it most likely has something to do with the yangming meridian;
- If the pain is on the sides, it very likely has something to do with the shaoyang meridian;
- If the pain is on the neck, it has something to do with the taiyang meridian;
- If the pain is on the top, it is likely related to the jueyin meridian.

Some diseases are accompanied with abnormal reactions at some acupuncture points, including changes of skin conditions, skin temperature, and skin electric resistance. These facts are also helpful in the diagnosis process.

For the third part about how the theory of meridians helps physicians to make efficient treatment plans, acupuncture treatment is based on the needling at various points on the body to dredge the meridian Qi so that all the organs can be nurtured and balanced with Qi, Xue, and various nutrients, and consequently, the disease can either be controlled or cured. The choice of acupuncture points is based on the definite analysis in second part above. Beside some local points, the choice of major acupuncture points is related to the distribution of the meridians. That is, when a meridian or an organ is sick, some points on this meridian or the meridians related to the organ will be used. Here, we omit some details. For interested readers, please consult (Qiu and Zhang, 1985).

From this brief, yet detailed enough exposition of Chinese traditional medicine, one can see that human body is indeed a whole, a system, with different parts and organs inseparably connected. Therefore, from what a system is meant to be (Lin, 2007; Lin, 2008b), it follows that each human being can be seen and studied as a spinning yoyo with its field expanding indefinitely into the space.

3.3 SYSTEM AND ITS ENVIRONMENT

As indicated by nature, no natural system exists in isolation. So, in this section, we study the concepts of environments, open and close systems, and interaction of systems.

3.3.1 Environments

Both the concepts of components and structures reflect the internal organization of systems, while interactions between systems and their environments reveal the external impacts on the systems by the environments. Only when one comprehends both the internal organization and the external impacts, he can acquire a relatively complete understanding of the behaviors of the systems.

Definition 3.6 For a given system S, the totality of all matters that are located outside the system S and that have certain kinds of associations to the system S constitutes the environment E of S. Symbolically using the ordered-pair Definition 3.4 of systems, the environment E is simply an another system that contains S as one of its elements and/or some elements of S are also elements of E.

Each system is born, develops, and evolves within certain environment. Its structure, state, attributes, and behaviors are closely related to and correlate with its environment. Hence, systems rely on their environments; interactions between environments and systems constitute the external condition for the development of systems. A system may possess different structures in different environments. That explains why in different environments, the same system may possess a varied set of behaviors and attributes. Therefore, the environment plays an important role in the emergence of the whole system. When the environment changes, the system has to develop a certain new set of holistic attributes in order to adjust itself to the changing external conditions. This end explains the fact that the development of complexities in the environment creates complexities in the system. Therefore, to investigate a system, one has to pay attention to the system's interaction with its environment.

Each environment has its objectiveness and relativeness.

Definition 3.7 The set of all objects that separate system S and its environment E is referred to as the boundary of the system S, denoted $B(S)$.

Proposition 3.1 If a given system S and its environment E can be written as $S = (M, R)$ and $E = (M_E, R_E)$, then $B(S) = M \cap M_E$.

Proof is straightforward and is omitted.

For each physical system, its boundary objectively exists. However, for some systems, their boundaries can be clearly defined, while some other systems, their boundaries cannot be easily located. For example, when civilizations are studied as systems, their boundaries at some geological locations and during certain historical moments can be surely drawn, while at some other geological locations and times, separating two cultures becomes an impossible task. So, according to the specifics of the problem of concern, one can select his desirable boundary and system. In such situations, the system will be defined as the totality of relevant events or matters that involve some of the research objects so that this totality is isolated out of the entire collection of interactive events and matters of the given task.

3.3.2 Open systems and close systems

When investigating a physical system, it is most likely that the system is artificially carved out from its environment. So, at the same time when one recognizes the classification certainty of the chosen system and its environment and definite difference between the internality of the system and the externality of the environment, he also needs to acknowledge the varied degrees of this certainty and definiteness and the relativeness of the classification between the system and its environment.

Definition 3.8 When a system does not interact with any external entity, the system is referred to as a closed system. Any system that is not closed is referred to as an open system. Symbolically, a system $S = (M, R)$ is closed, if and only if its environment E is simply the trivial system $(\emptyset, \emptyset)$ or $(\emptyset, \{\emptyset\})$.

For input-output systems, which are systems that take matters in from their environments and give matters off into their environments, openness helps make the systems viable and evolve. For physical systems, their openness can be either voluntary or forced. This end has been empirically evidenced by the events of the recent world history. On the other hand, closeness is a necessary condition instead of a negative factor for general systems to possess their individual identities. However, (Lin and Forrest, 2010a) shows that an expanded period of closeness is the sufficient condition for a civilization to experience internal turmoil and to suffer from external aggressions. So, for an input-output system, its viable and stable existence is guaranteed by a balance between its openness and closeness.

Using the relationship between systems and their respective environments, the totality of all systems can be classified into two classes: open systems and close systems.

3.3.3 Systems on dynamic sets

Because the usage of rigorous mathematical language in the discussion of properties of general systems is very advantageous, we will continue to employ this approach. In this section, we will not consider the case when a set is empty; and all sets are assumed to be well defined. That is, we do not consider such paradoxical situations as the set containing all sets as its elements. To this end, those readers who are interested in set theory and the rigorous treatment of general systems theory are advised to consult (Lin, 1999).

Let us now look at the difference between various parts of a system, that is, the differences between the system's objects and between the relations of the objects. We will employ the concept of attributes to describe such differences. By attribute, it means a particular property of the objects and their interactions and the overall behavior and evolutionary characteristic of the system related to the property.

If speaking in the abstract language of mathematics, the attributes of a system are a series of propositions regarding the system's objects, relations between the objects, and the system itself. Due to the differences widely existing between systems' objects, between systems' relations, and between systems themselves, these propositions can vary greatly. For example, a proposition might hold true for some objects, and become untrue for others. In this case, we say that these objects have the particular attribute, while the others do not. As another example, a proposition can be written as a mapping from the object set to the set of all real numbers so that different objects are mapped onto different real numbers. In this case, we say that the system's objects contain differences.

Additionally, each physical system evolves with time so that its objects, relations, and attributes should all change with time. So, when we consider the general description of systems, we must include the time factor. By doing so, the structure of a system at each fixed time moment is embodied in the object set, relation set, and attribute set, while the evolution of the system is shown in the changes of these sets with time.

Summarizing what is analyzed above, the system of our concern can be defined as follows:

Definition 3.9 Assume that T is a connected subset of the interval $[0, +\infty)$, on which a system S exists. Then, for $t \in T$, the system S is defined as the following order triplet:

$$S_t = (M_t, R_t, Q_t) \tag{3.5}$$

where M_t stands for set of all objects of the system S at the time moment t, R_t the set of the relations between the objects in M_t, and Q_t the set of all attributes of the system S. For the sake of convenience of communication, T is referred to as the life span and S_t the momentary system of the system S.

What needs to be emphasized is that we study not only the evolution of systems with time, but also the evolution of the system along with continuous changes of some conditions, such as temperature, density, etc. In such cases, T will be understood as one of those external conditions.

3.3.4 Dynamic subsystems

Just like each set has its own subsets, every system has subsystems, which can be constructed from the object set, relation set, and the attribute set of the system. In short, a system s is a subsystem of the system S, provided that the object set, relation set, and attribute set of s are corresponding subsets of those of S so that the restrictions of the attributes of S on s agree with the attributes of s. Symbolically, we have

Definition 3.10 Let $s_t = (m_t, r_t, q_t)$ and $S_t = (M_t, R_t, Q_t)$ are two systems, for any $t \in T$. If the following hold true:

$$m_t \subseteq M_t, r_t \subseteq R_t|_{s_t}, \quad \text{and} \quad q_t = Q_t|_{s_t}, \quad \forall t \notin T \tag{3.6}$$

then s_t is known as a subsystem of S_t, denoted

$$s_t < S_t, \ \ \forall t \notin T, \ \text{ or } \ s < S \tag{3.7}$$

where $R_t|_{s_t}$ and $Q_t|_{s_t}$ represent respectively the restrictions of the relations and attributes in R_t and Q_t on the system S_t and are defined as follows:

$$R_t|_{s_t} = \{r|_{m_t} : r \in R_t\} \text{ and } Q_t|_{s_t} = \left\{q|_{m_t \cup r_t} : q \in Q_t\right\}.$$

In this definition, the mathematical notation of less than is employed for the relationship of subsystems, because the relation of subsystems can be seen as a partial ordering on the collection of all systems. Similarly, when equ. (3.6) does not hold true, we say that s is not a subsystem of S, denoted $s_t \not< S_t$, $\forall t \notin T$, or $s \not< S$. Let the set of all subsystems of S be S, then we have

$$\mathrm{S} = \{\mathrm{s} : s < S\} = \{s_t : s_t < S_t, \forall t \in T\} \tag{3.8}$$

For any given system $S = (M, R, Q)$, where $M = \{M_t : t \in T\}$, $R = \{R_t : t \in T\}$, and $Q = \{Q_t : t \in T\}$, as defined by equ. (3.5), take a subset set of its object set $A = \{A_t \subseteq M_t : t \in T\}$. Then by restricting the relation set R and the attribute set Q on A, we obtain the following subsystem of S induced by A:

$$S|_A = \left\{s_t = \left(A_t, R_t|_{A_t}, Q_t|_{(A_t, R_t|_{A_t})}\right) : t \in T\right\} \tag{3.9}$$

where the restrictions $R_t|_{A_t}$ and $Q_t|_{A_t}$ are assumed respectively to be

$$R_t|_{A_t} = \left\{r|_{A_t} : \text{each element in } r|_{A_t} \text{ has the same length, } r \in R_t\right\} \tag{3.10}$$

and

$$Q_t|_{(A_t, R_t|_{A_t})} - \left\{q|_{(A_t, R_t|_{A_t})} : q|_{(A_t, R_t|_{A_t})} \text{ is a well defined proposition on } s_t, q \in Q_t\right\}. \tag{3.11}$$

Proposition 3.2 The induced subsystem $S|_A$ is the maximum subsystem induced by A.

Proof. Let $s = \left\{\left(A_t, r_{t,A}, q_{t,A}\right) : t \in T\right\} < S$ be an arbitrary subsystem with the entire A as its object set. According equ. (3.6) we have

$$r_{t,A} \subseteq R_t|_s, \ \text{ and } \ q_{t,A} = Q_t|_s = Q_t|_{(A_t, r_{t,A})}, \ \ \forall t \notin T$$

So, from equ. (3.10), it follows that $s \leq S|_A$only when $r_{t,A} = R_t|_s$, the equal sign holds true. QED.

Proposition 3.3 Assume that A and B are subsets of M satisfying that $B \subseteq A \subseteq M$, then $S|_B \leq S|_A$.

Proof. According equ. (3.9), we have $S|_B = \left\{ s_t = \left(B_t, R_t|_{B_t}, Q_t|_{(B_t, R_t|_{B_t})} \right) : t \in T \right\}$. $B \subseteq A$ implies that $B_t \subseteq A_t$, $\forall t \in T$. So, it follows that $R_t|_{B_t} \subseteq R_t|_{A_t}$ and consequently $R_t|_{B_t} \subseteq (R_t|_{A_t})|_{B_t}$ and $Q_t|_{(B_t, R_t|_{B_t})} = \left(Q_t|_{(A_t, R_t|_{A_t})} \right)\Big|_{(B_t, R_t|_{B_t})}$. Therefore, $S|_B \leq S|_A$. QED.

Proposition 3.4 Let S be a system. Then the collection of all subsystems of S forms a partially ordered set by the subsystem relation "<".

Proof. This result is a straightforward consequence of Proposition 3.3. QED.

Assume that S^1, S^2, and S are systems with the same time span such that $S^1 < S$ and $S^2 < S$. Let S^i, $i = 1, 2$, denote the set of all subsystems of S^i. Then each element in the set $\mathrm{S}^{i1} - \mathrm{S}^{i2} = \{ s : s < S^1, s \nless S^2 \}$ is a subsystem of S^1 but not a subsystem of S^2. Similarly, each element in the set $\mathrm{S}^{i2} - \mathrm{S}^{i1} = \{ s : s < S^2, s \nless S^1 \}$ is a subsystem of S^2 but not a subsystem of S^1. Let $\mathrm{S}^{i1} \Delta \mathrm{S}^{i2} = \{ s : s < S^1, s \nless S^2 \} \cup \{ s : s < S^2, s \nless S^1 \}$ be the union of the previous two sets of subsystems of either S^1 or S^2; and $\mathrm{S}^{i1} \cap \mathrm{S}^{i2} = \{ s : s < S^1, s < S^2 \}$ the set of all subsystems of both S^1 and S^2.

Proposition 3.5 Assume that $S^1 < S$ and $S^2 < S$. Then $\mathrm{S}^{i1} \cup \mathrm{S}^{i2} = \{ s : s < S^1 \} \cup \{ s : s < S^2 \}$ is a subset of $\mathrm{S}|_{M^1 \cup M^2}$.

Proof. $S|_{M^1 \cup M^2}$ is a maximal element in the partially ordered set $(\mathrm{S}|_{M^1 \cup M^2}, \subseteq)$, satisfying $\forall s \notin S|_{M^1 \cup M^2}, s < S|_{M^1 \cup M^2}$. On the contrary, $\forall s < S|_{M^1 \cup M^2}$, we have $s \in \mathrm{S}|_{M^1 \cup M^2}$. So, $\forall s \in \mathrm{S}^{i1} \cup \mathrm{S}^{i2}$, we have $s \in \mathrm{S}|_{M^1 \cup M^2}$. QED.

What this result indicates is that the union $\mathrm{S}^{i1} \cup \mathrm{S}^{i2}$ of the sets of subsystems of two subsystems S^1 and S^2 is a subset of the $\mathrm{S}|_{M^1 \cup M^2}$ of the subsystems of the induced system on the union $M^1 \cup M^2$.

Proposition 3.6 Given two arbitrary systems S^1 and S^2, there is always a system S^{12} such that $S^1 < S^{12}$ and $S^2 < S^{12}$.

Proof. Without loss of generality, assume that $S^1 = \{ (M_t^1, R_t^1, Q_t^1) : t \in T^1 \}$ and $S^2 = \{ (M_t^2, R_t^2, Q_t^2) : t \in T^2 \}$. To construct the system $S^{12} = \{ (M_t^{12}, R_t^{12}, Q_t^{12}) : t \in T^{12} \}$, $T^{12} = T^1 \cup T^2$, we first assume that the object sets M_t^1 and M_t^2 are disjoint, that is, $M_t^1 \cap M_t^2 = \emptyset$, for any $t \notin T^1 \cap T^2$. Then, the desired systemS^{12} is defined as follows: for $t \notin T^1 \cap T^2$,

$$\begin{cases} M_t^{12} = M_t^1 \cup M_t^2 \\ R_t^{12} = R_t^1 \cup R_t^1 \\ Q_t^{12} = Q_t^1 \cup Q_t^2 \end{cases} \tag{3.12}$$

for $t \notin T^1 - T^2$, define $M_t^{12} = M_t^1$, $R_t^{12} = R_t^1$, and $Q_t^{12} = Q_t^1$, and for $t \in T^2 - T^1$, define $M_t^{12} = M_t^2$, $R_t^{12} = R_t^2$, and $Q_t^{12} = Q_t^2$, and is denoted as $S^{12} = S^1 \oplus S^2$.

Now, if $M_t^1 \cap M_t^2 \neq \emptyset$, for some $t \notin T^1 \cap T^2$, we simply take two systems $^*S^1 = \{(^*M_t^1, ^*R_t^1, ^*Q_t^1) : t \in T^1\}$ and $^*S^2 = \{(^*M_t^2, ^*R_t^2, ^*Q_t^2) : t \in T^2\}$ with $M_t^1 \cap M_t^2 = \emptyset$, for any $t \notin T^1 \cap T^2$, such that $^*S^i$ is similar to S^i, $i = 1, 2$, where similar systems are defined in the same fashion as in (Lin, 1999, pp. 201). Then, we define $S^{12} =^* S^1 \oplus {}^*S^2$. Up to a similarity, the system S^{12} is uniquely defined. Therefore, it can be seen as well constructed such that $S^1 < S^{12}$ and $S^2 < S^{12}$, where the time spans T^1 and T^2 are seen as the same as T^{12} such that when $t \in T^1 - T^2$ (respectively, $t \in T^2 - T^1$), we treat S_t^2 (respectively, S_t^1) as a system with empty object set. QED.

From this proposition, it follows that when the interactions of some given systems are considered, these systems can always be seen as subsystems of a larger system. On the other hand, this proposition also shows that there is always some kind of interaction between two given systems, which is embodied in the fact that they are subsystems of a certain system.

3.3.5 Interactions between systems

When there is an interaction between two objects of a system $S = (M, R, Q)$ (of time span T), where $M = \{M_t : t \in T\}$, $R = \{R_t : t \in T\}$, and $Q = \{Q_t : t \in T\}$, it can be described by using a binary relation of the system. We say that objects m_1 and $m_2 \in M$, which means either $m_i = (m_t)_{t \in T}$ such that $m_t \in M_t$, for each $t \in T$, or $m_i = m_t \in M_t$, for a particular $t \in T$, $i = 1, 2$, interacts with respect to an attribute $q \in Q$, it means that the proposition q holds true for a binary relation $r \in R$ that contains either (m_1, m_2) or (m_2, m_1) or both. The idea of interactions between systems is a natural generalization of that between two objects. Based on the discussion of the previous section, we will discuss interactions of systems in the framework of subsystems. Assume that $S_i < S$, $i = 1, 2$. Now, let us look at how these subsystems could interact with each other.

Definition 3.11 The system S_1 is said to have a weak effect on the system S_2 with respect to an attribute $q \in Q$, provided that for any $r_2 \in R_2$ there is $r_1 \in R_1$ such that the proposition q holds true for the ordered pair (r_1, r_2). When no confusion is caused, we simply say that the system S_1 affects the system S_2 weakly without mentioning q. If the system S_2 also exerts a weak affect of S_1, then we say that these systems interact with each other weakly.

Definition 3.12 The system S_1 is said to have a strong effect on the system S_2 with respect to an attribute $q \in Q$, provided that for any $r_2 \in R_2$ and any $r_1 \in R_1$, the proposition q holds true for the ordered pair (r_1, r_2). When no confusion is caused, we simply say that the system S_1 affects the system S_2 strongly without mentioning q. If the system S_2 also exerts a strong affect of S_1, then we say that these systems interact with each other strongly.

From these definitions, it follows that strong interaction requires interactions between every ordered pair of objects, which is a more rigorous requirement than that of weak interactions. Also, to maintain the intuition behind the concepts of interactions, in Definitions 3.11 and 3.12, we only look at two relations $r_i \in R_i$, $i = 1, 2$. In order to capture the general spirit, these individual relations should be replaced by subsets $\{r_i \in R_i : \Phi_i(r_i)\}$, where $\Phi_i(\ )$ stands for the proposition that defines the set, for $i = 1, 2$. By doing so, what are discussed in Definitions 3.11 and 3.12 become special cases.

Definition 3.13 Given a subsystem $s = (m, r, q)$ of a system $S = (M, R, Q)$, the totality of all objects in $M - m$, each of which interacts weakly with at least one object in m, is known as the environment of the subsystem s in S, denoted E^S.

Next, let us see how man and nature have been recognized as two interacting systems for over two thousand years.

3.4 A GRAND THEORY ABOUT MAN AND NATURE

The purpose of this section is to look at how humans for over two thousand years have been treated as systems, which are nested in the grand system of nature. In particular, if each system is a spinning yoyo, then the universe will be an ocean of spinning currents.

3.4.1 Tao Te Ching: The classic

The concepts of Tao and Te were first introduced in the Chinese classic, named Tao Te Ching (Lao Tzu, unknown). This book is the most translated book in the world. Among many reasons for the superabundance of translations are the following:

1. This book is considered to be the fundamental text of both philosophical and religious Taoism. The Tao, or Way, is at the heart of *Tao Te Ching*, and is also the centerpiece of all Chinese philosophies and thoughts;
2. The brevity and the insights it offers make it among the few classics in the world. It is so short yet so packed with food for thoughts. It can be read and reread without exhausting the opportunity for obtaining new insights; and
3. It is supposed to be, in the word of the author himself, "very easy to understand." However, when one actually tries to read it, one will realize that it is extremely difficult to comprehend the book fully.

The book *Tao Te Ching* was written during Chou Dynasty (1030–207 B.C.) around 400–300 B.C. by a person known as Lao Tzu. Here, Tao means the Way that things should be and Te means integrity, which signifies the personal quality or strengths of an individual, one's personhood. Te is the moral weight of a person, which may be either positive or negative. Together, Tao Te means the overall character of a human being and his environment.

In the literature, there was very little to be found about the author concerning who he was and what profession he was in. According to many well-documented researches, it has been believed that Lao Tzu could actually be the editor of the collection, named *Tao Te Ching*, of many old sayings. In Chinese history, there are so many precedents of influential thinkers being named Tzu: K'ung Tzu (Confucius),

Meng Tzu (Mencius), Mo Tzu (Mecius or Macius) that one would naturally ask: What does the word "tzu" really mean? Here, tzu in Chinese has many meanings, including, "son," "pupil,", "man,", "scholar," etc. When combined with the word "lao", one of the many understandings could well be "father" in the sense of a family setting.

Together with the belief that the book is a collection of old sayings, it is reasonable to say that Lao Tzu (father) is the pseudonym of the editor of the collection; and whoever is reading *Tao Te Ching*, can simply believe that he is reading the words of wisdom of his own ancestors, and guided to be a good citizen and how to live harmonically with nature.

3.4.2 The purpose of Tao Te Ching

According to many scholars of both the ancient and modern times, *Tao Te Ching* is not just simply about the family teaching of children; more importantly, it was written as a handbook of ancient rulers. As for its philosophical stand(s), it is worth mentioning that there have been many different points of view about this end. The following are four main points of view (for more details, see (Wu, 1990)):

1. *Tao Te Ching* reflects the wishes of the then-peasants' class of private ownership;
2. *Tao Te Ching* represents the position of the then-growing class of farmers;
3. *Tao Te Ching* stands for the thoughts of the then-declining class of slave owners; and
4. *Tao Te Ching* is a book of the art of war, emphasizing on philosophical and logical education.

3.4.3 Modern systems research in Tao Te Ching

There is not only much in *Tao Te Ching* of a mystical and metaphysical quality, and it is also a bold combination of cosmic speculation and mundane governance. In this subsection, we focus on some of the ideas and concepts of modern systems theory either implicitly or explicitly contained in *Tao Te Ching*. Our discussion is based on V. H. Mair's recent translation (Lao Tzu, unknown) of the manuscript, discovered at Ma-Wang-Tui site in 1973. This new manuscript can be dated much closer, which means several hundred years earlier, to the supposed date when the classic was written than all other manuscripts used in various other translations.

In the light of modern systems research, *Tao Te Ching* is a theory of a special system, consisting of and about man and his environment. This system is named Tao-Te. The Tao or the Way is about how things including man and nature should be, and Te or Integrity is about the man itself. This is evidenced in Chapter 62 (Note: in Tao Te Ching, each chapter is about one page long. So, there is no need to cite the particular page numbers.),

> "Man patterns himself on earth, earth patterns itself on heaven, heaven patterns itself on the Way, the Way patterns itself on nature."

Here, it is believed that there is a single and overarching Way that encompasses everything in the universe.

Similar to modern axiom systems of scientific theories, the system of Tao-Te was introduced and studied without a definition given to either Tao or Te.

- "The Way is concealed and has no name" (Chapter 3).
- "The ways that can be walked are not the eternal Way; the names that can be named are not the eternal name. The nameless is the origin of the myriad of creatures" (Chapter 45).
- "The appearance of grand integrity is that it follows the Way alone. The Way objectified is blurred and nebulous. How nebulous and blurred! Yet within it there are images. How blurred and nebulous! Yet within it there are objects. How cavernous and dark! Yet within it there is an essence. Its essence is quite real; within it there are tokens" (Chapter 65).
- "There was something featureless yet complete, born before heaven and earth; silent – amorphous – it stood alone and unchanging" (Chapter 69).
- "The Way is eternally nameless" (Chapter 76).

Intuitively speaking, Te represents self-nurture or self-realization in relation to the cosmos. It is in fact the actualization of the cosmic principle in the self. Te is the embodiment of the Way and is the character of all entities in the universe. Each creature has a Te which is its own manifestation of the Tao. Tao represents cosmic unity, while Te stands for the individual personality or character. *Tao Te Ching* portrays the absorption of the separate soul into the cosmic unity, it describes the assimilation of the individual personality (Te) into the eternal Way (Tao). In simplest terms, Te means no more than the wholeness or completeness of a given entity. It represents the selfhood of every being in the universe. It also has a moral dimension in the sense of adherence to a set of values. From Tao, the vast variety of creatures and things in the world spring. Contrary to the existence or being of all the things or beings in the world, the Tao, their origin, is without existence. In terms of general systems, Tao represents the system, while Te stands for the attributes of each member of the whole; and the theory of the system Tao-Te depicts how the members and the whole are connected in such a way that the whole system could be in a chaos or in harmony.

One of the features of systems, as studied in our modern time, is the structure of layers. That is, an object B of a system A can be a system itself; an object C of the system B can be a system again, ..., see Section 3.1 or (Lin and Ma, 1987) for details. This kind of layer structure can be seen at many different places in *Tao Te Ching*. For example,

- "When the Way is lost, afterward comes integrity. When integrity is lost, afterward comes humanness. When humanness is lost, afterward comes righteousness. When righteousness is lost, afterward comes etiquette" (Chapter 1).
- "Preeminent is one whose subjects barely know he exists; the next is one to whom they feel close and praise; the next is one whom they fear; the lowest is one whom they despise" (Chapter 61).

Related to the concept of layers of systems, if one ignores the parts of members of the systems, then the structure of relations can be seen as stratifiable. This idea is contained in Chapter 62:

> "When the great Way was forsaken, there was humanness and righteousness; when cunning and wit appeared, there was great falsity; when the six family relationship lacked harmony, there were filial piety and parental kindness; when the state and royal house were in disarray, there were upright ministers."

In terms of systems, it says that if the high level structure is difficult to control or to study, a lower level will always be there.

The concept of process, which involves the notion of time, has been studied by many scholars, including some of the greatest minds in the history, such as Newton, Poincaré, Einstein, etc. In terms of systems, the concepts of time systems (Mesarovic and Takahara, 1974; 1989; Lin and Ma, 1987) and tree-like hierarchy of systems (Lin, 1988b; 1990) have been introduced and studied from many different angles. As a matter of fact, concepts, related to that of process, can be seen at various places in *Tao Te Ching*. For instance, Chapter 5 contains the following:

> "The Way gave birth to unity, unity gave birth to duality, duality gave birth to trinity, and trinity gave birth to the myriad creatures. The myriad creatures bear Yin on their backs and embrace Yang in their bosoms. They neutralize these vapors and thereby achieve harmony."

In the words of von Bertalanffy, the Whole was created first. By looking at the Whole, some objects and relations among the objects become dominating, which in turn give birth to "duality." Now, the whole, the collection of the objects, and the entirety of the relations constitute the "trinity." This Trinity is the fundamental structure of all the things and beings in the universe. In the words of ancient Greek atomists, Lavoisier, Einstein, or the laws of conservation, the Way is constant while all other things will vary at the level of the Trinity according to the observations of a third entity. Based on the research of pansystems, Lao Tzu had used the idea of general symmetric relations to describe the relation between gentle and dramatic changes, the general symmetric process of the gradual development from the chaotic Qi (vapor) to the creation of myriad creatures. For more details, see (Wu, 1990).

In the point of ruling a state, Lao Tzu's statement describes the fact that only with a clear view of the Way, there will be a stable, peaceful and prosperous country, which is called the unity. The stable and peaceful environment furnishes the prerequisite and the potentiality (the duality) for the development of myriad societal events. Even though the existence of the myriad societal activities or entities is mutually constrained or self-contradictory, it is the existence of these myriad mutually constrained and self-contradictory activities and entities that the whole or the country is in a balanced prosperity. In terms of systems research, only under the condition that a system exists, the entireties of objects and relations will possess their meanings. With the definite system, its objects, and relations, it becomes reasonable to study the myriad properties and structures of systems.

One of the important structures of general systems is the center. In 1956, Hall and Fagen (1956) introduced the concept of centralized systems, where a centralized system is a system in which one object or a subsystem plays a dominant role in the operation of the system. The leading part can be thought as the center of the system, since a small change of the part would affect in some way the entire system, causing considerable changes. As for how to form a centralized system, Tao Te Ching has the following teaching (Chapter 2):

- "Now, for this reason, Feudal lords and kings style themselves 'orphaned', 'destitute', and 'hapless'. Is this not because they take humility as their basis?" (Chapter 2);
- "That which all under heaven hate most is to be orphaned, and hapless. Yet kings and dukes call themselves thus" (Chapter 5).

That is to say, in order to form a centralized system, the "center" or the leading part must have harmonic relations or connections with the rest of the system. This conclusion coincides with the main result in (Lin, 1988a).

When more than one system appears in the study of general systems, one of the first things that need to be done is to compare the systems under consideration. The study of the concepts of functions and mappings in mathematics is a good example. In the research of general systems, the concept of mappings between systems can be seen in several different places, such as (Cornacchio, 1972; Lin, 1991). The need of comparison was phrased by Lao Tzu in the following way (Chapter 17):

> "Observe other persons through your own person. Observe other families through your own family. Observe other villages through your own village. Observe other states through your own state. Observe all under heaven through all under heaven."

The scientific history has shown the facts that systems are everywhere, and that nowhere has no systems. A great treatment on these facts is (Bunge, 1977). Accompanying these facts, in the past half century, the concept of systems has been applied to the study of all corners of human knowledge, see, for example, (Klir, 1970; Wu, 1990). Meanwhile, the research of general systems has shown some technical problems. In short, it is extremely difficult to develop some of the fundamental concepts in the theory of general systems, such as the concept of general systems. For details, see (Wood-Harper and Fitzgerald, 1982). All these phenomena had been vividly presented in *Tao Te Ching* as follows (Chapter 79):

- "When the Way is expressed verbally, we say such things as 'how bland and tasteless it is!'
- 'We look for it, but there is not enough to be seen.'
- 'We listen for it, but there is not enough to be heard'. Yet, when put to use, it is inexhaustible!"

R. Descartes and Galileo developed the following methods about scientific research and administration individually:

- Divide the problem under consideration into as small parts as possible, and study each of the isolated parts (Kline, 1972),
- simplify the complicated phenomenon into basic parts and processes (Kuhn, 1962).

In the history of science and technology, Descartes' and Galileo's methods have been very successfully applied. They guaranteed that physics had won great victories one by one (von Bertalanffy, 1972). Why are the methods so powerful? According to Lao Tzu, we have the following (Chapter 26):

> "Undertake difficult tasks by approaching what is easy in them; do great deeds by focusing on their minute aspects. All difficulties under heaven arise from what is easy, all great things under heaven arise from what is minute. For this reason, the sage never strives to do what is great. Therefore, he can achieve greatness."

Not only does it tell why Descartes' and Galileo's methods work, but also does it teach the relation between parts and the whole being a higher level structure than all parts.

3.4.4 Some final words

Combining what has been done in this chapter, it can be convincingly seen since the ancient times that each human being is a system, and it is organically connected to the environment, and how the universe is a huge system made up of many small systems. In other words, as a system, human body possesses the structure of spinning yoyo field, and the universe is indeed an ocean of eddy fields of different scales spinning at various intensities. These eddy fields push against each other, penetrate each other, and interfere the affairs of each other. That is, the state of motion of any yoyo field is determined by its own internal structure and by the interactions with other fields.

With these background results laid out, we are already to look at the systemic structure of the mind.

Chapter 4

The four human endowments

The totality of each human being is of four-dimensional: body, mind, heart, and spirit. It is physically made up of flesh, bone, blood, hair, and brain cells, and systemically of self-awareness, imagination, conscience, and free will (Covey, 1989, p. 70). By using self-awareness, he is able to examine his own thoughts and has the freedom to choose his response to whatever he comes across or whatever is imposed on him. With imagination, he is able to create a (fantasy) world in his mind beyond the present reality. With conscience, he is deeply aware of what is right and wrong, of the principles that govern his behavior, and a sense of the degree to which his thoughts and actions are in harmony with the principles. And he has free will to act based on his self-awareness, free of all other influences.

In this chapter, we use the systemic yoyo model (Lin, 2007) to provide new insights as for what the human endowments – self-awareness, imagination, conscience, and free will – are and to address some of the very important questions related to the phenomenon of these human endowments. Our work here and those contained in the following chapters of this book indicate that as is expected in (Lin, 2007), the systemic yoyo model can indeed be equally employed to study such "exact" science as physics and the study of such "inexact" science as social science and humanities. Not only so, discussions in these chapters also imply that the thinking logic of this model can be powerfully utilized to produce convincing results about human behaviors. This work provides a brand new theoretical foundation for resolving some of the very important age-old problems widely studied since the dawn of the Western philosophy. Some of the results obtained herein are expected to produce immediate real-life benefits in terms of designing educational programs. This chapter is mainly based on (Lin and Forrest, to appear 1 and 2).

More specifically, based on what is analyzed in the previous chapter, in Section 4.1, we investigate the phenomenon of self-awareness by addressing the following questions:

1 Where is human self-awareness or self-consciousness from?
2 Why does each human being have its very core of his/her own identity?
3 How is self-consciousness used to examine one's own thoughts and to choose appropriate actions in response to specific circumstances?
4 How are individual self-awareness and the cultural emphasis on the importance of self-consciousness related?
5 How can the systemic yoyo model be employed to explain the reason why individuals maintain different degrees of self-motivation and self-determination?

Among many interesting results, we show that since each human, seen as a systemic spinning yoyo, breaths in and out materials, it constantly battles with other systems by pushing against each other and attracting toward each other. The constant battles between human systems collectively make them aware of their own existence, the existence of others, be they human beings or objects, and their private thoughts. At the same time when individual self-awareness naturally exists with each human being, the cultural emphasis on the importance of self-consciousness in general influences how much its citizens focus on and act according to their individual understanding of various matters. Also, we clearly demonstrate the following important result: When the systemic structure of a person is uneven with its component materials, his yoyo structure will more likely spin on its own, and that the more uneven the yoyo structure is, the more driven, or more self-motivated, or more self-determined the person will be. That is, instead of being innate, self-motivation and self-determination are functions defined by hardships and knowledge of the opposite possibilities.

In Section 4.2 on imagination, we look at the following problems:

1 Is there any underlying mechanism over which the human imagination works and functions in its action and process to imagine, to form mental images or concepts of what is not actually present to the biological sense organs?
2 How does the innate ability to imagine help each of us form personal and individually different philosophical value and belief systems within our minds based on elements derived from sense perceptions of the shared physical world?
3 Is Albert Einstein (1987) right when he said, "Imagination ... is more important than knowledge"?
4 How does the workshop of the human mind – imagination – practically work to reassemble known ideas and established facts for innovative uses?
5 Why is imagination so powerful that it can convert adversities, failures, and mistakes into assets of priceless value, leading to the discovery of some of the underlying truths, known only to those who use their imagination?

Among other results, we show that imagination is the collection of all the conscious (meaning through the sense organs) and unconscious (meaning not through any of the sense organs) records of the interactions between the underlying yoyo fields. When one needs to establish an abstract image or concept, he relies on his various trainings to tap into his reservoir of imagination and make matches between some of the recorded patterns and the desired concept, product or service. And the way we form our personal, individually different philosophical value and belief systems, is similar to how a civilization forms its underlying assumptions and values of philosophy. When one experiences an adversity, a failure, or made a mistake, our analysis indicates that he experiences one of the following situations: facing great resistance, being completely stopped temporarily, or having taken a regretful choice. If a person is able to tap deeply into his imagination, then he can find out how to avoid the same situation from happening in the future and/or how to take advantage of similar situations in the future.

In Section 4.3 in terms of conscience, we address:

1 Where is conscience from? Why can it help distinguish whether an action is right or wrong?

2 Why is conscience deeply aware of the principles that govern one's behavior and a sense of the degree to which his thoughts and actions are in harmony with the principles?
3 How are the contents of his conscience affected by the culture?
4 Could the concept of world conscience possibly hold true and actually work so that one day in the future people can eventually make decisions based on what is beneficial to all people?

By using our systemic yoyo modeling, it is found that conscience is a partial function with two output values defined on some of the spin patterns and interactions of these patterns of imagination, where certain kinds of field flows and interactions of the flows are assigned the value of +, known as being right or moral, and some other kinds of flows and interactions assigned the value of −, known as being wrong or immoral. Based on this understanding, we show among others that the capacity for conscience is genetically determined, while the subject matter of conscience is learned. Our discussion demonstrates that conscience does not stands for the reason that addresses whether right or wrong, and explains why underneath the rich varieties of contents of individual consciences in a society, there is obvious commonalities when compared to other societies. We also show that the concept of world conscience will not ever become true.

In Section 4.4, we use the yoyo model to address several problems related to free will:

1 What is the systemic mechanism for the existence of the human endowment free will?
2 Does rational agent actually exist? If they do, when do they experience uncertainties in their decision-making?
3 Are all laws of nature causally deterministic? Is there such a thing as freedom in the reality?
4 Does moral responsibility require free wills of people? How are individuals morally responsible for their conducts?

It is shown that free will is the human ability to predict at least for the short term what one can or cannot accomplish and what choices are better or the best for the situation involved. Here, where his self is located, inside the domain of the ± function or outside the domain, determines how he would behave in terms of making and keeping promises. Using the commonly accepted definition of rational agents, we show in terms of whole evolutions that no human being or firm can be rational due to limitations in our knowledge and the development of technology, and that in the physical reality, there does not exist such a thing as freedom in absolute terms. As for moral responsibility, our work indicates that each individual or a small group of individuals has to be morally responsible for the well-being of their society, otherwise this individual or small group of individuals will be run over by other people or elements of the society mercilessly.

4.1 SELF-AWARENESS: THE FIRST ENDOWMENT

By self-awareness it means the human awareness that one exists as an individual and an entity that is separate from other people and objects with private thoughts and individual rights (Cooke, 1974, p. 106). It also includes the understanding that other people are similarly self-aware. Self-awareness is a self-conscious state in which attention focuses on oneself. It makes people more sensitive to their own attitudes and dispositions (Branden, 1969, p. 41).

At various circumstances, self-consciousness is used synonymously with self-awareness. It is credited only with an individual's development of identity. That is, in terms of epistemology, self-consciousness is a personal understanding of the very core of one's own identity. It is during the times when self-awareness or self-consciousness is awakened that people come to knowing themselves. Jean-Paul Sartre (June 21, 1905–April 15, 1980), a French existentialist philosopher and one of the leading figures in 20th century French philosophy (Gerassi, 1989), describes self-consciousness and self-awareness as being non-positional; it is not attached to any particular location.

4.1.1 The origin of self-awareness (self-consciousness)

With the utilization of self-consciousness, people examine their own thoughts and have the power to choose their responses to whatever situations they are faced with. That is why it is often observed that when someone loses himself/herself in a crowd, he/she may very well act totally untypical of the person. Self-consciousness or self-awareness is the basis for individuals traits, such as accountability, including responsibility, answerability, enforcement, blameworthiness, liability and other terms associated with the expectation of account-giving (Schedler, 1999), and conscientiousness, which stands for the trait of being painstaking and careful, or the quality of acting according to the dictates of one's conscience. Self-consciousness includes such elements as self-discipline, carefulness, thoroughness, organization, deliberation (the tendency to think carefully before acting), and need for achievement (Salgado, 1997).

With different degrees of ability to mobilize their self-consciousness, people are greatly affected by their own self-awareness, as some people scrutinize themselves more than others. And the importance of self-consciousness is emphasized differently from one culture to another. Even so, individuals maintain a varying degree of self-motivation and self-determination (Cohen, 1994, p. 136).

So, a natural question is where self-awareness is from? To this end, based on the systemic yoyo model, for any given person, his self-awareness is really a natural consequence of his underlying multi-dimensional systemic yoyo structure. Specifically, in theory, we can think of the totality of all materials that can be physical, tangible, intangible, or epistemological, and that all these matters are contained in the systemic yoyo of the person, if he is situated in isolation from other yoyo structures. That is, he is a whole being of his own.

However, in reality no man lives in isolation; systems are of various kinds and scales; and the universe can be seen as an ocean of eddy pools of different sizes, where each pool spins about its visible or invisible center or axis. To this end, one good example in our 3-dimensional physical space is the spinning field of air in a tornado. In the solenoidal structure, at the same time when the air within the tornado spins about the eye in the center, the systemic yoyo structure continuously sucks in and spits out air. In the spinning solenoidal field, the tornado takes in air and other materials, such as water or water vapor on the bottom, lifts up everything it took in into the sky, and then it continuously sprays out the air and water from the top of the spinning field. At the same time, the tornado also breathes in and out with air in all horizontal directions and elevations. If the amounts of air and water taken in by the tornado are greater than those given out, then the tornado will grow larger with increasing effect on everything along its path. That is the initial stage of formation of the tornado. If the opposite holds true, then the tornado is in the process of dying out. If the amounts of air and water that are taken in and given out reach an equilibrium, then the tornado can last for at least a while. In general, each tornado (or a systemic yoyo) experiences a period of stable existence after its initial formation and before its disappearance.

Because each person is a system with his own yoyo structure, he also constantly takes in and spits out materials. As influenced by the eddy (horizontal) spin, the meridian directional movement of materials in the yoyo structure is actually slanted instead of being perfectly vertical (Figure 4.1). In this figure, the horizontal vector stands for the direction of spin on the yoyo surface toward the reader and the vertical vector the direction of the meridian field, which is opposite of that in which the yoyo structure sucks in and spits out materials. Other than breathing in and out materials from the black hole (we will call it the south pole of the yoyo) and big bang (it will accordingly be named as the north pole of the yoyo) sides, the yoyo structure also takes in and gives out materials in all horizontal directions and elevations, just as in the case of tornadoes discussed earlier.

In the process of taking in and giving out materials, formed is an outside surface of materials that is mostly imaginary of our human mind; this surface holds within its boundary most of the contents of the spinning yoyo. The density of materials of this surface decreases as one move away from the yoyo structure. The maximum density is reached at the center of the eddy field. As the spin field, which is the field of the combined eddy and meridian fields, constantly takes in and gives out materials, there does not exist any clear boundary between the yoyo structure and its environment, which is analogous to the circumstance of a tornado that does not have a clear-cut separation between the tornado and its surroundings.

Now, this description of the general yoyo structure of a human being provides a fundamental theoretical explanation for where human self-awareness or self-consciousness is from. In particular, because each system breaths in and out materials throughout each part of its surface area, between different systems, there is a constant battle for

1 Pushing against each other when materials that are emitted outward are thrown against each other. This is how each person feels that he is separate from other people and objects. And

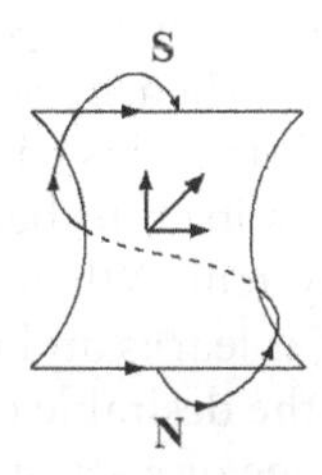

Figure 4.1 Slanted meridians of a yoyo structure.

2 Attracting toward each other when different systems try to absorb a piece of material of common interest. This is how the feeling of individual rights and entitlements is created.

These constant battles between systems collectively make them aware of their own existence, the existence of others, and their private thoughts, be they human beings or objects. Of course, the strength of such self-awareness is determined by the intensity of individual yoyo's spin field.

The reason why each person focuses on himself, is sensitive to his own attitudes and dispositions is because his very own viability is at stake, where his viability is determined by how strong his yoyo field spins and how much and how effectively the field can absorb materials from the environment.

4.1.2 Existence of core identity

The answer to the question of why each human being has its very core of his/her own identity lies in the fact that each person is the three-dimensional realization of a multi-dimensional field, which spins about its invisible axis, where the axis of spin stands for the center and core of his identity. Similar to the situation of identity search of civilizations (Lin and Forrest, 2010b; 2010c, Section 6.3), when person *N* feels inferior comparing to person *M*, sooner or later *N* will want to improve himself by adopting elements, such as a different way of looking at things, a specific philosophical point of view, or a new system of values, from the yoyo structure of *M*. As the change in *N* brings forward greatly wanted benefits, such as social status improvement, increased earning power, etc., the process of wanting to be just like *M* starts to reverse with the original sets of philosophical principles and values being revived. Further improvement in *N* weakens the influence of *M* and strengthens the commitment of *N* to its original being in two ways. At societal level, the economic benefits of personal improvement enhance *N*'s social status so that *N* gains confidence in himself and becomes socially assertive. At individual level, the voluntary improvement destroys original connections between different items in *N*'s value system, among his philosophical principles, and the way he used to connect with the outside world, leading to crises of identity. Because of the crises, *N* turns to religion. That is, person *N* selectively borrows items from other people and adapts, transforms, and assimilates them so as to strengthen and insure the survival of the core values of his own belief system.

4.1.3 Self-consciousness and selection of thoughts and actions

As what we have seen earlier, self-consciousness comes into being from the constant battles of the systemic yoyo structure of one person against all other spinning fields from the environment. In particular, when person *N* (Figure 4.2) battles with the spin field of another yoyo field *M*, *N* starts his engagement in the interaction with *M* gradually so that over time, *N* adjusts itself to better handle the interaction in order to maintain its own viability as a system. When *N* constantly interacts and battles with many systems at the same time, it learns and remembers how it adjusted itself to face various interactions to produce the desirable outcomes. So, when a new situation appears, *N* naturally associates it to one or a set of previous similar scenarios and tries to mimic the same reactions as before in hope of predicting similar desirable outcomes.

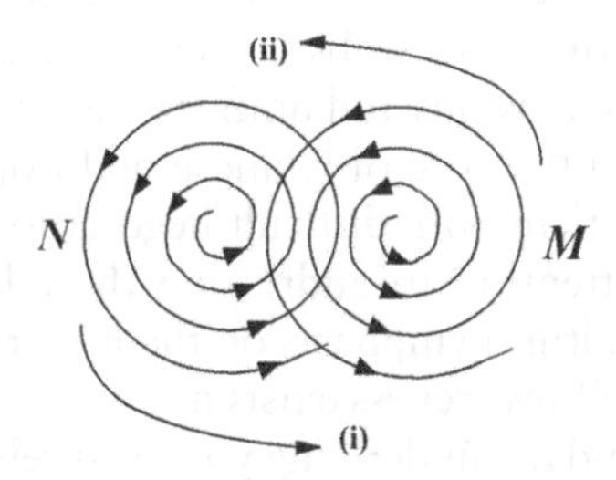

Figure 4.2 Acting and reacting spin fields of system *M* and a neighboring system *N*.

This systems modeling explains why when one suddenly faces a completely new difficulty or challenge, he tends to be panicky or experience anxiety. In such a situation, a more mature person would temporarily stop what he is working on at the very moment and organizes his thoughts and takes his time to come up with a most suitable reaction by utilizing his other three human endowments: imagination, conscience, and free will. What separates this more mature person from other relatively immature people is that through his various battles and interactions with other yoyo fields, he has gained a better understanding of himself and others. That is, he has accumulated a rich deposit of knowledge and experience and developed a better ability to command over the elements in the deposit so that he can refer back to them as freely and as many times as he needs to.

4.1.4 Self-awareness and cultural emphasis

To see how individual self-awareness and the cultural emphasis on the importance of self-consciousness is related, let us first look at how cultures are formed by treating each culture and every social organization as an abstract, high dimensional spinning yoyo. First, the geographic feature of each specific land always has its individuality and unshared characteristics. For example, Europe is mainly surrounded by oceans and has rivers well distributed throughout the land; and China is enclosed by deserts or extreme natural conditions in the north, cold, mountainous, and dry yellow-earth plateau in the west, the Himalayas in the southwest, lush jungles in the south, and open seas in the east. And throughout much of the land, people depend on one of the two river systems, the Yellow River and the Yangtze River, to survive. As what (Lin and Forrest, 2010a) discovered, different geographic characteristics of natural environments make people living on different lands hold different sets of philosophical principles and views of the world. For example, the openness in the geographic characteristics of Europe made Europeans accustomed to open thinking and freedom of movements (or individualism) due to the reason that they did not and will not have much need for difficult negotiation and compromise with anyone. On the other hand, because the geographic environment and condition of China more resembles that of the spinning dish used in Fultz's dishpan experiment with only the east open to large bodies of water. However, in ancient times without much capability to travel overseas, China was completely situated in a "dish with a solid periphery". So, the Chinese civilization is a perfect realization of a high dimensional spinning yoyo (of fluids), in which Chinese history is pretty much a social version of the periodic pattern changes observed in the dishpan experiment, where as a nation, it has gone through divisions and unifications alternately. So, throughout the history, collectivism has been the core of Chinese philosophical value, around which Chinese people have been patient with difficult negotiations and compromises.

Now, let us turn our attention to addressing the relationship between individual self-awareness and the cultural emphasis on the importance of self-consciousness. Because personal ability of self-awareness exists naturally with the underlying systemic yoyo structure of a person, and the underlying yoyo is surely situated in the spin field of the culture in which the person lives, since the very moment when the yoyo structure of the person is formed in the well controlled area between the spinning pools of his parents (Figure 4.3, where not all possibilities are shown), the content of his self-awareness and how he sees everything in the environment by using his self-awareness have been greatly affected by the value system of the family and the underlying philosophical principles and values of the culture. That is, individual self-awareness itself, as an ability and human endowment, naturally exists with each human being no matter where he lives. However, how the environment and world is seen within the self-awareness is greatly tinted by the fundamental philosophical principles and value system of the culture. That is, the cultural emphasis on the importance of self-consciousness in general influences how much the citizens focus on and act according to their individual understanding of various matters. For instance, in a culture that values individualism, the citizens would tend to focus more on their own roles in various affairs and the effects on their personal well beings of different events. On the other hand, in a culture that emphasizes on collectivism, the citizens would more likely place potentially organizational benefits over their personal gains, including the extreme case that they might have to sacrifice their lives to bring about the potential organizational benefits for others.

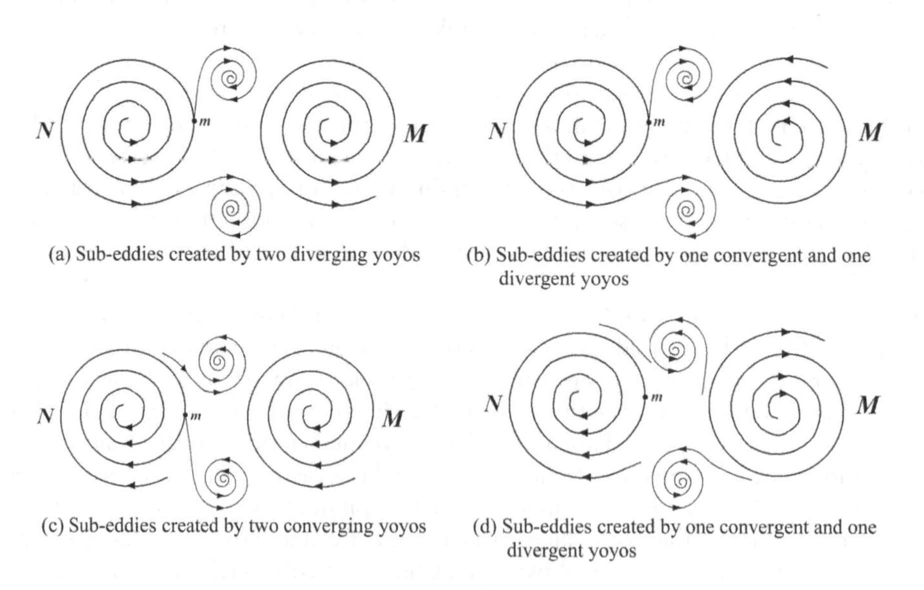

(a) Sub-eddies created by two diverging yoyos

(b) Sub-eddies created by one convergent and one divergent yoyos

(c) Sub-eddies created by two converging yoyos

(d) Sub-eddies created by one convergent and one divergent yoyos

Figure 4.3 *N* and *M* jointly produce sub-eddies in areas of their control.

4.1.5 Maintenance of self-motivation and self-determination

Over the years, many scholars from various angles have considered the following question: How can the systemic yoyo model be employed to explain the reason why individuals maintain different degrees of self-motivation and self-determination? However, as of this writing, still no satisfactory answer is derived. For details, see (Colvin, 2008) and references found there.

To address this problem, first let us model the meaning of self-motivation and self-determination as follows: If a yoyo spins on its own without much push or pressure (or no push or pressure at all) from the environment, then we say that the yoyo structure is self-motivated or self-determined. As for the degree of self-motivation and self-determination, it stands for the strength or intensity of the spin of the yoyo field. With this systemic modeling in place, the answer to the question of this subsection lies with the very reason why materials in the universe rotate in the first place.

According to the concept of uneven space and time of Einstein's theory of relativity (1987), we know that all materials have uneven structures. Out of these uneven structures, there naturally exist gradients. With gradients, there will appear forces. Combined with uneven arms of forces, the carrying materials will have to rotate in the form of moments of forces (Lin, 2008b, p. 31). With this reason on why materials in the universe rotate in the first place and continue to do so, we can readily see that when the yoyo structure of a person is made up of uneven materials, his yoyo structure will more likely spin on its own, and that the more uneven the yoyo structure is, the more driven, or more self-motivated, or more self-determined the person will be. This end in fact very well illustrates why people who grew up in difficult environments and knew the existence of better living conditions or who have lived in a rich variety of drastically different environments tend to have greater motivations or ambitions in life than those who grew up in monotonic comfort.

Our discussion here also indicates that for any chosen person, his self-motivation and self-determination of whatever degree is determined during his life time by the levels of difficulties he encounters and the accompanying knowledge of good quality of lives, especially during his childhood when minor discomforts can easily amount to major crises in life. So, instead of being innate, self-motivation and self-determination are functions defined by hardships and knowledge of the opposite possibilities.

4.2 IMAGINATION: THE SECOND ENDOWMENT

By imagination, it means the faculty of the human mind to imagine, to form mental images or concepts of what is not actually present to the biological senses, and the action or process of forming such images or concepts. With this faculty of the mind, one derives meaning to experience and understanding to knowledge. It is a fundamental human endowment through which people make sense of the world (Norman, 2000, p. 1–2) and plays a key role in the learning process (Egan, 1992, p. 50).

4.2.1 Mechanism over which imagination works and functions

We encounter everything in life only through imagination. Whatever we touch, see and hear coalesce into a mental picture via the faculty of imagination. The ability to imagine is the innate ability through which we form our complete personal philosophical value and belief systems within the mind based on elements derived from sense perceptions of the shared physical world (Harris, 2000, p. 94).

The common use of the term imagination is for the process of forming in the mind new images not previously experienced either partially or in different combinations. In this sense, imagination not being limited to the acquisition of exact knowledge is free from objective restraints. For example, the ability to imagine one's self in another person's place is very important to social interactions and mutual understanding. Albert Einstein (1987) said, "Imagination … is more important than knowledge. Knowledge is limited. Imagination encircles the world." The most famous technological/informational inventions and entertainment products in the entertainment business have been created from the inspiration of human imagination.

Imagination is the workshop of the human mind, in which known ideas and established facts may be reassembled into new combinations and put into new uses and it is known as the creative power of the human soul (Hill, 1928, p. 131). For example, to have a purpose in life, to establish self-confidence, to be proactive, and to take on leadership roles, one has to first create these qualities in his imagination and foresee the potential of him actually possessing them.

Any tangible achievement in life generally grows out of one's imagination. He first forms the thought in his imagination, then organizes the thought into realistic ideas and implemental plans in his mind, and then, and only then, physically transforms those mental ideas and plans into the reality. As soon as a thought germinates in the mind, the interpretative and creative nature of the imagination begins to examine relevant facts, concepts, and ideas; at the same time, it creates new combinations and plans out of these facts, concepts, and ideas. Through its interpretative ability, the imagination registers mental vibrations and thought waves, by which the physical body as dictated is put into motion by making use of the various outside available resources. Imagination is the only thing in the world over which we have absolute control. Even although one might be deprived of different privileges, rights, and properties, such as honors, individual freedom, or wealth, and he might be cheated upon in many different ways, there is no one in the world that can deprive him of the control and use of his imagination (Hill, 1928, p. 131).

When one properly uses his imagination, he will be able to convert his adversities, failures, and mistakes into assets of priceless value. It will lead him to the discovery of some of the underlying truths, known only to those who use their imagination. That is why it has been shown time and again that the greatest reverses and misfortunes in life often open the door to golden opportunities (Hill, 1928, p. 138). For instance, if you want to know what a certain fellow would respond to a specific circumstance, use your imagination, put yourself in his place, and find out what you would do under the same circumstance. That is the imagination in action (Hill, 1928, p. 142).

After learning the magnificent power of imagination, one might naturally ask: Is there any underlying mechanism over which the human imagination works and functions in its action and process to imagine, to form mental images or concepts of what is not actually present to the biological sense organs?

Going along with the systemic yoyo model of self-awareness, the human endowment – imagination – is also a natural consequence of the human yoyo structure.

Because biological sense organs are simply the body parts of the 3-dimensional realizations of the underlying multi-dimensional spinning fields of human yoyos, a great deal of the field structures of the world these sense organs cannot really pick up. From the fact that each human yoyo constantly interacts with other spin fields in the forms of pushing against each other or grabbing over materials of common interests, it follows that the experience and knowledge gained from the human field interactions are much richer than what is known based on the information collected by the sense organs. This end provides a systemic model for human imagination. That is, the so-called human imagination is in fact the collection of all the conscious (meaning through the sense organs) and unconscious (meaning not through any of the sense organs) records of the interactions between the underlying yoyo fields. When one needs to establish an abstract image or concept, he relies on his various trainings to tap into this reservoir of experience and knowledge collected both consciously and unconsciously from his field interactions with others. Here, the level of training on his self-awareness determines how deeply he can reach into this reservoir. Consequently, it determines how thought provoking his established concepts and images will be.

4.2.2 Formation of philosophical values and beliefs

It is shown in (Lin, 2008a) that the material world and human thoughts share the same structure, the yoyo field structure. So, when human imagination is called for action, it simply matches what is given, be it a difficult problem, a challenging situation, or a difficult task, with some of the relevant knowledge or experience from the reservoir. If the match is nearly perfect, the action will be considered successful. If the match is not good or no match is found, then new experience or knowledge will be added into the reservoir. Remember, this reservoir is composed of two parts. One part is the collection of all learned knowledge and experience through the sense organs; and the other the collection of all knowledge and information collected not through the sense organs. With adequate training on self-consciousness, one can "magically" apply a great deal of materials out of the reservoir of his imagination.

To understand how the innate ability to imagine helps each of us form personal and individually different philosophical value and belief systems within our minds based on elements derived from sense perceptions of the shared physical world, the discussion in Subsection 4.2.1 suggests that our ability to imagine, which is the ability to match a realistic situation with what is in the reservoir of imagination, is indeed innate. We were born with such a capability, just as we naturally knew we needed to eat when we were hungry. As for how through using imagination, we form our personal, individually different philosophical value and belief systems within our minds based on elements derived from sense perceptions of the shared physical world, it is similar to how a civilization, a human organization of the largest scale, forms its underlying assumptions and values of philosophy. In particular, at the time when a person is born, he lives in an extremely primitive condition and passively receives whatever is provided to him from the parents and the limited environment. From these initial contacts and interactions with the outside world, the newborn forms his elementary beliefs, basic values, and fundamental philosophical assumptions, such as "I am here for you to take care of me," "You have to satisfy my needs; otherwise, I blame you for the consequence," etc.

4.2.3 Imagination more important than knowledge?

Since the underlying yoyo field interactions are very basic and the parents nurture the newborn, the fundamental structure of the newborn's yoyo field is mainly formed through the interactions with the parents. Due to physical and mental limitations, minor obstacles of life make the newborn completely dependent on the parents. So, over time as he naturally formed a value system and a set of philosophical assumptions, on which he reasons in order to obtain what he wants, to explain whatever inexplicable, develops approaches to overcome hardships, and established methods to administrate his personal affairs.

As time goes on, he acquires more tools and knowledge from different sources in much greater environment. As all the learned (either consciously and unconsciously) knowledge that are from either sense organs or the invisible yoyo fields start to connect and form understandings of higher levels, some kinds of circulations of knowledge and information begins to form within the content of his self-consciousness and the reservoir of his imagination. As soon as a circulation appears, Bjerknes's Circulation Theorem guarantees the appearance of abstract, multi-dimensional eddy motions within the content of the person's self-consciousness and the reservoir of his imagination. That explains how imagination helps each of us form personal philosophical value and belief systems within our minds based on elements derived from the interactions between our spin fields and the fields existing in the shared physical world. As for why our value and belief systems are different from one another, it is because our spin field structures are not identical to each other so that the interactions of our individual fields with the fields of the shared world are different. Sometimes, the differences can be drastic.

This end illustrates why children growing up in different families tend to have different value and belief systems and why children growing up even in the same households turn out to be different from one another in their value and belief systems. In the latter case, although the children grow up in the same households, each of them experiences through quite different field interactions with the environments. In particular, in a household of two parents the oldest child grows up with two adults taking care of him; the second child with three beings running around about him with one of the three competing with him; ...

When comparing imagination and knowledge, Albert Einstein (1987) once commented: "Imagination ... is more important than knowledge". Is he correct? The answer to this question, based on the analysis earlier, depends on what Einstein means when he uses the words "imagination" and "knowledge." If to him imagination means one's ability to deeply reach into his knowledge base, which is composed of all the conscious and unconscious records of the interactions between the underlying yoyo fields, and to collect and organize all of the relevant information, and knowledge the collection of facts and theories gained through sense organs, then Einstein is correct. On the other hand, if both of the words "imagination" and "knowledge" stand for the same thing, the totality of all the conscious and unconscious experience and records of the interactions between the underlying yoyo fields, then Einstein did not say anything worthwhile discussing.

4.2.4 How imagination reassembles known ideas/facts for innovative uses

For each purposeful innovation or invention, the person involved generally has an idea about what he is looking for. Because the material world and human thoughts share the same structure – the yoyo field structure (Lin, 2008a), the level of training the person has gone through in the past helps him to make matches between the patterns stored in his imagination and the desired concept, product or service he is looking for. Here, the ability to reassemble different patterns available in one's head in various desired ways separate inventors and manufacturers.

Also, what is concluded here is that for an innovation to take place or an invention to be made there must be such a person that he is a visionary who can spot the need for something different and new. After he mentally goes through the idea numerous times either alone or collectively with others and is fully convinced that such an innovation is truly needed, he then passes on his mental image to another person, whom might be called a craftsman or engineer, who has the ability to materialize the imagined new concept, product, or service.

In earlier days, the visionary and the craftsman tend to be the same person. However, in our modern times where science and technology are highly advanced and sophisticated, the visionary and the engineer tend to be different entities. And each is often plural and contains a team of able bodies. When a team of engineers is involved, each member of the team contributes to the effort by providing his individual design. After the brainstorming, all the individual designs are pooled together to form the best and the potentially most realizable prototype. That is, the individual imaginations of the members of the team are combined together to produce the desired product at a much higher level than any of the original, individual designs.

For accidental inventions, the word "accidental" speaks the meaning. That is, the person involved in the process accidentally observes or recognizes a useful pattern out of the available information in his imagination. After that, engineers actualize the pattern into practically usable entity.

4.2.5 Imagination converts adversities, failures, and mistakes into assets

History has repeatedly shown that imagination is so powerful that it can convert adversities, failures, and mistakes into assets of priceless value, leading to the discovery of some of the underlying truths, known only to those who use their imagination.

The power of imagination is entailed in the fact that imagination is the reservoir of all the records of interactions of the spin field of a person with the fields of all other systems in the environment. In order to address this question well, let us first look at the systemic models for the concepts of adversities, failures, and mistakes, respectively.

Imagine that a systemic yoyo K interacts with another yoyo H (Figure 4.4). Because each of the spin field extends infinitely into the space, each of the yoyos is in actuality situated entirely in the spin field of the other yoyo. As indicated in the boxed areas, the interacting fields in these regions flow against each other. So, if a small yoyo m is flowing along the field of yoyo H in the left boxed area in Figure 4.4(a) or in the left side of the boxed area in Figure 4.4(b) in the upward direction, then yoyo m is experiencing an adversity, because its desire of moving upward is facing great resistance from yoyo

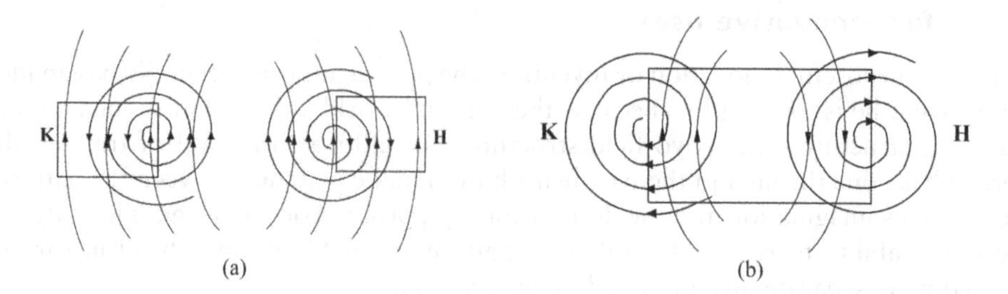

Figure 4.4 Interaction between two spin yoyos.

K. If the desired upward movement of m in one of the said area is completely stopped by the downward push of yoyo K, then m experiences a failure. If yoyo m in the said areas takes the new direction of flow of yoyo K because in the said region K seems to be more powerful, it is possible that later on when m is travelling in some other part of the field of K it realizes that it did not like the way it is forced to comply with a completely new set of rules. Then, this post-event feeling signals that m has made a mistake earlier in taking on a new direction in its pattern of movement.

When a person experiences an adversity, a failure, or makes a mistake, it simply means that he experiences one of the three scenarios just described: facing great resistance, being completely stopped temporarily, or taking a regretful choice. The reason why such an experience is seen as an adversity, a failure, or a mistake, is because in the reservoir of the imagination of the person, there is no record of any similar pattern. Now, if a person is able to tap deeply into his imagination, then he can take the experience to heart and find out how to avoid the same situation from happening in the future and/or how to take advantage of similar situations in the future. This end represents the priceless value of the adversity, the failure, and the mistake. Countless examples existing in the literature (Hill, 1928) evidence that all successful people are those who can in one way or another use their imagination to convert adversities, failures, and mistakes, no matter how damaging they are, into assets and lessons of priceless value in their future, more successful endeavors.

4.3 CONSCIENCE: THE THIRD ENDOWMENT

By conscience, it means the ability with which one distinguishes whether his actions are right or wrong, and is deeply aware of the principles that govern his behavior and a sense of the degree to which his thoughts and actions are in harmony with the principles. When one does things not in agreement with his moral values, his conscience will make him feel remorse; when his actions confirm to his moral values, his conscience brings him feelings of rectitude or integrity. Conscience also represents the attitude which informs a person's moral judgment before performing any act.

4.3.1 The functionality of conscience

Many authors from different angles, including religious views, secular views, and philosophical views, have studied the concept of conscience and its role in ethical decision-making. Religious views of conscience link this concept to an inherent morality, to the universe, and/or to divinity with many nuances. Modern day scholars in fields of ethology, neuroscience, and evolutionary psychology treat conscience as a function of the brain that developed to facilitate reciprocal altruism within societies (Buss, 2004; Pfaff, 2007; Tinbergen, 1951). It is found that conscience can prompt different people in quite different directions, depending on their beliefs. This observation suggests that while the capacity for conscience is probably genetically determined, its subject matter is probably learned, like language, as part of a culture. For example, while one person feels it's his moral duty to go to war under certain situations, while another person may very well feel a moral duty to avoid any war no matter what circumstance is concerned with. Also, numerous case studies of brain damage have shown that damage to specific areas of the brain can result in reduction or elimination of inhibitions with a corresponding radical change in behavior patterns. When the damage occurs to adults, they may still be able to perform moral reasoning; however, when the damage occurs to children, they may never develop that ability (Hare, 1970). In terms of philosophical views, the word "conscience" etymologically means with-knowledge. It stands for the reason that addresses questions of whether right or wrong and is accompanied with the sentiments of approbation and condemnation.

Currently, in the arena of international politics, there appeared a concept of world conscience. This concept is about the idea that with global communication possible and mostly available today we, as one people, will no longer be estranged from one another, whether it is culturally, ethnically, or geographically. Instead, we will approach the world as a place in which we all live, and with newly gained understanding of each other we will begin to make decisions based on what is beneficial for all people.

In (Lin and Forrest, to appear 1), we learned that self-awareness is one product of the constant battles between human yoyo structures. These battles make people aware of their own existence, the existence of others, be they human beings or objects. And, the strength of self-awareness is determined by the intensity of individual yoyo's spin field. In Section 4.2, it is concluded that the so-called imagination is in fact the totality of all the interactions between the human yoyo fields and between individual yoyo fields and the physical yoyo fields recorded either consciously or unconsciously. When one needs to utilize his imagination, he simply taps into this reservoir of records collected both consciously and unconsciously from his field interactions with others. Along with these systemic models, it can be seen that the so-called conscience is simply a partial function with two output values such that the partial function is defined on some of the spin patterns and interactions of these patterns stored in the reservoir of imagination. Here, certain kinds of field flows and interactions of the flows are assigned the value of +, known as being right or moral, and some other kinds of flows and interactions assigned the value of −, known as being wrong or immoral. The reason why this function is only partial is because other than the flow patterns and their interactions that are assigned either a + value or a value −, there are still plentiful of other flow patterns and interactions in the reservoir of imagination that are not assigned with any + or − value. For the sake of convenience of communication, let us call this partial function the ± function. By the domain of this function, we mean the totality of all the flow patterns and the flow interactions each of which has been assigned either a + or − value.

The assignments of + and or − values of certain flow patterns and interactions of flows start when the person is still an infant and continue throughout the entire life of the person. Initially, the domain of the ± function is the empty set. As the person grows older, the domain starts to expand. And, if some dramatic event happens during his lifetime, the assignments of the + and or − values may be altered.

As suggested by the formations of the sub-eddies in Figure 4.3 in (Lin and Forrest, to appear 1), some of the very first assignments of + values should be given to the flows heading in the same direction; and − values to currents flowing in opposite directions. It is because same directional flows feed into the existence of the sub-eddies, while opposite directional currents slow down or even attempt to stop the rotations of the sub-eddies. In daily language, the initial + values are given to behaviors that help strength the well being of all children involved, and the initial − values to the behaviors that damage or destroy the well being. For instance, when two children fight over a toy, most likely the parents would demand the children to share, reinforcing a + value on making everybody happy, even though the parents know very well which child actually owns the toy. When two little children are hitting each other, the parents would simply stop the fight forcefully, indicating the assignment of a − value to violence.

After a person reaches a certain age or level of maturity, his ± function will be quite well defined deeply in his head. So, whatever action he takes or thought he thinks of, he would unconsciously compare it to the elements in the domain of his ± function. If the action or the thought has a + value, he senses the feeling of rectitude as the consequence of being constantly appraised by adults for such occasions; if the action or thought has a − value, he feels remorse, since similar situations have been cursed regularly by grownups and he has let his care takers down one more time. If the action or the thought does not have a well assigned + or − value, the person will feel afraid of the potentially uncertain reactions from others. Afterward, he receives either a + or − value, or he will start his journey to explore the potential value for his specific action or thought by pushing this action further or thinking of the thought deeper until he reaches a certain outcome or he is attracted to some other more interesting, or urgent, or important topic before reaching any meaningful + or − value.

So, a person's capacity for conscience is genetically determined, because our discussion on conscience indicates that conscience is completely established on imagination, which in turn is dependent on self-awareness, while self-awareness is an innate ability. On the other hand, the subject matter of conscience, the specific content of the domain of the ± function and the assignments of the + or − values, is learned.

4.3.2 Conscience, behavior, thought and action

Conscience is deeply aware of the principles that govern one's behavior and a sense of the degree to which his thoughts and actions are in harmony with the principles. It is because our discussions in Subsections 4.2.2 and 4.3.1 imply that self-awareness senses the existence of philosophical value and belief systems and that of the ± function and its domain of definition in the reservoir of imagination. When the existing value and belief systems and the ± function are combined, meaning that the elements in these systems are also assigned either + or − values, the principles that govern one's behavior are formed. So, instead of conscience, it is self-awareness that it is deeply aware of these principles and creates a sense of degree to which one's thoughts and actions are in harmony with the principles.

What is discovered so far indicates that there does not exist any inherent morality, that conscience is indeed related to the common form of motion of materials – the yoyo structure – in the universe, and that conscience can be seen as a function of the brain developed through interactions of various yoyo fields to reciprocate altruism within societies, as studied in evolutionary psychology (Tinbergen, 1951; Pfaff, 2007; Buss, 2004). At the same time, our discussion also shows why when damages to specific areas of the brain occur to adults, they may still perform moral reasoning; and when the damages occur to children, they may never develop that ability. In particular, because conscience is a $\pm$ function with an ever-expanding domain over a person's lifetime, if the part of his brain that stores the information of the domain of the $\pm$ function is damaged, then he will either lose all the information of the domain or be unable to update the information of the expanded domain. In either case, if he is a child, whose knowledge of the $\pm$ function is nearly zero, he will never be able to gain any information on the $\pm$ function. That is why he might never be able to have the ability to perform moral reasoning.

4.3.3 How conscience is affected by culture

Our conclusions here agrees with the philosophical view on conscience as with knowledge, but not on that conscience stands for the reason that addresses questions of whether right or wrong. They also explain why conscience prompts different people in quite different directions. It is because slight differences in the environment in which one grows up lead to drastic variations in the content of his philosophical value and belief systems, and how his $\pm$ function is defined in his yoyo structure.

In particular, if one grows up in a family that is enemy centered, where the family's security is volatile, depending on the movements of its enemy, then it is likely that he will grow up always defining enemies and wondering what his enemies, which are those whose yoyo fields spins differently from his own, are up to and seeking self-justification and validation from the like-minded people, who are those their yoyo fields spin harmonically with his own. In this case, his judgment is narrow and distorted and he is defensive, over-reactive, and often paranoid. The power he has comes from anger, envy, resentment, and vengeance; it is a negative energy that shrivels and destroys, leaving little energy for anything else.

On the other hand, if one is born into and grows up in a family that is friend centered, where the family security is a function of the social mirror and highly dependent on the opinions of others, then it is likely that when he is fully grown, he constantly define who his friends are and makes decisions according to what his friends might think and is easily embarrassed by whatever he does. In this case, he sees the world through a self-made social lens and is limited by his social comfort zone. His actions are as fickle as opinions.

From the discussion in Subsection 4.1.4, it follows that the content of one's conscience is greatly affected by his family and the initial environment in which he grows up, while the family and the environment are part of a culture. So, his conscience is greatly affected by the culture. Here, by greatly affected by the culture, we do not

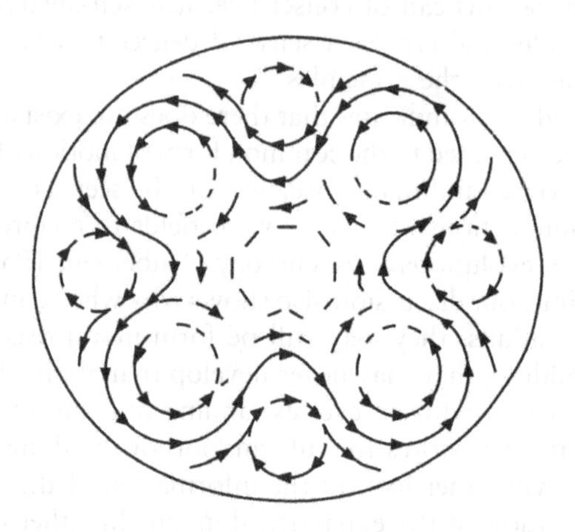

Figure 4.5 Existence of different spin fields in one dominating pool.

mean that the specific contents of an individual's conscience have to coincide with the norm of the culture.

Instead, as shown in Figure 4.5 (from Fultz's dishpan experiment), along with the rotation of the dish, periodically local eddies form. Although these local whirlpools spin along with the overall dish, their individual spin directions can be completely opposite to that of the dish. This laboratory experiment in fact explains why it is possible that children grow up and live adult lives completely opposite of what they were lectured to and that in any society, no matter what the societal norm is, there are always such people who are in odds with the majority of the population. On the other hand, in the rotating dish, no matter how different a temporary local eddy might spin, it still travels along the overall direction of the dish. That implies that underneath the rich varieties of contents of individual consciences in a society, there are obvious commonalities when compared to people of other societies.

4.3.4 Is world conscience possible?

The concept of world conscience stands for that one day in the future people from different parts of the world would make decisions based on what is beneficial to all people. Is this imagined state of world affairs possible to become true ever?

The answer to this question is definitely NO. It is because human societies are eddy fields formed by the natural geographical characteristics of various locations and by the distribution of natural resources (for details, refer to (Lin and Forrest, to appear 1)); and the idea of a world conscience means the appearance of a uniform flowing spin field without any local eddies existing, covering the entire globe. As evidenced by the existence of variously different patterns of airstreams from around the Earth, the idea of a world conscience can be likened to wishing all the airstreams from around the globe flow in the same direction without any local twists. It is simply impossible due to differences in temperature, humidity, pressure, and the unevenness of the surface of the Earth. As for human societies, due to differences in, for instance, the availability of natural resources from one region to another, people will have different economic interests and political desires. These and other naturally inherent differences will have to create conflicts (local eddies) between peoples. In particular, the individualism of the West and the collectivism of the East will not be entertained at the same time by any world-wide political or economic decision meaningful to all peoples' day-to-day lives.

4.4 FREE WILL: THE FOURTH ENDOWMENT

By free will, means the human ability to keep the promises one makes oneself and others. It is the human ability to make decisions and choices and to act in accordance with those choices and decisions. The extent to which personal free will is developed is tested in day-to-day lives in form of personal integrity. It stands for one's ability to give meaning to his words and walk the walk. It is an integral part of how much value is placed on oneself.

4.4.1 Systemic mechanism of free will

The concept of free will appears when people study the question of whether or not and in what sense rational agents exercise control over their actions and decisions. To address this question, one needs to understand the relationship between freedom and cause, and determine whether the laws of nature are causally deterministic. Various philosophical positions differ on whether all events are deterministic or indeterministic and on whether freedom can coexist with determinism or not – compatibilism versus incompatibilism. The existence of free will brings forward religious, ethical, and scientific implications. For example, in terms of religion, free will might imply that the omnipotent divinity does not assert its power over individual will and choices. In terms of ethics, it might imply that individuals can be held morally accountable for their actions. In terms of science, it might imply that the responses of the body and mind are not perfectly corresponding to stimuli or determined by physical causalities. The exploration on the concept of free will has been one of the central issues since the beginning of the Western philosophical thought.

In philosophy, the basic positions on free will can be categorized in terms of their answers to the following questions:

- Does determinism hold true?
- Does free will actually exist?

By determinism, is meant that all current and future events are definitely determined by past events (Wu and Lin, 2002). This point of view has been well represented by the so-called Laplace's demon, which is a hypothetical "demon" envisioned in 1814 by Pierre-Simon Laplace (Whitrow, 2001; Suppes, 1993) such that if it knew the precise location and momentum of every atom in the universe then it could use Newton's laws to reveal the entire course of cosmic events, past and future. (Please consult (Lin, 2008b) for some recent studies on Newton's laws of motions and their generalizations). Compatibilism holds the point of view that the existence of free will and the truthfulness of determinism are compatible with each other; that is opposite to incompatibilism, which is the point of view that there is no way to reconcile the beliefs that in a deterministic universe there is such thing as free will (Ginet, 1983).

In terms of moral responsibility, each society generally holds its people responsible for their conducts by praising or blaming for what they do individually or collectively. Therefore, many scholars believe that moral responsibility requires individual free wills of the people. This end leads to the question of whether or not individuals are ever morally responsible for their conducts; and, if so, in what sense. For more details on the studies along this line, one can refer to (Benditt, 1998; Dennett, 1984; Hume, 1740; Greene, 2004).

In the realms of science, many arguments for or against free will are based on an assumption about the truth or falsehood of determinism. Although scientific methods have historically held the promise of turning such assumptions into facts (recent studies (OuYang, et al. 2002) show that either of these assumptions are completely correct so that scientific methods in this case will not be able to turn these into any fact), such facts would still need to be combined with philosophical considerations in order to amount to an argument for or against free will. For example, if compatibilism is true, the truth of determinism would have no effect on the question of the existence of free will. On the other hand, a proof of determinism in conjunction with an argument for incompatibilism would add up to an argument against free will. In particular, researches in physics, biology, neuroscience, neurology and psychiatry, psychology, and other scientific areas have touched on the concept of free will. For instance, among others, Robert Kane (1996) capitalized on the success of quantum mechanics and the wide arrange acceptance of chaos theory in his defense of incompatibilist freedom in his *The Significance of Free Will* and other writings. Many biologists were involved in one of the most heated debates in biology about that of "nature versus nurture", concerning the relative importance of genetics and biology as compared to culture and environment in human behavior (Pinel, 1990). Despite of many important laboratory findings in neuroscience since the 1980s, scholars in the field did not interpret their experiments as evidence for the inefficacy of conscious free will by pointing out that although the tendency to press a button may be building up for 500 milliseconds, the conscious will retains a right to veto that action in the last few milliseconds (Libet, 2003).

In terms of the yoyo model, the so-called free will is the human ability to predict at least for the short term what one can or cannot accomplish and what choices are better or the best for the situation involved. In particular, with self-awareness, one forms his reservoir of imagination and his $\pm$ function of moral values with its constantly expanding domain. What matters here is where his self, as identified by his self-awareness, is located, inside the domain of the $\pm$ function or outside the domain.

If his identified self is inside the domain of his $\pm$ function with a specific value assigned to it, then the assignment of a $+$ value to his self will force him to make as accurate predictions as possible on what he can or cannot accomplish and what choices are better or the best for the situation involved. If he were uncertain with respect to a situation, his promise either to himself or others would be that he would try his best without any guarantee for success. On the other hand, if a $-$ value is assigned to himself, he will still force himself to make as honest predictions as possible on what he can or cannot accomplish and what choices are better or the best for the situation involved. However, in this case, most likely he will make promises opposite to what he foresees based on his imagination and his $\pm$ function. Now, when the third scenario occurs, where his identified self is not inside the domain of his ± function, then whatever appears, he would make his random promise, because no matter whether or not he keeps his promise, the outcome would not bear any conscientious consequence to him.

That is, in terms of the development of one's conscience, by placing him in an appropriate environment during his upbringing, the person's ability to keep promises can be drastically different from one kind of environment to another.

At this juncture, we like to spend a little time on predictions. All the current methods of prediction developed on modern science and technology have not produced forecasts with desired degree of accuracy other than live reports. To this end, predictions of major, zero-probability natural disasters can be seen as a supporting evidence. And predictions of stock market behaviors are another piece of evidence (Wu and Lin, 2002). On the other hand, the predictions one makes about what he can or cannot accomplish and what choices are better or the best for the situation involved can be quite accurate, because these predictions are made using structural analysis on the flow patterns and interactions of different flows. In particular, any change in a structure or an interaction of structures for the foreseeable future in general is preceded by internal changes in some of the structures involved. So, when one makes prediction on what he can or cannot accomplish, he basically studies his yoyo field structure and its possible interactions with other field structures in his environment. Such structural analysis involving rotations are more accurate than any method developed in modern science, because the later possibility currently is established on linear thought, while yoyo fields are nonlinear entities. For more detailed discussion to this end, please consult (Lin, 2008b).

This discussion on free will does not imply that the omnipotent divinity from the point of view of religion does not assert its power over individual will and choices. In fact, it more specifically indicates that no matter how one is brought up, whether he has a defined ± function with his self included in its domain or not, whether his assigned value to his self is + or −, all of which could be seen as directly under the powerful influence of the omnipotent divinity, his individual will and choices are still divinely determined by his ability to predict or how undefined his ± function is on his identified self.

In terms of ethics, our discussion implies that no matter whether or not one has a defined ± function and whether or not his self is in the domain of this ± function, he is still situated in an over-riding whirlpool, whose dominating pattern of spin dictates what specific value should be assigned to one's self. In terms of science, our discussion implies that the responses of human body and mind are not necessarily corresponding directly to stimuli or determined by external physical causalities, because the stimuli and physical causalities simply stand for environmental fields interacting with the yoyo structure of the human being; how the body and mind react to these external fields is determined by the person's prediction about what he can or cannot accomplish and what choices are better or the best for the situation involved from analyzing the patterns of flow and interactions of fields.

4.4.2 Rational agents and uncertainty

In areas of scientific studies, such as economics, game theory, decision theory, and artificial intelligence, anything that makes decisions, typically a person, firm, machine, or software, is referred to as a rational agent. Such an agent has a clear preference, models uncertainties via expected values, and always chooses to perform the action that results in the optimal outcome for itself from among all feasible actions. The concept of rational agents is also studied in the fields of cognitive science, ethics, and philosophy. In these studies, the action a rational agent takes depends on:

- The preference of the agent
- The agent's knowledge of its environment, which may come from past experiences
- The actions, duties and obligations available to the agent, and
- The estimated or actual benefits and the chances of success of the actions.

From our systemic yoyo modeling, it can be seen that no human being or firm can be rational in the long term or in terms of whole evolutions, because predicting its preference well into the future and how its preference will be affected by other fields interactions will be an impossible task. To this end, current studies in behavioral economics also show that rational human agents do not generally exist (Kahneman and Thaler, 2006).

Now, assume that for short-terms, some rational human agents do exist. That would mean that these agents can perfectly predict for the immediate future what they can or cannot accomplish and what choices are better or the best for the situation involved. In this case, they may still very well experience uncertainties when they process information due to uncertainties contained in the information. In particular, when the available information contains uncertainty, the conclusions drawn on the information will be naturally uncertain; and if the available information is perfectly correct, due to limitations of modern science and technology, the information may not be necessarily processed perfectly, leading to uncertain outcomes (Lin, 1998(guest editor)a; 1998). Now, even if we assume that modern science and technology can perfectly process the correct information, due to varied expectations, different people will behave differently based on their knowledge of the perfect information, leading to an uncertain future, for more details, please consult (Liu and Lin, 2006). As a matter of fact, what we see here also indicates that rational human agents cannot practically exist for either the short-term or the long-term in general except for some occasional, accidental cases.

4.4.3 Laws of nature and causal determinacy

In the study of blown-up theory, there is a concept known as equal quantitative effects. This concept was initially introduced by OuYang (1994) in his study of fluid motions and later, it is used to represent the fundamental and universal characteristics of all materials' movements (Lin, 1998(guest editor)a). By equal quantitative effects, it means the eddy effects with non-uniform vortical vectorities existing naturally in systems of equal quantitative movements, due to the unevenness of materials. In this definition, by equal quantitative movements, it means such movements that quasi-equal acting and reacting objects are involved or two or more quasi-equal mutual constraints are concerned with. For example, the relative movements of several planets of approximately equal masses are considered equal quantitative movements. In the current laboratories of physics, the interaction between the particles to be measured and the equipment used to do the measurement has been often seen as an equal quantitative movement. Many events in daily lives, such as wars, politics, economics, chess games, races, plays, etc., can also be seen as examples of equal quantitative effects.

More specifically, every time when a measurement uncertainty exists, one faces the effect of equal quantities. For example, when we observe an object, our understanding of the object is really constrained by our background knowledge (because both of us and the object are spinning fields interacting with each other), our ability, and limitation of human sense organs. When an object is observed by another object (two yoyos interact on each other), the two objects cannot really be separated apart. So, such well known theories as Bohr's principle, the relativity principle about microcosmic motions, von Neumann's Principle of Program Storage, etc., all fall into the category of equal quantitative effects.

Now, the systemic yoyo model well indicates that due to equal quantitative effects, which exist widely in the universe, at least some of the laws of nature would be neither about causalities nor deterministic. In particular, the enclosed areas in Figure 4.6 stand for the potential places for equal quantitative effects to appear. In these areas, the combined pushing and pulling is small in absolute terms. However, it is generally difficult to predict what will come out of the power struggle.

Next, let us turn our attention to address the question: Is there such a thing as freedom in the reality?

The term freedom has many different meanings. In philosophy, it has been given numerous interpretations by philosophers and schools of thought. In general, the protection of interpersonal freedom has been an object of a social and political investigation, and the metaphysical foundation of inner freedom is a philosophical and psychological question. In terms of politics, freedom means the absence of interference with the sovereignty of an individual by the use of coercion or aggression. The idea is that members of a free society should have full dominion over their public and private lives. The opposite of a free society is a totalitarian state, which highly restricts political freedom in order to regulate almost every aspect of behavior. In this sense 'freedom' refers solely to the relation of humans to other humans, and the only infringement on it is coercion by humans (von Hayek, 1991, p. 80, 81). The right for an individual to act according to his own will is known as liberty in political philosophy. Both individualist and liberal conceptions of liberty relate to the freedom of the individual from outside compulsion or coercion. On the other hand, a socialist perspective associates liberty with equality across a broader array of societal interests, such as reasonably equitable distributions of wealth. That is, without relatively

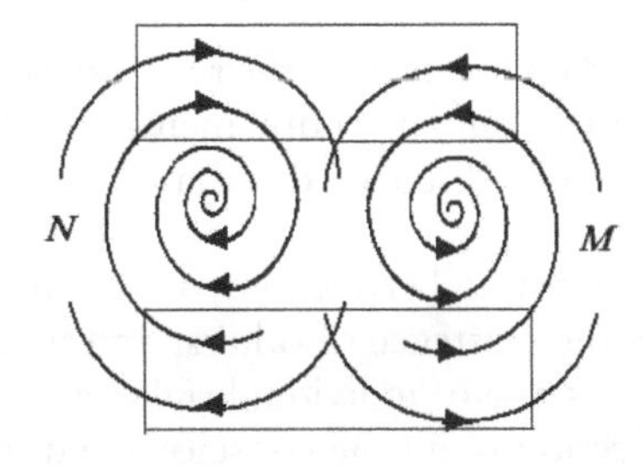

Figure 4.6 Structural representation of equal quantitative effects.

equal ownership, the subsequent concentration of power and influence into a small portion of the population inevitably results in the domination of the wealthy and the subjugation of the poor. Thus, freedom and material equality are seen as intrinsically connected. Meanwhile, the classical liberal argues that wealth cannot be evenly distributed without force being used against individuals which reduces individual liberty (Mill, 2003).

Our discussion in this part of the book indicates that in the physical reality, there does not exist such a thing as freedom in absolute terms. In particular, because each person exists alongside with many other spin fields, his way of physical movement is naturally constrained and influenced by other fields. On the other hand, in one's imagination, he can in theory freely combine different patterns of movement to form what he desires. However, his potential ways of putting patterns together are still dictated by the naturally existing patterns and interactions of the patterns.

4.4.4 Moral responsibility and free will

In terms of the relationship between moral responsibility and free wills of people and whether or not individuals are morally responsible for their conducts, as what we have seen, each society stands for a powerful over-riding spin field (a dish) (Figure 4.5). According to the dishpan experiment, when the periphery of the dish is solid, local eddies will form periodically. While at times these local eddies might rotate vigorously in a direction not harmonic to the powerful field of the society, these local eddies, when seen as individual entities, still have to travel along with the over-riding spin direction of the "dish." Now, what is implied here is that although people have their individual free wills and a degree of mental freedom of imagining various patterns of flow and various interactions that could realistically exist, they still have to comply with the laws, which are established to protect the well-being of the society. That is, moral responsibility does require free wills of people, where by moral responsibility we mean the spin direction of the society and free wills of individual people the formation of local eddies. On the other hand, free wills of people naturally exist in any society. So, considering the orderly operation and the well-being of the society, each individual or a small group of individuals, seen as a local eddy, has to be morally responsible for that the well-being of the society is not harmed; otherwise this individual or small group of individuals will be run over by other people or elements of the society mercilessly.

4.5 A FEW FINAL WORDS

In this chapter, we established on the basis of the recent development in systems science a systemic yoyo model for the human endowments – self-awareness, imagination, conscience, and free will. What is discovered is that

1 Due to constant battles existing between yoyo structures, humans become aware of their own existence, the existence of others, and their private thoughts. At the same time, they obtain a sense of individual rights and entitlements.
2 Imagination is the collection of all the conscious and unconscious records of the interactions between an underlying human yoyo field and other fields.

3 Conscience is a partial function with two output values defined on some elements of the imagination. For certain kinds of field flows and interactions, their function values are $+$, while for some other kinds of field flows and interactions, a $-$ value is assigned.
4 The so-called free will is the human ability to predict at least for the short term what one can or cannot accomplish and what choices are better or the best for the situation involved. With self-awareness, one forms his reservoir of imagination and his $\pm$ function of moral values with its constantly expanding domain. Where his self, as identified by his self-awareness, is located, either inside the domain of the $\pm$ function or outside, determines how one would make and keep his promises.

That is, self-awareness is a faculty of the mind. By purposeful utilization of one's self-awareness, meaning that he purposely gets involved in various situations so that his underlying yoyo structure will be more uneven, he can continuously cultivate, develop, extend and broaden all of his four endowments: self-awareness, imagination, conscience, and free will.

Chapter 5

Character and thought

Based on what has been investigated in the previous two chapters, this chapter is devoted to the study of the systemic structure of human character, thought, desire, and enthusiasm on a unified theoretical foundation (the systemic yoyo model and its figurative analysis) in order to establish tangible results that can be potentially useful in practice. Here, both desire and enthusiasm are two extremely important driving forces behind each and every personal and professional success ever recorded in history.

By using the systemic yoyo field models of the four human endowments – self-awareness, imagination, conscience, and free will, developed in (Lin and Forrest, to appear 1 and 2) or Chapter 4, we are able to model character and thought as specific yoyo field movements or formations. Because of these models, we are able to derive convincing results using logic reasoning. And then, we address some of the age-old problems related to desire and enthusiasm. This chapter is mainly based on (Lin and Forrest, to appear 3 and 4).

One of the several important implications of this work is that studies in social sciences and humanities can also be carried out in a similar fashion as those of natural sciences, where concepts and results are established in abstract, but real-life like spaces, such as Euclidean spaces. In our cases here, concepts and results can be developed by using the systemic yoyo model.

The framework of reasoning established in this chapter can be employed to illustrate many seemingly unrelated social and psychological phenomena. It shows how a unified framework can play the role of theoretical foundation for all the conclusions developed by many authors, such as Plato, Aristotle, Descartes, Immanuel Kant, David Hume, Georg Wilhelm Friedrich Hegel, to name a few, throughout the history since antiquity without reaching satisfactory conclusions. The theoretical and practical values of what follows in this chapter are implicitly contained in the presented discoveries so that a more tangible understanding of desire and enthusiasm is derived and that practically beneficial results can potentially be obtained by introducing appropriate educational programs in order to foster ambition, drive, desire and enthusiasm in any person of average intelligence.

The systemic model we use to base our discussions is that each human being is a system (Chapter 3); each general system can be seen as a high-dimensional rotational field (Lin, 2007; Chapters 1–3), whose realization in the three-dimensional space can be visualized as the (Chinese) yoyo (Lin and Forrest, to appear 1). That is, each human being has an underlying multi-dimensional yoyo field structure. His physical body is the realization of the multi-dimensional yoyo structure in the three-dimensional space;

and his behaviors and conducts are the realizations of specific forms of motion of the yoyo structure in our familiar material world. When we look beyond our bodies into the space, the universe is indeed an ocean of yoyos of different sizes and scales, spinning in their individual directions, intensities, and orientations.

This chapter is organized as follows: In Section 5.1, we investigate the following questions: What is character in terms of the systemic yoyo model? Why can it be counter-productive in a different society? Is it generally true that those who possess character have enthusiasm and personality sufficient to draw to them others who have character, as claimed by Hill (1928, p. 164)? What could be the laws, existing in the human dimension, that govern human effectiveness? How can such laws be potentially written in terms of the systemic yoyo model? How can breaking loose from deeply embedded, undesirable habitual tendencies, such as procrastination, impatience, criticalness, selfishness, be difficult, especially if one knows these tendencies are not good? Knowledge can be gained through learning; skills can be acquired through doing. How could one work on his desire?

Among others, we establish the following results: Using self-awareness, one develops his imagination along with his upbringing. Utilizing free will, he experiments ways to interact with external systems. Whatever the consequences, the experiments and their outcomes are stored in the reservoir of his imagination. Pressured by external fields, he adopts optimal patterns for his underlying field structure to achieve the best possible balance in its interaction with specific circumstances and to deal with different yoyo structures. As the person matures over time, his ways of handling often-seen situations (external yoyo fields) are repeatedly applied, leading to the same or similar outcomes. That is, if no new-event happens, one is expected to have a relatively stable system of traits, consisting of his optimal patterns of field flow and field interactions, that specifies how he would relate and react to others, to known kinds of stimuli, and the often-seen external field structures of the environment. This relatively stable system of traits is the character of the person. Because each character is developed in a specific culture, which represents a mighty field that constantly acts on the person's underlying yoyo, the very way of rotation of his underlying field structure carries some of the characteristics of the field of the culture. That is why when the structure of a person's character developed in one society is used in a different society without modification, it may be counter-productive, because the field of the new culture might spin in a direction not congruent to that of the original culture. Because a trauma one experiences stands for a severe shock to his systemic yoyo field such that the very structure of the field can be drastically reorganized, it explains why major trauma that occurs in one's life, even in his adulthood, can sometimes have a profound effect on the character structure of the person (Brunet, et al., 2007).

Personality is the subsystem of one's character that consists of all the three-dimensional realizations of his optimal patterns of field flow and field interactions that show how he related and reacted to others, to various stimuli, and to the environment. Enthusiasm is a state of intensive spin of the underlying yoyo field of one's mind and body that has a specific tilt in the axis of rotation. Comparing to self-motivation and self-determination, the intensity of spin of enthusiasm fluctuates and has a shorter lifespan. Detailed analysis implies that those who possess a certain kind of character with sufficient enthusiasm and personality will draw to them others who have the same kind of character.

In terms of the laws in the human dimension regarding effectiveness, it is shown that when two spin fields of the same scale and identical spinning intensity face off, they will eventually be locked up in an extreme hostility if no third party is involved. And, if one field is much mightier than the other, in the process of adjusting itself, the tiny yoyo will be sucked into the black hole of the mighty field.

In Section 5.2, we study the concept of thoughts by addressing such intriguing questions as: What is the systemic mechanism for thoughts to form so that they help humans model the world and to deal with what they face? Why are a person's acts always in harmony with the dominating thoughts of his mind? How could the dominating thoughts of the mind bring forward desirable outcomes according to the nature of the thoughts so that one may shape his worldly destiny according to his own liking? Why could the human mind be controlled, guided, and, directed to the desired ends? How can the systemic yoyo model be employed to explain the two creations of any material achievement– first in the mind and the second in the physical world? What is the mechanism underlying the phenomenon of mental inertia? How does human mental inertia work? Can anyone who pursues after his own desired success avoid from traveling along the bloody trail created by mental inertia? The results below highlight some of what we discover in this section. Each thought is a local eddy in the reservoir of someone's imagination. Such understanding explains why new thought generally is triggered by some hint from the outside world. Each thinking process is a process of utilizing the hint to generate a local eddy in the reservoir of imagination by pulling relevant information and knowledge together to form an organic whole. Pattern matches and the well-formulated ± function of the conscience work together to form judgments about newly experienced situations. That is why to insure sound judgments being made, intellects sort through relevant knowledge and experiences against the present situation, revealing the vital importance of quality education and of the richness of the living environment. Because thoughts are simply local eddy pools in the mind and can be manipulated by mixing, matching, sifting, and sorting concepts, facts, perceptions, and experiences, they have become the most highly organized form of energy known to man. Because any bodily movement is simply a 3-dimensional realization of the underlying multi-dimensional yoyo structure of man, it explains why every voluntary movement of human body is caused, controlled, and directed by thought through the operation of the mind.

Each field flow or interaction of the outside world can play the role of a stir for someone's mind. It is because no matter how well one knows a pattern of field flows, there is always a different, new angle from which he can uncover some surprising new insights. It is shown that we do not really have absolute control over our thoughts in reality and that as long as a dominating thought is derived on accurate information, bodily realization in the 3-dimensional space of the thought will become visible in no time. The formation of a local eddy in imagination is a feedback process with the initial stir being the input. Reacting to the stir, various groupings of relevant information are formed in the conscious mind and judged by using the ± function. Among all the groupings with + response values, one picks out the grouping with the most desirable + value. This feedback procedure makes people feel that they have control over their own thoughts. In this process, the original flow pattern of imagination experiences turmoil with local eddy pools formed and sense organs become keen to whatever supportive to individual groupings of information. When facing difficulties or realizing newly found resources, the chosen response will be modified. So, another feedback loop emerges. This loop creates the feeling that the human mind is guided and directed to the desired ends.

After a specific movement appears in a human field, that formation can be realized in the 3-dimensional space; and only after that, some tangible or visible material achievement can be possibly recognized. This sequential before and after of events explains why any material achievement has to be first accomplished in the mind and then in the physical world. At this juncture, the concept of duality of yoyo fields is employed to illustrate why as long as one has a definite goal to achieve, there is always a person or people who create various difficulties for the former person along his pursuit of his goal. In terms of the relationship between individuals or a small group of individuals and their society, the "well-being" of the society stands for the maintenance of the inertia the society. The relative stability means that the underlying yoyo field of the society rotates quite vigorously without any local pool spinning in a different direction. In such a uniformly spinning fluid, any rebellious local movement will surely be crushed by the inertia of the vigorous field. The dishpan experiment indicates that for one who pursues after his desired success to avoid traveling along the bloody trail created by the inertia of the society, he has to control his field movement in conformation with the acceptable form of motion of the greater culture. As he obtains support from others and causes many others to pursue after similar goals of success, the inertia of his greater environment will be altered by his effort.

In Section 5.3, we focus on the study of desire, an unsettled problem considered since antiquity, by addressing such questions as: What is desire in terms of the systemic yoyo model? Why is desire so powerful in terms of dictating one's thinking and physical conduct? Where is desire from? How can we determine the depth of one's desire that makes him appear to possess supper human powers in his pursuit of life-time goals? Can desire be artificially strengthened or deepened? Along with other interesting results, by using our systemic model, we discover that desire is how one wants his underlying yoyo structure to be in terms of one of or any combination of its attributes: direction, orientation, spin intensity, and scale. With self-awareness, he senses how his yoyo field could be potentially made uniform so that it does not have to be as it is. That is, desire comes from and is created by the differences naturally existing between human yoyo fields. With the ever expanding domain of his $\pm$ function, desirable changes are assigned with $+$ values, making his desire vary over time. So, along with the $\pm$ function, desire motivates one to act; and in the restless movement of field rotation and interactions, desire removes the antithesis between itself and its object, and that the object of immediate desire is a living thing and forever remains an independent existence. For the relationship between desire and fear, other than they share the same brain circuit (Berridge, et al., 2009), our work shows that they are really the different sides of the same coin. Desire describes what one wants for his yoyo structure, while fear how one does not want to lose what he already has within his field.

Desire also stands for a local eddy, similar to thought, in the reservoir of someone's imagination and is specifically caused by such stirs that are created from comparisons with and interactions between human field structures. That is why desire can create a thinking process to pull relevant information in the reservoir of imagination together in order to determine what is desirable and what is not. By the work of the $\pm$ function, each available stimulus is assigned a value of preference so that all available stimuli are ordered according to their desirability. As for the depth of desire, it can be measured using the intensity of the local eddy pool of the desire in the reservoir of imagination. It is derived that when the intensity of spin of the local pool reaches an extremely high level, an extraordinarily amount of materials of the imagination will be pulled into the back hole of the pool, and various innovative connections are established to bridge seemingly related information and facts. When these connections are materialized in the physical world, they become valuable guidance for where for one to look for potentially useful resources. This sort of guidance, invisible to others, makes the specific individual appear to possess super human power in his attainment of the objective of his deeply rooted desire.

As for whether or not desire can be artificially created and strengthened, our modeling indicates that the answer is YES. In particular, the main idea is to make one's yoyo field structure uneven either by designing and employing reward systems or by creating obstacles in his life while providing him with the knowledge that when an obstacle is conquered, something magnificent will appear, or by a combined use of these two methods.

Section 5.4 focuses on the investigation of enthusiasm by looking at the following questions: What is the systemic mechanism under the intense state of mind, known as enthusiasm, that explains why enthusiasm inspires and arouses a person to put action into the task at hand and makes originally monotonous works more enjoyable to finish? What is the connection between great leaders and their own enthusiasm? How is their own enthusiasm spread over to their followers? When one is situated in a difficult situation and a long way from realizing his definite goal in life, what is the underlying mechanism for him to be able to kindle the fire of enthusiasm in his heart and keep it burning? And why is that before very long the obstacles that now stand in the way of attaining one's definite goal will melt away, and he will find himself in possession of power that he did not know he possessed? The technique with which one can fix any idea he chooses in his mind is called self-suggestion. What is the theoretical foundation for this technique of self-suggestion to work? Why is self-control so important that it can direct enthusiasm to constructive ends, to build up instead of tear down?

Among other interesting results, we derive the following conclusions. It is derived that there are two ways for a field that spins in its inertial state to suddenly acquire the momentum to rotate at a higher level of intensity: a new unevenness in the structure of the field suddenly appears, or an ever-presenting field of the environment suddenly, unexpectedly disappears. In terms of suddenly causing new structural unevenness, one can simply discover either new strengths in himself or new patterns of interaction with other fields. When the field of increased intensity is realized in the 3-dimensional space, it makes people feel that his enthusiasm has inspired and aroused him to put action into the task at hand. And this realization makes the person himself feel that his enthusiasm makes originally monotonous works more enjoyable to finish. By leadership, is meant one's capability to adjust his underlying field structure so that many other neighboring fields would spin in similar fashion without much difficult readjusting. With this systemic model of leadership, all the relevant studies of leadership can be unified into an organic whole. It is argued that the leader creates his high degree of spinning intensity by making use of an emerging, overreaching, and overpowering conceptual field structure. His increased field intensity is his enthusiasm, which is spread over to his followers by utilizing the field influence of the conceptual field motion, when other fields are still in their individual inertial states of motion.

If a person can keep his underlying field spin at its newly acquired momentum and intensity, the First and Third Laws on State of Motion imply that all the other field structures, which used to resist the change in the person, will have to change their forms of motion accordingly in an accommodating way. That is why before long all the obstacles that now stand in the way of attaining the person's definite goal will melt away, and he will find himself in possession of power that he can change behaviors of others according to his likings.

At the end, Section 5.5 concludes this chapter with some relevant comments.

5.1 HUMAN EFFECTIVENESS

By character, it means a person's overall, relatively permanent system of traits that are manifested in specific ways on how he relates and reacts to others, to various kinds of stimuli, and the environment (Hergenhahn, 2005). Character develops over time in each individual to enable him/her to interact successfully within a given society and to help him/her to adapt to its mode of production and social norms. When the structure of a person's character developed in one society is used in a different society without modification, it may be counter-productive.

5.1.1 Systemic yoyo model of character

Character is developed along with one's natural growth. For example, if a child lives in a treacherous environment and interacts with adults who do not take the long-term interests of the child to heart, then it is very likely that the child will form a pattern of behavior that help protect himself from harms potentially existing in the malign social environment. Also, major trauma that occurs later in one's life, even in adulthood, can sometimes have a profound effect on the character structure of the person (Brunet, et al., 2007).

It is character that it represents what a person is and that it gives a person real and enduring power and great influence in life. One can obtain character only by building it through using his own thoughts and deeds. Those who possess character have enthusiasm and personality sufficient to draw to them others who have character (Hill, 1928, p. 164). In the studies of character ethics, it is assumed that human effectiveness are governed by laws, existing in the human dimension, that are as real, as unchanging as the laws of science in the physical dimension (Covey, 1989, p. 32).

Each person's character is a composite of his habits. Habits are powerful factors in any person's life. Because they are consistent and often unconscious patterns, they constantly express one's character and produce his/her effectiveness or ineffectiveness. Just as how gravity works in the physical world, habits also possess tremendous pulls. Breaking loose from deeply embedded habitual tendencies, such as procrastination, impatience, criticalness, selfishness, etc., which violate basic principles of human effectiveness, requires a strong willpower and a burning desire for change (Covey, 1989, p. 46). The gravitational pull of habits is a powerful force. It can be utilized effectively to create the cohesiveness and order necessary for one to achieve whatever goals he/she has in life.

According to Covey (1989, p. 47), a habit is the intersection of knowledge, skill, and desire, where knowledge is the theoretical paradigm, the what to be done and the why, skill the method of how to achieve what wants to be done, and desire the motivation, the want to do. That is, to create a desirable habit, one needs to work in all these three dimensions. Because to break through to new levels of achievements from an old paradigm that has been a source of pseudo-security for years one has to work on knowledge, skill, and desire simultaneously, the drive for such a potentially labor-intensive work has to come from the call of a higher purpose. Only so, one is willing to subordinate what he thinks he wants now for what he wants later.

To describe the concept of character in terms of the systemic yoyo model and to explain why it can be counter-productive in different societies, let us first look at the following law:

The First Law on State of Motion (Lin, 2007): *Each imaginable and existing entity in the universe is a spinning yoyo of a certain dimension. Located on the outskirt of the yoyo is a spin field. Without being affected by another yoyo structure, each particle in the said entity's yoyo structure continues its movement in its orbital state of motion.*

This law was initially introduce to address several open questions related to Newton's First Law of Motion, such as "if a force truly impresses on an object, the force must be from the outside of the object. Then, where can such a force be from"? "In their state of motion, all objects possess the so-called natural resistance to changes; then how can such a resistance be considered natural"?

For our purpose, what we need to emphasize from the First Law on State of Motion is the natural resistance to changes each system possesses. In particular, with the naturally existing self-awareness, one develops his reservoir of imagination along with his upbringing. By utilizing free will, he experiments different ways to interact with various external systems. No matter what happens, the processes and consequences of the experiments are well stored either consciously or unconsciously in the reservoir of his imagination. Pressured by external fields, such as the invisible and/or visible mandates of the culture, he adopts the optimal patterns for his underlying yoyo to rotate in order for his field structure to achieve the best possible balance in its interaction with various specific circumstances; also he develops the most efficient ways to deal with different yoyo structures. As he grows older, some of his unique ways of handling often-seen situations (external yoyo fields) are repeatedly applied, leading to the same or similar outcomes.

According to how character is defined (Hergenhahn, 2005) and the systemic yoyo modelings of self-awareness, imagination, conscience, and free will in (Lin and Forrest, to appear 1–2), the First Law on State of Motion implies that if no new field structure enters one's cycle of life (his spin field), then the overall pattern of spin of his underlying yoyo structure will continue its accustomed state of motion. That is, in his ordinary life without any new-event happening, one is expected to have a relatively stable system of traits, consisting of his optimal patterns of field flow and field interactions, that specifies how he would relate and react to others, to known kinds of stimuli, and the often-seen external field structures of the environment. This relatively stable system of traits is the character of the person.

Because each person develops his character in a relatively specific culture, an overriding spin field that constantly acts on the shape of the person's underlying yoyo, the very way of rotation of his underlying field structure carries some of the characteristics of the spin field of the culture, such as the unique rotation angle or the unique speed of spin. That explains why when the structure of a person's character developed in one society is used in a different society without modification, it may be counter-productive. For example, a school teacher, growing up in a society that emphasizes on individualism and democracy, tends to be very interested in testing various new methods of teaching and different ways to involve students. By behaving so, his unique individuality can be greatly enhanced and noted by others. On the other hand, in a different culture that focuses on collectivism, school teachers are more like to be prudent on using any novel method of teaching, because the unknown outcome may very well ruin the collective effort of producing their expected high quality students. Here, the seemingly depressed pattern of behavior is the natural result of the need for achieving group success. Now, if the former teacher ends up in the latter culture, he will have a difficult time to be accepted by any reputable school system there, because his interest in testing out new ideas will be frowned upon by his colleagues and school administrators. Equally, if a teacher from the latter culture teaches in the former teacher's setting, he will also suffer from various hardships due to his accustomed commitment to quality that his colleagues in the new setting do not even care about. As for which of the two school systems truly produce better quality students, it is beyond our discussion here.

Our systemic yoyo model analysis explains why character is developed along with one's natural growth; and our example shows why and how the environment in which one grows up bears the role of shaping his character. As for why major trauma that occurs in one's life, even in his adulthood, can sometimes have a profound effect on the character structure of the person (Brunet, et al., 2007), it is because each major trauma stands for such a shock to the whole systemic yoyo field of the person that the fundamental hierarchy of the field might be reorganized. It means that the reservoir of the person's imagination could be reshuffled; the $\pm$ function of his conscience redefined; and the domain of his $\pm$ function reshaped.

5.1.2 Attraction of character

Because the so-called character stands for the overall pattern and shape of one's underlying yoyo field structure, that explains why character represents what a person is and why it is an indicator for what enduring power and influence a person has in his life over others. Because character is the totality of how a person's underlying yoyo field spins under various, ever-presenting pressures and influences of other fields with the work of his free will, it explains the reason why if one wants to build a stronger character, he can achieve the goal by using his own thoughts and deeds.

Based on numerous case studies, Hill (1928, p. 164) concludes that those who possess character have enthusiasm and personality sufficient to draw to them others who have character. Based on the systemic yoyo model of character, established in the discussion of Subsection 5.1.1 above, we now know that by character, it means the relatively stable system of traits, consisting of one's optimal patterns of field flow and field interactions, that specifies how he would relate and react to others, to various known stimuli, and often-seen external field structures of the environment.

To make sense out of what Hill concluded a degree of satisfaction, we need to first understand the concepts of enthusiasm and personality. Enthusiasm signifies a whole-hearted devotion to an ideal, cause, study or pursuit, or merely being visibly excited about what one's doing (Tucker, 1972). It stands for intense enjoyment, interest, and/or approval. And, by personality, is meant one's aggregate conglomeration of decisions he has made throughout his life. There are inherent natural, genetic, and environmental factors that contribute to the development of one's personality (Allen, 2005).

So, based on the systemic yoyo model of character, the so-called personality is the subsystem of one's character that consists of all the three-dimensional realizations of his optimal patterns of field flow and field interactions that show how he related and reacted to others, to various stimuli, and to the environment. And, by using the systemic yoyo model, enthusiasm can be visualized as a state of intensive spin of the underlying yoyo field of one's mind and body that has a specific tilt in the axis of rotation. Here, the special tilt in the axis of rotation is emphasized to represent the person's specific whole-hearted devotion. When compared to the systemic modeling of self-motivation and self-determination, as discussed in Question 5.1.5 in (Lin and Forrest, to appear 1–2), enthusiasm is the same as that model except that its state of intensive spin of the underlying yoyo field of the mind and body does not last as long as that of self-motivation and self-determination and that throughout the lifespan of enthusiasm, the intensity of spin of the underlying yoyo field might change from strong to weak and vice versa alternatively.

This model of enthusiasm provides a detailed explanation for why enthusiasm is contagious to all with whom the enthusiast comes in contact (Hill, 1928, p. 153). In particular, assume that there are two people N and M. Initially, none of them is specifically enthusiastic about anything particular. That is their underlying yoyo fields are loosely acting on each other, Figure 5.1(a), where the fields of N and M overlap so that they act on each other without forcing each other to change their respective shapes. Now, for whatever reason, N becomes visibly enthusiastic over what he is doing so that its field suddenly spins at a much higher level of intensity so that the field of M is greatly affected. As shown in Figure 5.1(b), the originally circular spin field of M now becomes oval, where the left-hand side is helped by the field of N to travel faster and further than before, and the right-hand side faces much greater resistance than before from the field of N that is newly strengthened.

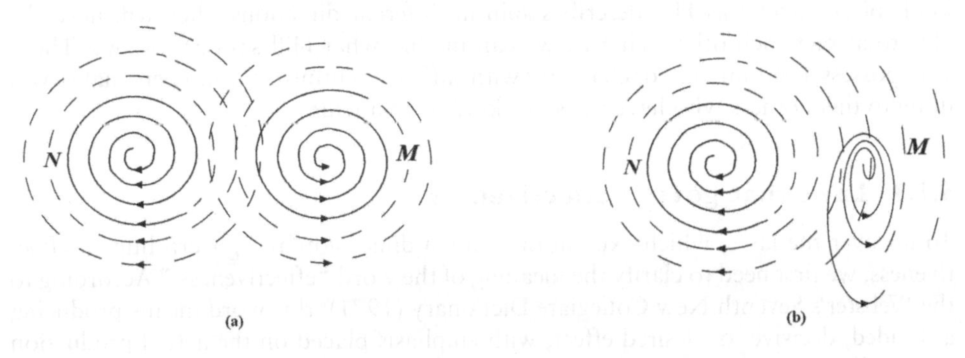

Figure 5.1 Imaginary acting force of an enthusiast *N* over *M*.

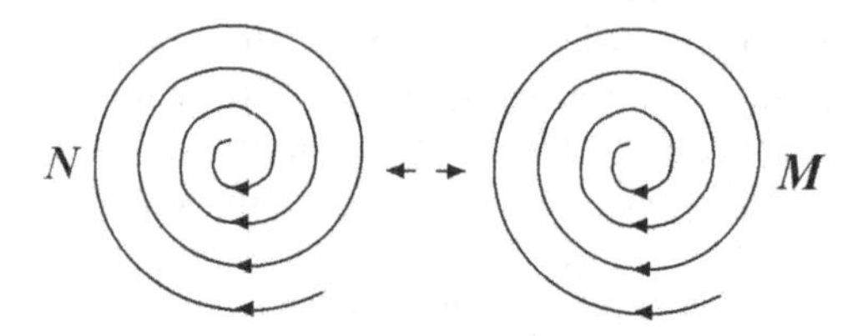

Figure 5.2 The fields of *N* and *M* spin harmonically and convergently.

This analysis indicates that what Hill claims is not completely correct. Specifically, as shown in Figure 5.1, if the yoyo fields of persons *N* and *M* spin in opposite directions, when one suddenly experiences a surge in its intensity of spin, the other person in fact suffers from the consequence. For instance, *M*'s territory is altered against his will because now he is forced to move along a new, elliptical orbit; while he is pressured to behave in certain ways in order to conform with the increased intensity of *N* (the left-hand side of *M*), he can no longer do whatever he likes to do as before without first fighting to obtain *N*'s approval (the right-hand side of *M*). On the other hand, if the yoyo fields of *N* and *M* are both convergent and spin harmonically, meaning that they spin in the same direction (Figure 5.2), then their original, naturally existing attraction will be greatly strengthened by the sudden increase in the field intensity of *N*. This is exactly what Hill describes: enthusiasm is contagious to all with whom the enthusiast comes in contact, if we assume that by "all with whom the enthusiast comes in contact," he means those who are also willing to associate with the enthusiast. By carefully analyzing different ways two yoyo fields could possibly interact with each other, it can be easily seen that the two scenarios we mentioned here are only two special cases of many possibilities. This end in fact explains why no matter what one is enthusiastic about, other than those who are genuinely supportive of him, there are countless many others who hold varied and quite opposite opinions about his newly found excitement.

So, from the previous analysis, we can conclude that what Hill (1928, p. 164) claims is about the specific kinds of characters that would bring one what he desires in life and that would create him the needed or desired beneficial influence over others. At the same time, as pointed out earlier, if the yoyo fields of people who possess the kinds of characters as Hill describes spin in different directions, they will never be able to attract each other. That is, we can modify what Hill says as follows: Those who possess a certain kind of character with sufficient enthusiasm and personality will draw to them others who have the same kind of character.

5.1.3 Laws that govern effectiveness

To uncover the laws, which exist in the human dimension and govern human effectiveness, we first need to clarify the meaning of the word "effectiveness." According to the Webster's Seventh New Collegiate Dictionary (1971), this word means producing a decided, decisive, or desired effect, with emphasis placed on the actual production of the effect when in use or in force. By using the yoyo model, this word specifically

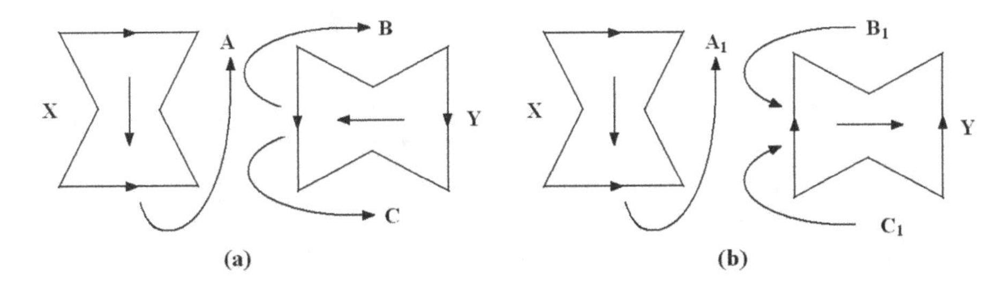

Figure 5.3 The tendency for yoyos to line up along their axes of rotation.

stands for one's capability to make other fields to flow in directions and ways he desires. In particular, if person N can make M spin at a desired orientation (or the tilt of M's axis of rotation), intensity, speed, and direction (Figure 5.1), then N is said to have an effect on M.

Based on the systemic yoyo structure underlying each and every human being, it can be naturally seen that any law existing in the human dimension that bears a degree of governing power over human effectiveness should be about special characteristics of a spin field and its interactions with other eddy fields of humans. To this end, let us first look at the simplest case with only two people, named X and Y, who live side by side in isolation from others. The reason in general why two people can live together is because they possess similar and complementing personalities and characters (Hendrix, 2001). To accommodate this requirement, let us assume without loss of generality that the yoyo fields of X and Y satisfy the following:

Left Hand Rule I: When holding our left hand, the four fingers represents the spinning direction of the eddy plane and the thumb points to the direction along which the yoyo structure sucks in and spits out materials along its axis of rotation (the narrow neck).

Now, let us see why yoyo structures of X and Y have the tendency to line up side by side with their axes of spin parallel to each other. As a matter of fact, if the yoyos X and Y are positioned as in Figure 5.3(a), then the meridian field A of X fights against C of Y so that both X and Y have the tendency to realign themselves in order to reduce the conflicts along the meridian directions. Similarly in Figure 5.3(b), the meridian field A_1 of yoyo X fights against B_1 of Y. So, the yoyos X and Y also have the tendency to realign themselves as in the previous case.

Assume that after some compromise, X and Y do line up from their initial position in Figure 5.3(a) with their axes of rotation parallel to each other except that their meridian fields flow in opposite directions (Figure 5.4(a)), then X and Y will further realign so that their meridian fields will be facing the opposite directions (Figure 5.4(b)). After reaching the situation in Figure 5.4(b), there is not much X and Y can do to reach further compromise without interference from outside of their two-men system.

If from their initial position in Figure 5.3(b), X and Y line up with their axes of rotation parallel to each other and their meridian fields flowing in the same direction

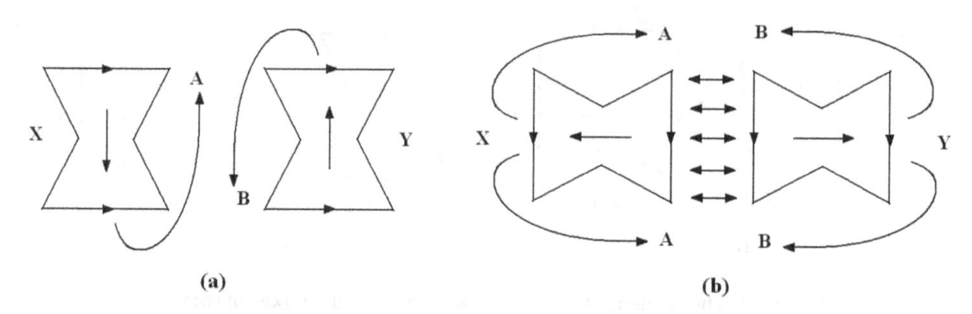

Figure 5.4 Realignment of X and Y making their meridian fields flow in the same direction.

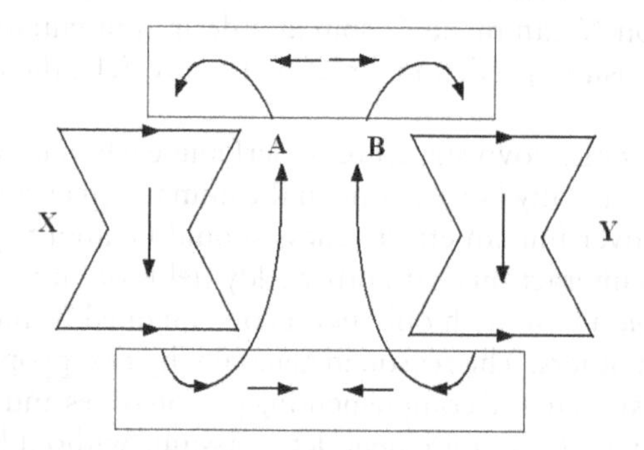

Figure 5.5 X and Y reach a complete compromise.

(Figure 5.5), then the black-hole sides tend to attract each other and the big-bang sides repel each other. So, the yoyos X and Y also tend to realign to a situation like the one in Figure 5.4(b). However, what is different of the situation in Figure 5.4(a) is that in this case, the attraction and repulsion in the enclosed areas in Figure 5.5 are much weaker than the intensity of fight against each other between X and Y in Figure 5.4(a). That is, the alignment of X and Y in Figure 5.5 represents the most stable positioning of these fields in terms of their interactions.

Now, consider two random people of similar social status who are thrown together by some external force, such as sharing a dormitory room in a university setting or an office in an employment environment, or assigned to work on a project. Similar to what is just analyzed, the yoyo fields of these two people also have the tendency to realign themselves so that their axes of rotation would be parallel to each other. So, the interaction between their eddy fields can take any of the scenarios in Figure 5.6.

By looking over the possibilities of interaction between yoyo fields N and M in Figure 5.6, we can see the following:

1 In scenarios (a) and (e), M attracts N, because N is divergent and M convergent. In these cases, M and N will never combine into one greater field.
2 In scenarios (d) and (h), N attracts M, because N is convergent and M divergent. So, M and N will never combine into one mightier yoyo.

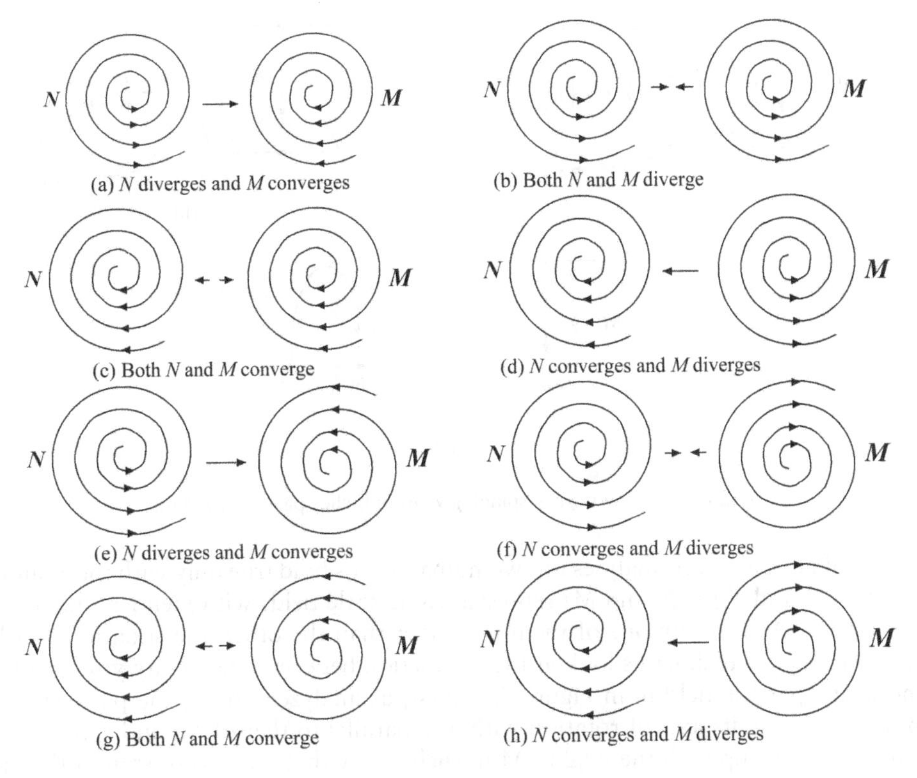

(a) *N* diverges and *M* converges
(b) Both *N* and *M* diverge
(c) Both *N* and *M* converge
(d) *N* converges and *M* diverges
(e) *N* diverges and *M* converges
(f) *N* converges and *M* diverges
(g) Both *N* and *M* converge
(h) *N* converges and *M* diverges

Figure 5.6 Interactions between two same-scale spinning yoyos.

3. In scenarios (b) and (f), *N* and *M* repel each other, where both *N* and *M* are divergent; and
4. In scenarios (c) and (g), *N* and *M* attract each other. In particular, in (c) *N* and *M* have a tendency to combine, while in (g), because the rotational directions of *N* and *M* are opposite of each other, they tend to combine and destroy each other.

Because the interaction between N and *M* in scenario Figure 5.6(c) has the tendency of combining the two spin fields, let us look at the other side of these two fields. The interaction of the opposite sides of N and M is depicted in Figure 5.6(b), where *N* and *M* repel each other. So, the dynamics between *N* and *M* in scenario Figure 5.6(c) can be described as follows: As *N* and *M* travel toward each other in their attempt to combine into a greater field, their closer distance makes the diverging fields of *N* and *M*, as shown in Figure 5.6(b), start to repel each other. The force of repellence comes from the opposite spinning directions of the fields of *N* and *M*. Under the influence of this force, *N* and *M* are pushed away from each other. When *N* and *M* travel away from each other to a certain distance, the attractions of the other sides of *N* and *M* (Figure 5.6(c)) start to once again pull them together. Such alternating effect of repulsion and attraction keeps *N* and *M* together and on their individual terms.

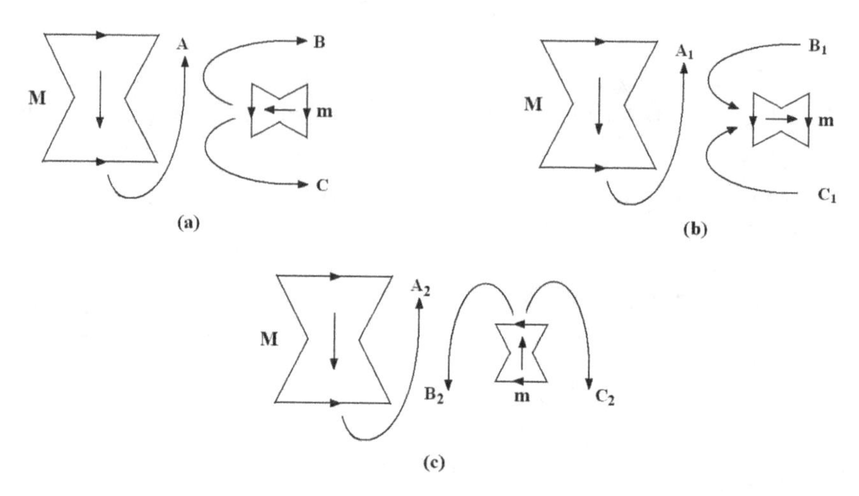

Figure 5.7 How mighty spinning yoyo M bullies particle yoyo m.

As of now, all these analyses on two human yoyos hold true only with the assumption that X and Y (or *N* and *M*) represent same scale fields with identical intensities of rotation. And, if any one of them is weaker than the other, the situation will be different. To this end, let us look at the interaction between one mighty yoyo field M and a tiny particle field m in Figure 5.7. First, as analyzed above, the particle yoyo m has to line up its axis of rotation with and parallel to that of M; and secondly, m is forced to line up with the mighty M in such a way that the axis of spin of the tiny m is parallel to that of M and that the polarities of m and M face the same direction. For example, Figure 5.7 shows how particle yoyo m has to rotate and reposition itself under the powerful influence of the meridian field of the much mightier and larger yoyo structure M. In particular, if the two yoyos M and m are positioned as in Figure 5.7(a), then the meridian field A of M fights against C of m so that m is forced to realign itself by rotating clockwisely in order to reduce the conflicts with the meridian direction A of M. If the yoyos M and m are positioned as in Figure 5.7(b), the meridian field A_1 of yoyo M fights against B_1 of m so that the particle yoyo m is inclined to readjust itself by rotating once again clockwisely. If the yoyos M and m are positioned as in Figure 5.7(c), then the meridian field A_2 of yoyo M fights against B_2 of m so that the tiny particle yoyo m has no choice but to reorient itself clockwisely to the position as in Figure 5.7(b). As what has been just analyzed, in this case, yoyo m will further be rotated until its axis of spin is parallel to that of M and its polarities face the same directions as M. Also, in the process when the tiny yoyo m adjusts itself with respect to the might M, the powerful meridian field A would have carried m to the entrance of its black hole side.

This end describes how realistically those who possess a certain character and have sufficient enthusiasm and personality can draw to them others who have the same kind of character (but weaker enthusiasm and personality), as claimed by (Hill, 1928, p. 164). This last statement can surely be seen as a law, written in terms of the systemic yoyo model, in the human dimension that governs at least some aspect of human effectiveness.

What we have done so far in this subsection only dealt with the interactions between two yoyo fields. When more fields are involved, we will have to consider situations of three-, four-, or n-body problems, for any natural number $n > 2$. These problems have been extremely difficult and unsettled since the dawn of modern science over 300 years ago. For more details, see (Lin, 2007) and references there.

5.1.4 Difficulty of breaking loose from undesirable habits

Based on what we have obtained earlier, each habit comes from consistent and unconscious repetition of a field flow pattern in one's underlying yoyo structure in his dealing with the spin fields of other human beings, situations, or objects. It belongs to his relatively stable system of traits, known as his character. Because over time each trait in the system has been tested, confirmed, and reconfirmed with various relevant, but different field flow patterns, it has been set in an inertial state of motion, as what is described in the First Law on State of Motion (Lin, 2007). So, when one knows that one of his habits is not good, meaning that he recently encountered a new pattern of field flow, leading to his realization of the deficiency of that specific habit, he sees the need to change or modify the habit (the pattern of flow). However, comparing to how well the flow pattern (the deficient habit) has worked over time, this newly acquired need for change in general is not very strong, because the recently encountered pattern has not appeared enough times. In other words, to truly make the needed change in the deficient habit, a new yoyo field has to present steadily and powerfully in order to provide a constant and influential appearance of the newly experienced pattern of flow, as suggested by the First Law on State of Motion. That end explains why in general it is difficult to break loose an undesirable habit without a powerful and steady interference of a new yoyo field.

5.1.5 Working with desire

Similar to self-motivation and self-determination, as discussed in (Lin and Forrest, to appear 1–2), for one to work on his desire, meaning that he works on making his underlying yoyo structure spin more on its own for at least a period of time, he has to, in theory, make his yoyo field not as even or uniform as before. It is because the newly created internal unevenness will naturally produce gradients, which in turn lead to moments of force so that the yoyo field will naturally rotate on its own. In particular, to create the necessary unevenness, one has to experience hardships and/or learn the existence of better possibilities. Because of this reason, it can be concluded that in general desires can be more naturally created by external forces than by reasons from within. So, one way to address the question of how one can work on his desire is to constantly compare his own state of affairs with those who are more successful.

5.2 THOUGHTS AND CONSEQUENCES

Thoughts (and thinking) are mental forms and processes. They are the most highly organized form of energy known to man. Thinking allows humans to model the world and to deal with what they face according to their objectives, plans, and desires. And, every voluntary movement of the human body is caused, controlled, and directed by thought through the operation of the mind. The presence of any thought or idea in one's consciousness tends to produce an associated feeling and to urge him to transform that feeling into appropriate muscular action that is in perfect harmony in the nature of the thought.

5.2.1 Formation of thought

The concept of thoughts is similar to that of imagination. Each thinking process involves mental manipulations of what is observed or received through the sense organs, as when abstract concepts are formed, problems are resolved, and decisions are made (Baum, 2004). The human brain makes patterns matching; in each moment of reflection, new situations and new experiences are judged against those learned or experienced from the past and judgments are then formed. To insure sound judgments being made, the intellect sorts through all relevant knowledge and past experiences against the present situation, while keeping the present situation distinct and separate from the past experiences. Here, the outstanding knowledge is generally forced upon him or acquired through his own volition under highly emotional conditions when his mind was most receptive. The three great, organized forces of passing on knowledge are: schools, press, and the church (Hill, 1928, p. 321).

With practice and adequate training, the intellect can readily mix, match, sift, and sort concepts, facts, perceptions, and experiences. This mental activity and process is known as reasoning. The self-awareness of this mental process of reasoning is referred to as an access to one's own consciousness (Block, 2007).

Thought is one thing over which one has absolute control; one has the power to select the material that constitutes the dominating thoughts of his mind. And, the dominating thoughts of the mind bring forward desirable outcomes according to the nature of the thoughts. That is, thought can be employed as an important tool, with which one may shape his worldly destiny according to his own liking. To this end, the so-called self-control is solely a matter of thought control. When one deliberately chooses his thoughts to dominate his mind and firmly ignores all outside suggestions, he exercises self-control in its highest and most efficient form (Hill, 1928, p. 183–184).

To achieve a definite purpose or goal in life, Napoleon Hill (1928, p. 245) suggested sowing the seed of the purpose in the subconscious mind, thinking accurately, and then utilizing creative thought of a positive, nondestructive nature, without hatred, envy, selfishness, and greed, to awaken the seed into growth and maturity. The reason why one has to think accurately in order to materialize his definite purpose in life is because the mind can be controlled, guided, and, directed to either creative and constructive ends or destructive ends. For instance, all the greatest of all achievements, whether in literature, art, finance, industry, commerce, transportation, religion, politics, or science are usually the results of thoughts formulated in someone's mind, and then transformed into reality by his followers through the combined use of their collective minds and bodies (Hill, 1928, p. 248).

Because any achievement in the material world is always first created in the conscious mind through imagination as a thought and then transformed into the tangible physical reality through the subconscious mind, it suggests on how concentration actually works and explains why the first step in achieving any goal in life is to create a definite mental picture for what is desired, then followed by concentrating on that picture until the subconscious mind translates it into the ultimate and desired form (Hill, 1928, p. 250).

Also, along the road of achieving success of any kind, one always needs to be aware of the widely existing phenomenon of mental inertia of human societies, even though almost every significant breakthrough in the field of scientific endeavor is first a break with tradition, with old ways of thinking, and with old paradigms (Kuhn, 1996). Because of this common but tragic mental inertia, Socrates sipped the hemlock, Stephen was stoned, Bruno was burned at the stake, Galileo was terrified into retraction of his starry truths. Forever would one desiring magnificent success in the physical world almost surely travel down that bloody trail in different degrees through the pages of history.

A thought is simply a local eddy in the pool of available knowledge that has been recorded either consciously (through sense organs) or unconsciously (not through any sense organ) in the reservoir of imagination. Thoughts are not imagination, but local spin fields in the big pool of various available patterns of flow and interactions of flows readily available in the reservoir of imagination. This systemic modeling of thoughts explains why a new thought in general germinates in the mind as triggered by some kind of hint collected from the outside world. The hint could be a new understanding of a well understood pattern or interaction, or a realization of a new combination of several relevant patterns or interactions, or a newly emerging pattern or interaction that has been unconsciously available before in the reservoir of imagination.

Each so-called thinking process (Baum, 2004) is a process of utilizing the hint from the outside world to generate a local eddy in the reservoir of imagination by pulling relevant information and knowledge together to form an organic whole. The working mechanism here is similar to the situation of a relatively calm pond of water that is suddenly interrupted by an object falling into the pond. The impact of the object creates waves propagating through the entire pond by mobilizing each water particle. Whichever "water" particle is useful and relevant, its essential meaning will be gathered to form a local eddy current in the reservoir of imagination in order to form a good pattern match. Pattern matches and the well-formulated $\pm$ function of the conscience work together to form judgments about newly experienced situations and experiences.

Because the definition of the $\pm$ function is formed and enriched throughout one's entire lifetime, when a new situation or experience does not match any element in the domain of the $\pm$ function perfectly, if the situation or experience is crucial to the well-being of the person, he will be forced to employ thinking process to decide what value, either $+$ or $-$, should be assigned to the situation or experience at hand. This process of value-assignment generally takes some time for mental manipulation for a quite while for relevant currently unknown facts to fall in places. This end explains why to insure sound judgments being made, intellects, those with solid education backgrounds, sort through all relevant knowledge and past experiences against the present situation, while keeping the present situation distinct and separate from the past experiences. It also explains the vital importance of quality education and that of the richness of the living environment in which one grows up. Our analysis entails the fact that by quality education, it means that other than forcing the fundamental information necessary for one to function effectively and efficiently in the society and skills on how to learn what is not taught to him on his own, it is more about how to form needed eddy pools to resolve problems that have never been seen before and to meet challenges. By the richness of living environments, it means such an environment that is filled with different activities and experiences that make one's yoyo pool uneven so that more or less the person will be self-propelled to achieve. As what we have studied earlier, structural unevenness is the reason for any yoyo structure to spin due to reasons from within.

Because thoughts are local eddy pools in the mind and can be manipulated by mixing, matching, sifting, and sorting concepts, facts, perceptions, and experiences, they have become the most highly organized form of energy known to man. This is especially true in our modern time with advanced technology and has been manifested time and again by the designs and manufactures of some of the most amazing products man has ever imagined, such as steam engines, airplanes, computers, etc.

As for why every voluntary movement of the human body is caused, controlled, and directed by thought through the operation of the mind is because any bodily movement is simply a 3-dimensional realization of the underlying multi-dimensional yoyo structure of man. Just as in the case where the first existence of materials and then followed by the appearance of time, which is simply a sign of eddy motion of materials, is argued (Wu and Lin, 2002), in this case, the existence of materials, the formation of local eddy currents, precedes the actual rotation of the materials. And only after materials start to rotate, the rotating materials can be possibly actualized in the 3-dimensional space for our sense organs to pick up.

The discussion above in this subsection in fact has addressed why a person's acts always in harmony with the dominating thoughts of his mind. As a matter of fact, one's acts, which are realizations of some eddy currents in his multi-dimensional field structure, take place after his relevant thoughts (the eddy currents) have been formulated. The reason why the eddy currents were formulated in the first place inside his reservoir of imagination is because he has gone through some mental manipulations, which requires consumption of energy, concentration, gathering of information, and making matches and adjustments. That explains why only dominating thoughts in general produce tangible bodily acts in the 3-dimensional space.

5.2.2 Thoughts and desirable outcomes

Based on daily experience, it is found that the dominating thoughts of the mind bring forward desirable outcomes according to the nature of the thoughts so that one may shape his worldly destiny according to his own liking (Hills, 1928). To make a theoretical sense for this observation, let us analyze as follows.

Each thought, as just discussed, is a local eddy pool in the reservoir of imagination, caused by a stir from an outside source. And, pretty much any field flow or interaction from the outside world one can focus on and make it a stir in his mind. In particular, if a flow or interaction has never registered in his reservoir of imagination, then he can surely ponder over the new phenomenon and see what new information or knowledge he could potentially extract and how personally he can benefit from this new discovery. If the flow or interaction is a simple repetition of what is well known, then he can use his imagination to look at it from a new angle for the purpose of creating a new twist to it. This mental manipulation is similar to looking at the familiar natural surrounding through various lenses of different colors. What might be surprising here is that when our eyes are covered with a lens of certain color, although the overall arrangement and layout of the surrounding looks the same, some of the previously invisible patterns may appear and become obvious. In terms of mental lenses, they are known as paradigms. As is well argued by Covey (1989), how facts are seen depends on in which paradigm one lives. The same facts can be interpreted very differently or even in contradictory ways. This fact of life alone indicates that no matter how well one knows a pattern or interaction of field flows, there is always a different, new angle from which one can uncover some surprising new lights about the known pattern or interaction.

As for whether or not one has absolute control over his thoughts, our systemic yoyo modeling suggests that it be not exactly so. It is because first of all, it is true that one has the power to select the material that constitutes a specific thought of his mind. However, to make the thought real-life relevant and realizable in the 3-dimensional space, the thought has to be accurate, derived on materializable field patterns or interactions, and developed on the currently available technology. For instance, let us imagine that we liked a remote area in a third-world country; the specific location was touched by neither modern industry nor currently advanced commerce. Based on what we knew about health and healthy living, we imagined our quality life in this godly environment, where the air was not even slightest polluted, water still tasted pure and sweet, all the food naturally grown. So, in this place we could live our lives in complete harmony with nature. However, after our initial wave of excitement, generated from actually living in the said place, was over, we had to face the day-to-day details of living. Very soon, we would withdraw from this special place and retreat back to our accustomed environment of comfort. This imaginary scenario has been played out in real-life by many. That is, the absolute control of thoughts does exist in theory. However, in reality, it does not hold up.

As for why dominating thoughts of the mind can bring forward desirable outcomes according to the nature of the thoughts, it is because when a thought in someone mind become dominating, he would have spent a lot of his time and energy on making it exciting and potentially materializable in our 3-dimensional space. As what we just studied about the systemic structure of thought, as soon as a stir appears to a person's self-awareness, the entire reservoir of his imagination is affected so that relevant information is gathered and related $\pm$ values assessed. The fact that each human being is a multi-dimensional spinning field implies that as long as the dominating thought is derived on accurate information, bodily realization in the 3-dimensional space of the thought will become visible in no time.

That is, to anyone, who is self-determined, meaning that his underlying yoyo spins on its own without any or much interference from the outside world, thinking is an important, must-have tool and thinking process the inevitable path for him to travel through in order to materialize a worldly destiny according to his own liking, which fits his unique way of spin.

This analysis well demonstrates why the so-called self-control is solely a matter of thought control; and when one deliberately chooses his thoughts based on accurate and realizable information to dominate his mind and firmly ignores all outside suggestions that are not congruent with his thoughts, he exercises self-control in its highest and most efficient form, as claimed by Napoleon Hill (1928, p.184).

5.2.3 Controlling, guiding, and directing the mind

The mind can be seen as fundamentally made up of the four human endowments: self-awareness, imagination, conscience, and free will. As discussed above, when a stir appears through the faculty of self-awareness, one takes in the new information and mobilizes one's entire reservoir of imagination to formulate a response by making use of one's conscience (the $\pm$ function) and free will (the ability to make at least short-term predictions). That is, the actual formation of a local eddy in his imagination is really a feedback process, in which the initial stir is the input. To respond to the stir, various groupings of relevant information and knowledge are formed in the conscious mind. These groupings of known data are judged by using the $\pm$ function. When a $-$ value is seen assigned to the response produced out of a specific grouping, the relevant combination of information and knowledge is disregarded. When a response does not have a definite $\pm$ value, more relevant data out of his reservoir of imagination are pulled into the specific grouping until a definite $\pm$ value can be possibly assigned. Among all the groupings with $+$ response values, the person orders the response values according to his preference or the instantaneous need or liking of the moment so that he could handily pick out the one response with the most desirable $+$ value. This feedback procedure makes people feel that they have control over their own thoughts. For a general treatment of feedback systems, please consult (Lin and Ma, 1990).

Now, with the mind fully activated toward formulating a desirable response or outcome, the originally relatively stable flow of the reservoir of imagination begins to experience turmoil and various local eddy pools start to form. Accordingly, the sense organs become keen to whatever in the surroundings that is supportive to individual groupings of information and knowledge. Along with the alert and excited sense organs, the entire underlying multi-dimensional yoyo structure of the person starts to act in various harmonic ways correspondingly. So, this purposeful movement, which might well be unconscious to the person, of the underlying structure is eventually actualized in the 3-dimensional space. When facing difficulties or realizing newly found resources in this 3-dimensioanl physical space, the originally chosen response will be voluntarily modified either to overcome the difficulty or to take advantage of the newly found resource. Once again, a feedback loop emerges. This feedback loop creates the feeling that the human mind is guided and directed to the desired ends.

This systemic explanation developed in Subsection 5.2.3 in fact illustrates that if one has a definite purpose or goal in life, in one way or another, he has been terribly stirred or troubled when his established world view and belief system was somehow turned upside down at least partially. Through the feedback mechanism, he is able to form his most desirable outcome in his conscious mind by making use of his self-awareness, imagination, the $\pm$ function of his conscience, and his ability of making predictions. If for whatever reason, the stir is truly severe, then he will naturally have a burning desire to materialize his imagined outcome in the physical world in order to even out the imbalance created by the stir in his reservoir of imagination and the associated $\pm$ function of his conscience. By making use of what we just obtained in the previous paragraph, in order to bring about bodily actions toward materializing the burning desire, one should concentrate on the chosen outcome, think accurately about all relevant information, knowledge, and available resources. Here, for one to achieve true concentration, he has to avoid wasting his energy, especially his mind power, on seemingly related, but in reality not necessary needs, such as hatred, envy, selfishness, and greed. Each of these unnecessary needs can easily consume all the available time, energy, and resources needed to meaningfully realize any noble goal in life.

5.2.4 Mental creation and physical materialization

To see in terms of the systemic yoyo model why any material achievement consists of two creations: first in the mind and the second in the physical world, let us look at the general fact that materials, be they tangible or intangible, exist prior to any form of movement of the materials (Lin, 1998b). Speaking more specifically, before a particular aspect of a human yoyo field can be realized in our 3-dimensional space, the relevant field has to first experience some sort of formation or movement. And only after a special aspect of the multi-dimensional field is realized in the 3-dimensional space, some tangible or visible material achievement can be possibly recognized. This sequential before and after of events provides the reason why any material achievement has to be first accomplished in the mind and then in the physical world.

At this juncture, it is necessary to note that due to the duality of yoyo fields (Lin, 1998b), meaning that yoyo fields exist in pairs, each of which contains two fields spinning in directions opposite of each other, when one forms an idea or a definite goal to reach with a burning desire, there will naturally be an opposite idea or definite goal out there in the immediate surroundings for someone else or people to accomplish. This latter person or people in real-life generally create various degrees of difficulties or hardships for the former person along his pursuit of his goal.

Assume that person A has a burning desire to reach a definite goal. To successfully materialize his dream, he will have to put in some kind of hard work into his pursuit. For instance, person A desires to possess a lot of wealth through conventional means, hard work combined with intelligence, without hurting anyone else. If he is lucky enough to eventually materialize his goal, then he will generally experience through the following stages of struggle.

1 In the beginning stage of forming his own very goal, many people within his immediate surroundings will discredit his idea by saying directly to him such things as "money is not everything in life," even though A never says and believes that money is everything.
2 As the goal begins to crystallize in A's mind, more people will start to discredit this idea by providing him similar reasons as "money is not everything in life." At the same time, these people will try to convince each other that A is about to make a terrible mistake in his life.
3 When A begins to carry out his plan of acquiring a lot of money, these same people plus additional more will communicate with A and among themselves that A is crazy and will never make it, because his plan is flawed.
4 As A starts to bring about initial signs of success, these people will begin to circulate various untrue stories of A just to calumniate on A's character.
5 As A enjoys his eventual success of owning of a lot of wealth, some of the other people will also have achieved what they unconsciously wanted: the service A provides, fame and some monetary awards, from trying to destroy A.

Now, let us see why these stages 1–5 generally play out as outlined. At stage 1, for whatever reason, person A's reservoir of imagination is stirred so that some special local eddy begins to form. That is, his underlying yoyo field starts to move differently from before. Along with this subtle change, all the yoyo fields in A's immediate surroundings begin to feel the pressure. As their natural resistance to the pressure, the existence of such resistance is guaranteed by the First Law on State of Motion, they fight back toward A. As A moves forward to crystallize his plan, his yoyo field inevitably places more pressure on the other yoyo fields, creating more intensive resistance from these fields. When A starts to carry out his plan, his crystallized goal is realized and becomes visible in the 3-dimensional world. This realization provides a tangible target for the other people to react to. As A's field actually sucks in wealth by providing what other people need in their lives, these people found more materials to give out from their fields: making and telling untrue stories about A. As A eventually reaches the great success as he envisioned from the very start, the high level of intensity of his spinning field structure inevitably has also helped several other fields to rotate at high intensities.

What is interesting here is that the word "wealth" can be replaced by almost any other word, and similar stages of struggle play out just the same.

5.2.5 Mental inertia

The previous discussions have described the mechanism underlying the phenomenon of mental inertia and how human mental inertia works. As for how a person, who pursues after his own desired success, can avoid from traveling along the bloody trail created by mental inertia, let us look into our systemic model in more details to see whether or not there is such a possibility that one yoyo can increase its intensity of spin without being drowned or totally wept out by others. To do this end, let us look at some of the representative examples from the past.

Historically, the reason why Socrates, a classical Greek philosopher who is credited as one of the founders of Western philosophy, sipped the hemlock is because his pursuit of virtue and strict adherence to truth clashed with the then-current course of Athenian politics and society (Brun, 1978). He lived during the time of the transition from the height of the Athenian hegemony to its decline with the defeat by Sparta and its allies in the Peloponnesian War. However, Socrates praised Sparta in various dialogues. And, his offense to the city of Athens was his position as a social and moral critic. Rather than upholding a status quo and accepting the development of immorality within his region, Socrates worked to undermine the collective notion of "might make right" so common to Greece during this period. And, his attempts to improve the Athenians' sense of justice may have been the source of his execution.

As for Stephen being stoned to death, it is a story from the Bible. Amidst the rapid development of the church in Jerusalem, it appeared that there was a hint of discrimination. The problem was serious; immediate attention was required. To resolve the problem, the congregation selected seven men of good reputation, full of spirit and wisdom among themselves to oversee the daily ministry. The first man chosen was Stephen, indicating that he was the first choice of people. However, this godly man was later stoned to death by a mob of angry Jews. What happened was that: Stephen visibly taught the people of Jerusalem what he believed. Consequently, some Jews were convinced that Stephen was a leading figure and drive force of the church. They decided to confront him in a highly visible venue, believing that they could easily discredit him and turn back the hearts of the people. However, the outcome infuriated them. After that, the Jews attacked Stephen with false witnesses and accusations. After false charges had been leveled against him, Stephen was given the opportunity by the Jewish high priest to make a public defense of himself. However, instead of a defense to save his own skin, Stephen went on an offense for what he believed. What happened did not sit well with these Jewish leaders, and that ultimately cost him his life.

Giordano Bruno (1548–February 17, 1600) was an Italian philosopher best-known as a proponent of heliocentrism and the infinity of the universe. He is often considered an early martyr for modern scientific ideas, because he was burned at the stake as a heretic by the Roman Inquisition, a system of tribunals developed by the Holy See during the second half of the 16th century, responsible for prosecuting individuals accused of a wide array of crimes related to heresy (Griggs, 1969).

Galileo Galilei (February 15, 1564–January 8, 1642) was an Italian physicist, mathematician, astronomer, and philosopher. He played a major role in the Scientific Revolution of the 16th–17th centuries. He also defended heliocentrism, claimed it was not contrary to those Scripture passages, and openly questioned the veracity of the Book of Joshua (10:13), where the sun and moon were said to have remained unmoved for three days to allow a victory to the Israelites. In 1616, Cardinal Bellarmine, acting on directives from the Roman Inquisition, delivered Galileo an order not to hold or defend the idea that the Earth moves and the Sun stands still at the center. For the next several years Galileo stayed well away from the controversy. When his book, *Dialogue Concerning the Two Chief World Systems*, was published in 1632, Galileo alienated many of his defenders in Rome, include the Pope, and was ordered to stand trial on suspicion of heresy in 1633. With the court sentence he spent the rest of his life under house arrest.

The bottom lines, indicated in all these and relevant historical stories, are the same: Each individual or a small group of individuals has to be morally responsible for the well-being of their society; otherwise this individual or small group of individuals will be run over by other people or elements of the society mercilessly, as concluded in (Lin and Forrest, to appear 1–2) when they address the question of how individuals are morally responsible for their conducts. Specifically, the "well-being" of the society stands for the maintenance of both the mental inertia and the inertia of the field motion of the society. Speaking differently, each relatively stable society has its well adopted system of beliefs and philosophical values as well as a fully developed hierarchy of social structure. The relative stability means that the underlying yoyo field of the society rotates quite vigorously without any local pool spinning in a different direction. In such a uniformly spinning fluid, any rebellious local movement will undoubtedly be crushed mercilessly by the inertia of the vigorously spinning field of the society. As indicated by the dishpan experiment (Hide, 1953; Fultz, et al., 1959; or see (Lin and Forrest, to appear 1–2) for a brief description of the experiment), for such a uniform spinning fluid to allow local eddy pools to exist, the local pools have to appear gradually at roughly the same time. So, when one pursues after his desired success, it is possible for him to avoid traveling along the bloody trail created by mental inertia of the society. In particular, the person has to control its field movement in conformation with the acceptable form of motion of the greater culture. As he obtains more support from others and causes many others to pursue after similar goals of success, the inertia of his greater environment will be altered by his effort. Just as what is well recorded in the history, this process will not be easy; it generally takes time; and it may well turn out to be the case that one gets such a trend of change started, and many others in the following generations devote their collective efforts and intelligence to eventually get the mental inertia of the greater environment altered.

5.3 DESIRE

On the back of all achievement, all self-control, and all thought control is there the magic something known as desire. It is the depth of this desire that it limits one from achieving high. When a person has strong desires, he/she would appear to possess supper human powers to achieve magnificently and to climb high (Hill, 1928, p. 185–186).

5.3.1 Desire in terms of systemic yoyo model

As a matter of fact, each and every personal and professional achievement starts from a strongly rooted desire. Desire is the seed of all magnificent achievement, the starting place, back of which there is nothing or at least there is nothing of which we have any knowledge. According to Hill's life time, intensive studies on achievements, it is discovered that each cycle of human achievement works more or less in the following fashion: First, form a picture and some objective in the conscious mind, derived on top of a definite purpose or goal based upon a strong desire; secondly, focus the conscious mind upon this objective by constant thinking of it and believing in its eventual attainment; thirdly, sooner or later the subconscious section of the mind takes up the picture or the outline of this objective; that will impel one to take the necessary physical actions to transform that picture and objective into the eventual physical reality (Hill, 1928, p. 250).

Because desire is so important for one to achieve magnificently in his personal and professional lives, then what is desire? To this end, there are several different studies. In terms of psychoanalysis, desire designates the impossible relationship that a person has with his objet petit a, meaning the unattainable object of his desire, which is sometimes also called the object cause of desire. According to French psychoanalyst and psychiatrist Jacques Lacan (1901–1981), desire proper (in contrast with demand) can never be fulfilled. Lacan argues that desire first occurs during a "mirror phase" of a baby's development, when the baby sees an image of wholeness in a mirror which gives him a desire for that being. As that baby grows older and matures as an adult, he still feels separated from himself by language, which is incomplete, and so the person continually strives to become the desired whole. He uses the term "jouissance" to refer to the lost object or feeling of absence which a person believes to be unobtainable (Evans, 1996).

In terms of philosophy, the concept of desire has been identified as a philosophical problem since antiquity. For example, in *The Republic* written in approximately 380 B.C., Plato (1991) argues that individual desires must be postponed in the name of the higher ideal. In Aristotle's *De Anima* (Polansky 2007), the soul is seen to be in motion, because animals long for various things and in their desire they acquire locomotion. Aristotle argues that desire is implicated in animal interactions and the propensity of animals to motion. At the same time, Aristotle acknowledges that desire cannot account for all purposive movement towards a goal. He posits that perhaps reason, in conjunction with desire and by way of the imagination, makes it possible for one to apprehend an object of desire, to see it as desirable. In this way, reason and desire work together to determine what is a good object of desire. This description resonates with that of desire in the chariots of Plato's Phaedrus, for in the Phaedrus (Plato, 2005) the soul is guided by two horses, a dark horse of passion and a white horse of reason. Here passion and reason, as in Aristotle, are also together. Socrates (Kofman, 1998) does not suggest the dark horse be done away with, since its passions make possible a movement towards the objects of desire, but he qualifies desire and places it in a relation to reason so that the object of desire can be discerned correctly, so that one may have the right desire.

In Passions of the Soul, Descartes (1649) writes of the passion of desire as an agitation of the soul that projects desire, for what it represents as agreeable, into the future. For Immanuel Kant (1790), desire can represent things that are not only the objects presently at hand but also absent; it is also the certain effects that do not appear and that what affects one adversely be curtailed and prevented in the future. The moral and temporal values are attached to desire in such a way that objects that enhance one's future are considered more desirable than those that do not. In 1740, David Hume suggests that reason is subject to passion. Motion is put into effect by desire, passions, and inclinations. It is desire, along with belief, that motivates action. Kant (1790) establishes a relation between the beautiful and pleasure. He says that "I can say of every representation that it is at least possible (as a cognition) it should be bound up with a pleasure. Of representation that I call pleasant I say that it actually excites pleasure in me. But the beautiful we think as having a necessary reference to satisfaction." And desire is found in the representation of the object. Georg Wilhelm Friedrich Hegel (1807) begins his exposition of desire with the assertion that "self-consciousness is desire." It is in the restless movement of the negative that desire removes the antithesis between itself and its object, "... the object of immediate desire is a living thing...," and object that forever remains an independent existence. Hegel's inflection of desire via stoicism, a school of Hellenistic philosophy founded in Athens by Zeno of Citium in the early third century B.C., considering that destructive emotions are the result of errors in judgment, and that a sage, or person of "moral and intellectual perfection," would not have such emotions (Stanford Encyclopedia of Philosophy), becomes important in understanding desire.

In terms of emotions, desire stands for a sense of longing for a person or object or hoping for an outcome. When a person has the desire for something or someone, his sense of longing is excited by the enjoyment or the thought of the item or person, and he wants to take actions to obtain his goal. The motivational aspect of desire has long been noted by philosophers. For example, Thomas Hobbes (1588–1679) asserted that human desire is the fundamental motivation of all human action (Macpherson, 1962). According to the early Buddhist scriptures, the Buddha stated that monks should generate desire for the sake of fostering skillful qualities and abandoning unskillful ones (Bhikkhu, 1996); and in one's training for his eventual liberation, he must work with motivational processes based on skillfully applied desire (Collins, 1982, p. 251).

While desires are often classified as emotions by many, psychologists describe desires as something different from emotions and argue that desires arise from bodily structures, such as the stomach's need for food and the blood needs oxygen, whereas emotions arise from a person's mental state. Berridge, Robinson, and Aldridge (2009) find that although humans experience desire and fear as psychological opposites, they share the same brain circuit. Kawabata and Zeki (2008) showed that the human brain categorizes each stimulus according to its desirability by activating three different brain areas: the superior orbito-frontal, the mid-cingulated, and the anterior cingulated cortices. Their work on pleasure and desire shows that reward is a key element in creating both of these states, and that a chemical called dopamine is the brain's "pleasure chemical". They also show that the orbitofrontal cortex has connections to both the opioid and dopamine systems, and stimulating this cortex is associated with subjective reports of pleasure.

Based on our systemic models of human endowments, character, and thought, developed in (Lin and Forrest, to appear 1 & 2) and in this chapter, desire can be modeled as how one wants his underlying yoyo structure to be in terms of one of or any combination of its attributes: direction, orientation, spin intensity, and scale. In particular, along with the development of one's self-awareness, he naturally builds his reservoir of imagination and his $\pm$ function with its ever-expanding domain. By knowing himself and others, he senses either consciously or unconsciously and either correctly or incorrectly how his yoyo field can be made uniform and even or different so that it does not have to rotate as vigorously, or flow in a certain direction, or orient in a certain way, or be a certain scale, as it has been. So, as the person matures, his desire changes over time due to his knowledge on how his field could be made uniform or different evolves.

This systemic model of desire explains why each desire designates an impossible relationship that a person has with the unattainable object of his desire or the object cause of desire. It is because as long as the person is alive, his underlying field has to rotate on its own without being constantly pushed by some external fields. For this end to hold, his yoyo field must have some degree of unevenness. Because one's initial desires appear along with the development of his self-awareness, they indeed start to occur during the mirror phase of a baby's development. However, instead of wanting to be whole, as claimed by Lacan (1901–1981) (Evans, 1996), the baby just wants unconsciously to be more even than he is in terms of his underlying field structure. As the baby grows older, he continues to feel not as even as he likes. As a matter of fact, as he matures into an adult and as his self-awareness becomes more awaken, he may very well feel much more uneven than when he was a child, because with the growth in age he generally experiences more constant comparisons with others so that some of the unevenness in his own field become more obvious with time.

As for why individual desires must be postponed in the name of higher ideal, as written by Plato (1991) in approximately 380 BC, our discussions earlier indicate that individual yoyo fields or local eddy fields have to be morally responsible to the well-being of the society; otherwise, other people and elements of the society would run over these individuals mercilessly. So, if an individual does not postpone his personal desire in order to entertain the need of the society to achieve or maintain its relative stability, before long this specific individual will no longer exist physically; its yoyo field will be completely destroyed by the powerful, overwhelming flow of the field of the society. Here, corresponding to Plato's dark and white horses of the soul, the viability of an individual's field depends on its self-awareness about its own existence and those around him and its $\pm$ function on which he senses what to do.

In our modeling, the soul has been implicitly identified with the spin field of an individual. So, the soul is in spinning motion because of its internal uneven structure and acquires its locomotion because in its desire it rotates more vigorously. This end is

reasoning behind this idea is that rewards provide the material and/or psychological indicators of recognition for progress toward the attainment of the desired goal; and the obstacles help to strengthen the internal drive for a local pool of desire to form and to spin vigorously.

The reason why Hill's (1928, p. 250) cycle of human achievement actually works can be analyzed as follows: With a strong desire (a fast rotating pool in imagination), if one has a firm and clear definition of success and how he can tell whether or not he has achieved his defined success, then his innovative connections of knowledge, formed on the basis of the strong desire, will guide him to the right direction to devote his attention, energy, and time for the eventual attainment of the defined success. When his progress is hindered by unexpected difficulties, this guidance will provide possible ways for him to get around or tell him how to overcome the unexpected. And equally important, these innovative connections of knowledge will help him locate available resources both from within his reservoir of imagination and from the outside world useful and valuable to attaining his desired goal. By constantly thinking of the defined success and believing in its eventual attainment, one in fact keeps his eyes open and mind active in searching for possibilities that can be employed to his advantage in his attainment of his goal. By actively exercising his sense organs and mind for potentially available resources, his human endowments – self-awareness, imagination, conscience, and free will – will eventually make his underlying field structure spin in a necessary form. When this form of motion is realized in the three-dimensional physical world, it means that he has taken the necessary physical actions to transform his mentally defined success into the eventual physical reality.

5.4 ENTHUSIASM AND STATE OF MIND

By enthusiasm, it means intense enjoyment, interest, and/or approval. It represents a state of mind that inspires and arouses a person to put action into the task at hand. It is contagious to all with whom the enthusiast comes in contact. Enthusiasm is the vital moving force that expels action. The greatest leaders are those who know how to inspire enthusiasm in their followers. When one mixes his works with enthusiasm, the originally tedious works will no longer feel stressful or monotonous. Enthusiasm in general energizes one's entire body so that he can get along with little rest and enables him to perform from two to three times as much work as he usually perform in a given period of time without experiencing fatigue. Enthusiasm is a vital force that one can harness and use for good purpose and with which one recharges his body and develops a dynamic personality.

5.4.1 Systemic mechanism of enthusiasm

Some people are blessed with natural enthusiasm while others must acquire it. The procedure of developing enthusiasm is simple. It begins at doing the work or rendering the service one likes best. If one is so situated that he cannot conveniently engage in the work that he likes best, then he can for the time being proceed along another line very effectively by adopting a definite purpose or goal in life that contemplates him engaging in that particular work at some future time. That is, one may be a long way from realizing his definite goal in life, but if he kindles the fire of enthusiasm in his heart, and keep it burning, before very long the obstacles that now stand in the way of attaining that goal will melt away, and he will find himself in possession of power that he did not know he possessed (Hill, 1928, p. 153–154).

To practically acquire enthusiasm for those who are not naturally blessed with the special state of mind, they only need to use the technique called self-suggestion (Hill, 1928, p. 164) for a period of time until a small still voice within themselves starts to affirm that they will positively realize their goals. More specifically, the technique of self-suggestion works as follows: 1. Completely write out the definite goal in clear and simple language; 2. Write out a relatively detailed plan on how to transform the goal into reality; 3. Read over the descriptions of the goal each night right before bed with full faith and enthusiasm in the ability to transform the definite goal into reality; and 4. While reading the description, imagine the full possession of the object of the goal.

Along with enthusiasm, one needs self-control to direct his enthusiasm to constructive ends. Enthusiasm is the vital force that arouses a person to action, while self-control is the balance that directs the action to build up instead of tear down. To this end, Napoleon Hill (1928, p. 175) studied 160,000 prisoners in the United States. He found that 92% of these men and women are in prison because they lacked the necessary self-control to direct their energies constructively. On the other hand, the records of those the world remembers as great indicate that each and every one of them possesses the quality of self-control.

We have established the systemic mechanism for enthusiasm in (Lin and Forrest, to appear 3), when we addressed the following question: Is it generally true that those who possess character have enthusiasm and personality sufficient to draw to them others who have character, as claimed by (Hill, 1928, p. 164)? To see why enthusiasm inspires and arouses a person to put action into the task at hand and makes originally monotonous works more enjoyable to finish, let us cite the relevant details here.

By enthusiasm, is signified a whole-hearted devotion to an ideal, cause, study or pursuit, or merely being visibly excited about what one's doing (Tucker, 1972). As what has been modeled in (Lin and Forrest, to appear 3), enthusiasm is a state of intensive spin of the underlying yoyo field of one's mind and body that has a specific tilt in the axis of rotation. When compared to the systemic model of self-motivation and self-determination, enthusiasm is the same as self-motivation and self-determination except that its state of intensive spin does not last as long as that of self-motivation and self-determination and that throughout the lifespan of enthusiasm, the intensity of spin might vary up and down alternatively.

In order to uncover the systemic mechanism of enthusiasm, we need to address why a field spinning in its inertial state could suddenly acquire the momentum to rotate at a higher level of intensity in terms of its speed and strength. Based on our systemic yoyo model, there are only two ways for this phenomenon to occur. One is that a new unevenness in the structure of the field suddenly appears; and the other an ever-presenting field in the immediate neighborhood suddenly, unexpectedly disappears, breaking the inter-field equilibrium while providing some advantage to the field of our concern.

There are many different ways to cause a new structural unevenness to appear suddenly. One category of methods includes those about discovering new strengths in oneself. Another category contains methods about discovering new patterns of interaction with other fields. Discovering new strengths within oneself can be about finding not previously known, but realizable combinations of well known patterns of flow or different ways a specific flow pattern can be directed. For instance, one used to eat his breakfast, consisting of a bowl of cereal, a cup of milk, and a cup of fruit mix, in the following order. He first prepares and eats his bowl of cereal; after placing his bowl in the sink, he then fills up his cup with milk from a bottle in the refrigerator, waits a few minutes for his milk to be heated up before he drinks the milk. After he throws the cup into the sink, he once again reaches into the refrigerator the second time to get the big container with fruit mix to fill up his second cup. When this procedure becomes a routine, this person will naturally repeat the sequence of operation daily without giving any second thought to it. Now assume that one day for whatever reason he has to hurry up in the morning, he unconsciously finishes his breakfast as follows: When he prepares his bowl of cereal from the pantry, he also conveniently takes out the milk bottle and the container of fruit mix from the refrigerator. When filling up his bowl with cereal, he also fills up his cups with milk and fruit mix, respectively. When he returns the box of cereal back into the pantry, he also conveniently put the milk bottle and fruit mix container back into the refrigerator. While he is eating his cereal and fruit mix, his cup of milk is heated up. On his way to put his bowl and cup into the sink, he finishes his cup of milk so that he put all of his table wares into the sink at the same time. This simple example illustrates how we can improve the efficiency in many areas of our lives by focusing on ourselves.

As for discovering new patterns of interaction with other fields, it includes finding out more about others so that one's own behaviors can be accordingly modified in order to take advantage of the new discovery and to achieve much improved effectiveness. For instance, in a sales situation, a large commercial customer bought many units of a certain product. That is great for the salesman! And to most salesmen, the story stops right there and when he receives his fat commission check. However, if a salesman has his eyes open for discovering new patterns of interaction, he might very well feel curious as to why the customer bought so many units of his product. After actually checking into the details with the representative of the commercial buyer, he may surprisingly find out how the buyer actually uses his products to benefit many other well-known companies. Now, equipped with this new knowledge, the salesman will undoubtedly feel very excited about his product and will surely brag about it to other potential buyers. This end in practice may create many more major sales for the salesman in the weeks, months, and even years to come.

So, when a new structural unevenness appears suddenly within one's yoyo field either by discovering new strengths within himself or by realizing new patterns of interaction with others, his underlying field obtains a boost in its intensity of spin. As one field suddenly spins at a greater strength, the following First Law on State of Motion implies that in comparison all other fields in the neighborhood will still be moving in their individual inertial states.

The First Law on State of Motion (Lin, 2007): *Each imaginable and existing entity in the universe is a spinning yoyo of a certain dimension. Located on the outskirt of the yoyo is a spin field. Without being affected by another yoyo structure, each particle in the said entity's yoyo structure continues its movement in its orbital state of motion.*

This fact implies that the strengthened field senses the potential of winning various scores against all other fields with the broken equilibrium. It is because its movement suddenly experiences fewer constraints from other fields. When the field of suddenly increased intensity is realized in the 3-dimensional space, it makes people feel that his enthusiasm has inspired and aroused him to put action into the task at hand. And this 3-dimensional realization makes the person himself feel that his enthusiasm makes originally monotonous works more enjoyable to finish. Here, both the task at hand and the originally monotonous works have been reflected in the form of spinning motion of the specific field before it suddenly gains its unexpected intensity. As a matter of fact, these feelings are truthful reflections of the state of affairs, because the fact that this specific field is spinning at a much increased intensity while others are still rotating in their individual inertial states implies that this fast moving field can suddenly accomplish many objectives that he could not easily accomplish before.

Similar analysis can be carried out for the situation when an ever-presenting field in the immediate neighborhood suddenly and unexpectedly disappears. It is because the sudden and unexpected disappearance of a neighboring field breaks the inter-field equilibrium of the neighborhood. This sudden lose in the inter-field balance in actuality creates an imbalance or unevenness within the said field. So, this external imbalance is immediately transformed into a situation of internal unevenness. What is a little different from what is just analyzed above though is that when such an inter-field imbalance appears unexpectedly, all the fields of the neighborhood are affected with expected appearance of internal unevenness. So, whichever field that does expect such a change from occurring would have its advantage over other fields.

This yoyo field modeling indeed explains vividly why enthusiasm stands for intense enjoyment, interest, and/or approval. The so-called intense enjoyment and interest come from one's sudden liberation from the constant constraints of other fields. As the field of increased intensity spins faster than before, as it would like to base on the unevenness of its internal structure, it does experience an ease, which can be expressed as an intense enjoyment and interest. Because suddenly the field suffers from less constraint from others, it feels like that its behaviors have been approved by these less constraining fields.

To address why enthusiasm is a vital force with which one recharges his body and develops a dynamic personality. Let us first look at the concept of personality. According to (Allen, 2005), personality stands for one's aggregate conglomeration of decisions he has made throughout his life. There are inherent natural, genetic, and environmental factors that contribute to the development of one's personality. The systemic model of personality, developed in (Lin and Forrest, to appear 3), says that the so-called personality is the subsystem of one's character that consists of all the three-dimensional realizations of his optimal patterns of field flow and field interactions that show how he related and reacted to others, to various stimuli, and to the environment. Combining this model with what we just discussed, enthusiasm indeed make one more dynamic, spinning at a increased intensity, in terms of how he would proactively relate and react to others, to stimuli, and to the environment.

5.4.2 Leadership and enthusiasm

To see the connection between great leaders and their enthusiasm, let us first look at what is meant by leadership. As a matter of fact, leadership is one of the most salient aspects of the organizational context and a difficult concept to define. It is defined as the process of social influence in which one person can enlist the aid and support of others in the accomplishment of a common task (Chemers, 2001), or ultimately about creating a way for people to work together and to make something extraordinary happen (Kouzes and Posner, 2007). In the research of leadership, many different theories have been developed by various authors.

For example, the trait theory, which was initiated by Thomas Carlyle (1841), attempts to identify the talents, skills, and physical characteristics of men that are associated with effective leadership (House, 1996). Kirkpatrick and Locke (1991) argue that key leader traits include: drive (a broad term which includes achievement, motivation, ambition, energy, tenacity, and initiative), leadership motivation (the desire to lead but not to seek power as an end in itself), honesty, integrity, self-confidence (which is associated with emotional stability), cognitive ability, and knowledge of the business. Facing criticism, recent studies of the trait theory identify leadership skills, not simply a set of traits, but as a pattern of motives suggesting that successful leaders tend to have a high need for power, a low need for affiliation, and a high level of self-control (McClelland, 1975).

(Spencer, 1841) argues that it is the times that produce the leaders and not the other way around. This theory assumes that different situations call for different leadership characteristics. According to this group of theories, no single optimal psychographic profile of a leader exists, and what an individual does, while acting as a leader, is in large part dependent upon characteristics of the situation in which he functions (Hemphill, 1949). It is found (Van Wormer, et al., 2007) that

1. The authoritarian leadership style, in which the leader makes decisions alone, demands strict compliance to his orders, and dictates each step taken, to a large degree future steps were uncertain to others, is approved in periods of crisis but fails to win the hearts and minds of the followers in the day-to-day management;
2. The democratic leadership style, in which collective decisions, assisted by the leader, are made, before accomplishing any task, perspectives are gained from group discussions, members are given choices and collectively decide the division of labor, and feedbacks are given by individual group members, is more adequate in situations that require consensus building; and,
3. The laissez faire leadership style, in which freedom is given to the group for policy determination without any participation of the leader, who remains uninvolved in work decisions unless asked, does not participate in the division of labor, and very infrequently gives praise, is appreciated by the degree of freedom it provides, but as the leader does not "take charge", he can be perceived as a failure in protracted or thorny organizational problems.

That is, leaderships and their styles are contingent upon the situation.

According to the functional theory, the leader is responsible for making sure whatever necessary to his group's needs is taken care of. So, he is detrimental for his group's effectiveness and cohesion (Wageman, et al., 2008). This theory has most often been applied to team leadership as well as organizational leadership. According to this theory, there are five broad functions a leader provides when he promotes his unit effectiveness. These functions include: environmental monitoring, organizing subordinate activities, teaching and coaching subordinates, motivating others, and intervening actively in his group's work.

A formal organization (Cecil, 1970, p. 884–89) is such a human hierarchy that it is established for achieving defined objectives. It consists of divisions, departments, sections, positions, jobs, and tasks so that the entire organization would behave impersonally in regard to relationships with clients and with its members. Based on merit or seniority employees are ranked so that the higher one's position is in the hierarchy, the greater his presumed expertise and social status. It is this bureaucratic structure that heads are appointed for administrative units and they are endowed with the authority corresponding to their positions. Now, other than the appointed administrative chief, a leader may still emerge informally, which underlies the formal organizational structure. He is the leader of the underlying informal organization that is made up of the personal objectives and goals of individual employees. The unchanging human needs – personal security, maintenance, protection, and survival – of the employees are met by the informal organization and its emergent leaders (Knowles and Saxberg, 1971, p. 884–89).

Each informal leader influences a group of employees toward obtaining a particular result. He does not depend on any title or formal authority. Instead, he is recognized by his capacity for caring for others, clear communication, and a commitment to persist (Hoyle, 1995). On the other hand, although each appointed manager has the authority to command and enforce obedience, he still has to possess adequate personal attributes to match his authority, because the authority is only potentially available to him. In the absence of sufficient personal competence, a manager may be confronted by an emergent unofficial leader, who can challenge the manager's role and reduce it to that of a figurehead. Since only authority of position has the backing of formal sanctions, it follows that whoever wields personal influence and power can legitimize this only by gaining a formal position in the organizational hierarchy with commensurate authority (Knowles and Saxberg, 1971, p. 884–89). Therefore, leadership can be defined as one's ability to get others to willingly follow. Every organization needs leaders at every level to achieve functionality and efficiency.

Now, let us look at the systemic model and analysis of the concept of leadership. Based on the definitions (Chemers, 2001, Kouzes and Posner, 2007), leadership is one's capability to adjust his underlying field structure so that many other neighboring fields would spin in similar fashions without much difficult readjustment. In particular, if one can utilize a process of social influence to obtain aids and supports of others in accomplishing a common task (Chemers, 2001), it implies that there has appeared a big whirlpool (the common task). This pool might initially be conceptual and physically invisible. However it does cover a large territory, within which many smaller fields (individual people) are located. Now, the leader is the person who can realign all the individual eddy fields in such a way that the conceptual large field becomes a visible reality. This exact analysis can be used to understand the definition of leadership of (Kouzes and Posner, 2007), where the initially invisible large field is the expected something extraordinary.

With this systemic model of leadership, we can unify all the relevant studies into one organic whole. For instance, to be a leader, it is indeed important for a person to possess some key elements, such as talents, skills, and physical characteristics, as claimed in the trait theory (House, 1996). Let us now see why the key elements include drive, leadership motivation, honesty, integrity, self-confidence, cognitive ability, and

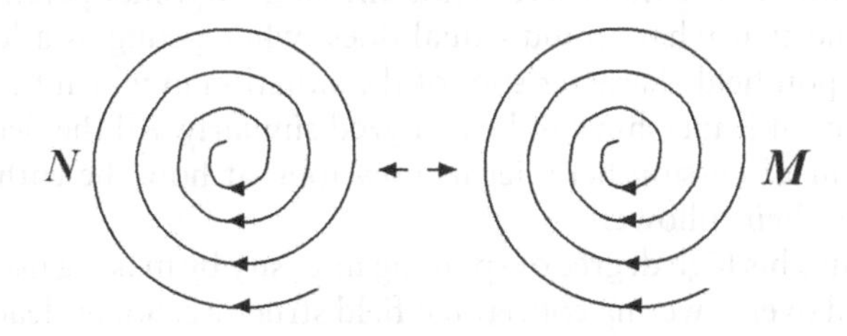

Figure 5.8 Convergent fields spinning harmonically.

knowledge of the business, as argued by (Kirkpatrick and Locke, 1991). First of all, according to what we have analyzed in (Lin and Forrest, to appear 1–2), drive stands for the intensity of spin powered from within, caused by the internal unevenness of one's field structure. Evidently, from the First Law on State of Motion, it follows that only such a field has the potential to drag other fields along to spin harder. As for why leadership motivation is a key, it is because when a conceptual large field appears, several existing yoyo fields might play the leadership role, meaning that each of these fields fits the form of motion of the conceptual field in one way or another. As implied in the study of centralized and centralizable systems (Lin, 1988b), these potential leaders have to work something out in order to form a cohesive whole either through fighting off those fields which are fundamentally different or through forming coalitions with each other. As analyzed in (Lin and Forrest, to appear 3), only when convergent fields spin harmonically (Figure 5.8), they have the tendency to compromise and to form a greater field; in all other possibilities, no fields can potential compromise with one another. That is, to be a leader, one needs to devote extra amount of time and energy to deal with other fields pleasantly and unpleasantly. That explains why leadership motivation is a key element for leadership.

As for honesty, integrity, self-confidence, cognitive ability, and knowledge of the business, they are involved with accurate thinking based on the available resources in one's reservoir of imagination, where the person's identified self is located, within or outside of the domain of his $\pm$ function, and his ability to apply his human endowment free will (or his ability to make short-term predictions).

Our modeling and analysis indeed indicate that patterns of field movement are the fundamental reason why a person would become a leader. His overreaching field influence on others (due to a high degree of intensity of spin) makes him seen as having a high need for power, a low need for affiliation (because he attracts others to him and not the other way around), and a strong self-control (that is thought-control, in which one deliberately chooses his thoughts based on accurate and realizable information to dominate his mind and firmly ignores all outside suggestions that are not congruent with his thoughts (Lin and Forrest, to appear 3)).

At the same time, our modeling and analysis explain why, as argued by (Spencer, 1841), it is the times that produce the leaders and not the other way around and that different situations call for different leadership characteristics. In particular, different situations stand for different forms of field motion. To fit a situation well, where the situation stands for the existence of a conceptual field structure, only those whose fields are in conformation with the conceptual field motion will potentially become the leader. This end verifies that there is not any single optimal psychographic profile for leadership, and that what an individual does, while acting as a leader, is in large part dependent upon field characteristics of the situation in which he functions.

Other theories of leadership can be analyzed similarly. All the details are omitted here. To finish our discussion here, let us now look at how the enthusiasm of great leaders spreads to their followers.

A leader creates his high degree of spinning intensity by making use of an emerging, overreaching, and overpowering conceptual field structure. So, his leadership ability is his enthusiasm. In other words, the leader's increased field intensity is the result of his structural consonance with the emerging large field and this increased intensity also stands for a newly acquired enthusiasm. This enthusiasm is spread over to the leader's followers by utilizing his increased intensity of spin combined the field influence of the conceptual field motion, when other fields are still in their individual inertial states of motion.

5.4.3 Kindling and maintaining the fire of enthusiasm burning

When a person is situated in a difficult situation and a long way from realizing his definite goal in life, how to kindle and maintain the fire of enthusiasm in his heart and keep it burning immediately become very important. And what is very interesting is that before very long the obstacles that now stand in the way of attaining one's definite goal will melt away, and he will find himself in possession of invaluable powers that he did not know he possessed.

First, when one has a definite goal in life, it means that in one way or another, he has been terribly stirred or troubled when his established world view and belief system was somehow turned upside down at least partially, for details see the discussion in Subsection 5.2.3. From that discussion, it follows that the definite goal is formulated more or less through the following procedure.

1 When one faces a troubling stir, one's self-awareness takes in the new information and mobilizes one's entire reservoir of imagination to formulate a response by making use of one's conscience (the $\pm$ function) and free will (the ability to make at least short-term predictions).
2 To respond to the stir, various groupings of relevant information and knowledge are formed in the conscious mind.
3 These groupings are judged by using the $\pm$ function of one's conscience. Among all the groupings with $+$ response values, one picks out the one response with the most desirable $+$ value.
4 Along with the fully activated mind, one's sense organs become keen to whatever in the surroundings that is supportive to the chosen response.
5 Accompanying the fully activated mind and stimulated sense organs, the entire underlying multi-dimensional yoyo structure of the person starts to act in various harmonic ways correspondingly.

Now, when this person feels that he is situated in a difficult situation and a long way from realizing his definite goal in life, it simply means that his newly adjusted yoyo structure is experiencing great resistance from the surrounding fields due to the inertia of their individual movements. If for whatever reason the external stir is truly troublesome and severe, then he will naturally have a burning desire to materialize his imagined outcome in the physical world in order to even out the imbalance created by the stir in his reservoir of imagination and the associated $\pm$ function of his conscience. This burning desire stands for how he wants his underlying yoyo structure to be in terms of its attributes: direction, orientation, spin intensity, and scale (see the discussion in Subsection 5.3.1 above). Now, the newly created (by the stir) severe unevenness in his field structure explicitly implies the appearance of a much increased intensity of spin. This state of intensive spin is exactly an enthusiasm the person recently acquired. So, there is no need for the person to kindle the fire of enthusiasm no matter how difficult situation he is in and what a long way he is from realizing his definite goal in life. As for how he can keep his enthusiasm burning in his heart, our analysis indicates that as long as he does not settle for less, meaning that he would not allow any state of field motion between the current chaotic state and his desired state to be acceptable, his fully activated mind and stimulated sense organs will continue to look out for whatever supportive to achieve his desired outcome.

If the person can keep his enthusiasm burning in his heart, it means that his underlying field is kept spinning at its newly acquired momentum and intensity, the First and the following Third Laws on State of Motion imply that all the other field structures, which used to resist the change in the person, will have to change their forms of motion accordingly in an accommodating way. This end explains why before very long the obstacles that now stand in the way of attaining one's definite goal will melt away, and he will find him in possession of power that he did not know he possessed. Here the power is that he can change behaviors of others according to his likings.

The Third Law on State of Motion (Lin, 2007): *When the spin fields of two yoyo structures N and M act and react on each other, their interaction falls in one of the six scenarios as shown in Figure 5.9(a)–(c) and Figure 5.10(a)–(c). And, the following are true:*

1 *For the cases in (a) of Figures 5.9–5.10, if both N and M are relatively stable temporarily, then their action and reaction are roughly equal but in opposite directions during the temporary stability. In terms of the whole evolution involved, the divergent spin field (N) exerts more action on the convergent field (M) than M's reaction peacefully in the case of Figure 5.9(a) and violently in the case of Figure 5.10(a).*

2 *For the cases (b) in Figures 5.9–5.10, there are permanent equal, but opposite, actions and reactions with the interaction more violent in the case of Figure 5.9(b) than in the case of Figure 5.10(b).*

3 *For the cases in (c) of Figure 5.9–5.10, there is a permanent mutual attraction. However, for the former case, the violent attraction may pull the two spin fields together and have the tendency to become one spin field. For the latter case, the peaceful attraction is balanced off by their opposite spinning directions. And, the spin fields will coexist permanently.*

5.4.4 The working of self suggestion

The technique with which one can fix any idea one chooses in one's mind is called self-suggestion. To see how and why this technique works, what the First Law on State of

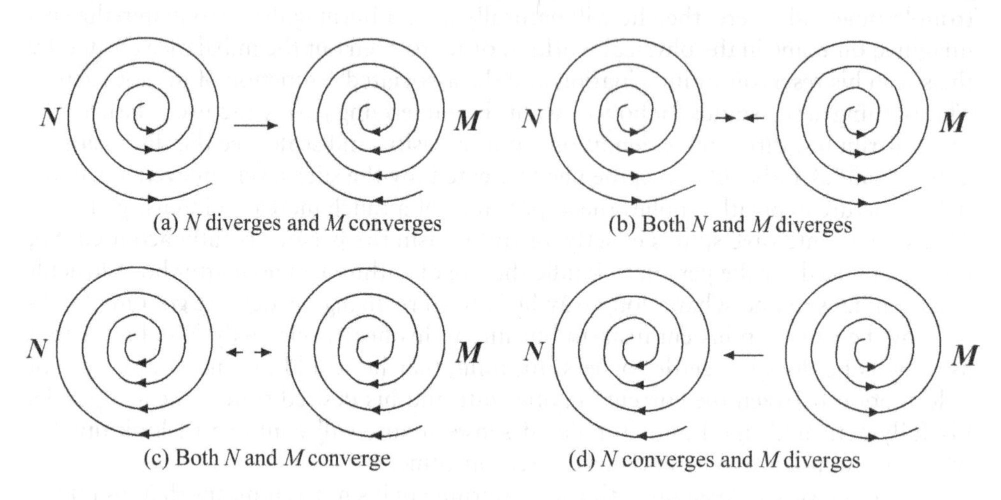

Figure 5.9 Same scale acting and reacting spinning yoyos of the harmonic pattern.

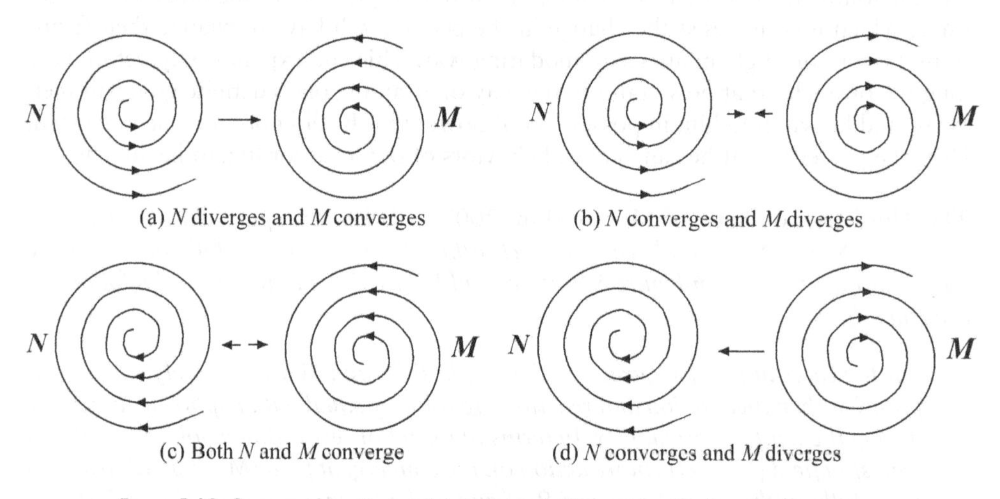

Figure 5.10 Same scale acting and reacting spinning yoyos of inharmonic patterns.

Motion suggests is that the mind, when seen as a part of the person's underlying yoyo structure, experiences inertia in its ordinary operation. It is occupied by accustomed thoughts, repeatedly used logic of reasoning, often visited facts, etc. So, to sow a chosen idea into the mind and make it the center of focus, one has to fight his very own field structure against all the dominating parts, which are in their individual inertia states of motion. Only after this new center of focus exerts a constant force on the other parts over a period of time, the states of motion of the other parts will alter their respective forms of movement in order to accommodate the existing change in their environment. Here, the key is that this new center of focus has to be established permanently and exerting constant pressures on the other parts of the mind first before the other parts would accept its existence and yield to its influence.

What Hill (1928, p. 164) suggests on how to apply self-suggestion in fact implies how one should make his predictions using his free will on the realization of his goal (by writing out a relatively detailed plan). If this plan is made on accurate thoughts, then the person would not have any trouble to form a specifically intensive spin in his reservoir of imagination. And as we have discussed previously, when such a specific spin is formed in his field structure, its realization in the 3-dimensional space will be recognized in no time.

5.4.5 Importance of self control

Self-control can direct enthusiasm to constructive ends, to build up instead of tear down. Because enthusiasm is a state of intensive spin of one's underlying yoyo field, it can of course be realized in the 3-dimensional space with one of the following three possibilities: a constructive consequence, no contribution to the society whatsoever, or a destructive end. Now, the so-called self-control is simply the thought-control, each thought a local eddy pool in the reservoir of imagination, and thinking process the process of generating a local pool by pulling relevant information and knowledge together to form an organic whole.

For a person, if his identified self is located within the domain of his ± function of his conscience with a + value, then assuring about a constructive consequence being realized in the 3-dimensional space out of his enthusiasm, he would constantly compare each newly formed component in his thought with the definition of his ± function. If an imagined pattern of field flow does not have an already assigned ± value, further thinking will be needed until a definite value is provided. If a pattern has an assigned – value, the pattern has to be either erased out of the local eddy pool of the thought or modified to insure a consequent + value.

As we have discussed in (Lin and Forrest, to appear 2), how one would react to a stimulus is dependent on where his identified self is located, either inside the domain of his ± function or outside, and even if his identified self is inside the domain, it is also dependent on what value is assigned to the self. So, the analysis in the previous paragraph will change from one person to another.

5.5 SOME FINAL COMMENTS

In this chapter, with the help of the systemic yoyo model, we have gained some brand new understanding on human desire and enthusiasm, two most important driving forces behind every great success ever recorded in history. What is most practically important of this work is that these driving forces can be purposely developed in any person of average intelligence. As of this point, what is still left open is how to specifically design programs to help people develop these vital forces in their personal as well as in professional lives.

Also, by using the systemic yoyo model, we have considered in this chapter many important, unsettled problems investigated by many scholars throughout the history since antiquity without reaching scientifically satisfactory results. All of our results and discoveries are derived on the basis of the systemic yoyo model. This model not only works as the common framework of our discussions and but also plays the roles of systematic reasoning and thinking logic. Because of the introduction of this rigorously proven methodology, we are able to establish the related key concepts on a unified foundation, overcoming the weakness of the relevant studies of the past of having no solid ground for relatively rigorous reasoning. As expected in (Lin, 2007) and as shown in this and previous chapters, this systemic yoyo model is indeed useful in the studies of social science and humanities.

The presentations in this and the previous chapters are based on (Lin and Forrest, 2010), where additional topics are also addressed.

Part 3

Mathematics seen as a systemic flow: A case study

Chapter 6

A brief history of mathematics

The purpose of this chapter is not to briefly and simply walk through the milestones in the history of mathematics. Instead, we look at this history as a system of human thoughts in the light of whole systemic evolution by focusing on several threads of thoughts that have existed since the beginning of recorded history. By doing so, we show the existence of the systemic yoyo structure in human thoughts so that the human way of thinking is proven to have the same structure as that of the material world.

By using parallel comparison we reveal the underlying structure that exists in human thoughts. After highlighting all the relevant ideas and concepts, which are behind each and every crisis in the foundations of mathematics, it becomes clear that some difficulties in our understanding of nature are originated from confusing actual infinities with potential infinities, and vice versa.

By walking through the history along the specified threads, we see clearly how the underlying systemic yoyo structure exists and how the spin field of this structure goes through the pattern changes as what is vividly described in the dishpan experiment. By doing so, we expect to show that in the thought system, called mathematics, another crisis is overdue and, as a matter of fact, such a crisis has arrived and appeared in the foundations of the system.

In this chapter, we assume that mathematics is an organic system of thoughts created on the bases of experience, intuition, and logic thinking. By sorting through the ideas appeared in and concepts studied in the history of mathematics, we discover that some ideas and concepts, such as those of infinitesimals, infinities, limits, etc., underlie each crisis in the foundations of mathematics. That is, through our systemic analysis of the history of mathematics, it is found that the systemic yoyo structure existing in human thoughts is indeed the same as that of the physical systems studied in (Lin, 2007) and as that of economic systems studied in (Lin and Forrest, D., 2008 A–C).

This chapter is organized as follows. Section 6.1 looks at how mathematics got started in the history. Then the first, second, and third crisis that appeared one after another in the foundations of mathematics are described in Sections 6.2–6.4.

A note about references. A lot of knowledge of mathematics is needed to understand the rest of this book. However, all these knowledge are assumed without particularly pointing to relevant reference books for the reader. So, let us list the necessary references here once for all. For the naïve and ZFC axiomatic set theories, (Lin, 1999) contains an introduction to both. Also, all the information regarding the history of mathematics in this chapter without mentioning the source is from either (Eves, 1992) or (Kline, 1972).

6.1 THE START

Archeological evidence that has been collected throughout history suggests that man has employed counting for as far back as 50,000 years ago. Since recorded history is much shorter than that, imagination has to be employed to fill in a huge gap here.

6.1.1 How basics of arithmetic were naturally applied

Because of the need for survival, humans started out with a sense of numbers and recognition of more and less. Along with the gradual evolution of human society, counting became imperative. Probably the earliest way of keeping count was by some simple tally method, employing the principle of one-to-one correspondence. An assortment of vocal sounds was developed as a word tally against the number of objects in a group. Then, an assortment of symbols was derived to stand for these numbers. After developing simple grouping systems for numbers, multiplicative groupings and positional numeral systems were established. Many of the modern-day computing patterns in elementary arithmetic appeared only in the fifteenth century due to a lack of plentiful and convenient supply of suitable writing medium in the earlier days.

6.1.2 Mathematics evolves with human society

Early mathematics was developed to meet practical needs and the evolution of more advanced forms of society. So, usable calendars, systems of weights and measures, methods of survey were invented. A special craft came into being for the cultivation, application, and instruction of this practical science. In such a circumstance, tendencies toward abstraction were bound to develop and the science was then studied for its own sake. It was in this way that algebra grew out of arithmetic and geometry out of measuration.

In the last centuries of the second millennium B.C., along with economic and political changes, some civilizations disappeared or waned and new people, including the Greeks, came to the fore. The alphabet was invented and coins were introduced. The new civilization made its appearance in towns along the coast of Asia Minor and later on the mainland of Greece, on Sicily, and on the Italian shore.

For the first time, men started to ask *why*'s in fundamental ways. The past empirical processes, sufficient for the question *how*, no longer sufficient to answer these more scientific inquiries of *why*.

6.1.3 Birth of modern mathematics

As man started to ask *why*'s in fundamental ways, mathematics, in the modern sense of the word, was born in this atmosphere of rationalism.

Thales is the first known individual with whom mathematical discoveries are associated. He used logical reasoning instead of intuition and experiment. The next outstanding Greek mathematician is Pythagoras. His philosophy rested on the assumption that whole numbers are the cause of the various qualities of man and matter. This led to an exaltation and study of number properties, and arithmetic along with geometry, music, and spherics.

Among many important works, Pythagoras independently discovered the Pythagorean theorem that the square of the hypotenuse of a right triangle is equal to the sum of the squares of the two legs.

6.2 FIRST CRISIS IN THE FOUNDATIONS OF MATHEMATICS

Historically speaking, the concept of integers arises from the need of counting finite collections of objects. The need to measure quantities, such as length, weight, and time, gives rise to the concept of fractions, because it is very seldom that a length, as an example, is an exact integral number of linear units. Thus, a rational number is defined as the quotient of two integers p/q, where $q \neq 0$. The system of rational numbers seems to be sufficient for practical purposes at the time.

6.2.1 A geometrical interpretation of rational numbers

Imagine that we are given a horizontal straight line and we mark two distinct points O and I on the line with I located to the right of O. Let us use that segment OI as out unit of length. If we understand the mark O as 0 and I as 1, then all the positive and negative integers can be presented by a set of points on the line spaced at unit intervals apart with the positive integers marked to the right of O and the negative integers to the left of O. The fractions with denominator q may be then represented by the points that divide each of the unit intervals into q equal parts. So, for each rational number, there is a corresponding point on the line. Due to the fact that to the early applications of mathematics, the system of rational numbers is closed with respect to the arithmetic operations, $+$, $-$, $\times$, $\div$, and meets the need of all the practical purposes, it is natural for the early scholars to believe that all the points on the line have been used up in the way just described.

6.2.2 The Hippasus paradox

In the fifth century B.C., unexpected discovery of irrational numbers, called the Hippasus paradox, together with other paradoxes, such as those constructed by Zeno, gives rise to the first crisis to the foundations of mathematics. This discovery is one of the greatest achievements of the Pythagoreans. This new finding implies that there are still points on the line not corresponding to any rational number. In particular, the Pythagoreans found that the diagonal of a unit square does not correspond to a rational number point on the line.

The rigorous proof for the existence of irrational numbers was surprising and disturbing to the Pythagoreans, since it was a mortal blow to the Pythagorean philosophy that all in the world depend on the whole numbers. Geometrically, this discovery was

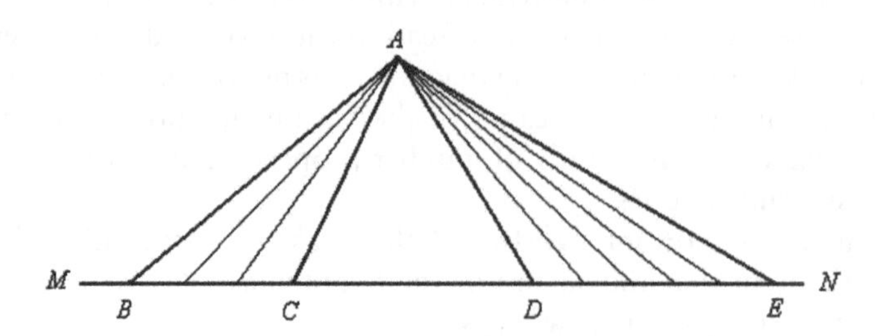

Figure 6.1 Triangles ABC and ADE.

startling, because who could doubt that for any two given line segments one is able to find some third line segment that can be marked off a whole number of times into each of the two given segments? Let the two segments be s and d such that s is the side of a square and d the diagonal. If there were third segment t that could be marked off a whole number of times into s and d, one would have $s = bt$ and $d = at$, where a and b are positive integers. Since $d = s\sqrt{2}$, we have $d = at = s\sqrt{2} = bt\sqrt{2}$ or $\sqrt{2} = a/b$, a rational number. This contradiction implies that there exist incommensurable line segments, meaning segments without a common unit of measure.

This discovery that like magnitudes may be incommensurable is proved to be extremely devastating. It caused some consternation in the Pythagorean ranks. Not only did it appear to upset the basic assumption that all in the world depend on the whole numbers, but because the Pythagorean definition of proportion assumed that any two like magnitudes are commensurable, all the propositions in the Pythagorean theory of proportion had to be limited to commensurable magnitudes, making the theory unsound, and their general theory of similar figures become invalid.

For quite a long time, $\sqrt{2}$ was the only known irrational number. Later Theodorus of Cyrene (ca. 425 B.C.) showed that $\sqrt{3}$, $\sqrt{5}$, $\sqrt{6}$, $\sqrt{7}$, $\sqrt{8}$, $\sqrt{10}$, $\sqrt{11}$, $\sqrt{12}$, $\sqrt{13}$, $\sqrt{14}$, $\sqrt{15}$, $\sqrt{17}$ are also irrational numbers. The final resolution of this crisis in the foundations of mathematics was achieved in about 370 B.C. by the Eudoxus.

6.2.3 How the Hippasus paradox was resolved

To get a sense about how the Eudoxus resolved the first crisis in the foundations of mathematics, let us look at the following:

Proposition 6.1 (1) *Triangles having equal bases and equal altitudes have equal areas.* (2) *Of any two triangles having the same altitude, that one has the greater area which has the greater base.*

Evidently, statement (2) above is a corollary of (1).

Assume that we have two triangles ABC and ADE, where the bases BC and DE lie on the same straight line MN, as shown in Figure 6.1. Without knowing the existence of irrational numbers, the Pythagoreans assumed implicitly that any two line segments are commensurable. So, BC and DE have some common unit of measure u so that

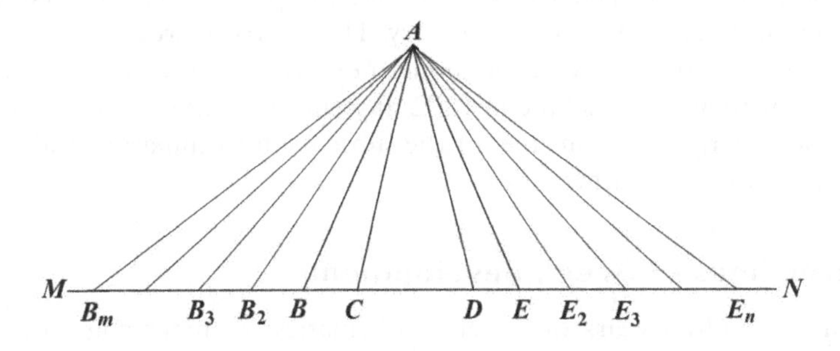

Figure 6.2 The Eudoxus' construction.

BC $= pu$ and DE $= qu$. That is, the unit u goes into BC p times and DE q times. Mark off these points of division on BC and DE and connect each of them with vertex A. So, the triangles ABC and ADE are divided, respectively, into p and q smaller triangles. From Proposition 6.1 (a), it follows that all these smaller triangles have the same area. So, $\Delta ABC : \Delta ADE = p : q = BC : DE$. That is, the Pythagoreans had shown the following result.

Proposition 6.2 *The areas of triangles having the same altitude are to one another as their bases.*

With the later discovery that two line segments do not have to be commensurable, this Pythagorean's proof, along with others, became incorrect. This is why the discovery of irrational numbers was extremely devastating to the Pythagoreans.

Here is how the Eudoxus resolved the crisis. He first redefined proportion as follows: Magnitudes are said to be in the same ratio, the first to the second and the third to the fourth, when if any equi-multiples whatever be taken of the first and the third, and any equi-multiples whatever of the second and the fourth, the former equi-multiples alike exceed, are alike equal to, or are alike less than the latter equi-multiples taken in corresponding order. In symbols, if A, B, C, and D are any four unsigned magnitudes, A and B being of the same kind (both line segments, or angles, or areas, or volumes) and C and D being of the same kind, then the ratio of A to B is equal to that of C to D when, for arbitrary positive integers m and n, $m\text{A} \geq, =,$ or $\leq n\text{B}$ according to as $m\text{C} \geq, =,$ or $\leq n\text{D}$.

Going back to the previous example, on CB produced, mark off, successively from B, $m - 1$ segments equal to CB and connect the points of division, $\text{B}_2, \text{B}_3, \ldots,$ B_m, with vertex A, shown in Figure 6.2. Similarly, on DE produced, mark off, successively from E, $n - 1$ segments equal to DE and connect the points of division, $\text{E}_2, \text{E}_3, \ldots, \text{E}_n$, with vertex A. Then, $\text{B}_m\text{C} = m(\text{BC})$, $\Delta \text{Ab}_m\text{C} = m(\Delta \text{ABC})$, $\text{DE}_n = n(\text{DE})$, $\Delta \text{ADE}_n = n(\Delta \text{ADE})$. Now, Proposition 6.1 implies that $\Delta \text{Ab}_m\text{C} \geq, =,$ or $\leq \Delta \text{ADE}_n$ according to as $\text{B}_m\text{C} \geq, =$ or $\leq \text{DE}_n$, That is, $m(\Delta \text{ABC}) \geq, =,$ or $\leq n(\Delta \text{ADE})$ according to as $m(\text{BC}) \geq, =,$ or $\leq n(\text{DE})$. Hence, by the Eudoxian definition of proportion, $\Delta \text{ABC} : \Delta \text{ADE} = \text{BC} : \text{DE}$. So, Proposition 6.2 is established. Here, neither commensurable nor incommensurable quantities are mentioned, since the Eudoxian definition applies equally to both situations.

The Eudoxian revised theory of magnitudes and proportions becomes one of the great mathematical masterpiece of the history. The Eudoxus' treatment of incommensurables coincides essentially with the modern exposition of irrational numbers that was first given by Richard Dedekind in 1872. It seems that this crisis in the foundations of mathematics is largely responsible for the subsequent formulation and adoption of the axiomatic method in mathematics.

6.2.4 Three lines of Greek development

The first three hundred years of Greek mathematics, commencing with the initial efforts at demonstrative geometry by Thales about 600 B.C. and culminating with the remarkable Elements by Euclid about 300 B.C., constitute a period of extraordinary achievement. Looking back to this period, we notice three important and distinct lines of development. The first was the development of the materials that ultimately was organized into the Elements by Euclid, began by the Pythagoreans and followed by Hippocrates, Eudoxus, Theodorus, Theaetelus, and others. The second line is the development of notions connected with infinitesimals and with limits and summation processes that did not attain final clarification until after the invention of calculus in modern times. The following Zeno paradoxes, the method of exhaustion of Antiphon and Eudoxus, and the atomistic theory associated with the name of Democritus belong to this line of development.

Zeno Paradox 1 (The Dichotomy): If a straight line segment is infinitely divisible, then motion is impossible, for in order to traverse the line segment, it is necessary first to reach the midpoint, and to do this one must first reach the one-quarter point, and to do this one must first reach the one-eighth point, and so on, ad infinitum. It follows that the motion can never even begin.

Zeno Paradox 2 (The Arrow): If time is made up of indivisible atomic instants, then a moving arrow is always at rest, for at any instant the arrow is in a fixed position. Since this is true of every instant, it follows that the arrow never moves.

These and some other paradoxes were devised by the Eleatic philosopher Zeno (ca. 450 B.C.) to address the question of whether a magnitude is infinitely divisible or it is made up of a very large number of tiny indivisible atomic parts. There is evidence that since the time of Greek antiquity, schools of mathematical reasoning developed their thoughts using each of these two possibilities. Zeno paradoxes above assert that motion is impossible whether we assume a magnitude to be infinitely divisible or to be made up of a large number of atomic parts. Because of these paradoxes, infinitesimals were excluded from Greek demonstrative geometry.

The third line of development is that of higher geometry or the geometry of curves other than the circle and straight line, and of surface other than the sphere and plane, originated in the continued attempts to solve three construction problems: (1) Construct the edge of a cube that has twice the volume of a given cube; (2) divide a given arbitrary angle into three equal parts; and (3) Construct a square that has the area equal to that of a given circle.

To summarize, based on Pythagorean theorem, it was found that for the right triangle with both legs of length 1, the hypotenuse $\sqrt{2}$ could be not represented by using whole numbers or ratios of whole numbers. The discovery fatally shook the foundation of the Pythagorean school's assumption and assaulted the common belief of Greek society. It made the scholars panic and intranquil and directly vacillated the foundations of mathematics of the time. The chaos created by this discovery and Zeno paradoxes, has been called the first crisis in the history of mathematics. However, it was because of the occurrence of this crisis that the history of mathematics turned from intuition and experience based arguments to deduction based proofs. The axiomatic geometry and formal logic were born and developed, leading to the formation of the concept of irrational numbers and the expansion of the number field.

6.3 SECOND CRISIS IN THE FOUNDATIONS OF MATHEMATICS

The seventeenth century is very conspicuous in the history of mathematics. Early in the century, John Napier revealed his invention of logarithms, Thomas Harriot and William Oughtred contributed to the notion and codification of algebra, Galileo Galilei founded the science of dynamics, and Johann Kepler announced his laws of planetary motion. Later in the century, Gerard Desargues and Blaise Pascal opened a new field of pure geometry. Rene Descartes launched modern analytic geometry, Pierre de Fermat laid the foundations of modern number theory, and Christiaan Huygens made his distinguished contributions to the theory of probability and other fields. Then toward the end of the century, after many other seventeenth century mathematicians had prepared the way, the creation of calculus was made by Isaac Newton and Gottfried Wilhelm Leibniz. It was during this century that many new and vast fields were opened up for mathematical investigation.

6.3.1 First problems calculus addressed

The first problems occurring in the history of calculus were concerned with the computation of areas, volumes, and lengths of arcs. In their treatment, one finds evidence of applying the two assumptions about divisibility of magnitudes. The famous Greek method of exhaustion is usually credited to the Eudoxus (ca. 370 B.C.) and can be considered as the Platonic school's explanation to Zeno paradoxes. The method assumes the infinite divisibility of magnitudes and has, as its foundation, the following proposition.

Proposition 6.3 *If from any magnitude there be subtracted a part not less than its half, from the remainder another part not less than its half, and so on, there will at length remain a magnitude less than any pre-assigned magnitude of the same kind.*

Of the ancients, it was Archimedes who made the most elegant applications of the method of exhaustion and who came the nearest to actual integration. In his treatment of areas and volumes, Archimedes arrived at equivalents of a number of definite integrals found in our modern-day calculus textbooks. With time, such scholars as Kepler, Cavalieri, Torricelli, Fermat, Pascal, Saint-Vincent, Barrow, and others, employed the ideas of infinitesimals and indivisibles and produced results equivalent to the integrations of such expressions as x^n, $\sin\theta$, $\sin^2\theta$, and $\theta\sin\theta$.

On the other hand, differentiation has originated in the problem of drawing tangents to curves and in finding maximum and minimum values of functions on the ideas set forth by Fermat in 1629. Isaac Barrow (1630–1677) was one main character who made important contributions to the theory of differentiation.

6.3.2 Fast emergence of calculus

Aided by analytic geometry calculus was the greatest mathematical tool discovered in the seventeenth century. It proved to be remarkably powerful and capable of attacking problems quite unassailable in earlier days.

It was the wide-range and astonishing applicability of the theory that attracted the bulk of the mathematics researchers of the day with the result that papers were published in great profusion with little concern for the unsatisfactory foundations of the theory. The processes employed were justified largely on the ground that they worked. It was not until the eighteenth century was almost elapsed, after a number of absurdities and contradictions had crept into mathematics, that mathematicians felt the need to examine and to establish rigorously the basis of their work.

6.3.3 Building the foundation of calculus

The effort to place calculus on a logically sound foundation proved to be a difficult task. Various ramifications occupied the better part of the next hundred years. A result of this careful work in the foundations of calculus was that it led to equally careful work in foundations of all branches of mathematics and to the refinement of many important concepts.

Chronically, calculus was created in the latter half of the seventeenth century. The eighteenth century was largely spent in exploiting the new and powerful methods of calculus. And, the nineteenth century was largely devoted to the effort of establishing calculus on a firm, logical foundation.

Since the early calculus was established on the ambiguous and vague concept of infinitesimals without a solid foundation, many operations and applications of the theory were reproached and attacked from various angles. Among all the reproaches, the most central and the ablest about the faulty foundation of the early calculus was brought forward by Bishop George Berkeley (1685–1753). In 1734, he published a book, entitled "The Analyst: A Discourse Addressed to an Infidel Mathematician". The "infidel mathematician" is believed to have been either Edmond Halley or Newton. This book was a direct attack on the foundation and principles of calculus, specifically on Newton's notion of fluxion and on Leibniz's notion of infinitesimal change. The following passage has been most frequently quoted from The Analyst:

> "And what are these fluxions? The velocities of evanescent increments? And what are these same evanescent increments? They are neither finite quantities nor quantities infinitely small, nor yet nothing. May we not call them the Ghosts of departed Quantities?"

To understand what Berkeley was talking about, let us look at an example from a paper by Newton on how to compute the area a region with a curved boundary. In his work, Newton claimed that he avoided using infinitesimals by going through the following steps: Give an increment to x, expand $(x+0)^n$, subtract x^n, divide by 0, compute the ratio of the increment in x^n over that in x. Then by throwing away the 0 term, he obtained the fluxion of x^n . Berkeley said that the variable was first given an increment, then let the increment to be 0. This involved the fallacy of a shift in the hypothesis. As for the derivative seen as the ratio of the disappeared increments in y and x, that is, the ratio of dy and dx, Berkeley called these disappeared increments "neither finite quantities nor quantities infinitely small, not yet nothing. May we not call them the Ghosts of departed quantities?" (Kline, 1972).

It was exactly because of the early calculus did not have a solid theoretical foundation that many criticisms seemed at the time reasonable so that conflicts between achievements of calculus in applications and the ambiguity in the foundation became more and more incisive. With the passage of time, a serious crisis in the foundations of mathematics became evident. The first suggestion to replace the sandy foundation of calculus on a more solid basis came from Jean-le-Rond d'Alembert (1717–1783), who observed in 1754 that a theory of limits was needed. It was Joseph Louis Lagrange (1736–1813) who was the earliest to attempt a rigorization of calculus. With his work, the long and difficult task of banishing intuitionism and formalism from calculus began.

6.3.4 Rigorous basis of analysis

In the nineteenth century, the theory of analysis continued to grow on an ever-deepening foundation due to the 1812-work of Carl Friedrich Gauss who set new standards of mathematical rigor. A great stride was made in 1821 when Augustin-Louis Cauchy (1789–1857) successfully executed d'Alembert's suggestion by developing an acceptable theory of limits and then defining continuity, differentiability, and definite integral in terms of the concept of limits.

Such examples as a continuous function that has no derivative at any point in its domain (by Karl Weierstrass, 1874) and a function that is continuous for all irrational points and discontinuous for all rational points in its domain (by Georg Bernhard Riemann) suggested that Cauchy had not resolved all the difficulties in the foundation of analysis. Accordingly, Weierstrass advocated and materialized a program to first rigorize the real number system and then establish the basic concepts of analysis in particular, and mathematics in general, on this system. Today, it can be stated that essentially all of existing mathematics is consistent if the real number system is consistent.

In the late nineteenth century, Richard Dedekind (1831–1916), Georg Cantor (1845–1918), and Giuseppe Peano (1858–1932) showed how the real number system and the great bulk of mathematics could be derived from a set of postulates established for the natural number system. Then, in the early twentieth century, it was shown that the natural numbers can be defined in terms of concepts of set theory and thus that the great bulk of mathematics can be made to rest on a platform in set theory.

Specifically, on the basis of Cauchy's theory of limits, Dedekind proved the fundamental theorems in the theory of limits using the rigorized real number theory. With the combined efforts of many mathematicians, the methods of ε–N and $\varepsilon - \delta$ became widely accepted so that infinitesimals and infinities could be successfully avoided. The mathematical analysis, developed along this line, has been called the standard analysis.

It has been a common belief that due to the satisfactory establishment of the theory of limits, calculus has ever since been constructed on a solid, rigorous theoretical foundation. So, in the history of mathematics, the second crisis in the foundations of mathematics is considered resolved successfully. In the following chapter in this book we will see that this belief cannot be further from the truth. That is, the Berkeley paradox did not really go away. And, more paradoxes are created in the theory of limits and in the foundations of mathematics.

6.4 THIRD CRISIS IN THE FOUNDATIONS OF MATHEMATICS

Georg Cantor (1845–1918) was interested in number theory, indeterminate equations, and trigonometric series. This last interest somehow inspired him to look at the foundation of mathematical analysis. In 1874, he commenced his revolutionary work on set theory and the theory of the infinite, where he developed a theory of transfinite numbers based on the actual infinite. Since so much of mathematics is permeated with set concepts, the superstructure of mathematics can actually be made to rest upon set theory as its foundation.

6.4.1 Inconsistencies of Naïve set theory

In 1897, Burali-Forti brought to light the first publicized paradox of set theory. Cantor two years later found a non-technical description of a very similar paradox. In his theory of set, Cantor proved that for any given set X, there is always another set Y such that the cardinality of X is less than that of Y. What is shown is similar to the situation that there is no greatest natural number. There also is no greatest transfinite number. Now, consider the set that contains all sets as its elements. Surely no set can have more elements than this set of all sets. But, if this is the case, how can there be another set whose cardinality is greater than the cardinality of this set of all sets?

In 1902, Bertrand Russell discovered another paradox, which involves nothing more than just the concept of sets itself. In particular, let the set of all sets that are members of themselves by M, and the set of all sets that are not members of themselves by N. Now, let us consider this question: Is N a member of itself N? There are only two possible answers to this question: Yes, N is; or No, N is not a member of itself. If N is a member of itself, then N belongs to M and not of N. So, N is not a member of itself. A contradiction. On the other hand, if N is not a member of itself, then N is a member of N and not of M. So, N is a member of itself. Another contradiction. That is, in either case, we are led to a contradiction.

Since the discovery of the above contradictions within Cantor's set theory, more paradoxes have been produced in abundance. These modern paradoxes of set theory are related to several ancient paradoxes of logic. For instance, Eubulides (the fourth century B.C.) is credited for the remark" "This statement I am now making is false." If what Eubulides said is true, then by what is meant, the statement must be false. On the other hand, if what Eubulides said is false, then it follows from what is implied that his statement must be true. That is, Eubulides' statement can be neither true nor false without entailing a contradiction. As an another example, Epimenides (the sixth century B.C.) is claimed to have made the following remark: "Cretans are always liars". A simple analysis of this remark easily reveals that it, too, is self-contradictory.

The existence of paradoxes in set theory, like those described above, surely suggests that there must be something wrong with that theory. However, what is more devastating is that now, the foundation of mathematics faces another major crisis, called the third crisis of mathematics, since these paradoxes naturally cast doubt about the validity of the entire fundamental structure of mathematics.

6.4.2 Confronting the difficulty of Naïve set theory

A close examination of the paradoxes considered above reveals the fact that in each case, a set S and a member m of S are involve, where the definition of the member m depends on S. That is, in a sense, a circular definition is involved. To avoid the paradoxes, it seems that one has to impose a restriction on the concept of set. That is why Russell introduced his Vicious Circle Principle: No set S is allowed to contain members m definable only in terms of S, or members m involving or presupposing S. However, this approach of disallowing circular definitions met with one serious objection, since many fundamental concepts of mathematics are introduced using such definitions. In 1918, Hermann Weyl undertook the task to find out how much of mathematical analysis can be constructed without using circular definitions. After successfully obtaining a considerable part of the analysis, he was stuck by being unable to derive the important theorem that every non-empty set of real numbers having an upper bound has a least upper bound.

To resolve the third crisis in mathematics, there have appeared three main schools of thought – the so-called logistic, intuitionist, and formalist schools. Each of these schools has attracted a sizable group of followers and generated a large body of associated literature. Russell and Alfred North Whitehead are the chief expositors of the logistic school. L. E. J. Brouwer led the intuitionist school, and David Hilbert developed the formalist school.

For the logistic school, its thesis is that mathematics is a branch of logic and that all mathematical concepts are to be formulated in terms of logical concepts and all theorems of mathematics are to be developed as theorems of logic. The notion of logic underlying all science dates back at least as far as the time of Leibniz. It was Dedekind (1888) and Frege (1884–1903) who were the first to actually reduced mathematical concepts to those of logic. And it was Peano (1889–1908) who was the first to rephrase the statement of mathematical theorems by means of a logical symbolism.

The logistic thesis was composed to push back the foundations of mathematics to as deep a level as possible. Since historically, the foundation of mathematics was established on the real number system, and then on the natural number system, and thence into set theory, and since the theory of classes is an essential part of logic, the idea of reducing mathematics to logic certainly arises naturally. The documental Principia Mathematica of Whitehead and Russell (1910–1913) purports to be a detailed reduction of mathematics to logic. It starts with "primitive ideas" and primitive propositions," corresponding to the "undefined terms" and "postulates" of a formal abstract development. These primitive ideas and propositions are not to be subjected to interpretation but are restricted to intuitive concepts of logic. They are accepted as plausible descriptions and hypotheses of the real world so that there is no need to prove the consistency of the primitive propositions. Then, Principia Mathematica develops mathematical concepts and theorems from these primitive ideas and propositions.

To avoid the paradoxes of set theory, Principia Mathematica sets up a hierarchy of levels of elements. The primary elements constitute those of type 0; classes of elements of type 0 constitute those of type 1; classes of elements of type 1 constitute those of type 2, and so on. So, one has such a rule that all the elements of any class must be of the same type. Since this rule precludes circular definitions, this new approach of constructing mathematics avoids the paradoxes of set theory. In the work of logistic school, the great difficulty is on how to obtain the circular definitions needed for certain areas of mathematics.

For the intuitionist school, its thesis is that mathematics is to be built solely by finite constructive methods on the intuitively given sequence of natural numbers. From this intuitive sequence of natural numbers, all other mathematical objects must be built in a purely constructive manner, employing a finite number of steps or operations. This school of thought was originated in about 1908 by Brouwer with some earlier ideas proposed by Kronecker (in the 1880s) and Poincaré (during 1902–1906).

For the intuitionists, the existence of an entity must be shown to be constructible in a finite number of steps. It is not sufficient to show that the assumption of the entity's non-existence leads to a contradiction. So, to the intuitionists, a set cannot be thought of as a ready-made collection, but must be considered as a law by means of which the elements of the set can be constructed in a step-by-step fashion. Because of this, such contradictory sets as the set of all sets" cannot exist.

Also, the intuitionists' finite constructability leads to the denial of the universal acceptance of the law of the excluded middle. For them, this law holds for finite sets, but should not be employed when dealing with infinite sets. Based on the intuitionist thesis, as of this writing, large parts of present-day mathematics have been successfully rebuilt with a great amount of work still needed. In short, intuitionist mathematics has turned out to be considerably less powerful than classical mathematics and in many ways it is much more complicated to develop. Based on what has been done, it is now a general conviction that the intuitionist methods do not leads to known contradictions.

For the formalist school, its thesis is that mathematics is about formal symbolic systems, in which terms are mere symbols and statements are formulas involving these symbols. The ultimate foundation of mathematics does not lie in logic but only in a collection of non-logical marks (symbols) and in a set of operations with these marks. From this point of view, the establishment of the consistency of the various branches of mathematics becomes an important and necessary part of the formalist program. David Hilbert, together with Bernays, Ackermann, von Neumann, and others, founded this school of thought after 1899 when he published his Grundlagen der Geometrie.

The fate of the formalist school hinges upon its solution of the consistency problems. To Hilbert, freedom from contradiction is guaranteed only by consistency proofs, in which one must prove by the rules of a system that no contradictory formula can appear within the system. However, in 1931, Kurl Godel showed that it is impossible for a sufficiently rich, formalized deductive system, such as Hilbert's system established for all classical mathematics, to prove consistency of the system by methods of the system. This result reveals an unforeseen limitation in the methods of formal mathematics. It shows "that the formal systems known to be adequate for the derivation of mathematics are unsafe in the sense that their consistency cannot be demonstrated by finitary methods formalized within the system, whereas any system known to be safe in this sense is inadequate (De Sua, 1956)".

To summarize what has been discussed above, it can be seen that because of the establishment of the theory of limits, calculus becomes a sound collection of powerful methods developed on a rigorous theoretical foundation. As to this point in time, the first and the second crises in the history of mathematics were successfully resolved. However, as a matter of fact, the rigorous theory of limits is established on the system of real numbers. And, to develop the system of real numbers, the theory of sets has to be employed as another deeper level of the foundation. However, in the maturing process of the naïve set theory, a series of paradoxes appeared one after another, constituting another even greater crisis in the system of modern mathematics.

6.4.3 Axiomatic set theory – an alternative

Until the year of 1900, most mathematicians believed that the publicized paradoxes in set theory were only some technical problems. As long as some necessary details were fine-tuned, the problem would be resolved. However, two years later, the well-known Russell paradox was publicized, astonishing the entire western communities of philosophers, logicians, and mathematicians. It was because a simple analysis of the Russell paradox suggests that as soon as the language of logic is used to replace that of set theory, the Russell paradox will directly touch on the entire theory of the foundation so that the common belief that mathematics and logic are two most rigorous and exact scientific disciplines is severely challenged. Facing the challenge, mathematicians and logicians were pressured to seriously treat and investigate the paradoxes of set theory. The chaotic situation, caused by the appearance of these set theoretic paradoxes, is adequately called the third crisis in mathematics. What needs to be pointed out is that to a certain degree, this third crisis is in fact a deepening evolution of the previous two crises, because the problems touched on by these paradoxes are more profound and involved a much wider area of human thought.

In the attempt to resolve the third crisis, mathematicians once thought of the possibility of giving up the naïve set theory as the theoretical foundation of mathematics, since the theory contains irresolvable antinomies, and identifying another theory as the foundation of mathematics. However, after a careful analysis, it was realized how difficult it would be to adopt this approach. So, instead, these mathematicians focused on modifying the naïve set theory in an attempt to make it a plausible theory.

As of the present-day, there have been two plans to remold the set theory. The first is to employ the theory of types, developed by Whitehead and Russell. The second is to continue Zermelo's axiomatic set theory. Walking along the idea of Russell's "theory of extensionality," Zermelo in 1908 established his system of axioms for his new set theory. After several rounds of modifications, Fraenkel and Skolem (1912–1923) provided a rigorous interpretation and formed the present-day ZF system. Since the ZF system accepts the axiom of choice, it is often written as ZFC system.

From the structures and the contents of the non-logical axioms of the ZFC system, it can be seen that the ultimate goal is still about establishing a rigorous foundation for mathematical analysis. The specific route of thinking and technical details can be expressed as follows: By introducing the axioms of the empty set and of infinity, the legality of the set of all natural numbers is warranted. The legality of the set of all real numbers is derived by using the axiom of lower sets. Then, the legality of each subset of those elements, satisfying a given property P, of the real numbers is based on the axiom of subsets. Therefore, as long as the ZFC system is consistent or contains no contradiction, the theory of limits, as a rigorous theoretical foundation of calculus, can be satisfactorily constructed on the ZFC axiomatic system.

6.4.4 Warnings of the masters

Up to this point in history, the success of the hard-working mathematicians is that in the ZFC system, various paradoxes of the two-value logic, which have appeared in the history, can be successfully interpreted; that is, these paradoxes will not reappear in the ZFC system. However, what remains open is that the community of mathematicians still cannot show in theory that no paradox whatever kind will ever be constructed in the ZFC system. That is, the consistency of the ZFC system or other presently available axiomatic set theoretic systems still cannot be shown. In other words, no theoretical guarantee exists that in the future no new paradoxes could be ever found in the current versions of modern axiomatic set theory.

Even so, of course, what is worth celebrating is that for over a century since the time when the ZFC system was initially suggested, it has been indeed fortunate that no new paradoxes were found in the ZFC system and other modern axiomatic set theoretic systems. To this end, let us not forget what Poincaré once said about the situation over a century ago:

> We set up a fence to protect our flock of sheep from potential attacks of wolves. However, at the time when we installed the fence, there might have already been a wolf in sheep's clothing being enclosed inside the flock of our sheep. So, how can we guarantee that in the future there will not be any problems?

We should notice that what Poincaré said is absolutely not a few sarcastic remarks. Instead, they represent this great master's intuitive judgments about the essence of the matter. The following Harsdorff's comment can be seen as another opinion which echoes with what Poincaré said. It was after several paradoxes were publicized in the naïve set theory that Hausdorff felt deeply graceful and reminded the community of mathematicians that:

> These paradoxes made people feel unease. It is not because of the appearance of the paradoxes. Instead, it is because we did not expect these contradictions would ever exist. The set of all cardinalities seems so empirically indubitable just as the set of all natural numbers is that naturally acceptable. So, the following uncertainty is created. That is, is it possible that other infinite sets, all of them, are this kind contradictory, specious non-sets? (Hausdorff, 1935)

However, what attracted the most attention and what was most surprising are what Robinson expressed in 1964 (Robinson, 1964):

> In terms of the foundations of mathematics, my position (point of view) is based on the following two main principles (or opinions): (1) No matter which semantics is

> applied, infinite sets do not exist (both in practice and in theory). More precisely, any description about infinite sets is simply meaningless. (2) However, we still need to conduct mathematical research as we have used to. That is, in our work, we should still treat infinite sets as if they realistically exist.

Without any doubt, Robinson's statement (1) is an-extremely deep, intuitive assessment instead of any irresponsible nonsense. As for his statement (2), it is due to some kind of inability on his part to alter anything existing at the time of his remark.

Chapter 7

Actual and potential infinities

After briefly looking at how the past three crises of mathematics had appeared and resolved in the foundations of mathematics in the previous chapter, this chapter focuses on the concepts of potential and actual infinite and relevant matters. In particular, the concept of infinite touches on those matters that seem to be beyond our finitary reach, no matter whether they are so small that we have to use our imagination to divide once and again without an end in sight, or they are so big that any chosen bound will be surpassed again and again forever, or that seem to fall in between the crack between parts and whole, where parts represent many while the whole a one. This last part actually provides an explanation for why systems science only started to appear as a systematic scientific investigation in the recent past, although the basic concepts of systems can be traced back as far as the recorded history can go and isolated studied on systems have been kept popping up at different times throughout the history.

After visiting some of the well-known paradoxes of infinite, this chapter looks at how different scholars throughout the history had debated over the difference between potential and actual infinite. Then, an analytical example, known as the vase puzzle, is given to show that not only temporally are the concepts of potential and actual infinite different, but also different analytically. After seeing so many intriguing and at the same time dreadful paradoxes of the infinite, a careful classification of paradoxes is presented to show that paradoxes of the past may very well become the catalyst for future development of thoughts. In other words, as indicated by the past development of science, other than cajoling, provoking, amusing, exasperating, and seducing, more importantly paradoxes arouse curiosity, stimulate, and motivate, and attract attention. They have inspired the clarification of basic concepts, the introduction of major results,and the establishment of various branches of inquiry.

After sufficient amount of discussion on the relevant aspects of potential and actual infinite is provided, the attention of this chapter turns to establish the descriptive definitions for the concepts of potential and actual infinite while introducing a set of necessary symbols. The chapter then considers how widely and pervasively these two different concepts of infinite have been employed at the same time in the entire spectrum of modern mathematics.

This chapter is organized as follows. Section 7.1 outlines several paradoxes that are related to the infinitely small, the infinitely big, and the scenarios involving many and one. After seeing how the concept of infinite can lead to difficulties in the conventional logical reasoning, Section 7.2 pays a close attention to look at how historically and analytically the great minds of the past had understood infinite. A classification of various paradoxes is presented in the last part of this section. Section 7.3 is devoted to the development of descriptive definitions of potential and actual infinite with a set of appropriate symbolism introduced. Section 7.4 provides a guide for the reader to walk through the system of modern mathematics in order to see how both potential and actual infinities are simultaneously employed in mathematics so that a major crisis in the foundations of mathematics is uncovered.

7.1 PARADOXES OF THE INFINITE

Historically, the concept of infinite has taken one of two forms in varying details. In one form, one finds boundlessness, endlessness, unlimitedness, and progressiveness. In the other form, one finds completeness, wholeness, and perfection. In the former case, there is always more to come without ending that conveys a sense of potentiality, while in the latter form, one can go beyond the completion of the involved progressive process so that it provides a sense of actuality. So, in this book, the former kind of infinite is referred to as potential infinite (respectively, infinity) and the latter as actual infinite (respectively, infinity). Because one of these concepts of infinite denotes temporarily a present progressive tense, while the other perfect tense, that explains why some scholars throughout history who have seen the infinite in terms of one form accused those who have seen it in terms of the other form of being incorrect.

No matter which form of infinite one applies, one has to face many paradoxes. In this section, we will look at some of these paradoxes in order for the reader to gain a sense of urgency of investigating the concept of infinite more closely.

7.1.1 The infinitely small

When one is interested in knowing the fine and finer details of the world, one is about to encounter such situations that one is fascinated and at the same time puzzled by nature's extremely fine intricacies.

Example 7.1 (The Paradox of Achilles and the Tortoise). This paradox that involves the concept of infinite is comes from Zeno, and supposes that Achilles can run twice as fast as the tortoise and they decide to have a race with the tortoise starting ahead a predetermined distance. Then before Achilles can overtake the tortoise, he must reach the point at which the tortoise starts. By that time, the tortoise will have advanced half the distance initially separating them. Achilles must now make up this distance. But by the time he does so, the tortoise will have advanced again. When this reasoning continues indefinitely, it seems that Achilles can never overtake the tortoise. On the other hand, with the given speeds and distances involved, we can compute exactly how long it takes Achilles to catch up with the tortoise from the start of the race.

Example 7.2 (The paradox of the divided stick). An imaginary stick of a fixed length is cut in half at some point in time. A half minute later, each of the halves is again cut in half. A quarter minute later, each of the quarters is again cut in half. If this process is allowed to continue, infinitely, what will remain at the end of the particular minute?

The paradox appears when one tries to answer this question. Is it infinitely many infinitesimally thin pieces of the stick? If so and intuitively it seems to be so, then there are two possibilities: (1) Each of the infinitesimally thin pieces has a width, and (2) Each of the pieces does not have a width. If possibility (1) holds true, then infinitely many such pieces would add up to an infinitely long stick. If possibility (2) holds true, then how could infinitely many of them make up a stick of any length at all?

7.1.2 The infinitely big

When one is embarking on a journey to know the grand and grander characteristics of the world, he will be astonished by the magnificence of the universe while feeling the potential reach of the unreachable. That is where the big becomes so very large that it becomes difficult to comprehend.

Example 7.3 (The paradox of the even numbers). If one pairs any chosen natural number n with $2n$, then mathematical induction implies that there are as many even numbers as natural numbers.

However, intuition seems to suggest that there are obviously fewer or, to be exact, half as many even numbers as natural numbers. At the same time, because each set here, the set of all even numbers and the set of all natural numbers, is infinite, there might be an urge for us to claim that there are as many even numbers as natural numbers. That is, no matter how we describe the situation, our intuition is in a state of turmoil.

Example 7.4 (The paradox of the hotel). Assume that a hotel with infinitely many rooms is totally occupied with guests. Then a new traveler arrives at the hotel and demands a room for the night. Without kicking anyone out of the hotel, this newcomer is accommodated easily in the first room as follows:

1 The guest who occupies the first room moves into the second room;
2 The guest who occupies the second room moves into the third room;
3 The gust who occupies the nth room moves into the $(n+1)$th room; ...

At first, this rearrangement of guests seems to be difficult to comprehend. However after pondering over the situation for a while, it does seem that many problems in the finite world could be resolved if we could find a way to push them indefinitely into the future, because tomorrow never arrives!

7.1.3 Many and one

When one employs his natural ability to collect many into a whole, he actualizes a one from the many. When this happens, the person will once again be puzzled by various unthinkable illogicalities.

Example 7.5 (Russell's paradox). In the naïve set theory Cantor defines a set to be a many which allows itself to be considered as a one. In other words, a set is a collection of elements. Now, consider such a set X that contains all those sets, each of which does not belong to itself as an element. The natural question that arises is that: Is X an element of X?

According to the definition of set X, we know that Y belongs to X if and only if Y does not belong to Y. So, in particular, X belongs to X if and only if X does not belong to X. What a blatant self-contradiction!

Going along with this Russell's paradox, there are many other paradoxes in the naïve set theory that is constructed on the very idea of considering one of many.

What the listed paradoxes above tell us is that when we look into the infinitely small, we run into intellectual difficulties; when we look at the infinitely big, we experience the uncertainty of whether or not we could finish the counting. Here, mathematical induction seems to have come to the rescue by putting the uncertainty to rest (Note: In the following chapters we will see that what the current versions of mathematical induction say is that actual infinite is the same as potential infinite. In other words, infinitely ongoing progressions are the same as those that can be actually finished.) And when we look at a many as a whole, we have to face the difficulty of accepting individualism and collectivism, two extremes of an uncompromising difficulty.

7.2 A HISTORICAL ACCOUNT OF THE INFINITE

In this section, which is based on (Forrest, Wen and Panger, to appear), we look at the infinite both historically and analytically using two different techniques of modeling, and then we see how paradoxes have been classified.

7.2.1 The infinite

Historically, Greek mathematicians did not directly study the infinite as a mathematical object, but they analogously addressed it in their work. For example, in their study of geometry, they acknowledged the existence of a line (Euclid, 1956) that goes on indefinitely, and thus, concluded that this indefinitely long line can be subdivided into infinitely many smaller line segments. Yet the Greeks never claimed this to be an infinity-based concept for they only studied finite lines with finite subdivided line segments. However, it is possible to make the claim that the understanding of an infinite framework was needed for the Greeks to perform these finite tasks. Many mathematicians do not claim the infinite to be a major part of mathematics (except set theorists, which will be explained later), but some acknowledge that they may be analyzing the infinite to the extent that they are working in an infinite framework (Kline, 1972). Aside from this idea of an infinite framework, infinity has explicitly pervaded the fields of mathematics and philosophy for hundreds of years.

Aristotle (1984) identifies the temporal distinction between actual and potential infinity, namely a point in time yields the infinitude of actual infinity whereas the infinitude of potential infinity exists over time. However, he claims infinity can never be expressed actually but only potentially. Furthermore, Aristotle uses the potential infinite with the mathematical infinite synonymously. He attempted to explain that infinity could be accepted solely on potential grounds, and any other attempts and corresponding failures of acceptance were based on the invalid concept of actual infinity.

Mathematically speaking, he defined the potential infinity to be understood by the never-ending process of counting, not by the idea of infinitely many numbers.

Descartes' (1984) contribution to the philosophy of the infinite provided the growing affiliation between the actual, mathematical infinite and the metaphysical infinite, which contrasted the Aristotelian view. Additionally, Descartes and his fellow rationalists believed the infinite took logical and epistemological precedence over the finite and did not require a sense of imagination to understand. Pierre Gassendi (1592–1655) believed (Descartes, 1984) that to understand the infinite really is to understand the negation of the finite. Since Descartes discounts the logical importance of the finite in understanding the infinite, he claims Gassendi's belief is not possible. The infinite is something real that is entirely larger than something that has an end. But, equivocating the infinite with the negation of the finite is incorrect because this ambiguously associates everything that is infinite is everything the finite is not.

Though Descartes did not entirely discount Aristotelian thought, Hegel did (Hegel, 1969, pp. 234–8), where Aristotle favored a mathematical infinite over a metaphysical one, Hegel claimed the opposite. His claim against the mathematical infinite stems from his belief that it never transcended the finite. Thus, he believed that the only true infinite was a metaphysical one, which carried many contradictions within itself. When these contradictions were analyzed, the results would be in mathematically infinite terms, or according to Hegel, in finite terms, which led to another contradiction in his concept of infinity. With this cyclical method of temporally fixing contradictions, Hegel could never conclude anything concrete.

Aside from claims made above, the mathematics of the nineteenth century focused heavily on set theory and not so much the philosophy of the infinite, and specifically, the concept of infinite sets to explain the existence of actual infinity. Bolzano (1950; Moore, 1990, p. 112) defined infinite sets as sets containing infinitely any members and utilized the set of points in time and the set of natural numbers as examples for such sets. Contrary to Aristotle, Bolzano believed that though endlessness can be understood in temporal terms, infinity was not meant to be conceived in these terms. In other words, understanding an infinite set does not require collecting all of its members but understanding its elements' pattern to yield the full set.

Nineteenth and twentieth century Dutch mathematician L. E. J. Brouwer (1983a; 1983b; Moore, 1990, p. 131–132) attempted to revive Aristotelian notions of infinity that met great discontent in Cantor's work. Cantor treated the set of real numbers was a completed whole, but Brouwer's disagreement entails that no totality can be achieved solely from mathematical experience. To Brouwer, what could be done mathematically was solely the identification of individual real numbers. Generally, many mathematicians tend to agree with Brouwer in this respect, which dissipates much of Cantor's claim explaining actual infinity. Specifically, a 20th century French mathematician Henri Poincaré identified Cantor's work as 'a perverse pathological illness that would one day be cured' (Dauben, 1979, pp. 1).

GottlobFrege, a 20th century analytic philosopher, questioned the existence of infinitely many things and turned to set logic to develop an explanation. A basic description of the Frege's argument is as follows (Frege, 1980, pp. 82–3):

> Whatever else exists, there must at least be the empty set, the set with no members. Then there is the set whose only member is the empty set. Then there is the set

whose only members are these two. And so on ad infinitum. Thus, there are infinitely many things.

Bertrand Russell, also an analytic philosopher of this time, showed Frege the paradoxical nature of sets. The root of this paradox was the concept of infinity, and Russell (1967) attempted to avoid this paradox by claiming a set was a different entity than its members. When Russell began to develop axioms for his set theory, he included Frege's idea of the existence of infinitely many things. However, by accommodating for the paradox in Frege's sets, he could not avoid all paradoxes.

There exists a series of paradoxes involving a single collection of infinitely many things. In other words, these paradoxes are based on the existence or non-existence of infinite sets. Historically, the natural numbers act as the best example for discussing and debating the existence of infinite sets. It is completely reasonable and trivial to consider the set of the natural numbers, and the set must be infinite because there is always the possibility of adding one to the previous natural number. However, it also seems reasonable to collect infinitely many things together and count them as a single entity – a set. This general discussion is one of the many examples indicating the paradoxical nature of infinite sets.

Once Cantor's (1955) theory of ordinals is considered, the paradoxes based on infinite sets clearly surface. For Cantor, his disinterest in the size of sets reflected his focus on his theory of ordinals. The theory of ordinals dealt with whether or not the members of a particular set were well-ordered or not, where a set is well-ordered if there is clearly a first element and if for any member of the set, that member has a clear predecessor and successor (unless no elements are left). Cantor's idea of the well-ordering of sets can be seen in the way he ordered the elements of the set of natural numbers. For example, Cantor would say '0, 1, 2, . . .' is a well-ordering because it satisfies the above conditions. In contrast, '. . ., −2, −1, 0, 1, 2, . . .' is not a well-ordering because there is no clear indication of which element is first. Furthermore, '0, . . ., 1, . . ., 2, . . ., 3, . . .' is not a well-ordering, even though 0 is clearly the first element, there is no clear successor of 0.

Ordinals can be defined as the shape of a well-ordering. Thus, we are able to identify a well-ordering just by knowing its corresponding ordinal. This identification is possible because ordinals are well-orderings of themselves that satisfy the following condition, 'for each set of ordinals (finite or infinite), there is another ordinal which is the first to succeed them all' (Moore, 1990, pp. 125). With explanation of Cantor's concept of well-ordering and ordinals, it is time to explain some of the paradoxes arising from this very concept.

The example of the Burali-Forti (1967) paradox provides a specific and clear example for where Cantor's work can be seen as problematic. The ordinals are mathematically similar to the natural numbers and real numbers, so it seems perfectly plausible to have a set containing all of the ordinals (like the set of all real numbers and the set of all natural numbers). The set of all ordinals would be considered a set of ordinals, and thus, there must be a first ordinal (not in the set) to succeed all of the ordinals. However, we presupposed the existence of a set containing all of the ordinals, so this is a contradiction. This contradiction is only one of the many resulting from the discussion of infinite sets.

In order to avoid these paradoxes in set theory, several Russellian and Cantorian revisions to set theory were carried out during the twentieth century. For example, it was discussed that the existence of the elements preceded the existence of the set due to their logical importance in set theory. This fixed Russell's question of a set being a member of itself. However, these claims did not settle all paradoxes. Mathematicians prefer building mathematics from a collection of foundational principles like axioms, and it is clear that attempts at developing axiomatic systems void of paradoxes have been made in set theory. However, this method of building mathematics from the axioms in set theory was proved to be problematic by two twentieth century mathematicians, namely the Austrian Kurt Godel (1967) and Norwegian Thoralf Skolem (1967). Both developed similar, extensive proofs against this method, which showed the impossibility of creating an axiomatic system in set theory that produced any truth about sets. Thus, the two mathematicians concluded that truth in set theory could not exist in the Euclidean paradigm.

Though the discrepancy between actual and potential infinity is seen in the debate between the existence and non-existence of infinite sets, the distinction can also be seen throughout the field of calculus. The concept of the infinite sum is vital to understanding calculus. Consider the following infinite sum, $1/2 + 1/4 + \cdots = 1$. Addition normally involves a finite number of summands, so this infinite amount of summands equaling a single number does not necessarily make sense. Also, the notion of infinity arises in the concept of limits in calculus. However, Leibniz claimed in a letter, "I am so much in favor of the actual infinite" (Leibniz, 1960-1, I., pp. 416). With Leibniz's impact on calculus, his favoritism for the actual infinite may explain the reason behind equating infinite sums and limits with finite numbers.

By recalling Aristotle's claim and applying it to the discussion of calculus, it would be consistent to say that the concept of actual infinity permeates calculus but receives very little address. Aristotle would identify calculus-based concepts involving actual infinity to be attempts resulting in an invalid application of infinity. That would be a case where Hegel and Aristotle agree because where calculus deals with infinite sums and other similar concepts, it is actually talking about a finite quantity. For example, the completion of an infinite number of tasks in a finite time is essentially the same concept as that of a limit in calculus. Thus, when studying calculus, actual infinity often arises but is not paralleled with potential infinity, resulting in an ambiguous blend of the two different types of the infinite in modern mathematics.

The overall problem in compartmentalizing all infinity that is actual versus all that is potential comes down to semantics, namely defining the term 'infinity'. There exists a consistent trend that every new concept of infinity results in a slight, or at times full, disagreement with a previous explanation of the term. However, each new result seems to generally oscillate between the concepts of endlessness and wholeness (or synonyms of these words). The problem arises from the fact that the concept of endlessness implies a sense of potentiality and lack of completion whereas the concept of wholeness implies a sense of actuality and full completion. With these two concepts of the infinite, we see the origins of the paradoxical nature of infinity. However, with an articulate distinction between the two types of infinity, important paradoxical aspects of mathematics and philosophy will dissipate. To this end will our effort in this chapter be directed.

7.2.2 The vase puzzle

Other than some scholars in the history, such as Aristotle, Descartes, Hegel, and others, who did not recognize both kinds of infinities at the same time, is there any

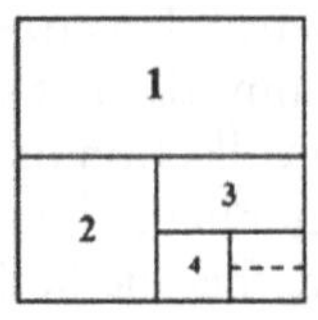

Figure 7.1 Obtain as many pieces of paper as needed out of a chosen area.

analytical evidence that these two kinds of infinities can in fact lead to different numerical consequences? The answer to this question is YES. For details, let us consider the following vase puzzle, which was first published in (Lin, 1999).

The Vase Puzzle: Suppose that a vase and infinitely many pieces of paper are available. The pieces of paper are labeled by natural numbers $1, 2, 3, \ldots$, so that each piece has one and at most one label on it. The following recursive procedure is performed:

Step 1: Put the pieces of paper, labeled from 1 to 10, into the vase; then remove the piece labeled 1.
Step 2: Put the pieces of paper, labeled from 11 to 20, into the vase; then remove the piece labeled 2.
...
Step n: Put the pieces of paper, labeled from $10n - 9$ through $10n$ into the vase; then remove the piece labeled n, where n is an arbitrary natural number $1, 2, 3, \ldots$

Question: After the recursive procedure is finished, how many pieces of paper are left in the vase?

Before we talk about the solution to this problem, some comments are needed here to make the situation practically doable. First, the vase need not be infinitely large; actually any size will be fine. Secondly, the total area of the infinite number of pieces of paper can also be any chosen size. For example, Figure 7.1 shows how an infinite number of pieces of paper can be obtained. Thirdly, the number labeling can be done according to the steps in the puzzle. Finally, the recursive procedure can be finished within any chosen period of time, the details of which are similar to that of any chosen size for the total area of the pieces of paper above.

Now, let us consider two different approaches of addressing this problem in the vase puzzle.

Approach I (an elementary modeling): To answer the question in the vase puzzle, let us define a function, based upon mathematical induction, by

$$f(n) = 9n,$$

which tells how many pieces of paper are left in the vase right after step n, where $n = 1, 2, \ldots$ Therefore, if the recursive procedure can be finished, the number of pieces of paper left in the vase should be equal to the limit of $f(n)$ as n approaches ∞. That is, infinitely many pieces of paper are left in the vase.

What we should note here is that this entire modeling process is based on the concept of potential infinite: the recursive procedure cannot be realistically finished. The step of taking limit only represents the progressive growth in the number of pieces of paper in the vase and the (imaginary) limit state.

Approach 2 (a set-theoretic modeling): Based on this modeling, the answer to the question of the vase puzzle is "no piece of paper is left in the vase." That of course contradicts the conclusion derived in the elementary modeling above. Here come the details.

For each natural number n, define the set M_n of the pieces of paper left in the vase right after the nth step of the recursive procedure:

$$M_n = \{x|x \text{ has a label between } n \text{ and } 10n + 1 \text{ exclusively}\}.$$

Then, after the recursive procedure is finished, the set of pieces of paper left in the vase equals the intersection:

$$\cap_{n=k}^{\infty} M_n, \text{ for } k \to \infty$$

That is, if x is a piece of paper left in the vase, x then has a label greater than all natural numbers. This contradicts the assumption that each piece of paper put into the vase has a natural number label. That is, the intersection $\cap_{n=k}^{\infty} M_n$, for $k \to \infty$, is empty. In other words, if after the recursive process is completed, there are still pieces of papers left in the vase, then we can pick one of them out. By assumption, this piece of paper must have a natural number label, say n. But, that is impossible, because at step n, this piece of paper had been taken out. A contradiction.

What we should note here is that this set-theoretic modeling is done on the basis of the concept of actual infinities. In particular, by using mathematical induction, we can show that for each natural number n, the piece of paper with label n has been taken out at step n. After that, an actual infinite is assumed to exist. That is, the existence of the sequence $\{1, 2, 3, \ldots, n, \ldots\}$ of all the pieces of paper that are taken out is assumed. As soon as this assumption is made, the conclusion, completely opposite to that of the elementary modeling, is inevitably produced.

This vase puzzle vividly shows the difference between actual and potential infinities. Of course, our example – the vase puzzle – is theoretical in nature. One might very well claim that in applications of mathematics, he will never run into such a problem. To this end, we should not be so sure, because in the study of quality control of the products out of an automated assembly line, for example, the concept of the infinite is involved. In particular, when studying quality, we draw a random sample of the products and use the sample statistics to make inferences on the continually expanding population, which is theoretically the collection of all the products that have been and will be produced from the assembly line. To make the inferences more reliable, we often treat the ever-expanding population as an infinite population, either actually infinite or potentially infinite. For more impacts brought forward by the vase puzzle, please consult (Lin, 1999), where a discussion about the connections between the vase puzzle and methodology, epistemology, and philosophy of science are respectively given. And some comments on the methodological indication of the fundamental structure of general systems, mathematical induction, and the knowability of the physical world can be found in (Lin and Fan, 1997).

7.2.3 A classification of paradoxes

What is presented in the previous subsection can be correctly referred to as a paradox, because the entire scenario seems to be reasonable according to the widely acceptable reasoning in mathematics and in science; however, the outcomes are contradictory to each other. Since the mid-1990s when the first author lectured around the world and presented this vase puzzle, among other topics, to different audiences, most of the young scholars were fascinated and greatly intrigued by such a pair of paradoxical and contradictory results. Even though these scholars were many in terms of numbers and widely spreading around the globe, no one had come up with a plausible explanation and found where the problem might be rooted.

By a paradox, it generally stands for such a statement that it sounds absurd and at the same time it has an argument to sustain it. When one scrutinizes the argument that sustains a paradox, he might discover the absurdity of either a buried premise or some preconception central to the classical physical theory or to the accepted thinking process. It has been more than often in history that each discovery of paradox has been the occasion for major reconstruction at the foundations of thought. In particular, studies of the foundations of mathematics in the past one hundred some years have been confounded and greatly stimulated by confrontation with two paradoxes, one propounded by Bertrand Russell in 1901–1902 and the other by Kurt Godel in 1931.

When paradoxes are compared to each other, it can be seen that some of them represent situations that seem absurd at first, but simple arguments sufficed to make us acquiesce them for good. For example, let us look at the following:

After traveling all day and feeling tired, three friends checked into a hotel. At the registration desk, they were told that the price was \$30 a night for them. So, each of them chipped in \$10 to cover the cost. Soon after they entered their room, the front desk called and informed them the correct price for the night should be \$25. So, bell boy from the front desk returned \$5 to the guests. While leaving \$2 tip for the bell boy, each of the friends kept \$1.00. So, the money transaction went as this: $\$(10-1)\times 3=\$27$ was the cost the three friends paid for the hotel stay, and \$2.00 went to the attendant for a total of $\$27+2=\$29$. Question: How could be one dollar missing?

In comparison, one might face with more difficult self-referential paradoxes. For example, Grelling paradox, derived by Kurt Grelling in 1908, is concerned with the heterological or non-self-descriptive adjectives. For example, the adjective "short" is short; the adjective 'English' is English; the adjective 'adjectival' is adjectival; the adjective 'polysyllabic' is polysyllabic. Each of these listed adjectives is self-descriptive or autological and is true of itself. Some other adjectives are heterological. For example, 'long' is not a long adjective; 'German' is not a German adjective; 'monosyllabic' is not monosyllabic. Now, Grelling's paradox arises as follows: Is the adjective 'heterological' autological or heterological?

To this inquiry, there are only two possibilities: 1) The adjective 'heterological' is heterological; and 2) The adjective 'heterological' is autological. If 1) is true, then it means by the definition of heterological adjectives that the adjective 'heterological' is autological. If 2) holds true, then the adjective 'heterological' is heterological. That is, the adjective 'heterological' is heterological if and only if the adjective 'heterological' is autological.

Similarly, Bertrand Russell (1967) in 1902 constructed his famous Russell's paradox. Its conclusion has been too absurd for the entire community of mathematicians to acquiesce in it in the past one hundred some years.

In other words, so-called paradoxes can be classified into three categories: veridical paradoxes, falsidical paradoxes, and antinomies, where

1. A veridical paradox contains a surprise. However, when one examines the argument, the surprise quickly dissipates itself;
2. A falsidical paradox also contains a surprise. However, as soon as one resolves the underlying fallacy, the surprise simply becomes a false alarm; and
3. A paradox is known as an antinomy, if it does contain a surprise and at the same time the surprise is accompanied by nothing less than a repudiation of part of the conceptual heritage.

As an example of falsidical paradoxes, let us look at the barber paradox, which is also due to Russell. In a village there is a barber, whose rule of conduct is that he shaves for those and only those men in the village who do not shave themselves. Now a paradox arises from the following question: Who shaves for the barber? There are only two possibilities: 1) The barber shaves himself; and 2) the barber does not shave for himself. If 1) is true, then the rule of conduct dictates that he does not shave for himself. If 2) holds true, then he shaves for himself. That is, he shaves for himself if and only if he does not shave for himself.

This unacceptable conclusion appears on the assumptions. We are given a story about a village and a man in the village who shaves those and only those men in the village who do not shave themselves. This self-rejection rule is the source of the trouble. So, the proper conclusion to draw has to be that there is no such barber. Our recourse in this comparable quandary is to declare a reduction of absurdity and conclude without any interim premise to disavow that there was no such barber. In this paradox, we arrive at the unacceptable conclusion by assuming the barber and deducing the absurdity that he shaves himself if and only if he does not. The self-contradictory conclusion simply a proves that no village can contain a man who shaves those and only those men in the village who do not shave themselves. So we acquiesce in the denial just as we acquiesce in the possibility, absurd on first exposure, of hotel expense scenario. In other words, any paradox, where is purportedly established is true, is called veridical or truth-telling paradox. For instance, the barber paradox is veridical provided that we take the proposition as being no such a village that contains such a self-contradicting barber.

A paradox is falsidical if its proposition not only seems at first absurd but also is false, where there is a fallacy in the purported reasoning. For instance, the hotel expense scenario represents such a falsidical paradox. Even so, falsidical paradoxes cannot be simply known as fallacies, because although fallacies most likely lead to false conclusions, they can also accidently produce true consequences and unsurprising conclusions as well as surprising ones. In a falsidical paradox, beside a fallacy always exists in the argument, the purportedly established proposition has furthermore seem absurd and to be indeed false.

Some of Zeno's paradoxes, when looked upon today, are falsidical paradoxes. For instance, let us look closely at the one Achilles and the tortoise. On these two fictitious characters, the paradox establishes the absurd proposition that so long as a runner keeps running no matter however slowly he is, another runner, no matter how fast he is, can never overtake him. The argument is that each time the pursuer reaches a spot where the pursued has been, the pursued has moved a bit beyond. When we try to make this argument more explicit, there appear several fallacies in this paradox. Firstly, it is the mistaken notion that any infinite succession of intervals of time has to add up to all eternity. With the currently knowledge of infinite series well established, we know that this notion is not true, because the infinite succession of intervals of time as chosen in the paradox satisfies the property that the succeeding intervals become shorter and shorter so that the whole succession actually takes a finite sum of time. That is, it is a question of a convergent series. Secondly, the paradox confuses a fictitiously constructed virtual scenario with the physical reality, where the concept of time is not fully explained and is misused.

In Grelling's paradox, as soon as the adjective 'heterological' is defined, we are asked whether or not it was heterological. As a matter of fact, along the same line of thoughts, one can establish the paradox just the same without introducing the particular adjective. Heterological is defined to mean not true of self, again a self-rejecting scenario, as pointed out earlier. Therefore, one can ask whether or not the adjectival phrase 'not true of self' is true or not. Then consequently, it can be found that it is true if and only if it is not true; hence that it is and it is not. So, the same paradox is established. When Grelling's paradox is seen in such a way, it seems to be unequivocally falsidical: its proposition is a self-contradictory compound proposition to the effect that the particular adjective is and is not true of self. However, it is worth noting that this paradox is very different of the falsidical paradoxes of Zeno, or any of the comical paradoxes of '$2 = 1$', in that one is at a loss to spot the fallacy in the argument.

Such paradoxes as Grelling's or Russell's are referred to as antinomies at least for the time being, because it is these antinomies that lead to crises in the commonly accepted thought and reasoning. Each antinomy produces a self-contradiction by employing the accepted ways of thinking. It evidences that fact that the tacit, tested, and trusted pattern of reasoning must be made more explicit; consequently, the accepted pattern of reasoning either needs be avoided or accordingly revised. Of course, as the history of science has shown, revision of a time-tested and trusted conceptual scheme is not unprecedented. In fact, it happens in a small way with each minor advance in science, and it occurs in a big way with each major advance in the contemporary way of thinking. For instance, the Copernican revolution and the shift from Newtonian mechanics to Einstein's theory of relativity are two incidences when the tacit, tested, and trusted belief systems had to be revised. And, there was such a time the theory that the earth revolves around the sun was known as the Copernican paradox, even by the same men who accepted it.

The previous classification of paradoxes is dynamic in the sense that with time, the classification of a particular paradox may change. For instance, in his days the falsidical paradoxes of Zeno had been seen as genuine antinomies. And then with the theoretical development of infinite series, a fallacy implicitly contained in these paradoxes is discovered: The notion that an infinite succession of intervals must add up to an infinite interval is no longer acceptable. Therefore, one man's antinomy is another man's falsidical paradox, give or take a few thousands or hundreds of years.

Although Russell used the barber paradox to illustrate his antinomy, the latter, however, has more than a hint of the former. Although in truth their parallel is exact, it is a simple point of logic that in no village there is a man who shaves those and only those men in the village who do not shave themselves so that he would shave himself if and only if he does not. The barber paradox was a veridical one showing that no such a barber can exist in reality. On the contrary, Russell's paradox is an antinomy instead of a veridical paradox where one can simply conclude that there is no class whose members are those and only those that are not members of themselves. The reason for this end is that the existence of such a class is without any doubt, while the non-existence of the barber is a sure possibility and consequence. The barber scenario barely qualifies as paradox in that people are mildly surprised at being able to exclude the barber on purely logical grounds by reducing him to absurdity; and even this surprise dissipates as one reviews the argument. On the contrary, Russell's paradox is a genuine paradox up to this point of our writing, because the criterion of class existence is so fundamental that we cannot easily give up. In a future century when the absurdity of that criterion become a commonplace and when a substitute criterion has enjoyed long enough tenure to be taken as the common sense, people will see Russell's antinomy as nothing more than a veridical paradox, showing that there is no such a class as that containing those and only those non-self-members. That is, with time one man's antinomy will become another man's veridical paradox, which in turn will become another man's platitude.

As of this writing, Russell's antinomy has made a more serious crisis in our modern day than did any other paradoxes. It strikes at very concept of classes of mathematics, where the concept is appealed to in an auxiliary way in most branches of mathematics, and increasingly so as passage of mathematical reasoning is made more explicit. The basic criterion for the existence of classes that is tacitly used at every place where classes are involved is precisely the criterion that is described by Russell's antinomy. Fortunately, the criterion for the existence of classes is only incidentally used in those branches of mathematics so that the progresses in these branches of mathematics are not adversely interrupted. That is why it has been possible for most branches of mathematics to go on blithely using classes as auxiliary apparatus in spite of Russell's and related antinomies. However, in comparison, the works presented in this book touches on every branch of mathematics as long as there is a constructive, existence proof using mathematical induction where 'for every' is unconsciously replaced by 'for all'. In other words, as soon as actual infinity is shown to be different of potential infinity, all these constructive proofs need to be reconstructed before the consequent existence results can be employed in the proofs of other theorems.

In short, the afore-mentioned classification of paradoxes, according to (Quine, 1976), is dynamic and time sensitive, because a past antinomy can become either veridical or falsidical with time, as what is shown above. Once again, at the conclusion of this section we like to mention that the discussion in this section is based on (Quine, 1976). For more details, please consult the original work.

7.3 DESCRIPTIVE DEFINITIONS OF POTENTIAL AND ACTUAL INFINITE

The exploration and research on potential and actual infinities are generally done simultaneously in many disciplines, such as philosophy, logic, computer science, mathematics, etc. From the angle of history, in this section, which is based on (Lin (guest editor), 2008, pp. 424–432), we analyze the concepts of potential and actual infinities descriptively with one point to start and two locations to cut in. By clarifying the difference and connection of these two different kinds of infinities on the level of mathematics, we introduce the symbolized, descriptive definitions for the concepts of potential and actual infinities.

7.3.1 A historical briefing

Historically speaking, Aristotle was the first scholar to clearly recognize the concept of potential infinity. He did not accept the existence of actual infinity. A little earlier than Aristotle, Plato was the first scholar in history who acknowledged the existence of actual infinity while against the existence of potential infinity. Over the history of more than two thousand years, scholars, who insisted on the concept of potential infinity, and those, who firmly stood for that of actual infinity, have been debating over which side is correct in the fashion of mutual rejection. Their debate has dealt with and touched on many scientific disciplines, including philosophy, logic, the theoretic foundation of computer science, mathematics, and others. For relevant references, please consult (Zhu and Xiao, 1991). These great minds, who were involved in this long-lasting debate, had left behind a great many elementary and unadorned descriptions about these two kinds of infinities. To this end, let us list a few of these descriptions.

Aristotle clearly recognized (Institute of Foreign History of Philosophy, Beijing University, 1962) that each infinite has to be a potential existence instead of an actual existence. He said that "because the process of division will never be finished (ongoing), it guarantees the potentiality of the existence of this kind of activity. However, it cannot guarantee the independent existence of infinite." Aristotle also stated (Institute of Foreign History of Philosophy, Beijing University, 1962) that "both time and space can be infinitely divided (ongoing), but have not been infinitely divided (done)."

Bayle held (Institute of Foreign History of Philosophy, Beijing University, 1962) that if we draw infinitely many line segments on a piece of material of the size of one square inch (done), we will make such kind of division. This kind of division would make Aristotle's infinite, which is only a potential (ongoing), become an infinite that is an actuality (done)."

Corresponding to what Bayle said, Lenin (1959) pointed out that the statement that "we draw an infinite many line segments on a piece of material of the size of one square inch (done)" means that we have completed the infinite division (done)!"

Weyl (1946) made the statement that "Brouwer has made this point clear: The sequence of natural numbers can pass on to the next number indefinitely so that it can go over any boundary which has been reached. Therefore, it opens the possibility of leading to infinity (ongoing). However, this possibility forever stays in the state of creation (generation) (ongoing) instead of a closed field of materials that exists in itself (done)."

7.3.2 An elementary classification

Based on these elementary and unadorned descriptions, we can clarify the differences between potential and actual infinities as in the following two points:

A From the angle of generation, each potential infinite is forever a present progressive tense (ongoing), while every actual infinite stands for a present or past perfect tense (done, finished).

B From the angle of existence, each potential infinite is dynamic and a possibility, while every actual infinite stands for a static state and a practical existence.

Therefore, the concept of potential infinite should possess the following two fundamental properties:

1 Not finitary. This property leads to the possibility of reaching actual infinity (ongoing).
2 Forever a present progressive tense. This property rejects the possibility of reaching the very end of the process (not done, but ongoing).

As for the concept of actual infinities, we can abstract the following two fundamental properties:

1 Not finitary. This property provides the possibility of leading to actual infinities (ongoing).
2 Definitely a perfect tense. That is, this property affirms the reachability of the very end of a process (done, finished).

For example, the symbol $[a, b]$ is usually used to represent a closed interval of real numbers from the number a to another number b in a Cartesian coordinate system. There are infinitely many points in this interval $[a, b]$. Let a variable x move in this interval along the positive direction of the x-axis toward the number b. Then, in this case, the variable x can not only approach the limit point b indefinitely (ongoing), but also eventually reach the limit point b (done). So, we can say that the variable x, in the interval $[a, b]$, approaches the limit point b in the fashion of actual infinities (or actual infinite). Additionally, we usually employ the symbol (a, b) to stand for the open interval of real numbers from the number a to another number b on the number line. Here, (a, b) contains once again infinitely many points while excluding the endpoints a and b. Now, same as before, let x be a variable which approaches b in the interval (a, b). Even though the point b is still the limit of the variable x, it is not included in the open interval (a, b). Therefore, the variable x can still approach the point b indefinitely from within the interval (a, b) (ongoing), but will never, ever reach the point b from within the interval (a, b) (forever ongoing). Hence, we can say that the variable x approaches the limit point b in (a, b) in the manner of potential infinities (or potential infinite).

As a second example, in mathematics, the symbol O_∞ is conventionally applied to represent a point on the real-plane that is indefinitely far away from the origin. When a variable x approaches this indefinitely far away point O_∞, the afore-described properties of potential infinities become:

(1′) Not finitary. So, it points to the possibility for the variable x to approach the imaginary point O_∞ (ongoing).
(2′) Forever a present progressive tense. So, it rejects the possibility for the variable x to reach the imaginary point O_∞ (not done, will forever ongoing).

At the same time, the afore-described properties of actual infinities in this case become:

(1′) Not finitary. So, it provides the possibility for the variable x to approach the imaginary point O_∞ (ongoing).
(2′) Definitely a perfect tense. That is, it affirms the fact that the variable x definitely reaches the infinitely far-away point O_∞ (not ongoing, but finished).

In short, each actual infinity materializes a present progressive tense (ongoing) to a perfect tense (done, finished), while every potential infinity strengthens a present progressive tense (ongoing) to a forever present progressive tense (ongoing). And, no matter whether it is a potential infinity or an actual infinity, it stands for a nonfinitary process. Therefore, we have the following:

Nonfinitary process {
- Actual infinity: definitely reach the end of process
- Potential infinity: never reach the end of process (ongoing)

7.3.3 A second angle

Now, let us further discuss and comprehend the concepts of potential and actual infinities with a different point of incision. When faced with our research subjects and phenomena, if the procedure of dealing with or judging on each and every subject and scenario is employed, we refer this method of research as to the method of listing or the procedure of listing. Here, we clearly define the concept of listing, which is different from that of enumeration in set theory. As for specific examples of using the method of listing, there are plenty of them in various fields of scientific studies. In the following, let us look at a few of the relevant cases.

Example 7.6 Let us look at the historically well-known statement about cutting space and time. That is, we draw a line segment on a material of the size of one square inch. After drawing one segment, we then a second one, then a third one, ... If this procedure is continued indefinitely and the line segments are drawn one after another, as long as we keep this process of drawing line segments ongoing (not ever finished), then what is faced in this concrete procedure of listing is a potential infinite. However, as what Bayle said earlier in the previous paragraphs, “if we draw infinitely many line segments on a piece of material of the size of one square inch," or as what Lenin stated, "if we have completed the infinite division", it means that if the said, specific procedure of listing is completed (done), or in other words, if the specific procedure of listing is exhaustively done (finished), then what this procedure of listing faces is no longer a potential infinity. Instead, it is a finished actual infinity. Hence, this indefinite, never-ending procedure of listing is a present progressive tense (ongoing), while completing exhaustively a procedure of listing is a perfect tense (done). In short, the transformation from listing to exhausting is a transformation from a present progressive tense to a perfect tense and one from a potential infinity to an actual infinity.

Example 7.7 Let us imagine an ideal empty cup G with a given precise predicate P defined on an infinite background world. When an object x is found to satisfy P, we throw this object x into the cup G. When another object y is found to satisfy P, we again place y into G. Since P is defined in an infinite world, this procedure of looking for objects and placing the found objects in G can of course be continued indefinitely. However, as long as this procedure of listing is not completed and is still in the process of listing (ongoing), then what is faced is a potential infinity. If this procedure of listing is exhausted, that is, if one can and has found all the objects that satisfy P and thrown all of these objects in G (done), then what is faced becomes an actual infinity.

Example 7.8 Based on the ideologies of Democritus, an ancient atomist, and Cantor, the founder of the naïve set theory, one knows that each infinite set is an actual infinite set. Now, given an infinite set A and a criterion f of choice defined in an infinite background world, we then can based on f pick an element from A, then another element, ... This procedure of selecting elements one after another from A according to f is also a specific process of listing.

Example 7.9 In Cantor's naïve set theory, the concept of one-to-one correspondence, when applied to infinite sets, is also a procedure of listing. It is because for any two infinite sets A and B and a one-to-one correspondence f from A to B, for each arbitrarily chosen element a in A, through f there must be a uniquely determined element b in B that corresponds to the element a in A. For another element a_1 in A, once again through f, one can find a unique element b_1 in B that corresponds to the second element a_1 in A; and vice versa. Because both A and B have infinitely many elements, this procedure or process can be continued indefinitely without an end. Therefore, it is also a procedure of listing.

Example 7.10 The concept of infinity of a sequence, defined by the $\varepsilon - N$ method in the theory of limits, and the thinking logic behind the concept are the same as what was used behind how natural numbers were constructed by Brouwer. Both thinking logics provide a procedure of giving a bound and going beyond this bound; then giving another bound and going beyond this new bound. And no matter how great the bound N one takes, he has a way to go beyond this bound. So, the procedure of giving and going beyond and giving again and going beyond again can be continued indefinitely without an end. That is, this is another concrete and typical process of listing widely studied in mathematics.

There are many varieties of procedures of listing in the scientific literature. However, based on the discussions of the previous examples, it can be readily seen that each non-terminating procedure of listing is about a potential infinite. Only when an infinite procedure of listing can be completed, that is, only after the procedure of listing is exhausted, one will face with an actual infinite. Conversely, in terms of actual infinities, since each of them must be a perfect tense, it means that the underlying procedure of listing must be exhausted, while in terms of potential infinities, since each of them must be a present progressive tense, it means that one has to reject the possibility that the underlying procedure of listing can be completed or exhausted. Therefore, we have

Not finite procedure {
Actual infinity: definitely exhuasted the listing process (done)
Potential infinity: the listing process cannot be completed

In short, no matter how we alter the way we look at and comprehend the concepts of potential and actual infinities, the essence of these concepts stays the same. That is, each actual infinite must be equivalent to a perfect tense, while each potential infinite a present progressive tense. In other words, no matter whether it is "completed" or "exhausted", it is always the same as a perfect tense. And no matter whether it is "never reachable" or "never exhaustible", it is always the same as a present progressive tense. Because any present progressive tense (ongoing) is different of a perfect tense (done), listing is not the same as exhausting. "Never reachable" is different of "reachable". Therefore, potential infinities and actual infinities, no matter whether we look at them either from the angle of generation or from the angle of existence, are two different concepts that can't be equaled or identified with each other under any circumstance. Otherwise we were not truthful with ourselves.

7.3.4 Summary of the highlights

Our discussion above can at least lead to the following conclusions: On the level of philosophy, we have clarified the difference and connection between potential infinities and actual infinities. To summarize, we have discovered one starting point and two angles for comprehending these concepts. The following summarizes the relevant details.

1 The starting point:

Not finite {
Actual infinity: it must be a perfect tense (done)
Potential infinity: it is forever a present progressive tense (ongoing)

2 Angle number 1:

Not finite {
Actual infinity: definitely reach the very end of the process
Potential infinity: impossible to reach the end of the process

3 Angle number 2:

Not finite processure	Actual infinity: definitely exguasted the processure of listing Potential infinity: impossible to exhaust the procedure

It can be seen that at the level of formal logic, there do not exist these concepts of potential infinities and actual infinities. It is because the formal logic only deals with possible subjects instead of existing subjects and both potential and actual infinities are all such concepts that deal with existence. That is, they are about the potential existence (potential infinities) as what Aristotle talked about and the actual existence (actual infinities) as what Plato recognized. Therefore, formal logic does not have any direct connection with the concept of infinite except some indirect connections. It is because formal logic can provide the tool of reasoning for the study of existent subjects, just like the case that the manipulation of the two-value logic has played the roles of foundation and tool of reasoning for the modern axiomatic set theory. As it is well known that in formal logic, there is a universal quantifier ∀, seen or read as "for each" or "for all". In our following studies, we will introduce such a quantifier E for the situation of listings, which will be understood as "for each". And we will limit the meaning of the existing universal quantifier ∀ to that of "for all". After strictly distinguishing the meanings of "for each" (listing) and "for all" (exhausted) in our studies of infinities, the listing quantifier E will become one of the tools of logic useful for the study of potentially infinite subjects, while the universal quantifier ∀ will be useful for the study of actually infinite subjects. The quantifier "E" on the level of logic corresponds exactly to the procedures of listing at the height of philosophy. And the quantifier "∀" on the level of logic corresponds exactly to the exhausted processes at the height of philosophy.

In terms of at the level of mathematics, it has been a common practice to directly apply the concepts of potential and actual infinities. For example, in the theory of limits, in order to avoid the Berkeley paradox, the definitions of limits, both infinite and finite, using $\varepsilon - N$ and $\varepsilon - \delta$ methods, are completely based on the thinking logic of potential infinite. Also, in Brouwer's mathematics of intuitionistic construction, the concept of actual infinities is completely excluded. Scholars in this system of mathematics completely refuse to recognize the convention that the totality of all natural numbers can make up a closed set. Therefore, from the start of how the totality of natural numbers is an intuitive, non-terminating process of generation to the construction of the continuum of the expanded form, the entire system of Brouwer's mathematics is established on the foundation of non-terminating processes and potential infinities. However, in the naïve set theory and the modern axiomatic set theory, from Cantor to Zermelo, the existence and construction of infinite sets have been established on the concept of actual infinite and perfect tense. For instance, for a given predicate P for constructing a set in an infinite background world under the meaning of Cantor and Zermelo, there exists a unique infinite set $A = \{x|P(x)\}$ determined by P. This set A consists of all such elements satisfying the predicate P and is produced on the basis of completely exhausting all the elements x that satisfy P. Therefore, in both the naïve set theory and the modern axiomatic set theory, each infinite set is constructed on the concept of completed actual infinities.

As a matter of fact, the creation of Cantors' naïve set theory indicates the fact that the development of mathematics had gone through the transition from studying finite objects and potentially infinite objects to the research of actually infinite objects. Before Cantor's time, many mathematicians believed in the idea of potential infinities. For example, Gauss expressed in a formative tone his position in a well-documented letter to Heiurich Shumacher: "I object the use of infinities as completed, actual entities. In mathematics, this is absolutely not allowed. Infinite is only a way of communication when dealing with limits" (Dauben, 1988; 1979). That was why in Joseph Dauben's book, entitled "Georg Cantor: His Mathematics of Philosophy of the Infinite", it was clearly pointed out that Cantor recognizes that his transfinite numbers and transfinite set theory face the objection of the traditional beliefs. One of his purposes of writing the book "Foundation" is to argue that there is no basis for people to object the concept of completed, actual infinities. He wished to answer in an irrefutable way to the suspicions of such mathematicians as Gauss, such philosophers as Aristotle, and such theologists as Thomas Aquinas. Cantor believed that it was a mistake to object the usage of actual infinities in mathematics, philosophy, and theology because of a well-accepted convention.

Finally, in terms of the theory of computer science, it is most important to recognize the direct application of the concept of potential infinities and the fundamentality of this concept. It is because in computer science, the possibility of practical performance is especially emphasized. As is well known, in computer science one comment about the historical role of the school of intuitionism is completely affirmed. That is (Zhu and Xiao, 1996), "when the development of computational mathematics is concerned with, the opinion and method of constructability of the school of intuitionism possess a very important meaning. Their requirement for operability is especially important for practical purposes. It is because when computers are employed, one has to think about operability." As discussed earlier, in terms of various beliefs of infinite, the school of intuitionism and its constructive system of mathematics still insist completely on the concept of potential infinities without leaving any elbowroom for the existence of actual infinities.

As a matter of fact, quite some scholars in the current scientific arena have contributed to the exploration and study of the concepts of potential and actual infinities, see, for example, (Hailperin, 2001; Engelfriet and Gelsema, 2004; Wang and Xu, 2003; Corazza, 2000; Kanovei and Reeken, 2000). For a more detailed review of current literature, please go to Section 7.3.6.

7.3.5 Symbolic preparations

In this subsection, on the basis of introducing a series of symbolic representations, we will establish the descriptive definitions for potential and actual infinities. In the establishment of formal systems, our symbolic representations can also be introduced as the corresponding "operators".

1 The symbol "$\uparrow$" is called an "open progressive word". Its meaning can be read as:

   (a) A present progressive tense;
   (b) For each ________ $=_{df}$ E________;

(c) List ________ $=_{df}$ *enu*________;
(d) Indefinitely approaches ________ $=_{df}$ *ina*______;
(e) Without bound ________ $=_{df}$ *kne*________.

2 The symbol "⊤" is referred to as a "positive completion word." Its meaning can be read as:

(a) Affirmed perfect tense;
(b) For all ________ $=_{df}$ ∀________;
(c) Exhausting ________ $=_{df}$ *exh*________.

3 The symbol "⊤̸" is referred to as a "negative completion word", its meaning can be read as:

(a) Negation of a perfect tense;
(b) Negation of for all ________ $=_{df}$ ¬∀________;
(c) Negation of exhausting ________ $=_{df}$ ¬*exh*________;
(d) Will never reach ________ $=_{df}$ ¬*rea*________.

Also, we will write "*fin*" for "finite", "*inf*" for "infinite", "*poi*" for "potential infinite", "*aci*" for "actual infinite". That is, we have $fin =_{df}$ "finite", $inf =_{df}$ "infinite", $poi =_{df}$ "potential infinite", and $aci =_{df}$ "actual infinite".

As is known, in the naïve set theory and the modern axiomatic set theory, a set A is finite, provided that there is a natural number $n \in N$ such that a one-to-one correspondence between A and n can be established. That is, if $\exists n(n \in N \wedge A \sim n)$, then A is called a finite set, denoted $A[fin]$. Now, a set A is infinite, provided A is not finite, denoted $A[inf]$. That is, $A[inf] =_{df} \neg A[fin]$. And, each infinite set, that is, each not finite set, is a completed, actually infinite set, because any infinite set $A[inf] = \{x | P(x)\}$ consists of all such elements x that satisfy the specified predicate P. So, in this background of mathematical modeling, "infinite" is defined as "not finite". Therefore, in this system of mathematics, one insists on either the bisecting principle of "not finite" and "actually infinite" or the thinking principle of not distinguishing between "potential infinite" and "actual infinite" due to their blurred differences. In our work, we will stay away from such a background of mathematical modeling and will define the concepts of "potential infinities" and "actual infinities" on the high level of abstraction as follows:

$$aci =_{df} \neg fin \wedge \uparrow \wedge \top; \quad \text{and}$$

$$poi =_{df} \neg fin \wedge \uparrow \wedge \not\top.$$

That is, no matter whether it is a "potential infinity" (*poi*) or an "actual infinity" (*aci*), firstly, it should not be finite (¬*fin*) and it should have entered a present progressive tense (↑). This is a common characteristic of "potential infinite" and "actual infinite", which is the possibility of leading to an infinity (¬*fin*∧ ↑). Secondly, on this common background, the difference between "potential infinities" and "actual infinities" is spelled out clearly. This difference is: For each actual infinite, a perfect tense must be affirmed (⊤: ∀, *exh*, *rea*). For each potential infinite, a negation of all perfect tenses must be cleared (⊤̸: E, *enu*, *ina*). Since, in this case, a negation of all perfect tenses is declared, the present progressive tense is strengthened to a forever present progressive tense so that one will forever be situated in the process of "for each" (E), "listing" (*enu*), and "indefinitely approaching" (*ina*).

7.3.6 A literature review of recent works

For the completeness of our presentation in this chapter, let us look at some other scholars who have done works in recent years on related concepts, such as actual and/or potential infinities. Limited by space, we will only mention a few pieces of relevant works. The interested readers can either consult the papers cited here and the references listed in these papers or do a more thorough literature search along all the lines outlined here.

T. Hailperin (1992) of Lehigh University (USA) established a formal characterization of a potential infinite sequence as a rule-generatable sequence. In (Hailperin, 1997), in his study of ontologically neutral (ON) language Q and the concept of models for an ON language, Hailperin employed the concept of actually infinite sets in the definition of verity functions. On this basis, he showed that any assignment of truth values to the atomic sentences can be extended so as to be a verity function on Q^{cl}, the closed formulas of Q. And, based on this infinitistic assumption and validity defined to be all possible assignments of truth values to the atomic sentences, he provided a completeness proof along traditional lines. To improve his completeness proof for ON logic, he (Hailperin, 2001) introduced the concept of potential infinite verity functions on potentially infinite sequences, leading to the notion of potential infinite models, and proved the following fundamental theorem of potential infinite semantics along with several other important results.

Theorem 7.1: (Hailperin, 2001) *Let M be a potential infinite model with potential infinite limit L for an ON language* $Q(p_1, \ldots, p_r)$. *Then,*

(i) *there is a unique function* V_M *which extends M from atomic sentences so as to be a potential infinite verity function on closed formulas of Q;*
(ii) *for any closed formula* $(\wedge i)\varphi$ *(or* $(\vee i)\varphi$*) the sequence of values* $V_M(\varphi\ 0), \ldots, V_M(\varphi\ (v)), \ldots$ *ends with a tail of all 0's or all 1's;*
(iii) *for any closed formula* ϕ, *its* V_M *value is equaled to that of its L-finitization, that is,* $V_M(\phi) = V_M(S_L(\phi))$.

J. Engelfriet and T. Gelsema (2004) of Leiden University (The Netherlands) studied the middle congruence, a notion of structural equivalence of processes, in which replication of a process is viewed as a potential infinite number of copies of the process in the sense that copies are spawned at need rather than produced all at once. Their work is based on the standard congruence studied in (Milner, 1992) and the extended congruence first introduced in (Engelfriet, 1996). They introduced additional laws to model !P as an "actual" infinite number of copies of P and their notion of middle congruence shares with standard congruence the suggestion that replication is only "potentially" infinite. They proved that middle congruence has the same desirable properties as extended congruence: it is decidable and has a concrete multiset semantics, leading to the conclusion that these properties do not depend on the distinction between potential and actual replication.

After (Lin (guest editor), 2008) showed that the 4th crisis in the foundations of mathematics has appeared, a plan of resolution of this new crisis is provided by addressing two questions: select an appropriate theoretical foundation for modern mathematics and computer science theory; and choose an interpretation so that the known achievements of mathematics and computer science can be kept in their entirety. Related to this attempt, we have the internal set theory (Nelson, 1977), the non-standard set theory (Hrbacek, 1978; 1979) and the Kawai (1981) set theory. Then, V. Kanovei of Moscow State University (Russia) and M. Reeken of Bergische Universitat (Germany) (2000) studied which transitive ∈–models of the ZFC can be extended to models of a chosen nonstandard set theory. The spirit of all these works is that non-standard models of natural and real numbers can be used to interpret the basic notions of mathematical analysis of the 17th and 18th century, including infinitesimals and infinitely large quantities.

7.4 POTENTIAL AND ACTUAL INFINITIES IN MODERN MATHEMATICS

The purpose of this section, which is mainly based on (Lin (guest editor), 2008, pp. 433–437), is to pay a closer look at the concepts of potential and actual infinities and how they are employed in the system of modern mathematics.

It is a common belief that both Cantor and Zermelo employed completely the thinking logic of actual infinite in both naïve and modern axiomatic set theory, and that Cauchy and Weierstrass applied completely that of potential infinite in the theory of limits. However, when we explore in depth the essential intensions of both potential and actual infinite, and after we sufficiently understand the differences and connections between these infinites and revisit the realistic situations on how the concept of infinities has been employed in the modern system of mathematics, we discover that in set theory, the thinking logic of actual infinities has not been applied consistently throughout, and that in the theory of limits, the idea of potential infinities has not been utilized consistently throughout, either. As for those subsystems involving the concepts of infinities of modern mathematics, they generally contain both kinds of infinities at the same time. As a matter of fact in modern mathematics and its theoretical foundation, one only need analyze slightly and dig a little deeper, he will be able to see the reality that the thinking logics and method of analysis of employing both kinds of infinities are implicitly everywhere.

7.4.1 Compatibility of both kinds of infinities

For each system of scientific knowledge, it is generally developed and established on a theoretical foundation, just like a large building which definitely has its basic foundations and supporting walls. For the discipline of mathematics, it is no exception to this rule. Since the time when Cantor created the naïve set theory in the 19th century, people have found that any concept of mathematics can be established and introduced on the basic concepts of set theory, and that each mathematical theorem can always be derived from the axioms or the fundamental postulates of set theory. In short, with set theory developed, the totality of mathematics can be established on the basis of the set theory. Because of this reason, scholars commonly recognize the fact that set theory can be employed as the theoretical foundation of the mathematical sciences.

However, very unfortunately, some self-contradictory propositions were discovered in the naïve set theory. That is, some so-called paradoxes were discovered. So, if the entirety of mathematics is seen as a tall and large building that is constructed on top of the naïve set theory, then in the foundations and the supporting walls of this mathematical mansion, people discovered some major cracks. These discoveries made everyone feel unsafe and anxious to continue to stay within the inside of the mathematical mansion. Consequently, mathematicians modified the naïve set theory and established the modern axiomatic set theory wishing that no paradox would appear in this new set theory.

Constructed on top of the modern axiomatic set theory, the new foundation of the mathematical mansion,modern mathematics has been believed to be standing on relatively solid supporting walls. However, the modern axiomatic set theory has several different versions. The commonly applied version is known as the ZFC system, which was initially created by Zermelo in 1908 and then further polished by Frankel.

Specific to the field of study known as mathematical analysis, because of the ambiguity of the concept of infinitesimals at the initial stage of development of calculus, Berkeley pointed out a contradiction, the historically well-known Berkeley's paradox, existing in mathematical analysis. In order to clear up the Berkeley paradox, mathematicians established and developed the theory of limits; and this theory of limits became the theoretical foundation of calculus. And, the theoretical foundation of the theory of limits is the ZFC system. So, if we use N, C, and Z to denote respectively calculus, the theory of limits, and set theory, then the mathematical system, which consists of calculus and its theoretical foundation, can be simply written as $N \cup C \cup Z$. Now, let us illustrate the fact that $N \cup C \cup Z$ is exactly such a typical mathematical system that both potential infinities and actual infinities coexist side by side.

Firstly, it has been known earlier that set theory emphasizes the existence of actual infinities. In particular, because each infinite set $A = \{x|P(x)\}$ consists of all such elements x that satisfy the proposition P, it means that a perfect tense must be employed here. That is, if a procedure of listing all elements x that satisfy the proposition P is used to generate this infinite set A, it is only after this process of listing is exhausted, the set A can be generated. Therefore, no matter how A is constructed, its existence means a perfect tense. That is, the existence of A implies the existence of an actual infinity. However, on the other hand, in order to avoid the Berkeley paradox, the theory of limits employs the $\varepsilon - \delta$ and $\varepsilon - N$ methods to define infinities and infinitesimals of sequences. So, these methods are completely established on the concept of potential infinities by indefinitely giving bounds and going beyond the bounds. Consequently, the direct employment of both actual infinities and actual infinitesimals is successfully avoided. Therefore, when calculus and its theoretical foundation $N \cup C \cup Z$ are seen as a mathematical system, at the very elementary stage, as we just described, it has already contained the concepts of both potential and actual infinite.

Not only as what we have just discussed, even inside the modern axiomatic set theory, it is not like what was expected by Cantor and Zermelo that the axiomatic system involves only purely actual infinities without dealing with potential infinities.

As a matter of fact, what was expected by Cantor and Zermelo is impossible. As a matter of fact, from Cantor to Zermelo, the principle of one-to-one correspondence appeared everywhere. Just as what is pointed out in example 7.9 in Section 7.3 above that when applied to infinite sets, the principle of one-to-one correspondence is also a procedure of listing, and before the procedure is exhausted, the process is forever a present progressive tense (going). Therefore, what is facing the procedure must be a potential infinity. Secondly, it is known that the set$N$, consisting of all natural numbers, has an infinite number of elements. In mathematics, the symbol ∞ or ω is often used to represent an infinity. Therefore, we say that N has ω many elements. That is, the set N of all natural numbers contains an infinite number of elements. Let us order the natural numbers in N from the smallest to the larger as in the following sequence λ: $\{1, 2, 3, \ldots, \text{n}, \ldots\}$, and let k be a variable representing a natural number varying from one natural number to another, increasing indefinitely. Then, the variable k can increase without a bound and approach ω indefinitely closely. However, in mathematics, it is a conventional fact that for each natural number, its magnitude is finite. That is, any chosen natural number is less than the infinity ω. So, ω is not a natural number. Even though the variable k can increase without any bound and approach ω indefinitely, k can never reach ω. Hence, the process for the variable k to approach ω is forever a present progressive tense (ongoing), facing a potential infinity surely. What we just discussed is more than enough to indicate that the modern axiomatic set theory itself is a system compatible with the thinking logics of both kinds of infinite, instead of a system completely implementing the concept of actual infinite without touching on that of potential infinite.

Now, let us look at the theory of limits, on which calculus and mathematical analysis are based. In this theory, even though the concepts of infinities and infinitesimals are defined using the methods of $\varepsilon - \delta$ and $\varepsilon - N$, as discussed earlier, in the process of defining these concepts, the non-terminating present progressive tense (ongoing) of choosing a bound and going beyond the bound and choosing another bound and going beyond this new bound is used. So, all these definitions face directly with the thought of potential infinite. Especially in the theory of limits, when a variable x approaches its limit x_0, the convention is not to mention whether or not x can reach the limit point x_0. More than that, what is clearly written is $0 < |x - x_0| < \varepsilon$. That is, the variable x can indefinitely approach its limit point x_0 and will forever not reach the point x_0. So, the concept of the variable x approaching its limit point x_0 is forever a present progressive tense (ongoing) instead of a perfect tense. Here, there is no doubt that the thought of potential infinite is heavily implemented. Now, the question is: Can the theory of limits truly succeed in terms of completing avoiding actual infinities and 100% applying the concept of potential infinite? As a matter of fact, the answer to this question is: No, it is not possible! It is because in the theory of limits, one has to deal with the concept of irrational numbers, such as π or $\sqrt{2}$, etc. However, the analytic representation of any chosen irrational number must face or deal with actual infinities. It is because behind the decimal point, each irrational number contains an infinite number of digits and the concept of countable infinities is definitely a typical example of actual infinities. That is, for any irrational number θ, the totality of all digits $p_i, i = 1, 2, 3, \ldots, n, \ldots$, behind the decimal point of θ, contains such an index set N, containing all the positional ordinal numbers, that is the actually infinite set of all natural numbers. This end is completely contradictory to the intuitive construction used in defining irrational numbers. Beside all of these, if one needs to establish and develop calculus on the basis of the theory of limits, then he will have to employ such actually infinite field as the set of all real numbers, the set of all irrational numbers, etc. That is, what we have shown is that the theory of limits has to be a theoretical system established on the thoughts of both potential and actual infinite.

Summarizing what has been discussed in the paragraphs above, it can be seen that not only are modern mathematics and its theoretical foundation a theoretical system containing the thoughts of both potential and actual infinite, but also have its subsystems, involving the concepts of infinities, to be such systems that the thoughts of both kinds of infinities co-exist. Otherwise, the systems of mathematical theories would not be seen as a branch of modern mathematics.

7.4.2 Some elementary conclusions

In Section 7.3, we have mainly clarified differences and connections between potential and actual infinite. The core content of our work there contains two aspects: 1) Each actual infinity has to be a present progressive tense (ongoing), which is transited into a perfect tense (done); 2. Each potential infinity must be a forever present progressive tense (ongoing) strengthened from a present progressive tense (ongoing). Then, from two different angles, one is either "eventually reaching" or "never reaching", and the other either the "procedure of listing" or "exhausted procedure of listing", we provided two concrete models or specific realizations for present progressive tense and perfect tense. With these discussions in place, in this section, we provided evidence showing the fact that modern mathematics and its theoretical foundations employ the thoughts of both kinds of infinities. Furthermore, in the first subsection of this section, we detailed the following two points: (1) The totality of modern mathematics and its theoretical foundation is a theoretical system employing both kinds of infinities; (2) Some subsystems of the whole system in (1), involving the concepts of infinities, are also systems based on both kinds of infinities.

Summarizing our discussions, it can be readily seen that in modern mathematics and its theoretical foundation, the thinking method of allowing both kinds of infinities and related contents are employed everywhere. Therefore, there does not exist such a mathematical subsystem that completely implements the concept of actual infinities without touching on potential infinities, or the other way around. This conclusion indicates that employing both kinds of infinities and related methods of analysis is an intrinsic attribute of modern mathematics and its theoretical foundation. That provides the basis of legality for mixing up the two different kinds of infinities and applying the resultant methods of analysis within the system of mathematics.

Chapter 8

Are actual and potential infinity the same?

What has been discussed so far in this book indicates that the concepts of actual and potential infinite have been studied since antiquity and the question of whether or not they are the same has been debated since the very early part of the recorded history of Western civilization. Therefore, the history can be roughly divided into two segments: The early period that lasted more than two thousand years, and the later period that took place in the past one hundred plus years. During the former period, scholars mostly agreed that actual and potential infinite are different at least in temporal terms, while during the latter period, the boundary between these two kinds of infinite seemed to be blurred. However, the vase puzzle once again rekindled the light of asking whether actual and potential infinite are the same, because the puzzle seems to suggest they are different not only in temporal terms but also in analytical terms. This indication of course is fundamentally different from the purely temporal arguments existing during the first segment of history. In this chapter, we will show that the answer to this age old question is both yes and no, constituting a new crisis in modern mathematics.

This chapter is organized as follows. In Section 8.1, we look at how modern mathematics has treated actual and potential infinites as the same by mathematical induction, its history, and what is implied temporally by the inductive step. In Section 8.2, we look at how modern mathematics has treated actual and potential infinites as different from each other by going over three paradoxes that involve the concept of infinite. Then in Section 8.4, a rigorous logical analysis is given to show that in the system of modern mathematics, the conventions of both that the potential and actual infinites are the same and that they are different are widely employed as the situation of concerns calls for. In reconfirming the need for paradoxes in the development of mathematics, Section 8.4 looks at how paradoxes throughout history have played the role of catalysts that underlay the foundation for the constantly energetic growth of mathematics. Then, the problem of how to modify mathematical induction is considered and the consequent expected future of mathematics is outlined.

8.1 YES, ACTUAL AND POTENTIAL INFINITE ARE THE SAME!

First let us look at a situation where the assumption that actual infinities are equal to potential infinities is held in the system of modern mathematics. Specifically, let us look at mathematical induction.

8.1.1 Mathematical induction

Mathematical induction is a method of mathematical proof typically used to establish that a given statement is true of all natural numbers. It is done by proving that the first statement in the infinite sequence of statements is true, and then proving that if any one statement in the infinite sequence of statements is true, then so is the next one. However, it should not be misconstrued as a form of inductive reasoning, which is considered non-rigorous in mathematics. In fact, mathematical induction is a form of rigorous deductive reasoning. Symbolically, mathematical induction is given in one of the following two versions:

Mathematical Induction (version 1): Assume that $\Phi(n)$ is a proposition regarding natural number n. If the following two steps can be shown,

Initial step: $\Phi(k_0)$ holds true for a certain natural number k_0,
Inductive step: If $\Phi(n)$ holds true for $n = k \geq k_0$, then $\Phi(k+1)$ holds true,

then the proposition $\Phi(n)$ holds true for all natural numbers $n \geq k_0$.

Mathematical Induction (version 2): Assume that $\Phi(n)$ is a proposition regarding natural number n. If the following two steps can be shown,

Initial step: $\Phi(k_0)$ holds true for a certain natural number k_0,
Inductive step: If $\Phi(n)$ holds true for all natural number n satisfying $k_0 \leq n \leq k$, then $\Phi(k+1)$ holds true,

then the proposition $\Phi(n)$ holds true for all natural numbers $n \geq k_0$.

If we imagine that the proposition $\Phi(n)$ stands for a ladder of infinite length, where each step is labeled by one and only one natural number, then the intuition behind the mathematical induction is that if firstly we can somehow get on the ladder at step k_0, and if we have climbed the ladder successfully up to step k ($\geq k_0$), then we can without any trouble climb to the step $k+1$, then we can finish climbing the infinitely long ladder $\Phi(n)$.

In the previously stated mathematical inductions, if in the initial steps we have $k_0 \neq 1$, then we can simply use the substitution $m = n - k_0 + 1$ to make the counting of the steps of the proposition $\Phi(n) = \Phi(m + k_0 - 1)$ to start at $m = 1$. That is why in most textbooks, the initial steps of the mathematical induction starts with $n = 1$.

8.1.2 A brief history of mathematical induction

In 370 B.C., Plato's Parmenides (Cornford, 1939) may have contained an early example of an implicit inductive proof. The earliest implicit traces of mathematical induction can be found in Euclid's proof that the number of primes is infinite.

Theorem 8.1: There are infinitely many prime numbers.

Proof. Suppose that p is a prime number. Now let us consider the following number

$$n = 2 \cdot 3 \cdot 5 \cdot 7 \cdot \ldots \cdot p + 1 \tag{8.1}$$

which is defined as the product of all prime numbers less than or equal to p plus 1. If we can show this number n is prime, then this fact will mean that there is a prime that is greater than p.

To this end, suppose by contradiction that n is not prime. Then there is a number m such that $1 < m < n$ and that m divides n exactly. Let q be the smallest such number. If q were not prime, there then would be a divisor $r > 1$ of q such that $r < n$ and r divides n exactly. That implies that q would not be the smallest such number as assumed to be. Thus, q has to be prime.

Now, if $q \leq p$, then q is one of the primes used in the definition of n above. So n/q would equal an integer plus $1/q$, which is impossible, since q divides n exactly. Thus if n is not prime then there is a prime q greater than p.

Thus no matter whether n is prime or not, there is a prime greater than p. Thus for any prime number $p \geq 2$, there is a larger prime number. Hence there are infinitely many prime numbers. QED.

Another earliest implicit trace of mathematical induction can be found in Bhaskara's, an Indian mathematician and astronomer, cyclic method (Cajori, 1918, pp. 197).

Example 8.1 (Bhaskara's cyclic method). Let us look at the following quadratic equation

$$x^2 = Ny^2 + 1 \tag{8.2}$$

for minimum x and y. If an ordered triple (a, b, k) satisfies the equation

$$x^2 = Ny^2 + k \tag{8.3}$$

then it can be combined with the trivial solution $(m, 1, m^2 - N)$ of equ. (8.3) to obtain the new solution $(am + Nb, a + bm, k(m^2 - \mathrm{N}))$ of equ. (8.3), for any m.

In general, if two solutions (x_1, y_1, k_1) and (x_2, y_2, k_2) of equ. (8.3) are given, then

$$\begin{aligned} k_1k_2 &= (x_1^2 - Ny_1^2)(x_2^2 - Ny_2^2) \\ &= x_1^2x_2^2 - Nx_1^2y_2^2 - Ny_1^2x_2^2 + N^2y_1^2y_2^2 \\ &= (x_1^2x_2^2 + 2x_1x_2y_1y_2 + N^2y_1^2y_2^2) - N(x_1^2y_2^2 + 2x_1x_2y_1y_2 + y_1^2x_2^2) \\ &= (x_1x_2 + Ny_1y_2)^2 - N(x_1y_2 + x_2y_1)^2 \end{aligned}$$

That is, the ordered triple $(x_1x_2 + Ny_1y_2, x_1y_2 + x_2y_1, k_1k_2)$, known as the composition of the original solutions (x_1, y_1, k_1) and (x_2, y_2, k_2) of equ. (8.3), is also a solution of equ. (8.3).

For any given solution (a, b, k') of equ. (8.3), by making an adjustment we can easily obtain another solution (x, y, k') of equ. (8.3) such that $\gcd(x, y) = 1$. Specifically, let $p = \gcd(a, b)$. Then, we can define $x = a \div p$, $y = b \div p$, and $k' = k \div p^2$. Assume that we now start with a solution (a, b, k) of equ. (8.3) such that $\gcd(a, b) = 1$. Then the equation $a^2 = Nb^2 + k$ implies

$$a^2m^2 = Nb^2m^2 + km^2$$

which in turn implies

$$a^2m^2 + N^2b^2 = N(Nb^2 + k) + Nb^2m^2 + km^2 - kN$$

Then we have

$$a^2m^2 + N^2b^2 = Na^2 + Nb^2m^2 + km^2 - kN$$

from which we produce

$$a^2m^2 + 2amNb + N^2b^2 = (Na^2 + 2abmN + Nb^2m^2) + km^2 - kN$$

and

$$(am + Nb)^2 = N(a + bm)^2 + k(m^2 - N)$$

and

$$\left(\frac{am + Nb}{|k|}\right)^2 = N\left(\frac{a + bm}{|k|}\right)^2 + \frac{m^2 - N}{k} \tag{8.4}$$

That is, from the assumption that (a, b, k) of equ. (8.3), it follows that $\left(\frac{am + Nb}{k}, \frac{a + bm}{k}, \frac{m^2 - N}{k}\right)$, known as scaling (a, b, k) down by k, is also a solution of equ. (8.3).

When a natural number m is chosen satisfying that $(a + bm)/k$ is an integer, the other two numbers in the scaled triple $\left(\frac{am + Nb}{k}, \frac{a + bm}{k}, \frac{m^2 - N}{k}\right)$ will be integers, too. Among all such natural numbers m, choose one that minimizes the absolute value of $m^2 - N$. Then the initial solution (a, b, k) of equ. (8.3) is replaced by the scaled triple $\left(\frac{am + Nb}{k}, \frac{a + bm}{k}, \frac{m^2 - N}{k}\right)$. When this process is repeated, it will always terminate with a solution to equ. (8.1), as shown by Lagrange in 1768 (Stillwell, 2002, pp. 72–76). Optionally, when k equals ± 1, ± 2, or ± 4, the process can terminate right away. QED.

To see how the afore-described procedure works, let us look at the following example.

Example 8.2 Determine an integer solution to the following equation

$$x^2 = 61y^2 + 1 \tag{8.5}$$

Solution. To get started, let us make things easy by assuming $y = 1$. So, we can conveniently let $x = 8$ and $k = 3$. That is, the ordered triple $(8, 1, 3)$ satisfies the following equation:

$$x^2 = 61y^2 + k \tag{8.6}$$

Compositing this solution $(8, 1, 3)$ of equ. (8.6) with the solution $(m, 1, m^2 - 61)$ of equ. (8.6) produces a new solution of equ. (8.6) as follows:

$$(8m + 61, 8 + m, 3(m^2 - 61)) \tag{8.7}$$

Now, we scale equ. (8.7) down by 3 to produce

$$\left(\frac{8m + 61}{3}, \frac{8 + m}{3}, \frac{m^2 - 61}{3}\right) \tag{8.8}$$

For $(8 + m)/3$ and $(m^2 - 61)/3$ to be minimal, let us choose $m = 7$ so that the solution of equ. (8.6) is $(39, 5, -4)$, where the new k value is -4. By scaling this solution of equ. (8.6) down by 2, we obtain the rational solution $(39/2, 5/2, -1)$ of equ. (8.6). Compositing this solution with itself twice produces the solutions $(1523/2, 195/2, 1)$ and $(29718, 3805, -1)$ of equ. (8.6). Now, by compositing the solution $(29718, 3805, -1)$ with itself once gives the minimal integer solution $(1766319049, 226153980, 1)$ of equ. (8.5). QED.

An implicit proof by using mathematical induction for arithmetic sequences was introduced in the *Al-Fakhri fi'l-jabr wa'l-muqabala* (*Glorious on algebra*) written by al-Karaji, a 10th century Persian Muslim mathematician and engineer, around 1000 AD. He used an implicitly assumed principle of mathematical induction to prove the binomial theorem and properties of Pascal's triangle. However, neither he nor other ancient scholars explicitly stated the inductive hypothesis. Similarly, Francesco Maurolico in his *Arithmeticorum libri duo* (1575) used the technique of mathematical induction to prove that the sum of the first n odd integers is n^2.

The first explicit formulation of the principle of induction was given by Blaise Pascal (June 19, 1623–August 19, 1662) in his *Traité du triangle arithmétique* (1665). Pierre de Fermat (August 17, 1601–January 12, 1665) made ample use of a related principle, known as indirect proof by infinite descent. The inductive hypothesis was also employed by Jakob Bernoulli (December 27, 1654–August 16, 1705). From then on, mathematical induction became well known. The modern rigorous and systematic treatment of the principle came only in the 19th century with George Boole (November 2, 1815–December 8, 1864) (Grattan-Guinness and Bornet, 1997, pp. 40–41), Charles Sanders Peirce (September 10, 1839–April 19, 1914) (1881, pp. 85–95), Giuseppe Peano (August 27, 1858–April 20, 1932) (the Peano axioms, see his book of 1889, entitled *The Principles of Arithmetic Presented by a New Method*), and Richard Dedekind (October 6, 1831–February 12, 1916) (Cajori, 1918, pp. 197).

8.1.3 What is implied temporally by the inductive step

In theory, what the inductive step in the mathematical induction says is that as long as a proposition $\Phi(n)$ holds true for natural number n, one can show that the proposition $\Phi(n)$ also holds true for $n+1$. That is, what is implied by mathematical induction must be a present progressive tense – a potential infinite in the temporal terms. However, when mathematical induction is used, the positive conclusion is always drawn as that the proposition $\Phi(n)$ holds true for all natural numbers n – an actual infinite, which is a perfect tense in the temporal terms. If, as we know, a present progressive tense is different of any perfect tense, then how can the proof based on the reasoning of mathematical induction, a potential infinite, lead to a conclusion of an actual infinite? Such a jump from potential infinite to actual infinite can only be materialized if mathematicians have implicitly admitted that potential and actual infinities are the same. Since such applications of mathematical induction appear all over the entire system of mathematics, we can conclude that in modern mathematics, the convention that potential and actual infinites are the same has been implicitly assumed and widely accepted.

8.2 NO, ACTUAL AND POTENTIAL INFINITE ARE DIFFERENT!

Based on what has been presented earlier, in this section, which is based (Forrest, Wen, and Panger, to appear), we look at three well-known paradoxes that heavily involve the concept of infinite. By strictly separating actual infinite and potential infinite, these antinomies of yesterday become either veridical or falsidical paradoxes of the present day.

8.2.1 The Littlewood-Ross paradox

(Bollobás, 1986, pp. 26; Byl, 2000) states that each of infinitely many available balls are labeled by one and only one natural number, and is put into an empty urn as follows:

Step 1: At one minute before noon, the balls numbered 1 to 10 are put in, and the number 1 is taken out;
Step 2: At 30 seconds before noon, the balls numbered 11 to 20 are put in and the number 2 is taken out;
Step 3: At 15 seconds before noon, the balls numbered 21 to 30 are put in and 3 is taken out;
……
Step n: At 2^{1-n} minutes before noon, the balls numbered $10(n-1)+1$ to $10n$ are put in and n is taken out.

Question: How many balls are in the urn at noon?

The reason why this situation is of a paradoxical nature is due to the fact that there are an infinite number of steps taken between 11:59 am and 12:00 pm, every ball that is placed into the urn should have been removed at 12:00 pm. Thus there should be zero balls left in the urn at noon. However, in contrast to this solution, the urn gains nine balls every step that is taken, which means that there will be infinitely many balls in the urn at noon. The reasoning is similar to that of the vase puzzle discussed earlier.

At this juncture, a natural question is which answer should be more correct? According to Paul Benacerraf (1962), although the balls and the urn are well-defined at every step-moment prior to noon, no conclusion can be made about the state of the urn at noon. Thus, all that is known is that at noon the urn just somehow disappears. Hence, Benacerraf believes that the progressive procedure in this paradox is under-specified. Additionally, in the eyes of Jean Paul Van Bendegem (1994), the progressive procedure in this paradox is ill-posed, because according to what is stated, infinite many steps have to be performed before noon, and then one wants to know the state of affairs at noon. However, when infinitely many steps have to take place sequentially before noon, then noon becomes a moment that can never arrive. Additionally, to ask how many balls will be left in the urn at noon is to assume that this particular noon actually arrives. Therefore a contradiction is implicitly contained in the statement of the problem, and this contradiction is the assumption that one can somehow complete an infinite progression.

The reason why debate ensues about this paradox is because several concepts are implicitly involved here. The first of such concepts is time, which is used to provide such a feeling for the reader that the moment of completing an infinite progression is actual and as real as the arrival of noon on each day. The second of such concepts is the confusion of actual and potential infinities. In particular, because all steps take place in one-minute period of time, let us consider the intervals $(1, 1]$ and $(0, 1)$, where the first interval says that noon is the endpoint, or in other words, an infinite number of steps will actually be taken and noon will be reached. In this case, we see the occurrence of actual infinity because we acknowledge that noon will be reached after an infinite number of tasks are accomplished. Because there are infinitely many steps between 11:59 am and 12:00 pm, and at each step a ball is taken away from the urn, each and every ball that is placed into the urn is eventually removed. In this specific description, the balls placed into the urn are numbered. So, in order to explain how every ball is removed, consider the second step. Ten balls are added into the urn, namely those labeled 11 through 20, and the ball with label 2 is removed. Running this process infinitely many times before noon is reached means that balls with labels $1, 2, 3, \ldots,$ will eventually be removed from the urn. Thus we are left with no balls in the urn at noon.

For the case of the interval $(0, 1)$, there does not exist an endpoint of time. Here, the symbol ")" indicates that one approaches noon infinitely but will never actually get there as what we have seen in the previous paragraph. Thus, an infinite number of steps occur progressively and noon will not be reached in this case. If we observe Step 1, where balls labeled 1 through 10 are placed inside the urn and ball 1 is taken out, we see that the total amount of balls in the urn is 9, named those labeled 2 through 10. When we consider n steps as n runs to infinity, the amount of balls that remain inside the urn at the end of Step n is $9n$. Therefore, we can equate the number of balls that will still be in the urn at noon by setting it to be the limit of $9n$ as n approaches infinity. Because we know that noon will never arrive from within the interval $(0, 1)$, and each step results in 9 additional balls in the urn. So, when we are asked, "How many balls are in the run at noon?" the answer is infinitely many because 12:00 pm will never arrive during this interval.

According to (Lin, 2008b), time is merely a measurement of movement of materials. When time is employed in the way as in this paradox, one has to confront with the problem of separating what is abstract thinking and what is practically doable, or what is imaginary and what is practical. This end leads to what Van Bendegem's conviction that the particular noon can never arrive. By assuming that this particular noon actually arrives, as pointed out by Van Bendegem, an important contradiction has been implicitly contained in the statement of the problem. On the other hand, if one only allows potential infinity to hold true, he will be led to Benacerraf's conclusion that the progressive procedure in this paradox is underspecified. In short, if we clearly separate what is imaginary and what is practically operable, this Littlewood-Ross paradox disappears naturally. That is, since the particular noon never arrives, the question about what is the state of affairs at the noon no longer arises.

However, if we kick the concept of time out of the picture, the Littlewood-Ross paradox reduces into an equivalent version of the vase puzzle (Lin, 1999). In other words, when what is abstract thinking and what is practically doable, and what is imaginary and what is practical are intertwined together, the so-called Littlewood-Ross paradox appears; and when time is out of the concern, one operates entirely in the imaginary world along with the abstract thinking. So, instead of asking how many balls are in the urn at noon, it would be natural to inquire how many balls are in the urn when the infinite progression finishes, as in the case of vase puzzle. This end represents the fundamental difference between the Littlewood-Ross paradox and the vase puzzle. In particular, when the concept of time is involved, as in the case of the Littlewood-Ross paradox, one is buried in the fight of separating what is imaginary and what is practically possible, when the same situation is phrased without confusing with any realistic matter, as in the case of the vase puzzle, one is naturally led to the profound conclusion that actual and potential infinite (respectively, infinities) are different.

8.2.2 The paradox of Thompson's Lamp

Another well-known infinity-based paradox that has many variations is known as Thompson's Lamp Paradox. According to Geoffrey C. Berresford (1981, pp. 1), the paradox can be explained in the following manner:

> Suppose ... that [a] lamp is off, and I succeed in pressing the button an infinite number of times, perhaps making one jab in one minute, another jab in the next half-minute, and so on, ... After I have completed the whole infinite sequence of jabs, i.e. at the end of the two minutes, is the lamp on or off?

The following analysis shows that although the terminology of this specific description is sufficient, the presented solutions are dependent upon the language of the paradox.

First of all, let us take the time to explain the differences between Thompson's Lamp Paradox and the Littlewood-Ross Paradox. The latter emphasizes that the balls were numbered, and in turn, the ultimate concern is to obtain the number of balls in the urn at a certain time. However, in the case of Thompson's Lamp Paradox, we are not dealing with numbers; instead, the state of the lamp at the end of two minutes is the focus. Even though both paradoxes involve an infinite number of tasks, their differences show that each is unique. The solutions to each paradox are relatively similar, because applying actual and potential infinity to nearly all types of infinity-based paradoxes will result in similar solutions. Even so, the reasons why it is important to pinpoint out the differences is to show that actual and potential infinity will be useful as solutions to a multitude of different infinity-based paradoxes.

As we return to Thompson's Lamp Paradox, the following question naturally arises: which answer is correct (on or off or the lamp is broken before the two minute period is up)? If the third possible outcome is out of consideration, according to William McLaughlin (1997), the process of 'counting the real numbers' is impossible, which shows that 'Thompson's lamp is dysfunctional' as a paradox. He then further indicates that 'one cannot know unlimited natural numbers'. In other words, the human mind cannot accurately fathom the entire collection of natural numbers. If one cannot know unlimited natural numbers, then how can one ever decide at what point to turn the lamp on or off as he or she runs this process to the infinite eternality? That presents a good point; however, this paradox is theoretical in nature; and both theoretically and physically, the process of enlisting natural numbers does not end. His last claim ends with the concept that as two minutes is nearly reached, 'the lamp's behavior is not observable (even if, in some way, it were to be emitting pulses of light'). However, in both reality and theory the lamp's behavior is entirely observable when the time is infinitesimally close to the limit of two minutes. Actually, the state of the lamp stabilizes at a dimmed middle state at this time. To clarify this claim, consider a hummingbird in front of our eyes. The hummingbird's wings are moving so quickly that we observe a blurred span, although the exact position of the wings within that blurred span is unknown. Similarly, the switch is moving so quickly when one is infinitesimally close to two minutes that the exact state of the lamp is unknown. Instead, what is observed is a blurred (dimmed) middle state. In other words, McLaughlin attempts to integrate the real world with the theoretical world, which we will discuss the problems with this later in this paper. With the help of McLaughlin, we can see that this paradox involves two concepts, namely an infinite number of tasks and the distinction between the real and theoretical world.

With this understanding in place, let us now present our solutions to Thompson's Lamp Paradox. Since all steps are taken in two minutes, let us consider the intervals $(0,2]$ and $(0,2)$. The half-closed interval $(0,2]$ says that two elapsed minutes is the endpoint. Or in other words, an infinite number of steps will happen, and two elapsed minutes will be reached. In this particular interval, one observes the application of the concept of actual infinity because we are acknowledging that two elapsed minutes will be reached after an infinite number of tasks is accomplished. Because the lamp's state depends upon the 'jab' of the switch, the state of the lamp is unable to be determined by the time two minutes have elapsed. Even so, in this interval $(0,2]$, we can ensure that it will be either on or off at the end of two minutes (it cannot be both). With respect to McLaughlin's argument, there must be an endpoint of our knowledge if we cannot know infinitely many real numbers. Thus, McLaughlin would probably indicate that the limit of our knowledge results in an eventual end, i.e. the two minutes are reached. The reason this is the case is because we are acknowledging two minutes have elapsed after an infinite number of tasks is completed; so at the end of these two minutes the state of the lamp must be on or off.

Now let us consider the open interval $(0,2)$. In this case, there does not exist any endpoint, where an infinite number of steps will occur while the two-minutes time period will not be exhausted. At this point, we acknowledge that McLaughlin is thinking in terms of the real world. However, this paradox involves the theoretical concepts of infinity as well. Let us now consider step one in the description. At one minute, the lamp will be on. At thirty seconds toward the end of the time frame, the lamp will be off. At fifteen seconds toward to end of the time frame, the lamp will be on, and so on. If this process runs to infinitely many steps, then the state of the lamp will approach a dimmed middle state as it infinitely approaches two elapsed minutes. In other words, the lamp will never be on or off, and we will observe this dimmed middle state as we watch this process run to infinity. Thus, the state of the lamp, in this open interval, stabilizes at a dimmed middle state because this process will run through infinitely many 'jabs' , and two minutes will not elapse.

The reader may ask the difference between the results from the cases of $(0,2]$ and $(0,2)$ because both resulted in an indeterminacy of the state of the lamp. To clarify, in the case of the half-closed interval $(0,2]$, the state of the lamp is indeterminate between either on or off. We are simply unable to determine which of the states of either on or off the lamp will be at when the two minutes time moment is reached; but it must be one of the two because two minutes have elapsed. In the case of the open interval $(0,2)$, the state of the lamp is changing infinitely many times, and two elapsed minutes will never be reached. Thus, in the open interval $(0,2)$, when we approach the end of the time frame indefinitely, the state of the lamp is determinable exactly at the middle between on and off, because as we approach two minutes, the lamp shows a stabilizing dimmed brightness which is neither of the theoretical possible outcomes (on or off).

However, one of the solutions mentioned above is entirely theoretical whereas the other is occurring in reality. The solution on the half-closed interval $(0,2]$ attempts to compare theoretical concepts to real life situations. It presupposes that both an infinite number of tasks will be completed in a finite amount of time. However, that statement contains an explicit contradiction. This comparison between reality and the theoretical world (as seen in McLaughlin's paper) proves to be problematic. Aside from the theoretical solution above, the fallacy of this comparison proves Thompson's Lamp is no longer a paradox, as well. This fallacy of the comparison stems from the fact that time is a part of reality, and time forever continues. If time forever continues (in other words, two minutes will be reached, and so will three minutes, and so on), how can an infinite number of tasks be realistically completed knowing two minutes will be reached? Imagine these two minutes are being timed on a stopwatch, and at each half of the remaining time, the state of the lamp is changed. Once the stopwatch beeps at two minutes, the lamp will either be on or off, and there lies the answer. When that time is reached, we can simply count the jabs up until that point. This is the case because infinitely many 'jabs' is a theoretical concept that cannot be applied to real life situations. As we stay away from this abstract, theoretical world, the universal solution is that the state of the lamp will be either on or off once two real life minutes are reached. But is this truly the case? The general answer is no. To elaborate, the second solution on the open interval $(0,2)$, a stabilizing dimmed middle state, occurs in the real world. As the time infinitely approaches two minutes, the jabs will be occurring at such a high speed that the observer will see a dimmed middle state. In this case, we are acknowledging that these two minutes will not be reached on this interval, while the result is a dimmed middle state as we get infinitely close to the end of the two-minutes time period. Thus, we have just provided two unique solutions, one that is indeterminate in the theoretical world and the other that is determinant and that has a firm foundation in what will actually happen in the real world.

8.2.3 The paradox of Wizard and Mermaid

The final infinity-based paradox that we will evaluate is known as the Paradox of Wizard and Mermaid (Butler, 2006). Although this paradox has some variations, for the sake of this paper, Tetyana Butler presents an adequate, all-encompassing definition as follows:

> An infinitely rich wizard has a mermaid in the pond of his garden. He likes to play with her and gold coins. Every minute he throws two coins in the pond and she throws him back one coin after a half minute. When this game is allowed to run forever, how will the money be distributed? We are allowed to put a label on all of the coins, numbering them 1, 2, 3, 4, etc.

Before we present the solutions to the paradox, it is important to make some preemptive notes. The Mermaid is allowed to place a label upon the coins, which the other player will know because he is a wizard. If the Mermaid chooses not to label the coins, then she will get confused and not be able to obtain as many coins as possible and keep those she desires. However, if the Mermaid is smart, she would label the coins resulting in the opportunity for possessing as many coins as she is interested in, which will be explained in detail later. In the rest of this analysis, we assume that the Mermaid labels the coins and the Wizard knows how she labels the coins and plays along by giving the Mermaid the minimally labeled coins at each step without reusing previously used coins. Also, in all of the following scenarios we assume that the Mermaid start out moneyless; thus, the first minute (the Wizard's first step) results in a two coin gain for the Mermaid (except in Situation 2.1, seen later).

To start off our discussion, let us look at the paradox from two different angles: 1) At each minute, the Wizard gives the Mermaid two coins and in each thirty seconds between whole minutes, the Mermaid returns one coin; and 2) At each minute, the Wizard gives two coins to the Mermaid, and at *every* thirty seconds the Mermaid returns a coin.

Situation 1: In this scenario, we assume that at each minute, the Wizard gives the Mermaid two minimally labeled coins without reusing previously used coins, and at every thirty seconds between whole minutes, the Mermaid returns one coin. That is, the Mermaid only returns one coin at every $n/2$ minutes, where n is an odd natural number but does not equal 1.

Let us assume the Mermaid wants to keep the coin labeled k. Because when this game continues indefinitely, the net gain of coins for the Mermaid increases with the number of the rounds of giving and returning played, there will be such a step in which the wizard has to hand the Mermaid two coins, one of which is labeled k. Then in this case, at each half minute, the Mermaid will only return the coins that are not labeled with k. In particular, if $k = 1$, then the Mermaid receives coins labeled 1 and 2 at the

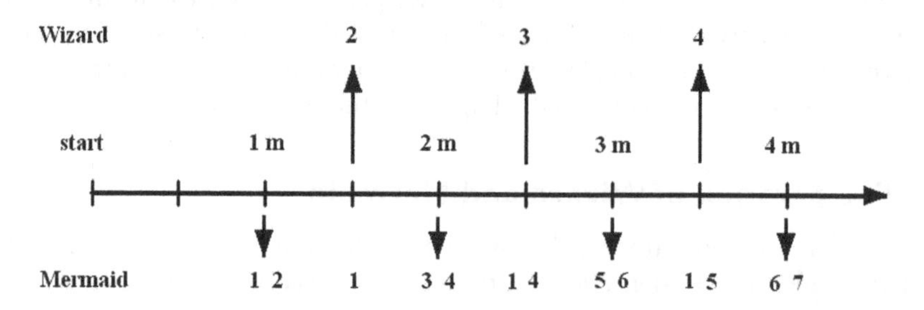

Figure 8.1 How the coins change hands between the Wizard and the Mermaid.

very first minute. She then will return coin 2 while keeping coin 1. At the following step, the Mermaid will obtain two coins neither of which is labeled with 1. Then the Mermaid returns any of the coins in her hands that is not coin 1 at the next step. When this strategy continues, the Mermaid will always possess the coin labeled with 1.

If $k = 2$, then the previous description will apply by simply switching the roles of the coins 1 and 2. Thus, the Mermaid can definitely keep coin 2 as she desires. Now, let us assume $k > 2$. If k is even, then coin k will be handed to the Mermaid at step $k/2$; if k is odd, then coin k will be handed over to the Mermaid at $(k+1)/2$. So, as soon as coin k is handed to the Mermaid, she has the choice to keep that coin and give away one of the other coins she has collected; and she will never run out of coins to give back to the Wizard. Thus, for any coin k, the Mermaid is able to keep it. Figure 8.1 shows the general exchange of coins between the Wizard and the Mermaid.

Next, let us analyze Solution 1 with respect to actual and potential infinities, respectively.

Solution 1.1. Here we assume that the infinite process of the play can actually finish. That is, the process of the coins changing hands will eventually end. In this case, the Mermaid can end up with almost any subset of the Wizard's coins.

Case 1.1.1. Assume that the Mermaid only wants to keep coin k and nothing else. Then this case has been partially explained in the section entitled Situation 1. And as soon as the Mermaid obtains coin k, she can then return the coins in her possession one by one along the order of their labels except k. By doing so, when the play eventually ends, the Mermaid will have coin k only.

Case 1.1.2. Assume that the Mermaid does not want to keep any coin for herself. In this case, all she needs to do is to return the coins in her possession along the natural order of their labels. In particular, at step 1, after coins 1 and 2 are given to the Mermaid, she returns coin 1. At step 2, when coins 3 and 4 are given to the Mermaid, she returns coin 2. At step 3, coins 5 and 6 are given to the Mermaid, she returns coin 3. Then if this process continues indefinitely and eventually ends, then every coin k that has been given to the Mermaid will eventually be given back to the Wizard at step k.

Case 1.1.3. Assume that the Mermaid wants to keep all coins with odd-number labels. In this case, the Mermaid can simply follow the following strategy. At step 1,

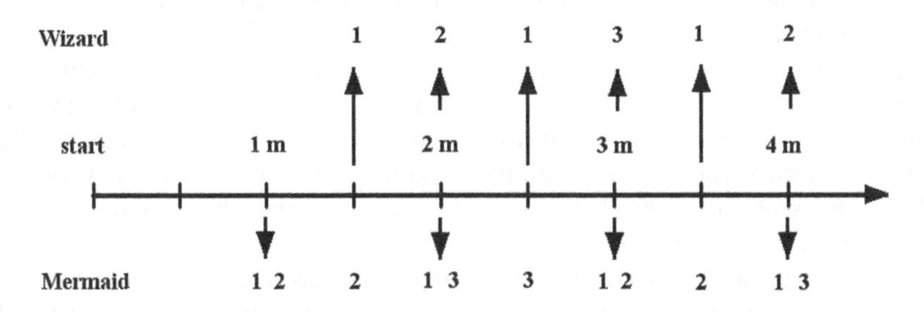

Figure 8.2 Distribution of three coins in the play.

coins 1 and 2 are given to the Mermaid, she then returns coin 2. At step 2, coins 3 and 4 are given to the Mermaid, she then returns coin 4 ... At step $k+1$, when coins $2k+1$ and $2k+2$ are handed over to the Mermaid, she then returns coin $2k+2$. When this play continues progressively and eventually finishes, the Mermaid will end up with all the coins with odd-number labels.

Case 1.1.4. Assume that the Mermaid wants to keep a particular subset X of all the coins N the Wizard starts off the game with such that if $x \in X$, then $x-1$, $x+1 \notin X$. That is, no pair of consecutive natural number labels belongs to X. For the sake of convenience of our communication, assume $N = \{1, 2, 3, \ldots, n, \ldots\}$, the set of all natural numbers, and $X = \{n_1, n_2, \ldots, n_k, \ldots\}$, a subset of N, such that $n_1 < n_2 < \ldots < n_k < \ldots$ Then, from the discussion in Case 1.1.1, it follows that the Mermaid can simply return coins in the natural order of their labels by skipping the coins with labels $n_1, n_2, \ldots, n_k, \ldots$

Solution 1.2. Here we assume that the play is a potentially progressive process, which can never be completed. That is, the process of the coins changing hands will never end. In this case, similar to what is discussed in the vase puzzle, the Mermaid will collect infinitely many coins as the play continues.

Situation 2. At each minute, the Wizard gives two coins, which can be reused in later steps, to the Mermaid and the Mermaid returns a coin *every* thirty seconds.

Case 2.1. Assume that no swap is allowed. For example, giving coin 2 and returning coin 2 simultaneously are not permitted. In this case, the Wizard does not have to be infinitely rich to play this game. Instead he only needs to have three coins to play the game indefinitely. So, if the game is interrupted at any moment in time, the Mermaid would end up with either one or two coins. In particular, the Mermaid could end up with coin 1, or coin 2, or coin 3 or one of the coin combinations of $\{1, 2\}$, $\{1, 3\}$, and $\{2, 3\}$, see Figure 8.2. However, if the game continues progressively either with or without an end, in other words, either an actual or potential infinite is assumed, the Mermaid will not know for sure what might be in her possession. That is, no matter if we assume the actual infinity or the potential infinity, the answer regarding the distribution of coins is indeterminate.

This situation is similar to the case of Thompson's Lamp Paradox with the following differences: 1) This current process can never actually finish unless additional conditions are imposed because no fixed time moment is chosen in the original description (we are allowing this process to run forever); 2) There is no middle state for the infinite process to converge to. Additionally, if the coins are not labeled and the process runs to infinity, the Mermaid can only end up with either one or two coins of her hands. In this scenario, even if the Wizard is infinitely rich, he does not have to use every coin to play this game.

Case 2.2. Assume that the exchange conditions are the same as in Situation 2, however, swaps occur. So, at the Wizard's first step, he will give the Mermaid two coins, while the Mermaid simultaneously returns one. At the Mermaid's next exchange, halfway between whole minutes, she will return the second coin. This results in the Mermaid having zero coins in her possession at the time of a minute and a half. This process will repeat and leave the Mermaid with zero coins. However, depending on when the process ends, assuming it does, the Mermaid will have either zero coins or one coin in her possession. In this case, no matter whether we assume actual infinity and potential infinity, the answer regarding the distribution of coins is indeterminate.

In summary, what is discussed above indicates that the concept of actual infinite in general is different of that of potential infinite.

8.3 A PAIR OF HIDDEN CONTRADICTIONS IN THE FOUNDATIONS OF MATHEMATICS

The purpose of this section, which is based on (Lin (guest editor), 2008a, pp. 438–445), is to show the existence of a deeply hidden pair of contradictions in the system of modern mathematics. It is indicated in Section 7.4 that it is an intrinsic attribute of modern mathematics and its theoretical foundation to mix up the intensions and methods of two different thoughts of infinities. That provides the basis of legality for using the methods of analysis, produced by combining the two kinds of infinities, in the study of the modern mathematical system. In this section, by precisely employing the method of analysis of mixing up potential and actual infinities, we card the logical and non-logical axiomatic systems for modern mathematics. The outcome of our carding implies that in modern mathematics and its theoretical foundation, some axioms implicitly assume the convention that each potential infinity is equal to an actual infinity, while some other axioms implicitly apply the belief that "each potential infinity is different of any actual infinity." That is, by using the concepts of potential and actual infinities, we uncover two contradictory thinking logics widely employed in the study of mathematics.

8.3.1 Implicit convention #1 in modern mathematical system

As is well known, the modern axiomatic set theory, which is usually denoted briefly as ZFC, is the theoretical foundation of the entire modern mathematics. It plays the role of the supporting walls of the mathematical mansion. The set of axioms, used to establish the ZFC system, can be grouped into two categories: One category contains those axioms called logical axioms. These axioms come from the manipulation system of logic introduced as the tools of reasoning in set theory. And, generally, the manipulation system of logic is divided into two blocks: manipulation of propositions and that of predicates. The second category of ZFC axioms contains those axioms called non-logical axioms. These axioms are about how to construct sets or the existence of certain specific sets. Therefore, in terms of the starting points and the conventions of unproven beliefs of the modern mathematics and its theoretical foundation, one should look at the contents in the following several areas:

1 Logical axioms: They are those axioms, laws of reasoning, and conventions on how to interpret quantifiers, widely employed in the manipulations of propositions and predicates;
2 Non-logical axioms: They are those postulates on how sets are constructed and what sets exist in the theoretical system of set theory; and
3 Techniques and methods of reasoning in deductions: They include mathematical inductions, transfinite inductions, proof by contradiction, etc. Of course, the validity of these techniques and methods of reasoning can also be derived within the ZFC system.

These logical, non-logical axioms, and techniques and methods of reasoning are all clearly established and stated inside the system. However, in this section, we will base our exploration on the two compatible thinking modes and methods of analysis of both kinds of infinities; and we will illustrate several implicit conventions on the thought deeply hidden in the axioms and methods as listed above.

Firstly, let us look at the law of introducing the universal quantifier ($\forall+$) and the convention on how to interpret this universal quantifier ($\forall$).

In the manipulation system of predicates of the ZFC, there are two quantifiers. One of them is called the universal quantifier, denoted $\forall$. Its interpretation is "for all" or "for each", and for each object$x$studied in our discourse, the phrase "for each x ..." is identical to "for all x ...". The symbol $\forall$ for the quantifier came from the first letter in the word "all".

The second quantifier in the manipulation system of the ZFC is called the existential quantifier, denoted $\exists$. This quantifier is interpreted as "there is" or "there are". The symbol $\exists$ of this existential quantifier came from the first letter of the word "exist".

In the manipulation of predicates, there is a law of reasoning, known as the law of introducing universal quantifiers, denoted ($\forall+$). If we use the language of logical manipulation, we have:

$$\Gamma \vdash A(a),$$

and if a does not appear in Γ, then we have

$$\Gamma \vdash \forall x A(x).$$

What do all these expressions mean? In each chosen text book on logic, the meaning of ($\forall+$) would be given briefly as follows. The symbol ($\forall+$) implies the thinking logic of reasoning employed in manipulations: If for any chosen entity a out of the universe of a scientific discourse, under a certain premise one can always deduce the fact that a satisfies the property A, then he can conclude that under the same premise, every entity in the universe satisfies the property A. Of course, it has to be emphasized that the choice of the entity a in the universe of discourse is arbitrary without any restriction. Especially, the choice of the entity is not constrained by the premise Γ with which it was shown that the entity satisfies the property A. That is, in ($\forall$+), it is clearly assumed that "a does not appear in Γ."

For example, when proving the proposition that "all points on the line perpendicular to a line segment and passing through the midpoint of the segment are equal distant from the endpoints of the segment", one chooses an arbitrary point on the line, proves it is equal distant from the endpoints of the segment, and then concludes that all points on the line, which is perpendicular to the segment and passes through the midpoint of the segment, are equal distant from the endpoints of the segment. Of course, as for the choice of the arbitrary point, other than the condition that it has to be on the specified line, there is no other restriction. Therefore, controlled by this thought of reasoning, it is inevitable for one to interpret the universal quantifier $\forall$ as either "for each" or "for all". And, consequently, "for each x, a certain conclusion holds true" now is equivalent to "for all x that certain conclusion holds true."

At this juncture, we need to point out: When one says to select an arbitrary entity a out of an infinite universe of discourse, or he talks about each entity from the universe, as we have discussed in Section 7.3, he is talking about a procedure of listing. And, any non-terminating procedure of listing, before it is finished, is forever a present progressive tense (ongoing). Or in other words, only when a certain procedure of listing is completed, one has a perfect tense (done). But, when a procedure of listing is finished, one will not talk about "for each" or "for any". Instead, he will talk in perfect tense about "for all" or simply "all".

That is, the phrases "for each" or "for any" are used before a procedure of listing is exhausted, where what one faces is only a potential infinite. However, the phrase "for all" in the interpretation of the ($\forall$+) reasoning and the quantifier $\forall$ of course must be about the perfect tense after a procedure of listing has been finished. So, in this case, what one faces must be an actual infinite. That is, the convention of reasoning applied in ($\forall$+), where the result true for an arbitrary entity holds true for all entities in the universe of discourse, and that where "for each" equals "for all" as in the interpretation of $\forall$, must implicitly carry on such a law of thought that each present progressive tense is equal to a perfect tense. Or in other words, each potential infinite is equal to an actual infinite.

Secondly, let us look at the concepts of cardinalities and ordinal numbers of infinite sets under Cantor-Zermelo's meaning. For an arbitrarily given infinite set $A = \{x|P(x)\}$, because it is made after collecting all the objects satisfying the property P, the existence of the set A must stand for a perfect tense. Even if one generates A by going through the procedure of checking each and every object x to see if it satisfys the property P, he must also first exhaust all such objects x before producing A. Therefore, what the existence of A faces must be an actual infinity. That is why all people know that the cardinality of any infinite set in set theory is a cardinality of actual infinite.

The so-called cardinality of a set is simply defined as how many elements the set A contains. Because of this reason, the set of all natural numbers $N = \{x|n(x)\}$, where $n(x)$ stands for the property that "x is a natural number", should not be an exception. That is, N is an actual infinite set; and what Nfaces is an actual infinity. The set $N = \{x|n(x)\}$ has infinitely many of elements.

In mathematics, the symbols ∞ or ω are often used to represent an infinity. So, we say that N has ω elements, which means that the set N of all natural numbers has infinitely many elements. Now, we order all the natural numbers in N from the smaller to the larger as the following natural number sequence λ: $\{1, 2, 3, \ldots, n, \ldots\}$, and let k be a variable that is defined on N and increases without bound over N. Then, the variable k can increase and approach ω without bound. Since in mathematics it is assumed that the magnitude of each natural number is finite, that is, each natural number is less than ω, we conclude that ω is not a natural number. Therefore, the process for the variable k to approach ω is forever a present progressive tense (ongoing), and what faces the natural number sequence λ is a potential infinite, as analyzed here. Since λ is nothing but an ordered list of the natural numbers in N, both N and λ represent the same set. However, as just discussed, when looking at the cardinality of N, it means an actual infinity. When looking at the order in λ, it means a potential infinity. That is, in studying the set of all natural numbers, what is implicitly implied is the thinking convention that each potential infinite is equal to an actual infinite.

What needs to be pointed out is that in our discussion above, the set N of all natural numbers is used as our platform. This choice of the set N is only for the purpose of making our discussion easier to follow and ready to be understood. As a matter of fact, since the ZFC system accepts the axiom of choice, the theorem of well-ordering any given set holds true, and for any well-ordered set A and any $x \in A$, the ordinality of $\{y|y < x\}$ is less than the ordinality of A. So, as long as one has the relevant knowledge, he can generalize our discussion above about N to any actual infinite set $A = \{x|P(x)\}$. That is to say, all the conventions and thoughts related to the concepts of cardinalities and ordinalities of infinite sets, as studied or employed in the modern axiomatic set theory, imply implicitly the thinking convention that each potential infinite is equal to an actual infinite.

Thirdly, let us once again look at mathematical induction. As is known, the inductive step in mathematical induction means that as long as a criterion a or property P holds true for n, then one can prove in theory that the criterion a or the property P holds true for $n+1$. This fact implies that for the matter of judging whether or not the criterion a or the property P holds true in general, one only needs to go through such a process that no matter what boundary is chosen, he can always go beyond this boundary. And, no matter how great the boundary N is, he can definitely go beyond this boundary N. That is, what's implied by this inductive step must be a present progressive tense (ongoing) and a potential infinite. However, by applying mathematical inductions, one concludes in general based on the inductive step that the criterion a or the property P holds true for all $n \in N$. Evidently, the conclusion like that the criterion a or the property P holds true for all $n \in N$ should only be drawn after the procedure of transferring from n to $n+1$ is completely exhausted. So, what the conclusion drawn on mathematical induction faces must be an actual infinity of the perfect-tense kind. Therefore, we have to ask: If, as usual, we know that each present progressive tense (ongoing) is not the same as any perfect tense (done) and each potential infinite is different of any actual infinite, then from the inductive proof reasoning on the basis of potential infinite, how can an inductive conclusion of actual infinite be drawn?

Only if scholars have implicitly admitted that each present progressive tense (ongoing) is equal to a perfect tense (done), or that each potential infinite is the same as an actual infinite, then they can materialize such a jump from an inductive proof of potential infinite to an inductive conclusion of actual infinite. Therefore, the legality and the wide-spread application of the method of reasoning as implied in mathematical induction are also supported by the thinking convention of implicitly accepting the convention that potential infinities are the same as actual infinities.

Similar thinking conventions implicitly employed can still be detailed. However, for our purpose, what we have done above is more than enough. Any theoretical system, if it can truly throughout hold on to the thinking convention as described in the previous paragraphs or applies such a thinking convention as an axiom without running into difficulties, can surely be called a self-sustaining system or a consistent system. However, in the following section, we will show that in the system of modern mathematics and its theoretical foundation, the implicit thought convention that potential infinities are the same as actual infinities, as discussed above, is not implemented consistently and throughout.

8.3.2 Implicit convention #2 in modern mathematical system

In the beginning of the 1980s before we constructed our medium logic system ML and the medium axiomatic set theoretic system MS, we once studied the thought convention of that as outlined in the "principle of the excluded middle" by tracing back the history as far as we could go. For our purpose here, there is a need to highlight some of our discoveries along this line.

Since the time of Aristotle (1984), formal logic has distinguished contrary opposites and contradictory opposites. If two concepts both contain their own affirmative contents and in a higher level concept of the same intension, there exists the largest difference between the two concepts, then these two concepts are seen as contrary opposites. For instance, good and evil, beautiful and ugly, men and women, true proposition and false proposition, potential infinities and actual infinities, etc., are all examples of contrary opposites. When the intension of one concept is the negation of the intension of another concept, these two concepts are known as contradictory opposites. For instance, working and not working, capital and not capital, men and not men, true proposition and not true proposition, actual infinities and not actual infinites, beautiful and not beautiful, good and not good, etc., are all examples of contradictory opposites.

Let us now assume P is a predicate (a concept or a property). If for any object x, the object x either satisfies P or does not satisfy P completely, that is, there does not exist such an object that partially satisfies P and partially does not satisfy P, then we say that P is a distinguishable predicate, denoted briefly dis P. And, if for another predicate P, there exists a certain object x that both satisfies and does not satisfy P partially, then P is called a fuzzy predicate, denoted fuz$P$. We will name the symbol $\sim$ as a fuzzy negation, which is interpreted as "partially". So, $\sim P(x)$ stands for the fact that the object x partially possesses the property P, while $P(x)$ means that the object x possesses the property P completely. The symbol ╕ we will name for opposite negation, which is interpreted as "opposite of", and the contrary opposite of P is denoted by ╕P. That is, we will employ the symbols P and ╕P to represent a pair of contrary opposite concepts, while the symbols P and $\neg P$ are employed to represent a pair of contradictory opposite concepts. As is well known, in the classical two-value logic, the formal symbol $\neg$ is known as the negation and is interpreted as "not".

Now, let us have a pair of arbitrary P and $\urcorner P$. If an object x satisfies $\sim P(x) \wedge \sim\urcorner P(x)$, then the object x is called a medium object of P and $\urcorner P$. This concept reflects what's often said in philosophy that "also this also that", where "this" means P and "that" $\urcorner P$. For example, the dawn is the medium for the dark night to transfer into daylight; 0 is the neutral number, neither positive nor negative; semi-conductor a medium between conductor and insulator, among others. As is known, in epistemology, there is such a basic principle that between the opposites there is always a medium object. The concept of opposites here stands for contrary opposite concepts P and $\urcorner P$.

However, in the classical two-value logical and classical mathematics, other than fuzzy properties and fuzzy concepts, which exist and are recognized universally in our daily lives, excluded in their research consideration the existence of medium objects, especially under suitable restrictions placed on the universe of discourse. Furthermore, in their chosen universe of discourse, the contrary opposites $(P, \urcorner P)$ and contradictory opposites $(P, \neg P)$ are treated as identical so that $\neg P$ is the same as $\urcorner P$. That is where such conclusions as "not beautiful is ugly," "not good is evil," etc., are from. Therefore, under the dominant influence of the thought of two-value logic, it is natural to have the thinking method as $\urcorner P \equiv \neg P$. For example, a true proposition P and the false proposition $\urcorner P$ are themselves a pair of contrary opposite concepts $(P, \urcorner P)$. But, in the propositional universe of discourse of the two-value logical thinking, there has existed such a belief that a not true proposition must be a false proposition. In the universe of discourse of all human beings, there appeared such a conclusion that not a man must be a woman. Similarly, when the concept of infinities is concerned, there must appear such conclusions that not an actual infinity must be a potential infinity, and that not a potential infinity must be an actual infinity. Speaking abstractly, within the framework of the two-value logical calculus, when a concept on a higher level of the contrary opposites $(P, \urcorner P)$ is restricted, in the universe of discourse, there must exist the thought convention that $\neg P \equiv \urcorner P$. For more detailed discussion on medium principle, please consult (Zhu and Xiao, 1996).

Modern mathematics has laid its foundation on the top of the modern axiomatic set theory. And in the modern set theory, no matter whether it is in the ZFC system or the NGB system, the two-value logical calculus is employed as the tool of deduction. So, all the rules of deduction and axioms of the classical two-value logical calculus are inherited by the modern mathematics and its fundamental research. Therefore, the rule of proof by contradiction $(\neg)$ of the two-value logical calculus is validated. Symbolically, the rule of proof by contradiction can be represented as:

$$\text{If } \Gamma \cup \{\neg A\} \vdash B, \quad \text{then } \neg B \Rightarrow \Gamma \vdash A,$$

or

$$\text{If } \Gamma \cup \{A\} \vdash B, \quad \text{then } \neg B \Rightarrow \Gamma \vdash \neg A,$$

where A is the proposition one wishes to either disprove or prove, and Γ the set of statements or the premises on which the discussion takes place, which could be, for example, the axioms of the theory one is working in, or earlier theorems he can build upon. The practical essence of the rule of proof by contradiction is the often-used method of deduction, known as proof by contradiction, in deductive reasoning. Within the system of two-value logical calculus, starting from the rule of proof by contradiction ($\neg$), one can prove the principle of the excluded middle: $\vdash A \vee \neg A$, and the principle of no contradiction: $\vdash \neg(A \wedge \neg A)$. Also, from the previous discussion, it follows that within the framework of two-value logical calculus, both contrary opposites ($P, \urcorner P$) and contradictory opposites (P, $\neg P$) coincide and become one. So, $\urcorner P$ is identical to $\neg P$. Therefore, the potential infinities (*poi*) and actual infinities (*aci*), as the opposites in a pair of contrary opposites ($P, \urcorner P$), must be a pair of contradictory opposites (P, $\neg P$). So, in the framework of the two-value logic, both potential infinities (*poi*) and actual infinities (*aci*) must represent the affirmative (A) and the negative aspect ($\neg A$) of a concept or fact and have to satisfy the principles of the excluded middle and no contradiction. That is, we have

$$\vdash poi \vee aci \quad \text{and} \quad \vdash \neg(poi \wedge aci).$$

That is, we can show $poi \neq aci$ in the same way as that if A and $\neg A$ have to satisfy both the principle of the excluded middle $\vdash A \vee \neg A$ and the principle of no contradiction $\vdash \neg(A \wedge \neg A)$, then it is impossible to derive $A \equiv \neg A$. What is discussed here indicates that in modern mathematics and its theoretical foundation, from the rule of proof by contradiction ($\neg$), it follows that the thought convention that "potential infinities are different of actual infinities" has to be implemented implicitly.

8.3.3 Some plain explanations

In this section, by employing the analysis method of allowing both kinds of infinities, we carded the axioms that are both logical and non-logical and the commonly applied deduction method system in modern mathematics and its theoretical foundations. The results of our carding indicate that some of the axioms and methods have implicitly employed the thought convention that "potential infinities are the same as actual infinities", while some other axioms carried on the convention that "potential infinities are not the same as actual infinities". This result of our analysis means that there is a need to study at a higher level the problem of compatibility between modern mathematics and its theoretical foundation. And, since this deeply hidden problem is revealed at the technical level of mathematical logic, a resolution should be eventually produced to solve this problem. However, what needs to be pointed out is that this kind of contradiction, deeply hidden in mathematical logic, is impossible to appear in the calculus system of the classical two-value logic. It is because no matter whether one starts with pure logical axioms or directly with non-logical axioms to describe the hidden contradiction as what is explained in this section, he in fact has touched on a series of impure logical concepts, such as potential infinities, actual infinities, progressive tense, perfect tense, reachability and non-reachability, listing and exhausting, etc. This end implies that we should explore the theory of limits and set theory in more depth in order to find other possible hidden contradictions. Also, our results obtained in this section essentially show that even though in the system of logic, it is often possible to construct proofs for consistency, soundness and completeness, in systems of impure logic, it is as of now still difficult to accomplish these tasks.

At this juncture, we should point out that the thinking logic and the reasoning method of coinciding potential infinities (*poi*) with actual infinities (*aci*) are inherently part of modern mathematics, for details, see Section 7.3, and that our discussion in the previous section tells us the fact that the thought convention of potential infinities (*poi*) being not the same as actual infinities (*aci*) is also an inherent part of modern mathematics. On the other hand, these results mean that the thought and method of equating *poi* and *aci* and the thought convention of *poi* ≠ *aci* are not artificially added into the system of modern mathematics and its theoretical foundation. Therefore, when we further study the problems of validity and consistency of the modern system of mathematics, we have to acknowledge the fact that there exists evidence within the system to warrant the thought convention of *poi* ≠ *aci* and that it is beyond any reproach to question the thinking logic and method of reasoning that equate *poi with aci.*

8.4 ROLE OF PARADOXES IN THE DEVELOPMENT OF MATHEMATICS

At this juncture, because of the logical reasoning and examples regarding the difference between actual and potential infinites in the previous sections, it is clear by now that these two kinds of infinites should be both temporally different and analytically distinct. However, in the modern system of mathematics, they have been treated indifferently, constituting a new crisis in the foundations of mathematics. To this end, two natural questions arise: How wide ranging is this crisis within mathematics? Will such a crisis affect mathematics adversely? In this section, we will address these two questions one by one, first showing the fact that paradoxes throughout the history have been one of the catalysts that stimulate the development of mathematics, and then showing by listing two examples on how mathematical induction has been widely employed in the constructive proofs of existence throughout the entire spectrum of mathematics.

8.4.1 Paradoxes in the history of mathematics

As what has been discussed in the previous sections, other than cajoling, provoking, amusing, exasperating, and seducing, more importantly paradoxes arouse curiosity, stimulate, and motivate, and attract attention. As a matter of fact, throughout the history of mathematics, paradoxes have inspired the clarification of basic concepts, the introduction of major results, and the establishment of various branches of inquiry. Based on the discussion in the previous section, the term "paradox" in section means, in a broad sense, an inconsistency, a counter-example to some widely held notion, a misconception, a true statement that seems to be false, or a false statement that seems to be true. It is in these various senses, for more details, see (Kleiner and Movshovitz-Hadar, 1994), that paradoxes have played an important role in the evolution of mathematics. Indeed, it has been the misunderstandings and seemingly unresolvable difficulties of the past that have always provided the opportunities of future development for mathematics (Bell, 1945, pp. 283), even though the misunderstandings and difficulties might seem paradoxical and challenging at the time; and one of the endlessly alluring aspects of mathematics is that its thorniest paradoxes have a way of blooming into beautiful theories (Davis, 1964, pp. 55).

In the rest of this subsection, we will look at some of the paradoxes that have had helped to shape the concept of numbers, curves, and continuous function, the meaning of logarithms of negative and complex numbers, properties of power series, equi-decomposition of volumes, etc.

8.4.1.1 Paradoxes related numbers

Throughout the development of numbers, there appeared the following paradoxes among others:

Number Paradox # 1: Negative numbers are greater than the positive infinity.

In the 17th century, Wallis questioned the concept of negative numbers as follows (Nagel, 1935): "[How can] any magnitude ... be less than nothing, or any number fewer than none?" He then proved the conclusion in Number Paradox # 1 as follows. Because for any chosen positive number a, $a/0 = +\infty$, $a \div$ a negative number has to be $> \infty$, because decreasing the denominator increases the fraction.

Number Paradox # 2: $\frac{1}{-1} \neq \frac{-1}{1}$.

In a letter to Leibniz, Arnauld, a 17th-century mathematician and philosopher, proved the inequality in Number Paradox # 2 as follows: The ratio of a greater quantity to a smaller quantity has to be greater than the ratio of a smaller to a greater. Leibniz agreed with the argument. At the same time, he asked for the tolerance of negative numbers, because they are useful; and, in general, they lead to consistent results (Cajori, 1913, pp. 39–40).

Number Paradox # 3: Real numbers can be written as meaningless (in the 16th century sense) complex numbers.

During the 16th century, as one of the great mathematical achievements of the time, Cardan established the solution of the cubic equation $x^3 = ax + b$ as follows:

$$x = \sqrt[3]{\frac{b}{2} + \sqrt{\left(\frac{b}{2}\right)^2 - \left(\frac{a}{3}\right)^3}} + \sqrt[3]{\frac{b}{2} - \sqrt{\left(\frac{b}{2}\right)^2 - \left(\frac{a}{3}\right)^3}}$$

Because applications of this formula on equations like $x^3 = 15x + 4$ introduces square roots of negative numbers, which he refused to recognize as legitimate mathematical entity, Cardan denied the possibility of his formula being applied to these cases.

Even so, Bombelli applied the formula to the specific equation $x^3 = 15x + 4$ and obtained $x = \sqrt[3]{2 + \sqrt{-121}} + \sqrt[3]{2 - \sqrt{-121}}$. At the same time, Bombelli noted by inspection that x = 4 is a solution of $x^3 = 15x + 4$ and that the other two solutions $-2 \pm \sqrt{3}$ are also real. Therefore, Number Paradox # 3 appeared.

"The whole matter seemed to rest on sophistry rather than on truth," noted Bombelli (Leapfrogs, 1980, pp. 2). To resolve this sophistry, he consequently developed the theory of complex numbers, although it took two and half centuries for complex numbers to be accepted as genuine mathematical entities.

8.4.1.2 *Paradoxes related logarithms*

In terms of the meaning of logarithms of negative and complex numbers, the early paradoxes arose in the first part of the 18th century in connection with integration.

Logarithm Paradox # 1: $\log x = \log(-x)$. In particular, $\log(-1) = \log 1 = 0$.

This difficulty appeared along with the following integration by using an analogy with the real case. In particular, Johann Bernoulli integrated $1/(x^2 + a^2)$ as follows:

$$\int \frac{dx}{x^2 + a^2} = \int \frac{dx}{(x + ia)(x - ia)} = -\frac{1}{2ai} \int \left(\frac{1}{x + ia} - \frac{1}{x - ia} \right) dx$$

$$= -\frac{1}{2ai} [\log(x + ia) - \log(x - ia)] = -\frac{1}{2ai} \log \frac{x + ai}{x - ai}$$

Then, in an exchange of letters with Leibniz, both men argued about the meaning of $\log[(x + ai)/(x - ai)]$ and whether or not $\log(-1)$ is real or imaginary. While Bernoulli asserted that $\log(-1)$ was real, Leibniz claimed it is imaginary.

To support his point, Bernoulli argued that because

$$\frac{dx}{x} = \frac{d(-x)}{-x}, \quad \int \frac{dx}{x} = \int \frac{d(-x)}{-x}$$

The conclusions in Logarithm Paradox # 1 had to be true.

Logarithm Paradox # 2: $\log(-1) = -2 - 4/2 - 8/3 \ldots$ must be imaginary.

Against Bernoulli's previous argument, Leibniz believed that $\log(-1)$ must be imaginary based on the following reasoning:

(i) Because the range of the function $\log a$, for $a > 0$, is all real numbers, it follows that $\log a$, for $a < 0$, must be imaginary, because all the real numbers have already been used up.
(ii) Because $\log i = \log(-1)^{1/2} = ½\log(-1)$, if $\log(-1)$ were real, then so would be $\log i$. But this is clearly absurd, as alleged Leibniz. Now,
(iii) Letting $x = -2$ in the expansion

$$\log(1 + x) = x - \frac{x^2}{2} + \frac{x^3}{3} - \ldots$$

yields $\log(-1) = -2 - 4/2 - 8/3 \ldots$ Since the series on the right diverges, it cannot be real, hence must be imaginary.

These two paradoxes were eventually resolved by Euler in one of his 1749 papers, while he noted that "for a long time … [this paradox has] tormented me" (Marchi, 1972, pp. 72). The key for Euler to have finally resolved this paradox was his formula $e^{i\theta} = \cos\theta + i\sin\theta$, which implies that $e^{i(\pi+2n\pi)} = \cos(\pi+2n\pi) + i\sin(\pi+2n\pi) = \cos\pi + i\sin\pi = -1$. So, one has $\log(-1) = i(\pi+2n\pi)$, where $n = 0, \pm1, \pm2, \ldots$ That is, $\log(-1)$ is multi-valued and all its values are complex.

8.4.1.3 Paradoxes related continuity

In terms of the concept of continuous functions, Euler (Edwards, 1979, pp. 301) defined in the 18th century that a continuous function was one that was given by a single expression (formula). So, one has the following

Continuity Paradox # 1: The absolute value function $f(x) = |x|$ is discontinuous, although one could draw the graph of the function without lifting his pen.

The reason for this paradox to appear is that the absolute value function $f(x)$ is defined as follows by two expressions:

$$f(x) = |x| = \begin{cases} x, & x > 0 \\ -x, & x \leq 0 \end{cases}.$$

Continuity Paradox # 2: The function $f(x) = 1/x$, comprising the two branches of a hyperbola, is continuous, although the graph of the function contains two disconnected branches.

It is because this function is written in a single expression.

These paradoxes did not become a problem until the work on Fourier series showed the untenability of this Euler's definition of continuous functions. In particular, the following function, defined by using three expressions,

$$g(x) = \begin{cases} -1, & -\pi < x < 0 \\ 0, & x = 0 \\ 1, & 0 < x < \pi \end{cases}$$

could be represented by a single expression, namely its Fourier series, as follows:

$$g(x) = \frac{4}{\pi}\left(\frac{\sin x}{1} + \frac{\sin 3x}{3} + \frac{\sin 5x}{5} + \ldots\right), \quad \text{for all } x \in (-\pi, \pi)$$

Hence, this function $g(x)$ was both continuous and discontinuous in the 18th-century sense of the concept. In an effort to reappraise and reorganize the foundation of the 18th-century calculus, in an 1821 work, Cauchy defined continuity of functions essentially the same way as we learn it today, where the now-accepted $\varepsilon - \delta$ formulation was later given by Weierstrass in the 1850s. However, as it turned out, the concept of continuity was very subtle and not fully understood even by Cauchy and his contemporaries of the early to mid-19th century. For example, Cauchy proved the following:

Continuity Paradox # 3: Each infinite sum of (a convergent series of) continuous functions is a continuous function (Bottazzini, 1986, pp. 110).

However, in the 1820s Abel constructed a counterexample essentially showing that the function

$$\sum_{n=0}^{\infty} \frac{\sin(2n+1)x}{2n+1}$$

is discontinuous at $x = k\pi$, $k = 0, \pm1, \pm2, \ldots$ Looking back to Cauchy's original proof of Continuity Paradox # 3, it can be seen that where Cauchy made a mistake is that he failed to distinguish between convergence and uniform convergence of a series of functions. As a matter of fact, such distinction was not made until the 19th century (Remmert, 1991, pp. 97).

Continuity Paradox # 4: Continuous functions are differentiable and not differentiable at the same time.

Those continuous functions as defined respectively by Euler and Cauchy turned out to be differentiable (except possibly at some isolated points) (Volkert, 1989). That explained why it was astonishing for the community of mathematicians when Weierstrass in the 1860s constructed an example of the following continuous function that is nowhere differentiable,

$$f(x) = \sum_{n=1}^{\infty} b^n \cos(a^n \pi x)$$

where a is an odd integer, b a real number in $(0, 1)$, and $ab > 1 + (3\pi/2)$. This example showed for the first time in history that the concept of continuity, as defined by Euler and Cauchy, is considerably broader than that of differentiability.

On the other hand, in the 1940s, Schwartz and Sobolev showed that every continuous function is indeed differentiable, if the concept of continuity is defined differently. In particular, if the function $f(x)$ is given as follows,

$$f(x) = \begin{cases} 1, & x > 0 \\ \frac{1}{2}, & x = 0 \\ 0, & x < 0 \end{cases}$$

then its derivative is given by

$$f'(x) = \begin{cases} 0, & x \neq 0 \\ \infty, & x = 0 \end{cases}$$

which is the Dirac delta function $\delta(x)$. What this example also shows is that even some discontinuous functions in the Cauchy's sense are differentiable in the Schwartz/Sobolev sense. What a shocking realization would it have been for the mathematicians of the second half of the 19th century, if they were to see our modern day development of mathematics!

8.4.1.4 Paradoxes related power series

In the 17th and especially 18th century, power series were treated as a powerful tool of calculus. They were dealt with just as if they were polynomials without any attention paid to the question of whether or not they were convergent. Consequently, beyond the impressive results such obtained, paradoxes appeared in abundance.

Power Series Paradox # 1: It is true that $-1 > \infty > 1$ so that ∞ must be a type of limit between the positive and negative numbers; and in this sense it resembles 0.

This paradox was due to Euler (Kline, 1972, pp. 447). In the power series expansion $(1+x)^{-2} = 1 - 2x + 3x^2 - 4x^3 + \cdots$, letting $x = -1$ produces

$$\infty = 1 + 2 + 3 + 4 + \ldots$$

And in the power series expansion $(1-x)^{-1} = 1 + x + x^2 + x^3 + \cdots$, substituting $x = 2$ leads to

$$-1 = 1 + 2 + 4 + 8 + \cdots \tag{8.9}$$

Because each term on the right-hand side of the second equation is greater than or equal to the corresponding term on the right-hand side of the first equation, one gets that $-1 > \infty$. Because one clearly has $\infty > 1$. Therefore, we have $-1 > \infty > 1$. From this outcome, Euler inferred that the infinity ∞ must be some kind of limit between the positive and negative numbers; and in this sense it resembled 0.

Of course, the later development of convergence of power series resolved this paradox.

Power Series Paradox # 2: $\log 2 = 0$.

This is another example of how the 17th and 18th century mathematics involved the art of series manipulation. In the power series expansion $\log(1+x) = x - x^2/2 + x^3/3 - \cdots$, letting $x = 1$ yields

$$\log 2 = 1 - \frac{1}{2} + \frac{1}{3} + \frac{1}{4} + \cdots$$

Now, rearranging the right-hand side as follows

$$\left(1 + \frac{1}{3} + \frac{1}{5} + \cdots\right) + \left(\frac{1}{2} + \frac{1}{4} + \frac{1}{6} + \ldots\right) - 2\left(\frac{1}{2} + \frac{1}{4} + \frac{1}{6} + \cdots\right)$$
$$= \left(1 + \frac{1}{2} + \frac{1}{3} + \frac{1}{4} + \frac{1}{5} + \cdots\right) - \left(1 + \frac{1}{2} + \frac{1}{3} + \frac{1}{4} + \frac{1}{5} + \cdots\right) = 0$$

establishes the fact that $\log 2 = 0$.

Riemann resolved this paradox in the mid-19th century by proving the fact that the sum of a conditionally convergent series can assume, upon rearrangement, any value. That is, "the discovery of this apparent paradox contributed essentially to a re-examination and rigorous founding ... of the theory of infinite series" (Remmert, 1991 , pp. 30).

8.4.1.5 Paradoxes related geometric curves

In the direction of geometry, the notion of curves is fundamental. To Euclid a curve is simply a "breadthless length". In his days, the collection of known curves was small – the conic sections, the conchoids, the crissoid, the spiral, the quadratrix, and a very few others. With the invention of analytic geometry in the 17th century, the notion of curves was expanded drastically: any equation in two variables came to represent a (plane) curve, although 17th century mathematicians did not have a widely accepted definition of curves (Bos, 1981, pp. 296). In the following three centuries, the investigation of curves attracted some of the best mathematicians using respectively the geometric, analytic, algebraic, arithmetic, and topological means. Then the 19th-century construction of the 'pathological' functions raised questions about the nature of curves. In response, Jordan in 1887 introduced the first formal definition of a curve to be $\{(f(t), g(t)|f, g: [0,1] \to \mathbf{R}$ are continuous functions$\}$, which represents the path of a continuously moving point.

Curve Paradox # 1: Jordan's definition of curves is both too broad and too narrow.

In particular, Peano in 1890 published his famous example of a 'space-filling curve', where a continuous function is constructed to map the unit interval onto the interior of a square. That is, according to Jordan's definition, the square is made into a curve. What an extremely undesirable state of affairs! In other words, Jordan's definition of curves is too broad.

On the other hand, one would surely want the graph of $y = \sin\frac{1}{x}$ and its limit points on the y-axis, which is symbolically

$$\{(x, \sin\tfrac{1}{x}) : x \in (-\infty, 0) \cup (0, \infty)\} \cup \{(0, y) : -1 \leq y \leq 1\})$$

to be called a curve. But this set of points is not the image of any continuous moving point. That is, Jordan's definition of curves is also too narrow.

A seemingly satisfactory resolution of this paradox was achieved by Menger and Uryson in the 1920s. To this end, one first had to clarify the notion of dimension (Menger, 1943); and then a curve was defined as a one-dimensional continuum (Whyburn, 1942). However, this definition of curves was only adequate until the 1970s when Mandelbrot introduced a class of new curves – his fractals, where the curves' dimensions are fractions (Gardner, 1976).

8.4.1.6 Axiom of choice

Banach and Tarski proved in 1924, by assuming the axiom of choice, that a pea and the sun are equi-decomposable. In other words, the volume of the pea may be cut up into finitely many non-measurable pieces; and a purposeful rearrangement of these pieces may become as voluminous as the sun. That is the well-known Banach-Tarski paradox (Wagon, 1985). In fact, what Banach and Tarski had shown is that any two bounded sets in the Euclidean space $\mathbf{R}^n$ are equi-decomposable as long as each of them contain interior points and if $n > 2$ (Blumenthal, 1940, pp. 351).

Similarly, Laczkovich in 1988 showed (Gardner and Wagon, 1989), by assuming the axiom of choice, that a given circle can be decomposed into finitely many non-measurable pieces; and these pieces can be purposefully rearranged together into a square of the same area as the original circle.

In short, what is presented in this subsection, for more details, please consult (Kleiner and Movshovitz-Hadar, 1994), shows that paradoxes, be they inconsistencies, counter-examples to some widely held notions, misconceptions, true statements that seem to be false, or false statements that seem to be true, have had very substantial and positive impacts on the healthy development of mathematics through the refinement and reshaping of concepts, the broadening of existing theories, and the creating of new concepts and theories. Additionally, what is exciting is that this process of development is ongoing not only in mathematics but also in other areas of scientific investigations.

8.4.2 Modification of mathematical induction and expected impacts

One of the conclusions our previous discussions point to is that in the conclusion statement of mathematical induction: "the proposition $\Phi(n)$ holds true for all natural numbers $n \geq k_0$," for details see Subsection 8.3.1, should be changed to the statement that "the proposition $\Phi(n)$ holds true for every natural number $n \geq k_0$". In other words, "for all" and "for every" should be different. The former stands for a perfect tense, while the latter a present progressive tense; and quantitatively, these different temporal tenses could lead to different answers, as shown in the previous sections. Now, a natural question is: Will such a one word alteration affect mathematics much? The answer is YES; all the constructive proofs of existence established throughout the entire spectrum of mathematics need to be re-established. In this subsection, we will look at two examples, one from real analysis, and the other from set theory, to why these constructive proofs need to be re-established.

Proposition 8.1 Any bounded sequence of real numbers has a convergent subsequence.

To make our point clear, let us look at the following version of the proof for this result. Assume that $\{a_i\}_{i=1}^{\infty}$ is a given, bounded sequence of real numbers. So, there are numbers m and M such that

$$m \leq a_i \leq M, \quad \text{for } i = 1, 2, \ldots$$

Now, let us cut the interval $[m, M]$ into two equal halves $\left[m, \frac{m+M}{2}\right]$ and $\left[\frac{m+M}{2}, M\right]$. Then, one of these subintervals must contain infinite many terms of the sequence $\{a_i\}_{i=1}^{\infty}$. Assume that the subinterval $\left[m, \frac{m+M}{2}\right]$ is such an interval that contains infinite many terms of $\{a_i\}_{i=1}^{\infty}$. Let us choose the minimum subscript k_1 such that $a_{k_1} \in \left[m, \frac{m+M}{2}\right]$.

Assume that for natural number n, we have picked $a_{k_1}, a_{k_2}, \ldots, a_{k_n}$ out of $\{a_i\}_{i=1}^{\infty}$ such that for any $1 \leq j \leq n$, the subscript $k_j > k_1, k_2, \ldots, k_{j-1}$, is the minimum such that $a_{k_j} \in [c_j, d_j] \subset [m, M]$, where $d_j - c_j \leq (M-m)/2^j$ and $[c_j, d_j]$ contains an infinite number of terms of $\{a_i\}_{i=1}^{\infty}$.

Now, we cut the interval $[c_n, d_n]$ into two equal halves $\left[c_n, \frac{c_n+d_n}{2}\right]$ and $\left[\frac{c_n+d_n}{2}, d_n\right]$. Then one of these two subintervals must contain an infinite number of terms of the sequence $\{a_i\}_{i=a_{k_n+1}}^{\infty}$. Assume that $\left[c_n, \frac{c_n+d_n}{2}\right]$ is such an interval and we pick the least subscript k_{n+1} such that $a_{k_{n+1}} \in \left[c_n, \frac{c_n+d_n}{2}\right]$. Then, we have successfully constructed $a_{k_1}, a_{k_2}, \ldots, a_{k_n}, a_{k_{n+1}} \in \{a_i\}_{i=1}^{\infty}$ satisfying $k_1 < k_2 < \cdots < k_n < k_{n+1}$.

By using mathematical induction, we have constructed a subsequence $\{a_{k_j}\}_{j=1}^{\infty}$ such that it is convergent. (For our purpose here, we omit the rest of the proof).

Since we now know the descriptive definitions of actual and potential infinites, all that are proven above by using mathematical induction is that for any natural number j, a term a_{k_j} of the sequence $\{a_i\}_{i=1}^{\infty}$ can be picked to satisfy a set of desirable conditions. So, the non-terminating process of getting one more term a_{k_j} out of $\{a_i\}_{i=1}^{\infty}$ can be carried out indefinitely. That is, we proved the existence of a potential infinity. Right after that, what is claimed is that the actual infinite, the subsequence $\{a_{k_j}\}_{j=1}^{\infty}$, has been obtained. That is, in real analysis, we assume that actual infinite is the same as potential infinite.

Proposition 8.2 The set of all real numbers has an uncountable cardinality.

Proof. Assume the opposite is true. In particular, assume that the set of all real numbers in the open interval $(0, 1) = R$ is countable. That is, there is an one-to-one correspondence between R and the set $N = \{1, 2, 3, \ldots\}$ of all natural numbers. In other words, the elements of R can be labeled by using natural numbers as in Figure 8.3, where each p_{ij}, $i, j \in N$, stands for a digit.

Now, we construct a real number as follows:

$$\theta = 0.p_1 p_2 p_3 \ldots p_n \ldots$$

where p_i, $i \in N$, is a digit such that $p_i = p_{ii} + 1$, and when $p_{ii} = 9$, $p_i = 0$. It is evident that $\theta \in (0,1)$ is a real number. On the other hand, because θ is different from each $\theta_m \in R$, it means that the list of all the elements in R as shown in Figure 8.3 does not include this very real number θ. It is a contradiction. So, the assumption that the set of all real numbers in the open interval $(0, 1) = R$ is countable is incorrect. QED.

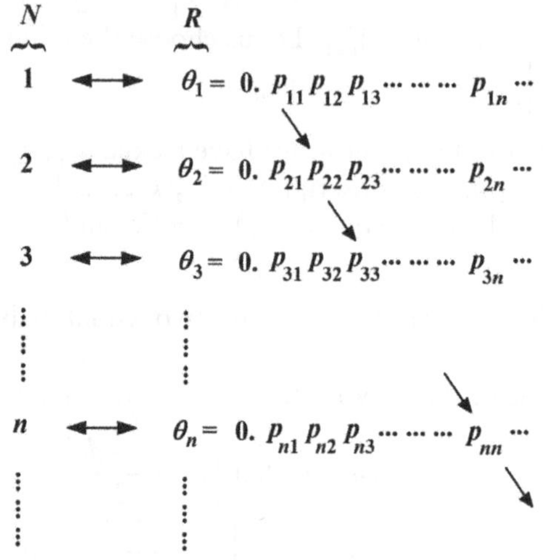

$$
\begin{array}{ccl}
\overbrace{N} & & \overbrace{R} \\
1 & \longleftrightarrow & \theta_1 = 0.\, p_{11} p_{12} p_{13} \cdots \cdots \cdots\, p_{1n} \cdots \\
2 & \longleftrightarrow & \theta_2 = 0.\, p_{21} p_{22} p_{23} \cdots \cdots \cdots\, p_{2n} \cdots \\
3 & \longleftrightarrow & \theta_3 = 0.\, p_{31} p_{32} p_{33} \cdots \cdots \cdots\, p_{3n} \cdots \\
\vdots & & \vdots \\
n & \longleftrightarrow & \theta_n = 0.\, p_{n1} p_{n2} p_{n3} \cdots \cdots \cdots\, p_{nn} \cdots \\
\vdots & & \vdots
\end{array}
$$

Figure 8.3 Cantor's diagonal method.

Now if we distinguish the concept of actual infinite with that of potential infinite and only accept the existence of potential infinites, then the previously constructed real number θ does not exist so that the relevant proposition cannot be established.

Chapter 9

Inconsistencies of modern mathematics

For over half a century, it has been commonly believed that the establishment and development of modern axiomatic set theory have provided a method to explain Russell's paradox. On the other hand, even though it has not been proven theoretically that there will not appear new paradoxes in modern axiomatic set theory, it has been indeed a century that no one has found a new paradox in modern axiomatic set theory. However, when we revisit some well-known results and problems with the thinking logic of allowing two kinds of infinite, the potential and actual infinite, we discover that various infinite sets in the framework of ZFC set theory, widely studied and employed in modern axiomatic set theory, are all specious non-sets.

For several hundred years now since the time when calculus was initially introduced, the common belief has been that the theory of limits had provided an explanation for the Berkeley paradox. However, when we revisit some of the age-old problems using the thinking logic of allowing both the concepts of potential and actual infinite, we find surprisingly that the shadow of the Berkeley paradox does not truly disappear in the foundation of mathematical analysis. At the same time, if we clearly distinguish the concepts of potential and actual infinite from each other, the fact is that the Berkeley paradox of the 18th century has already returned.

This chapter is organized as follows. In Section 9.1, we look at an inconsistency that exists with the concept of countable infinite sets. Then what is obtained in this section is generalized to uncountable infinite sets under the ZFC framework in Section 9.2. The phenomena of Cauchy theaters are discussed in Section 9.3, providing a second proof for the existence of inconsistencies with respect to the concept of infinite sets. Section 9.4 investigates the relationship between Cauchy theaters and the well-known diagonal method initiated by G. Cantor. By distinguishing the concepts of potential and actual infinite, Section 9.5 shows how Berkeley paradox returns as a new challenge to the foundations of mathematics. This chapter concludes with yet another proof for the existence of inconsistency in the system of natural numbers in Section 9.6.

9.1 AN INCONSISTENCY OF COUNTABLE INFINITE SETS

The purpose of this section, which is mainly based on (Lin (guest editor), 2008, pp. 446–452), to show the fact that countable infinite sets are self-contradictory non-sets.

Since about the start of the 20th century, it has been commonly believed that the establishment and development of modern axiomatic set theory have provided a method to explain Russell's paradox. On the other hand, even though it has not been proven theoretically that there will not appear new paradoxes in modern axiomatic set theory, it has been indeed a century that no one has found a new paradox in modern axiomatic set theory. However, when we revisit some well-known results and problems under the thinking logic of allowing two kinds of infinities, the potential and actual infinite, we discover that various countable infinite sets, widely studied and employed in modern axiomatic set theory, are all specious non-sets.

9.1.1 Terminology and notes

The descriptive definitions for potential infinite (*poi*) and actual infinite (*aci*) have been established in Chapter 7, where the names and interpretations of the symbols "$\uparrow$", "$\top$", and "$\not\top$", are also provided. In this section, corresponding to the mathematical background and concrete models needed for this section, we will select the appropriate interpretations for these symbols and construct a group of new expressions. In details, let us have:

$a \uparrow b =_{df}$ "variable *a* approaches the limit *b* indefinitely";
$a \top b =_{df}$ "variable *a* reaches the limit *b*";
$a \not\top b =_{df}$ "variable *a* can never reach the limit *b*";
$a \uparrow b \wedge a \top b =_{df}$ "variable *a* approaches indefinitely and reaches the limit *b*";
$a \uparrow b \wedge a \not\top b =_{df}$ "variable *a* approaches indefinitely and never reaches the limit *b*".

Additionally, let the symbol @ be a certain kind of quantity. It can represent a finite quantity or an infinite quantity. Specifically, @ can stand for a natural number n, or the symbol ∞, or various transfinite cardinalities, such as $\aleph_0, \aleph_1, \ldots, \aleph_n, \ldots$, or even the cardinality c of the continuum. On this basis, let us introduce the following abbreviation:

$S\,\overline{|k}\,@ =_{df}$ "the ideal container *S* holds @ many objects".

Note (I): In the discussion below, we will implement and employ the method of reasoning of allowing two kinds of infinities and the thought convention that potential infinities are different of actual infinities ($poi \neq aci$). This premise for our presentation that follows is made based on rational grounds, because it has been sufficiently proven (see (Lin (guest editor), 2008, pp. 433–445), or Sections 7-3 and 7-4 in this book) that the analysis method of allowing two kinds of infinities and the thought convention of $poi \neq aci$ are both inherent parts of the system of modern mathematics, whose theoretical foundation has been artificially added.

Note (II): In set theory, cardinalities and ordinalities are two different concepts. The cardinality of a set is uniquely determined, while its order type is not, because generally each non-empty set can be ordered in different ways. For example, there are infinitely many ways to order the elements in the set N of all natural numbers, such as

(1) $1, 2, 3, \ldots, n, n+1, \ldots;$
(2) $2, 3, 4, \ldots, n, n+1, \ldots, 1;$

(3) $1, 3, 5, \ldots, 2n+1, \ldots, 2, 4, 6, \ldots, 2n, \ldots;$
(4) $\ldots, n+1, \ldots, 3, 2, 1.$

Corresponding to the orderings in (1), (2), (3), and (4), the produced order types are respectively denoted as ω, $\omega+1$,$\omega+\omega=2\omega$, and ω^*. However, the cardinality of the set N is uniquely given as $\overline{\overline{N}}$. Also, in set theory, the cardinality of an well-ordered infinite set is named by aleph, its order type is known as an ordinal number or simply ordinality. So, after the set N of all natural numbers is well-ordered, its cardinality is $\aleph_0$, which is uniquely determined. And, the order types (1), (2), (3), that is, ω, $\omega+1$, 2ω, stand for different ordinalities of N after it is being well ordered. Therefore, cardinalities and ordinal numbers are two different concepts. However, they can also be seen as being the same when the universe of discourse and the method of ordering sets are specifically restricted. For example, if we only allow the set N of natural numbers being ordered as the λ-sequence, then the cardinality $\overline{\overline{N}}$ and order type $\overline{N}$ will and can be seen as the same. Additionally, the symbol "|" is often applied to represent a restriction. That is, $\models_{df}$ "restriction". So, we have

$$N = \{x \mid n(x)\} \mid \lambda(1, 2, 3, \ldots, n, \ldots) \Rightarrow \overline{\overline{N}} = \overline{N}.$$

This can also be written briefly $N \mid \lambda \Rightarrow \aleph_0 = \omega$.

Especially, the von Neumann scheme is one of the methods on how to construct the system of natural numbers. Its guiding principle is to start on the basis of the empty set $\emptyset$ so that each natural number is equal to the set of all smaller natural numbers. In this way, each natural number is equal to either the cardinality of the set of all smaller natural numbers or the ordinality of this set. This end is of course a phenomenon of finite ordinal numbers, which might not hold true when transfinite situations are concerned without adopting certain very special restriction, such as $N \mid \lambda \Rightarrow \aleph_0 = \omega$. This is actually the reason why people often use the statement that the set N of natural numbers has ω elements instead of using the more appropriate sentence that N has $\aleph_0$ elements.

Note (III): Let S be an ideal container and consider the following five criteria:

(α) There are @ D – atoms inside S;
(β) There are not @ D – atoms inside S;
(γ) The number or quantity of D – atoms inside S has reached @;
(ξ) The number or quantity of D – atoms inside S has not reached @; and
(η) The number or quantity of D – atoms inside S will never reach @.

Among these five criteria, we can clearly determine the following five relationships and conclusions:

(I) (α) and (β) cannot hold at the same time. That is, $\neg[(\alpha) \wedge (\beta)]$.
(II) (γ) and (ξ) cannot hold at the same time. That is, $\neg[(\gamma) \wedge (\xi)]$.
(III) (α) and (γ) are equivalent. That is, (α) iff (γ).
(IV) (β) and (ξ) are equivalent. That is, (β) iff (ξ).
(V) (η) implies (ξ). That is, $(\eta) \vdash (\xi)$.

Let k be such a variable that represents the quantity of D – atoms and that increases indefinitely and approaches @. Then, the previous abbreviation can be employed to rewrite Conclusion (III) as the following symbolic expression:

$$S|\overline{k}@ \text{ iff } k \uparrow @ \text{ and } k \top @. \tag{9.1}$$

Now, we use the language of set theory to translate our discussion above using the set N of all natural numbers. That is, our ideal container S will correspond to the set N, D – atoms the elements (the natural numbers) in N, and @ the cardinality $\aleph_0$ of N. If we impose the restriction that N can only be ordered from smaller natural numbers to larger ones, as in the λ-sequence, then the statement $N|\lambda \Rightarrow \aleph_0 = \omega$ in Note (II) above implies that ω can be used to replace $\aleph_0$. So, we have the following corresponding criteria:

(α') The number of natural numbers in N is ω;
(β') The number of natural numbers in N is not ω;
(γ') The number of natural numbers in N has reached ω;
(ξ') The number of natural numbers in N has not reached ω; and
(η') The number of natural numbers in N will never reach ω.

Based on the intensions of these five criteria, we have the following conclusions:

(I′) (α') and (β') cannot hold at the same time. That is, $\neg[(\alpha') \wedge (\beta')]$.
(II′) (γ') and (ξ') cannot hold at the same time. That is, $\neg[(\gamma') \wedge (\xi')]$.
(III′) (α') and (γ') are equivalent. That is, (α') iff (γ').
(IV′) (β') and (ξ') are equivalent. That is, (β') iff (ξ'). and
(V′) (η') implies (ξ'). That is, $(\eta') \vdash (\xi')$.

Let k' be a variable that represents the number of natural numbers and that increases indefinitely and approaches ω. Let us introduce the following abbreviation:

$\lambda|\overline{k'}\omega =_{df}$ "the set N of natural numbers contains ω many pairwisely different natural numbers."

Then, from this abbreviation and conclusion (III′) above, the following important result can be seen:

$$\lambda|\overline{k'}\omega \text{ iff } k' \uparrow \omega \wedge k' \top \omega. \tag{9.2}$$

9.1.2 An inconsistency with countable infinite sets

Now, we are ready to prove the following theorem:

Theorem 9.1: Each countable infinite set is a self-contradicting non-set.

Proof. Step 1. We first show that the set, made up of exactly all natural numbers, $N = \{x \mid n(x)\}$, where $n(x) =_{df}$ "x is a natural number", is a self-contradicting non-set. First, order the elements in $N = \{x \mid n(x)\}$ from smaller numbers to larger ones as follows:

$$\lambda : \{1, 2, 3, \ldots, n, \ldots\} \mid \omega.$$

Then, in terms of this λ-sequence, we have the following well-known theorems:

Proposition 9.1 Each natural number is a finite ordinal number. That is, $\forall n(n \in \mathrm{N} \rightarrow n < \omega)$.

Proposition 9.2 There are infinitely many natural numbers. That is, $N = \{x \mid n(x)\}$ contains ω many pairwisely different elements (see Note (II) above for details).

So, from Proposition 9.1, it follows that we can specifically represent the λ-sequence, as the following countable infinite sequence of ω many pairwisely different inequalities:

$$N^< = \{1 < \omega, 2 < \omega, 3 < \omega, \ldots, n < \omega, \ldots\}.$$

Let us introduce the following abbreviations:

$Ine =_{df}$ "inequality",
$n(in)Ine =_{df}$ "n is the unique natural number determined by some inequality in $N^<$",
$Inek =_{df}$ "the kth inequality in $N^<$",
$n(in)Inek =_{df}$ "n is the unique natural number contained in the kth inequality in $N^<$".

Let $\forall =_{df}$ "for all", $E =_{df}$ "for each". (For a more detailed discussion on the interpretational convention of the universal quantifier of the two-value logical calculus, please refer to (Zhu, Xiao, Song, and Gu, 2002) According to the interpretation of the universal quantifier in the classical two-value logic, we should have $\forall = E$. So, we have the following conclusion:

Proposition 9.3 In any circumstances, E and $\forall$ are exchangeable.

So, in $N^<$, we have

$$EnEk(n(in)Inek \rightarrow n = k). \tag{9.3}$$

For example, the uniquely determined value of the natural number contained in the 9th (respectively, kth) inequality in $N^<$ must be 9 (respectively, n). Not only this, by using mathematical induction, we can prove the following result:

$$\forall n(n(in)Inen). \tag{9.4}$$

That is, for each inequality in $N^<$, it is always true that $n = k$. Now, from Proposition 9.3 above, by replacing E by $\forall$ in equ. (9.3), it follows that:

$$\forall n \forall k(n(in)Inek \rightarrow n = k). \tag{9.5}$$

So, the following important conclusion is proven:

Proposition 9.4 $\forall n \forall k(n(in)Inek \to$ the appearances of n and k are exchangeable).

Let k' be a variable that represents the counting number and approaches ω indefinitely; and let us apply the following abbreviation:

$\lambda\overline{|k}\,\omega =_{df}$ "The set N of natural numbers contains ω many pairwisely different elements."

Therefore, from Note (III) and important conclusion in equ. (9.2), it follows that:

$$\lambda\overline{|k'}\ \omega \text{ iff } k' \uparrow \omega \wedge k' \top \omega.$$

Now, Proposition 9.2 above implies that $\lambda\overline{|k'}\,\omega$ holds true. So, the following proposition is true:

Proposition 9.5 $(k' \uparrow \omega) \wedge (k' \top \omega)$.

Similarly, since $N^<$ has ω many pairwisely different inequalities, when we use k to represent the variable that counts the number of inequalities in $N^<$ and increases indefinitely, the variable k can not only approach ω indefinitely, but also reach ω. Therefore, similar to our discussion about the λ-sequence above, when we are concerned with the number of inequalities in $N^<$, we would also have the true proposition: $(k \uparrow \omega) \wedge (k \top \omega)$.

Because $(k \uparrow \omega) \wedge (k \top \omega)$ take the truth value TRUE, the truth-value table of implication tells us that

$$\forall n \forall k(n(in)Inek \to (k \uparrow \omega) \wedge (k \top \omega)).$$

Since Proposition 9.5 above holds true, we can replace k in the expression by n, producing

$$\forall n \forall n(n(in)Inen \to (n \uparrow \omega) \wedge (n \top \omega)).$$

This expression is the same as

$$\forall n(n(in)Inen \to (n \uparrow \omega) \wedge (n \top \omega)).$$

So, we have

$$\forall n(n(in)Inen \to \forall n((n \uparrow \omega) \wedge (n \top \omega)) \tag{9.6}$$

However, on the other hand, in terms of the values n of the natural numbers contained in the inequalities of $N^<$, even though n can also approach ω indefinitely, it can never reach or is not allowed to reach ω. Otherwise, it would contradict Proposition 9.1 above. That is, it would contradict the well-accept conclusion: $\forall n(n \in N \to n < \omega)$.

So, when we focus on the value of the natural number contained in an inequality in $N^{<}$, we can only have the conclusion that $(n \uparrow \omega) \wedge (n \mathrel{\overline{\uparrow}} \omega)$. That is, we have

$$\forall n(n(in)Inen \rightarrow (n \uparrow \omega) \wedge (n \mathrel{\overline{\uparrow}} \omega)),$$

which implies

$$\forall n(n(in)Inen) \rightarrow \forall n((n \uparrow \omega) \wedge (n \mathrel{\overline{\uparrow}} \omega)). \tag{9.7}$$

From equs. (9.4) and (9.6) and applying the rule of separation, it follows that

$$\forall n((n \uparrow \omega) \wedge (n \mathrel{\overline{\uparrow}} \omega)). \tag{9.8}$$

Similarly, from equs. (9.4) and (9.7), it follows that

$$\forall n((n \uparrow \omega) \wedge (n \mathrel{\overline{\uparrow}} \omega). \tag{9.9}$$

We have to admit that equs. (9.8) and (9.9) contradict each other. This end implies that $N^{<}$ is not consistent, which means that the λ-sequence and $N = \{x \mid n(x)\}$ are self-contradicting non-sets.

Step 2. Let G be an arbitrary countable infinite set in the framework of ZFC. Then, the elements of G can be indexed by natural numbers. So, the conclusion that $N = \{x \mid n(x)\}$ is a non-set can be used directly to see the fact that G is also a self-contradicting non-set. QED.

9.2 UNCOUNTABLE INFINITE SETS UNDER ZFC FRAMEWORK

In this section, which is based on (Lin (guest editor), 2008, pp. 453–457), we will show that all uncountable infinite sets are self-contradictory non-sets.

After showing the fact that the set $N = \{x \mid n(x)\}$ of all natural numbers, where $n(x) =_{df}$ "x is a natural number", is a self-contradicting non-set in the previous section, in this section, we prove that in the framework of modern axiomatic set theory ZFC, various uncountable infinite sets are either non-existent or self-contradicting non-sets. Therefore, it can be astonishingly concluded that in both the naïve set theory and the modern axiomatic set theory, if any of the actual infinite sets exists, it must be a self-contradicting non-set.

9.2.1 An inconsistency of uncountable infinite sets

On the basis of the result proven in the previous section, we now show the following result:

Theorem 9.2: Under the premise that $N = \{x \mid n(x)\}$ is a non-set in the modern axiomatic set theory ZFC, all the various kinds of the so-called uncountable sets, as studied in the modern axiomatic set theory ZFC, are either non-existent or also self-contradicting non-sets.

Proof. It is a known fact that the rule of constructing sets in Cantor's naïve set theory is that of summarization, and another fact that by employing this rule of summarization unconditionally, paradoxes are produced in abundance. Among the various paradoxes is the following famous Russell's paradox that is the most basic and central: Let Σ be the set consisting of all such sets that are not elements of their own as elements. If we assume that Σ does not contain itself as an element, then we can prove Σ is an element of Σ. If we assume that Σ is an element of itself, then we can derive that the set Σ does not contain itself as an element. That is, no matter how we make the assumption, the consequence is the opposite. So, a contradiction is resulted.

By modifying the rule of summarization, the modern axiomatic set theory avoids the appearance of all the known paradoxes. As is well known, among the non-logical axioms in the ZFC, the central rules developed for constructing sets are:

(1) The axiom for the empty set;
(2) The axiom of infinity; and
(3) The power set axiom.

Here, axioms (1) and (2) can be employed to construct countable infinite sets and stipulate directly without any added conditions the existence of some sets. And, on the basis of (1) and (2), combined with the replacement and other axioms, it can be shown that the set of natural numbers $N = \{x \mid n(x)\}$ exists. After that, on the basis of (1) and (2) and through the use of the power set axiom, uncountable sets are constructed. Then, combined with other axioms, the set of real numbers $R = \{x \mid r(x)\}$ is constructed, where $r(x) =_{df}$ "x is a real number". With this end, the foundation of calculus is laid (Zhu and Xiao, 1991; 1996a).

In ZFC, it can be shown that Σ, as defined above, is a non-set (not a set), or not a ZFC-set. Because of this reason, one can no longer question whether Σ is a set of all those sets that contain themselves as elements or a set of all those sets that are not an element of themselves. Therefore, the potential appearance of the Russell's paradox is avoided. Also, under the premise that Σ is a non-set, we can no longer ask what sets are subsets of Σ in the framework of ZFC (for details, please consult Sections 7.3, 7.4, and 8.3 in this book), or what sets are the power set of Σ, etc. According to Figure 9.1, it follows that in ZFC, there does not exist any initial element of its own, and the following important conclusion:

Proposition 9.6 For any ZFC set $A \neq \emptyset$, each element of A must be a ZFC set. Any subset of A and the power set of A must be ZFC sets. And, only if A is a ZFC set, the power set $P(A)$ of A can be constructed using the power set axiom.

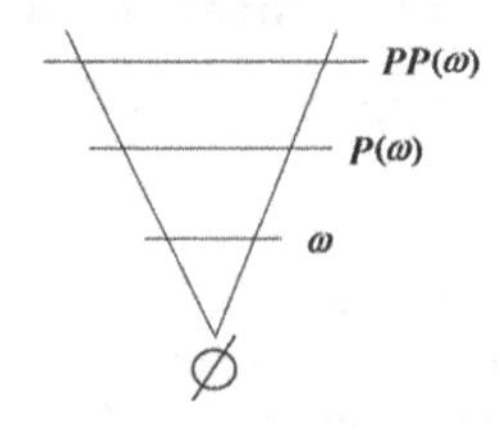

Figure 9.1 The hierarchy of sets.

Next, our discussion continues in two cases:

(I) If at the initial time when we constructed the ZFC framework it were found that $N = \{x \mid n(x)\}$ is a non-set, and that any countable infinite set is a non-set, according to (Lin (guest editor), 2008, pp. 446–452) or Section 9.1 in this book, then it would be impossible for us to employ the power set axiom to construct the power set of any countable infinite set. It is because from Proposition 9.6 above, it follows that only when a ZFC set is identified, the power set axiom of the ZFC will be used to construct the power set of the given set. Hence, since it is known that $N = \{x \mid n(x)\}$ is a non-set, there will no longer be any reason to construct $P(N)$, just like the situation that when we know Σ is a non-set, we will no longer consider $P(\Sigma)$. So, in this case, the so-called uncountable sets in the framework of ZFC do not exist at all.

(II) However, the history was not as what we imagined in case (I) above. It is long after when the ZFC system was initially established and been applied that we discovered the fact that $N = \{x \mid n(x)\}$ and any infinite countable set are non-sets. So, we have to discuss the true historical development in two possibilities.

Firstly, for an arbitrary ZFC uncountable set A, let us apply the axiom of choice to select a countable infinite subset A_C out of A. Of course, we have $A_C \subset A$. Now, we know $A_C$is a non-set. From Propositiŏn 9.6, it follows that any subset of a non-empty ZFC set must be a ZFC set. Now, the uncountable set A has a self-contradicting non-set A_C as its subset. This end is not allowed in ZFC system. So, A can no longer be a ZFC set. Or in other words, the uncountable set A is a non-set in the framework of ZFC.

Secondly, assume that $P(B)$ is the power set of a ZFC countable set B before it is found that each countable infinite set is a non-set. Then, in the ZFC system, we must have $B \mathit{in} P(B)$. Now, we assume we know that the countable set B is a non-set. From Proposition 9.6, it follows that each element of a ZFC set must be a ZFC set. Since now we know B is a non-set, the power set $P(B)$ contains a non-set B as its element. It means that the power set $P(B)$ has to be a self-contradicting non-set.

In short, under the condition that in ZFC, $N =, \{x \mid n(x)\}$ is a non-set, all various kinds of the so-called uncountable sets in ZFC either do not exist or are also self-contradicting non-sets. QED.

Because in the framework of ZFC, any actual infinite set is either countable or uncountable, from our theorem above and the result in (Lin (guest editor), 2008, pp. 446–452) or Section 9.1 in this book we have the following conclusion:

Proposition 9.7 In ZFC, any infinite set is a self-contradicting non-set.

9.2.2 Several relevant historical intuitions

Both the result in (Lin (guest editor), 2008, pp. 446–452) (or Section 9.1 in this book) and what is proven in this section are historically inevitable and had been felt by the great minds of the history. All we did here is to prove mathematically what these scientific masters had felt was correct. For example, Leibniz (Kline, 2005) once said,

"The wording of the count of all whole numbers is self-contradicting. It should be abandoned."

After there appeared the Russell's paradox in naïve set theory, Hausdorff (Hausdorff, 1962) was very emotional and reminded everybody, "the intranquility, caused by this paradox and widely felt, is not about the contradiction produced. Instead, it is about how we did not expect there could be a contradiction. The set, consisting of all cardinal numbers, seems to be so priori indubitable, just like how the set of all natural numbers is that naturally believable. Because of this, the following uncertainty is created: Is it possible that other infinite sets, that is, all infinite sets, are these kinds of non-sets which are apparently right but actually wrong and contain self contradictions?"

However, what attracted the most attention and astonishment was the opinion published by Robinson (Robinson, 1964): "As for the foundation of mathematics, my position (opinion) is based on the following two main principles:

(1) No matter what a meaning of words is used, infinite sets do not exist. (They do not exist in practice or in theory). More specifically, any statement on infinite sets is simply meaningless.
(2) However, we should still conduct mathematical works and activities as usual. That is to say, when we work, we should still treat infinite sets as if they truly existed."

At this juncture, we strongly believe that Robinson's statement (1) is a very significant intuitive judgment instead of an irresponsible nonsense. As for his statement (2), it could be out of his helplessness at the time. We believe that the main result in this section is consistent with Robinson's straight-forward words and the intuitions of the great minds of the past.

With our result in place, a natural question is: If the concept of actual infinite sets $A = \{x \mid p(x)\}$ has to be given up in the framework of ZFC, due to the serious problems coming along with the concept, would all the mathematical theories (that is the entire modern mathematics) developed on the set theory involving $\{x \mid p(x)\}$, have to be thrown out of the window?

The answer to this question is NO. As a matter of fact, the entire modern mathematics can be completely kept on the foundation of the mathematical system developed on the concept of potential infinities. It is because even though Cantor, Zermelo, and their followers constructed various kinds of actual infinite sets of the kind $\{x \mid p(x)\}$, when they deal with properties of elements, relationships between elements, structures of elements, and various kinds of functions defined on elements, the thinking logic of potential infinities have been essentially employed. That is, in their works, the method of reasoning of potential infinities has been used in place of the thinking principle of generation of actual infinities. So, as long as the mathematical system of potential infinities still employs the classical two-value logic and calculus as its tool of deduction, and as long as the universal quantifier $\forall$ is adjusted appropriately, we will be able to redevelop the entire modern mathematics in terms of the current intension.

At this very moment, another natural question is: Will this mathematical system based on potential infinities be a refurbished version of the constructive mathematics of the intuitionism? The answer to this question is NO. The fundamental differences include:

1 The intuitionistic logic is the tool of reasoning for the constructive mathematics of the intuitionism. And for the mathematical system of potential infinities, the classical two-value logical calculus is still its tool of deduction.
2 The starting point of the constructive mathematics of the intuitionism is not the ordinary set theory, the naïve set theory or the axiomatic set theory, but the natural number theory under the dual intuition of Brouwer's objects (Zhu and Xiao, 1991; 1996a).
3 The constructive analysis of the intuitionism is based on the continua of Brouwer's expanded form, while the mathematical analysis of potential infinities will be on the foundation of elastic power sets defined in an elastic natural number system.

In short, both systems of mathematics have their own individual starting points, their own individual methods of constructing the mathematical mansion, and their degrees of intensional riches will greatly differ from each other.

9.3 THE PHENOMENA OF CAUCHY THEATER

This section, which is based on (Lin (guest editor), 2008, pp. 458–468), employs a different approach to show that the countable infinite sets are self-contradictory non-sets.

The concept of infinities in the theory of countable sets was discussed in Section 9.1 by employing the method of analysis of allowing two different kinds of infinities. What was obtained is that various countable infinite sets, studied in the naïve and modern axiomatic set theories, are all incorrect concepts containing self-contradictions. In this section, we provide another argument to prove the same conclusion: Various countable infinite sets studied in both naïve and modern axiomatic set theories are all specious non-sets. The argument is given from a different angle on still the same premise of allowing two different concepts of infinities.

9.3.1 Spring sets and Cauchy Theater

For the predicate P: "natural number", it is well recognized that it is a predicate on an infinite background world, completely different of such predicate as "all the students who are currently in this classroom", defined on a finite background world. However, each infinite background world can be classified into two categories: One that contains all actual infinite worlds and the other of potential infinite worlds. Here, Cantor believed that the predicate P: "natural number" belongs to an actual infinite background world. So, he could construct the actual infinite set $N = \{x \mid n(x)\}$. However, Brouwer, an intuitionist, fundamentally rejected the concept of actual infinities. To him, natural numbers can only be situated in the process of being constructed one by one (ongoing). Even though new natural numbers can be constructed beyond any chosen bound, the collection of all natural numbers has to stay in the state of being generated (ongoing) and can never form a finished, closed field (done).

As is known, in the study of the limit theory, in order to avoid the Berkeley paradox, Cauchy and Weierstrass applied $\varepsilon - N$ and $\varepsilon - \delta$ methods to define infinities and infinitesimals of sequences. They also adopted Brouwer's thinking method of potential infinities implied in his forever unfinishable construction of natural numbers. That is, choose a bound, then go beyond that bound; choose another bound, and then go beyond this new bound. And, no matter how large the bound N is or how small the bound ε is, one can always pass beyond this bound. That is, they implemented totally the concept of potential infinite of endlessly establishing bounds and passing beyond the bounds. By doing so, the usage of the concepts of actual infinite and actual infinitesimals is avoided. Now, let us use the symbol $N = \{x \mid n(x))$ to represent the process of endlessly constructing natural numbers. That is, the ending parenthesis ")" is employed to indicate the process that more and more natural numbers are being continually constructed (ongoing), while the ending brace "}", as used by Cantor originally, means that the collection of natural numbers has formed a closed field, a perfect tense (done). For our purpose, we will specifically call the actual infinite set $N = \{x \mid n(x)\}$ as a rigid set and the potential infinite set $N = \{x \mid n(x))$ as a spring set. Another form of representation completely equivalent to this spring set is the following Cauchy Theater. Assume that

$$\lambda_n : \{1, 2, 3, \ldots, n\}$$

is the set of the first n natural numbers, ordered according to their magnitudes, starting with 1 and ending with n, a pre-determined natural number. Now, let the ending element n in λ_n vary. That is, let n start to increase indefinitely but forever satisfy $n < \omega$. Doing so constitutes a potentially infinite process (ongoing), where n increases indefinitely but is forever finite. Let us express this ongoing process as follows:

$$\text{C-N: } \{1, 2, 3, \ldots, \vec{n}),$$

and call it the Cauchy theater. Imaginarily, each natural number in C-N we can see as the numbering of a seat in the Cauchy Theater. So, our Cauchy Theater is a potentially infinite process (ongoing) with its number of seats, numbered using natural numbers, increasing continually without any bound. That is, our Cauchy Theater is exactly the spring set $N = \{x \mid n(x))$ of the present progressive tense with which Cauchy and Weierstrass continually constructed the natural numbers based on the idea of potential infinite by endlessly establishing bounds and surpassing the bounds.

The concepts of the so-called spring and rigid sets, as introduced above, are only representations for either the endless process of constructing natural numbers or the special case of the entire collection of all natural numbers. In fact, we can define these concepts for general cases. As pointed out at the end of Section 7.3, in order to construct the descriptive definitions for potential infinities (*poi*) and actual infinities (*aci*), we once introduced the symbols $\uparrow$ for open progressive word, $\overline{\top}$ for the negative completion word, and $\top$ for positive completion word, and various ways to interpret these symbols. Now, let us select the appropriate interpretations, further combine new abbreviations, and introduce their interpretations.

$x(In)A =_{df}$ "the object x is included into the set A";
$x(\neg In)A =_{df}$ "the object x is not included into the set A";

$ExP(x) \uparrow x(In)A =_{df}$ "any object x satisfying the predicate P can always be included in set A without any restriction";
$\forall xP(x) \top x(In)A =_{df}$ "all objects x satisfying the predicate P can be affirmatively included in set A";
$\forall xP(x) \overline{\top} x(In)A =_{df}$ "negation of the statement that objects x satisfying the predicate P can all be included in set A";
$Ex\neg P(x) \top x(\neg In)A =_{df}$ "definitely any or each object x not satisfying the predicate P is not included in set A".

It should be noticed that in the proof of the main theorem in Section 9.1, we introduced the listing quantifier E and interpreted it as "for each" or "for any" instead of "for all". And, the universal quantifier $\forall$is interpreted as "for all" and cannot be seen as "for each".

For any given predicate P and set A out of an infinite background world, if the following condition holds true:

$$(ExP(x) \uparrow x(In)A) \wedge (\forall xP(x) \overline{\top} x(In)A) \wedge (Ex\neg P(x) \top x(\neg In)A),$$

then the set A is called the spring set of the predicate P, denoted $A = \{x \mid P(x))$.

If a predicate P and a set A out of an infinite background world satisfy the following condition:

$$(ExP(x) \uparrow x(In)A) \wedge (\forall xP(x) \top x(In)A) \wedge (Ex\neg P(x) \top x(\neg In)A),$$

then the set A is called the rigid set of the predicate P, denoted $A = \{x \mid P(x)\}$.

Especially, for the given predicate n: natural number, and a set N, if the following holds true:

$$(Exn(x) \uparrow x(In)N) \wedge (\forall xn(x) \top x(In)N) \wedge (Ex\neg n(x) \top x(\neg In)N),$$

then the set N is called the rigid set of natural numbers, denoted $N = \{x \mid n(x)\}$.

For the predicate n: natural number, and a set N, if the following holds true:

$$(Exn(x) \uparrow x(In)N) \wedge (\forall xn(x) \overline{\top} x(In)N) \wedge (Ex\neg n(x) \top x(\neg In)N),$$

then the set N is called the spring set of natural numbers, denoted $N = \{x \mid n(x))$. This is exactly the Cauchy Theater discussed above:

$$\text{C-N} :\{1, 2, 3, \cdots, \vec{n}).$$

In fact, the above Cauchy Theater C-N or the spring set N satisfies two fundamental properties:

(a) The variable n can increase or be constructed indefinitely. So, this end clearly reflects the non-finitary property of C-N or N.
(b) In the process of n increasing infinitely, it is emphasized that $n < \omega$. So, it clearly rejects the possibility for n to reach ω, reflecting C-N's and N's characteristic of a "potential infinity".

That is, the number of seats in this Cauchy Theater, on one hand, increases without bond, and also, on the other hand, satisfies the following important conclusion:

Proposition 9.8 This Cauchy Theater can never have ω (that is, an actual infinite) seats.

This conclusion implies that the increase in the number of seats stays forever in the process (ongoing) of increasing without any chance of finishing. This fact indicates that when we look at this C-N theater from the angle of generation, it is forever a present progressive tense (ongoing). And, when we look at this C-N theater from the angle of existence, there is no doubt that this theater is dynamic, potential and unascertained.

At this juncture, let us conveniently point out that the reason why this so-called theater is named as Cauchy Theater and briefly denoted as C-N theater instead of using the name of Brouwer is because we like to clearly point out that our current research and discussion belong entirely to the non-constructive part of the system of modern mathematics. In our work, we neither touched on any part of intuitionistic logic nor established any connection with the constructive mathematics of the intuitionism. On the other hand, if the name of Brouwer appeared too often in our work, it might create the wrong feeling that our discussion had entered the category of intuitionistic constructive thinking.

9.3.2 Special Cauchy Theater in naïve and modern axiomatic set theories

Let us now observe the process for all the guests attending a movie reception to enter the theater. As shown in Figure 9.2, it is an ordinary 500-seat movie theater. And, on a special day, a movie reception will be held in this theater for all military officers of the rank of second lieutenant or higher who are stationed in or near the city. Before entering the theater, all the guests holding their tickets gather and wait in the rest area.

The theater has two entrances which are respectively watched by two attendants A and B. Here, the responsibility of A is to check the identification of each guest and that of B is to make sure no more than 500 people are allowed inside the theater, since no one is permitted to watch the movie without a seat. So, before anyone is allowed to enter the theater, A checks on all the guests in the rest area. After making sure everybody is a military officer of at least the rank of second lieutenant and stationed

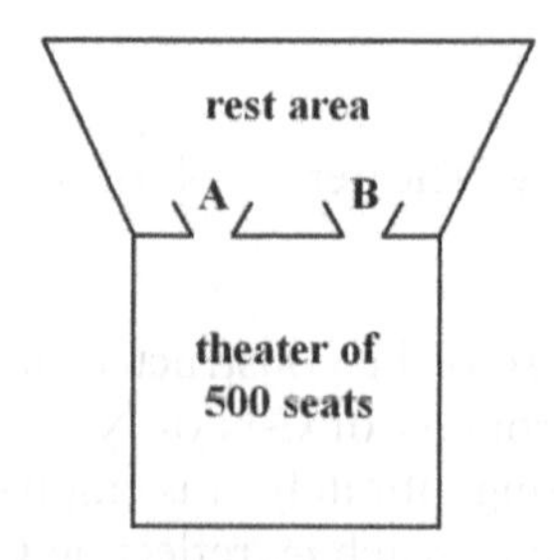

Figure 9.2 The layout of the movie reception venue.

in or near the city, A decides that the guests can enter the theater. However, B now is required to count the number of guests attending the reception. As soon as he knows that each guest holds exactly one ticket and there are no more than 500 guests, everyone is allowed to enter the theater. ... And the reception is a great success.

In the following, we will hold a special movie reception for all natural numbers in the previous Cauchy Theater. But, before the reception, let us review two well-known theorems.

Proposition 9.9 There are countable infinitely many natural numbers. That is, the set N of natural numbers has ω elements.

Proposition 9.10 Each natural number is a finite ordinal number. That is, if n is a natural number, than $n < \omega$.

As a matter of fact, these two well-known theorems can be combined into the following conclusion:

Proposition 9.11 There are ω many pairwisely different, finite valued natural numbers.

Now, as shown in Figure 9.3, we will hold a movie reception in our Cauchy Theater. When seen in the static framework, the Cauchy Theater is not any different from the ordinary theater that we just looked at earlier. It also contains a rest area, two entrance doors watched by two attendants A and B respectively, where the attendants' responsibilities are the same as before: A checks each guest's identification and B makes sure that the theater is not overloaded so that each guest has a seat when watching the movie. However, when seen as a dynamic entity, the number of seats in the Cauchy Theater can increase without a bound. Even so, at any specified moment of time, the number of seats in the theater is fixed and finite. And, the rest area is also open so that an infinite number of guests can enjoy their time there without any problem of space being limited. Now, each of the natural numbers holds one ticket, relaxing in the rest area and waiting to enter the theater. So, A checks the identifications of the guests in the rest area. When it is confirmed that all the guests are natural numbers, A decides without any hesitation that it is time for the guests to enter the theater. With

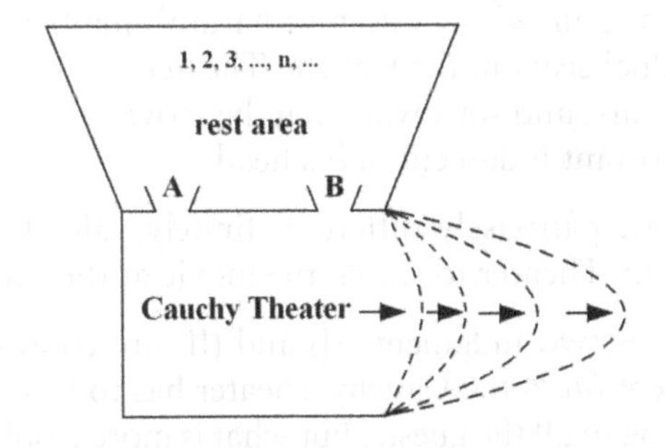

Figure 9.3 The dynamics of the cauchy theater.

this green light in place, B checks on the number of guests waiting in the rest area. When he finds out that there are ω of them, B immediately declares that the Cauchy Theater is not big enough to hold this many guests. Therefore, no one is allowed to enter the theater. Otherwise, the structure of the C-N theater would be damaged. This decision is made because B knows well the discussion in Section 9.3.1 above and the important conclusion in Proposition 9.8. That is, the Cauchy Theater will never have ω (an actual infinity) seats. The reasoning is that the Cauchy Theater stands for a process of potential infinite. Specifically, the increase in the number of seats forever stays in the process of getting greater (ongoing), and will never finish. So, the first judgment B makes in his head is that:

(I) The total of "ω many, pairwisely different, and finite-valued natural numbers" cannot enter the theater to watch the movie at the same time.

However, on the other hand, after B formed his judgment (I) and refused the guests (all natural numbers) waiting in the rest area to enter the theater, he started to question himself about the situation. It is because he remembers Proposition 9.10 above and reasons as follows: First, for any chosen natural number m, Proposition 9.10 implies that $m < \omega$. Since the C-N theater has at least $m+1$ seats, he cannot find a guest without sitting on a seat. Since there is no guest without a seat, then all the guests (all natural numbers) have their own individual seats.

Also, either mathematical induction or the method of proof by contradiction can be employed to prove that all the guests have a seat for each of them. For example, let us look at the proof by contradiction as follows.

Proof. First, order the natural numbers, seen as the guests, in their increasing order:

$$\lambda : \{1, 2, 3, \ldots, n, \ldots\},$$

so that λ is a well-ordered set. Now, we assume by contradiction that there exists such a natural number in λ that does not have its own seat in the Cauchy Theater. Let S be the set of all those natural numbers in λ who do not have their own seats in the Cauchy Theater. Then, S is a subset of λ. Since λ is well-ordered, the set S has the minimum element, denoted m. Since $m \in S$, m does not have his own seat in the Cauchy Theater. On the other hand, from the structure of the Cauchy Theater, it follows that the theater has at least $m+1$ seats. So, m must have his own seat. A contradiction. Therefore, the assumption by contradiction does not hold true. Hence, all natural numbers have their own seat in the Cauchy Theater. QED.

Since these ω many, pair wisely different natural numbers (all natural numbers) have their own individual seats in the Cauchy Theater, the guests can of course enter the theater, take their seats, and start watching the movie. Because of this, B draws the following second, important judgment in his head:

(II) "These ω many, pairwisely different, finitely valued natural numbers" can entirely enter the Cauchy Theater to watch the movie at the same time.

Because the previous two judgments (I) and (II) are contradicting to each other, the planned movie reception in the Cauchy Theater has to be cancelled. Not only this cancellation is so baffling to all the guests, but what is more troublesome is that the well accepted fact in set theory that "there are ω many, pairwisely different, and finitely valued natural numbers" does not at all hold true. Or in other words, in the naïve set theory and modern axiomatic set theory, the well-recognized actually infinite rigid set, consisting exactly of all the natural numbers, $N = \{x \mid n(x)\}$, is a self-contradicting non-set, which feels apparently right but actually wrong.

Now, let G be an arbitrary countable infinite set in the framework ZFC of the modern axiomatic set theory. Then, the elements in G can be labeled using all natural numbers using the concept of one to one correspondences. So, with the proven fact that $N = \{x \mid n(x)\}$ is a non-set in place, one can directly deduce the fact that G is also a self-contradicting non-set. Summarizing our discussion above, we have employed the phenomenon of Cauchy Theater to prove the following theorem.

Theorem 9.3: Any countable (infinite) set in the naïve set theory and modern axiomatic set theory is an apparently right but actually wrong non-set. QED.

9.3.3 Transfinite spring sets and transfinite Cauchy Theaters

The purpose of this and the next subsection, which is based on (Lin (guest editor), 2008, pp. 465–468), is to show using a different method that any uncountable set is a self-contradictory non-set.

In Section 9.2, it is shown that in the framework of ZFC, various countable infinite sets are all self-contradicting non-sets. In this section, we will generalize the concept of Cauchy Theater, as introduced above in this section, and establish the concept of transfinite Cauchy Theaters. After that, by employing a new method, different of that in Section 9.2, we will prove that various uncountable infinite sets, as studied in the naïve set theory and modern axiomatic set theory, are also self-contradicting non-sets.

In particular, as is known, in the system of modern mathematics, the natural-number based mathematical induction has been generalized to the transfinite induction based on transfinite ordinal numbers. And, this generalized induction has been widely applied in the deductions and reasoning of mathematics. Now, accordingly, we will generalize our concept of Cauchy Theater and establish transfinite Cauchy Theaters, denoted C-N-I.

Let $A[wos]$ be an uncountable, well-ordered set of some transfinite ordinal numbers such that $\overline{A[wos]} = \Omega$. That is, we assume that the ordinality of the uncountable set $A[wos]$ is Ω. On the other hand, in both the naïve and modern set theories, we have the following two known theorems:

Proposition 9.12 For each well-ordered set A, it cannot be isomorphic to any segment A_α, and $\overline{A_\alpha} < \overline{A}$, for $\alpha \in A$.

Proposition 9.13 The ordinality of the well-ordered set w_α, that consists of all ordinal numbers less than α, is the ordinal number α. That is, $\overline{w_\alpha} = \alpha$.

Let us now establish and define the concept of transfinite Cauchy Theaters on the basis of $\overline{A[wos]} = \Omega$ as follows.

Assume that Ω is an uncountable ordinal number. Then, let $A[wos]$ be the well ordered set consisting of all ordinal numbers less than Ω. From Proposition 9.13, it follows that $\overline{A[wos]} = \Omega$. And from Proposition 9.12, it follows that for any ordinal number η in $A[\overline{wos}]$, we always have $\eta < \Omega$. Let $\vec{\eta}$ stand for the ordinal number variable defined on $A[\overline{wos}]$ satisfying that $\vec{\eta}$ can arbitrarily increase but stays forever less than Ω. Then, we will use the following notation

$$\text{C-N-I: } \{1, 2, \ldots n, \ldots, \omega, \omega+1, \ldots \omega \cdot 2, \ldots, \omega^2, \ldots, \vec{\eta}).$$

to represent the transfinite Cauchy Theater we want to establish. Two fundamental properties of this C-N-I theater are:

(a′) The variable η stands for an ordinal number and increases without bound. And, no matter whether η is a non-limit number with a successor and a predessessor or a limit ordinal without a predessessor, such as ω, $\omega \cdot n$, ω^2, $\omega^n, \ldots, \omega^{\omega^{\cdot^{\cdot^{\cdot}}}}, \ldots$, etc., the variable η can always be increased by adding 1 to it. So, if the ordinal numbers in a C-N-I theater are seen as the labels of seats in the C-N-I theater, then for the seat labeled by an arbitrary label η, we can always increase the number of seats in the C-N-I theater in the form of $\eta + 1$. And we can do so indefinitely.

(b′) Even though the ordinal number η can increase without bound, it will forever stay smaller than Ω, that is, $\eta < \Omega$. This restriction is reasonable, since Propositions 9.12 and 9.13 above imply that

$$\forall \eta (\eta \in A[\overline{wos}] \rightarrow \eta < \Omega).$$

To make difference, the Cauchy Theater C-N defined on the set of all natural numbers will be referred to as the special Cauchy Theater, and the previously defined transfinite Cauchy Theater C-N-I as the general Cauchy Theater. If we employ the concept of spring sets, then C-N is exactly $N = \{x \mid n(x))$. And if we let

$$O(x) =_{df} \text{“}x \text{ is an ordinal number”},$$

then the C-N-I can also be written as the following transfinite spring set:

$$O_\Omega = \{x \mid O(x) \wedge x < \Omega).$$

The fundamental difference between the special C-N and the general C-N-I is that C-N does not contain any limit ordinal number, while the general C-N-I does and contains various levels of the present progressive tense (ongoing) of potential infinities and the perfect tense of actual infinities (done).

9.3.4 Transfinite Cauchy Theater phenomena in the ZFC framework

Based on the preparation above, let us now prove the following theorem:

Theorem 9.4: Each uncountable, actual infinite set in the framework of ZFC is a self-contradicting non-set.

Proof. Let $A = \{x \mid p(x)\}$ be an uncountable set in the framework of ZFC. Since ZFC accept the axiom of choice, the well-ordering theorem holds true. That is, any non-empty set can be well-ordered. So, the given uncountable set $A = \{x \mid p(x)\}$ is not an exception. Let us now first order A into a well-ordered set $A[wos]$. Since $A[wos]$ is well ordered, there is an ordinal number α such that $\overline{A[wos]} = \alpha$. Assume $\gamma \in A[wos]$, then Proposition 9.12 above implies that the segment $A_\gamma[wos]$ cannot be isomorphic to $A[wos]$ and $\overline{A_\gamma[wos]} = \gamma < \alpha$. Now, let us use the ordinal number $\gamma = \overline{A_\gamma[wos]}$ to replace the element γ in $A[wos]$ and construct the set $A[\overline{wos}]$ of ordinal numbers $\gamma = \overline{A_\gamma[wos]}$, for $\gamma \in A[wos]$, corresponding to $A[wos]$. Of course, $A[wos]$ and $A[\overline{wos}]$ are isomorphic to each other. So, we have $\overline{A[wos]} = \overline{A[\overline{wos}]} = \alpha$, and the following conclusion:

Proposition 9.14 $\forall \eta (\eta \in A[\overline{wos}] \rightarrow \eta < \alpha)$.

Since $A[\overline{wos}]$ is not countable and consists of all ordinal numbers less than α, Proposition 9.13 above implies that $\overline{A[\overline{wos}]} = \alpha$. Based on the concept of generalized Cauchy Theater in Section 9.3.3, we first establish the following transfinite Cauchy Theater for our purpose here:

$$\text{C-N-I: } \{1, 2, \cdots, n, \cdots, \omega, \omega + 1, \cdots \omega + n, \cdots, \omega \cdot 2, \cdots, \omega \cdot n, \cdots, \omega^\omega, \cdots \vec{\eta}),$$

where η is an ordinal number, which can increase indefinitely while forever stays smaller than $\overline{A[\overline{wos}]} = \alpha$. That is, $\eta < \alpha$ holds true forever. We now see the ordinal numbers in C-N-I as the labels of the seats in the C-N-I. So, on one hand, since η can increase indefinitely, the number of seats in C-N-I can always be increased in the form of $\eta + 1$; and on the other hand, since $\forall \eta (\eta \in A[\overline{wos}] \rightarrow \eta < \alpha)$, it clearly points out the fact that η can never reach α. Therefore, we have the following conclusion:

Proposition 9.15 C-N-I can never have α seats.

Now, let us organize a movie reception for all the ordinal numbers in $A[\overline{wos}]$ in the C-N-I as the venue. So, we have the following conclusion:

Proposition 9.16 Not all ordinal numbers in $A[\overline{wos}]$ are allowed to enter C-N-Ito watch the movie, assuming that each ordinal number in C-N-I must have its own individual seat.

In fact, it is because Proposition 9.15 implies that C-N-I does not have α seats. Or, let $P(\eta) =_{df}$ "η has one and only one individual seat in C-N-I that belongs to himself." Then, this previous proposition implies that

$$\neg \forall \eta (\eta \in A[\overline{wos}] \rightarrow p(\eta)) \tag{9.10}$$

Otherwise, if we assume $\forall \eta (\eta \in A[\overline{wos}] \rightarrow p(\eta))$, then since $A[\overline{wos}]$ has α elements, we conclude that C-N-I must have α seats. This end contradicts the conclusion in Proposition 9.15 above.

However, on the other hand, for each $\xi \in A[\overline{wos}]$, since $\xi < \alpha$, it follows that when η increases indefinitely in C-N-I, C-N-I can always be made to have $\xi + 1$ seats, no matter whether ξ is a limit ordinal number or not. Therefore, in C-N-I ξ has its own individual seat, which is uniquely his. That is, we have

$$E\eta(\eta \in A[\overline{wos}] \rightarrow p(\eta)).$$

Since in the classical, two-value logical calculus, it is assumed that $E = \forall$, we now can exchange the roles of E and $\forall$. So, the previous equation becomes

$$\forall\eta(\eta \in A[\overline{wos}] \rightarrow p(\eta)). \qquad (9.11)$$

As a matter of fact, equ. (9.11) indicates that all ordinal numbers in $A[\overline{wos}]$ have their own, unique, individual seats. Therefore, they can all enter C-N-I at the same time to watch the movie. Since equs. (9.10) and (9.11) are mutually contradicting with each other, it implies that $A[\overline{wos}]$ is a self-contradicting non-set.

Also, as stated earlier, well-ordered sets $A[wos]$ and $A[\overline{wos}]$ are order-isomorphic and satisfy $\overline{A[wos]} = \overline{A[\overline{wos}]} = \alpha$. So, if f is an order-isomorphism from $A[wos]$ onto $A[\overline{wos}]$, the elements in $A[\overline{wos}]$ can be indexed using the ordinal numbers in $A[\overline{wos}]$ through the order-isomorphism f. That is, from the fact that $A[\overline{wos}]$ is a self-contradicting non-set, it follows that the well-ordered set $A[wos]$ is also a self-contradicting non-set. And, since $A[wos]$ is obtained by ordering the elements in $A = \{x \mid P(x)\}$, in terms of sets, the sets $A[wos]$ and A are the same. Therefore, the uncountable set $A = \{x \mid P(x)\}$ is a self-contradicting non-set. QED.

9.4 CAUCHY THEATER AND DIAGONAL METHOD

The purpose of this section, which is based on (Lin (guest editor), 2008, pp. 469–473), is to show that a class of well-used methods in set theory is invalid.

Based on the concept of Cauchy Theater introduced in Section 9.3, it is proven that the argument employing the diagonal method for the fact that the set of all real numbers $R = \{x \mid r(x)\}$, where $r(x) =_{df}$ "x is a real number", has an uncountable cardinality, is not valid. The diagonal method has appeared in both the naïve and modern axiomatic set theories.

In particular, as is known, in the naïve set theory, the fact that the set R of all real numbers has an uncountable cardinality is proven by using the diagonal method as follows:

Prove by contradiction. Assume that the number of real numbers in the open interval $(0, 1) = \mathrm{R}$ has cardinality ω. Then, the elements in R can be labeled using natural numbers. Or in other works, between R and the set of all natural numbers $N = \{x|x$ is a natural number$\}$ an one-to-one correspondence can be established as in Figure 9.4, where p_{ij}, $i, j \in N$, is a digit.

Now, construct a real number

$$\theta = 0.p_1p_2 \ldots p_n \ldots,$$

where p_i, $i \in N$, is a digit, satisfying that $p_i = p_{ii} + 1$; when $p_{ii} = 9$, take $p_i = 0$. Since $\theta \in (0, 1)$, it is evidently that θ is a real number. On the other hand, since θ is different

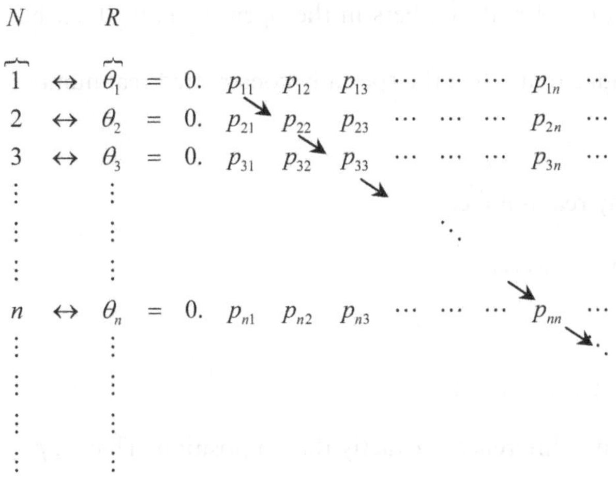

$$
\begin{array}{ccccccccccc}
N & & R & & & & & & & & \\
1 & \leftrightarrow & \theta_1 & = & 0. & p_{11} & p_{12} & p_{13} & \cdots\ \cdots\ \cdots & p_{1n} & \cdots \\
2 & \leftrightarrow & \theta_2 & = & 0. & p_{21} & p_{22} & p_{23} & \cdots\ \cdots\ \cdots & p_{2n} & \cdots \\
3 & \leftrightarrow & \theta_3 & = & 0. & p_{31} & p_{32} & p_{33} & \cdots\ \cdots\ \cdots & p_{3n} & \cdots \\
\vdots & & \vdots & & & & & & \ddots & & \\
n & \leftrightarrow & \theta_n & = & 0. & p_{n1} & p_{n2} & p_{n3} & \cdots\ \cdots\ \cdots & p_{nn} & \cdots \\
\vdots & & \vdots & & & & & & & & \ddots
\end{array}
$$

Figure 9.4 Imagined 1-1 correspondence between N and R.

of each θ_m in R, there is a finite position label $m \in N$ such that $p_m \neq p_{mm}$. Because of $E\theta_m(\theta_m \in R \to \theta \neq \theta_m)$, in the naïve set theory, it is equivalent to say that

$$\forall\theta_m(\theta_m \in R \to \theta \neq \theta_m).$$

If we let $E =_{df}$ "for each" and $\forall =_{df}$ "for all", then according to the convention of interpreting the universal quantifier $\forall$, in any given circumstance E and $\forall$ can be exchanged. Therefore, $\theta \notin R$. This end implies that R did not contain all real numbers in the open interval $(0, 1)$. That surely represents a contradiction. This means that the assumption that the set R, consisting of all the real numbers in $(0, 1)$, has a countable cardinality does not hold true. Therefore, the set of real numbers $R = \{x \mid x \in (0, 1) \wedge x$ is a real number$\}$ has an uncountable cardinality, denoted c.

In the following, we will point out that after we introduce the concept of spring set $N = \{x \mid n(x))$ or the Cauchy Theater C-N in the naïve set theory, it can be seen that the previous proof about that the set R has cardinality c is not reliable. It is because in the naïve set theory, the "finite positional difference principle of unequal real numbers" has some problem. Here, the details of the principle are: For two real numbers θ and θ':

$$\theta = 0.p_1p_2 \ldots p_n \ldots,$$

and

$$\theta' = 0.p'_1p'_2 \ldots p'_n \ldots,$$

where p_i and p'_i, $i \in$ N, are digits, if $\theta \neq \theta'$, then there must be that $\exists m(m < \omega \wedge p_m \neq p'_m)$. And, the inverse also holds true. That is,

$$\theta \neq \theta' \text{ iff } \exists m(m < \omega \wedge p_m \neq p'_m),$$

where $iff =_{df}$ "if and only if."

Let us now look at the process of using the diagonal method to prove that the set R, consisting of all real numbers in the open interval $(0, 1)$, has an uncountable cardinality.

When we judge that when the specially constructed real number

$$\theta = 0.p_1p_2 \ldots p_n \ldots,$$

is different of any real number

$$\theta_i = 0.p_{i1}p_{i2} \ldots p_{in} \ldots,$$

in the rigid set

$$R = \{\theta_1, \theta_2, \theta_3, \ldots, \theta_n, \ldots\},$$

the finite positional difference is exactly the ith position. That is, $p_i \neq p_{ii}$, or

$$\theta \neq \theta_i \text{ iff } \exists i(i < \omega \wedge p_i \neq p_{ii}).$$

Now, let us choose two arbitrary real numbers θ_i and θ_j from the set R. As long as $i \neq j$, the positional difference number i for judging $\theta \neq \theta_i$ is different of the positional difference number j for judging $\theta \neq \theta_j$. So, we have the following conclusion:

Proposition 9.17 The finite position numbers, used to judge the difference between θ and an element in R, are pairwisely different. That is, $\theta_i \neq \theta_j$ *iff* the difference position $i(\theta \neq \theta_i$ and $p_i \neq p_{ii}) \neq$ the difference position j $(\theta \neq \theta_i$ and $p_j \neq p_{jj})$.

Since the real number set

$$R = \{\theta_1, \theta_2, \theta_3, \ldots \theta_n, \ldots\}$$

has ω elements, as assumed in the proof by contradiction, the so-called telling the real number

$$\theta = 0.p_1p_2 \ldots p_n \ldots,$$

is different of all real numbers in R is to cheek whether or not θ is different from all the ω real numbers in R. Now, from Proposition 9.17 above, it follows that there must be ω many pairwisely different difference positions in order for one to be sure that θ is different of all the ω real numbers in R. As a matter of fact, under the principle of finite positional differences, in the process of employing the previous diagonal method to prove that the set R, consisting of all real numbers in $(0, 1)$, has an uncountable cardinality, we should have the following conclusion:

Proposition 9.18 "Determine that the real number θ is different from all the ω real numbers in R" *iff* "there exist ω many, pairwisely unequal, finite difference positions."

Now, as long as we hold a special reception in the Cauchy Theater for the aforementioned "ω many, pairwisely distinct, finite difference positions", we will know the conclusion that "there exist ω many, pairwisely unequal, finite difference positions" cannot hold true. Just like when we discussed about the phenomenon of the Cauchy Theater in Section 9.3, both the door keepers A and B have reached the following mutual understanding: The conclusion that "there exist ω many, pairwisely different, finite ordinal numbers" cannot hold true. Therefore, from Proposition 9.18, it follows that under the principle of finite positional differences, the entire process of argument using the diagonal method for that the real number θ is different from all the ω many real numbers in R cannot hold true.

At this juncture, what should be noticed is that since the conclusion that "there exist ω many, pairwisely distinct, finite difference positions" cannot hold true, Proposition 9.18 implies that the conclusion that "the real number θ is different of all the ω real numbers in R" cannot hold true. Therefore, the conclusion that "all real numbers in the set R have a finite positional difference" with real number θ cannot hold, either. However, on the other side, for each or for any real number in R, it must have a finite positional difference with θ. That is to say, even though we have

$$E\theta_i(\theta_i \in R \rightarrow \exists i(i < \omega \wedge p_i \neq p_{ii})),$$

we do not necessarily have

$$\forall\theta_i(\theta_i \in R \rightarrow \exists i(i < \omega \wedge p_i \neq p_{ii})).$$

That is, if we define a predicate P as "some real number θ_i in R has a finite difference position i with the real number $\theta(p_{ii} \neq p_i)$," then our discussion above implies the following:

$$\text{(II) } E\theta_i(\theta_i \in R \rightarrow P(x)) \Rightarrow \forall\theta_i(\theta_i \in R \rightarrow P(x))$$

does not hold true. However, the following statement:

$$\text{(I) } \forall\theta_i(\theta_i \in R \rightarrow P(x)) \Rightarrow E\theta_i(\theta_i \in R \rightarrow P(x))$$

holds true in any circumstance. That is to say, even though from "for all x, so and so" we can always produce "for each x, so and so," the opposite, "for each x, so and so", might not necessarily lead to "for all x, so and so." This fact implies that the thought convention of the well recognized "for each" equals "for all" (that is, $E = \forall$) in the classical two-value logic has serious limitations in the method of reasoning and deduction when both the concepts of potential and actual infinite are distinguishingly allowed.

9.5 RETURN OF BERKELEY PARADOX

For a long time now, the common belief has been that the establishment and development of the theory of limits had provided an explanation for the Berkeley paradox. However, when we revisit some of the age-old problems using the thinking logic of

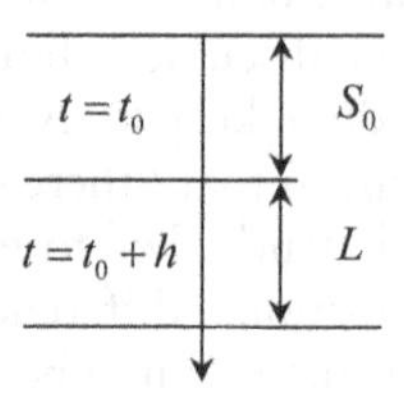

Figure 9.5 Falling distances of the free falling object.

allowing both the concepts of potential and actual infinite, we find surprisingly that the shadow of the Berkeley paradox does not truly disappear in the foundation of mathematical analysis. At the same time, if we clearly distinguish the concepts of potential and actual infinite from each other, the fact is that the Berkeley paradox of the 18th century has already returned.

This section is based on (Lin (guest editor), 2008, pp. 474–481).

9.5.1 Calculus and theory of limits: A history recall

During the 17th and the 18th century, because of the appearance of calculus and its wide range applications in various areas of learning, the theory of calculus was developed rapidly. However, during this time period, the theory of calculus was established on the ambiguous and vague concept of infinitesimals. Since the foundation was not solid, the theory of calculus was criticized severely from many different angles. The most central criticism is the famous Berkeley paradox. To illustrate this paradox, let us use the example of computing the instantaneous speed of a free falling object at the time moment t_0. The well-known distance formula for a free falling object is given by $S=\frac{1}{2}gt^2$. When $t=t_0$, the distance the object has fallen through is $S_0=\frac{1}{2}gt_0^2$. When $t=t_0+h$, the distance of falling is $S_0+L=\frac{1}{2}g(t_0+h)$. This end implies that during the h seconds, the object has fallen through the distance L, as shown in Figure 9.5, that is given by

$$L=\frac{1}{2}g(t_0+h)^2-S_0=\frac{1}{2}g(t_0+h)^2-\frac{1}{2}gt_0^2=\frac{1}{2}g(2t_0+h)h.$$

So, within the h seconds, the average speed of the falling object is

$$V=\frac{L}{h}=\frac{\frac{1}{2}g(2t_0+h)h}{h}=gt_0+\frac{1}{2}gh.$$

Evidently, the smaller the time interval h is, the closer the average speed is to the momentary speed at $t=t_0$. However, no matter how small h is, as long as $h\neq 0$, the average speed is not the same as the speed at $t=t_0$. When $h=0$, there is no change in the falling distance. So, $V=\frac{L}{h}=gt_0+\frac{1}{2}gh$ becomes the meaningless $\frac{0}{0}$. So, it becomes impossible to compute the instantaneous speed of the falling object at the time moment $t=t_0$.

Both Newton and Leibniz once provided several explanations to get rid of this difficulty.

a) Assume that $h$is an infinitesimal. So, $h \neq 0$ and the ratio

$$\frac{L}{h} = \frac{\frac{1}{2}g(2t_0 + h)h}{h}$$

is meaningful. And, this ratio can be simplified to $L/h = gt_0 + \frac{1}{2}gh$. Since the product of the infinitesimal h and a positive, bounded value can be ignored, the term $\frac{1}{2}gh$ can be erased, so that $gt_0 + \frac{1}{2}gh$ becomes gt_0. This is the instantaneous speed of the falling object at $t = t_0$. That is, $V|_{t=t_0} = gt_0$.

b) Just let the final ratio of

$$\frac{L}{h} = \frac{\frac{1}{2}g(2t_0 + h)h}{h} = gt_0 + \frac{1}{2}gh$$

be gt_0, which is the instantaneous speed at $t = t_0$.

c) Claim the limit of $\frac{L}{h} = gt_0 + \frac{1}{2}gh$ to be gt_0.

d) Both Newton and Leibniz also explained that when $h \to 0$, that is, it is not before h becomes zero 0 or after h becomes 0; at the exact moment when h becomes 0, the value of $L/h = gt_0 + \frac{1}{2}gh$ is gt_0.

None of these explanations could satisfactorily make people feel that the following contradiction has been resolved:

$$* \begin{cases} \text{(A) To make } \frac{L}{h} \text{ meaningful, one must have } h \neq 0; \\ \text{(B) To obtain } gt_0 \text{ as the instantaneous speed at } t = t_0, \text{ one must assume } h = 0. \end{cases}$$

In the process of solving a fixed, chosen problem, how can the same quantity h be not equal to 0 and be equal to 0 at the same time? This is the Berkeley's paradox in the history of mathematics.

Morris Kline (1979), in his book, entitled "Mathematical Thoughts in Ancient to Modern Times", pointed out that the derivative is the ratio of increments of the disappeared y and x, that is, the ratio of dy and dx. Berkeley factually pointed out that they are neither finite quantities nor infinitesimals; but they are something. These rates of changes are nothing but the apparitions of disappeared quantities.

The history of mathematics characterizes the chaotic situation existing in the community of mathematics in the 18th century after the birth of calculus as the second crisis of mathematics.

This chaotic situation in history forced mathematicians to look at and to treat this Berkeley paradox seriously so that the second crisis of mathematics could be successfully resolved. This effort directly led to the era of Cauchy–Weierstrass. During this era, Cauchy carefully and systematically established the theory of limits. Dedekind proved fundamental theorems of the theory on the basis of real numbers. Cantor and Weierstrass devoted their efforts to lay down the theoretical foundation for calculus, developed the $\varepsilon - N$ and $\varepsilon - \delta$ methods, so that the proposals and applications of the concepts of realistic infinitesimals and infinities are avoided. That is the present-day mathematical analysis.

9.5.2 Abbreviations and notes

In (Lin (guest editor), 2008, pp. 424–432) or Section 7.3 in this book, after introducing a set of abbreviations, we established the descriptive definitions of potential infinities (*poi*) and actual infinities (*aci*). There, the meanings and interpretations of the symbols: "$\uparrow$", "$\top$", and "$\not\top$", were also given. In this section, corresponding to the mathematical background and specific models of our study in this section, we will select some of the interpretations appropriate for our purpose. And, on this basis, we will construct a set of new abbreviations and their corresponding interpretations.

$a \uparrow b =_{df}$ "the variable *a* approaches the limit *b* indefinitely";
$a \top b =_{df}$ "the variable *a* reaches the limit *b*";
$a \not\top b =_{df}$ "the variable *a* never reaches the limit *b*";
$a \uparrow b \wedge a \top b =_{df}$ "the variable *a* approaches and reaches the limit *b*";
$a \uparrow b \wedge a \not\top b =_{df}$ "the variable *a* approaches, but never reaches the limit *b*".

Definition 9.1 If we have $a \uparrow b \wedge a \top b$, then we say that the variable *a* approaches the limit *b* in the fashion of actual infinities (done).

Definition 9.2 If we have $a \uparrow b \wedge a \not\top b$, then we say that the variable *a* approaches the limit *b* in the fashion of potential infinities.

It can be seen that for any variable *a*, it approaches its limit *b* either in the fashion of actual infinities or in the fashion of potential infinities. Besides, limit expressions usually are equations involving the limit symbol *lim*. For example, $\lim_{x \to x_0} f(x) = A(0, \omega)$, $\lim_{x \to 0} f(x) = A(0, \omega)$, $\lim_{x \to \infty} f(x) = A(0, \omega)$, etc. The two sets of variables in each of the limit expressions are $x \to x_0$ and $f(x) \to A$ with the set of variables $x \to x_0$ located in the subscript of the limit expression as the first variable, and the other set of dependent variables $f(x) \to A$ as the second variable.

Notes: No matter whether $a \uparrow b \wedge a \top b$ or $a \uparrow b \wedge a \not\top b$ one has, the common characteristic of the two possibilities is $a \uparrow b$. The inconsistency of the two is the perfect tense $a \top b$ (done) and the present progressive tense $a \not\top b$ (ongoing). However, in the theory of limits, all the relevant definitions of limits based on the $\varepsilon - \delta$ and $\varepsilon - N$ methods do not deal with and distinguish such concepts as $a \top b$ and $a \not\top b$. Instead, they are all unified by the concept of $a \uparrow b$. That is, they only consider a variable approaching its limit. As for how the variable approaches its limit, none of $a \top b$ and $a \not\top b$ is considered. As a matter of fact, these two fashions of approaching limits have a common characteristic: $a \uparrow b$. This characteristic is already enough for the definition of limits using the $\varepsilon - \delta$ and $\varepsilon - N$ methods. But for now, we will introduce to the theory of limits such concepts as a variable approaching its limit in the fashion of actual infinities and a variable approaching its limit in the fashion of potential infinities. And, we will strictly separate the cases for a variable to approach its limit either in the fashion of actual infinities or in the fashion of potential infinities.

It should be clear that when one studies problems in the theory of limits, he has valid supporting evidence to back up his need to employ the thinking logic and method of deduction of allowing both potential and actual infinities and the recognition that potential infinities are not the same as actual infinities ($poi \neq aci$). It is because in

Section 7.4, we have obtained the following conclusion: The thinking logic and method of analysis of allowing these two classes of infinities are an intrinsic attribute of modern mathematics and its theoretical foundation. Especially, they are an intrinsic attribute of the theory of limits. Also, we obtained in Section 8.3 the following conclusion: The thought convention that potential infinities are not the same as actual infinities ($poi \neq aci$) is also an intrinsic attribute of modern mathematics and its theoretical foundation.

Therefore, to the theory of limits we introduce such concepts as a variable approaching its limit in the fashion of actual infinities and a variable approaching its limit in the fashion of potential infinities, and clearly separate the fashions on how a variable approaches its limit. Other similar activities will not be externally and artificially added to the theory of limits. Instead, they are merely parts of the intrinsic attribute, thinking logic, and method of reasoning.

9.5.3 Definability and realizability of limit expressions

In terms of the first variable $x \to x_0$ in the limit expression $\lim_{x \to x_0} f(x) = A$, if the variable x approaching its limit x_0 in the fashion of actual infinities causes trouble or leads to contradictions, that is, $x \uparrow x_0 \wedge x \top x_0 \vdash B, \neg B$, then the variable x has to approach its limit x_0 in the fashion of potential infinities. That is, it must be that $x \uparrow x_0 \wedge x \bar{\top} x_0$.

Example 9.1 For the limit expression $\lim_{x \to 0} 1/x = \omega$, if $x \uparrow 0 \wedge x \top 0$, there must be a contradiction. It is because $x \top 0$ stands for the situation of $x = 0$ to actually become a reality. However, mathematically speaking, 0 cannot be used as a divisor. So, $x \top 0$ clearly means that the function $1/x$ will eventually become meaningless instead of ω. In fact, the function $1/x$ is not defined at $x = 0$. In short, for the first variable $x \to 0$ in the limit expression $\lim_{x \to 0} 1/x = \omega$, the variable x can only approach its limit 0 in the fashion of potential infinities. That is, only $x \uparrow 0 \wedge x \bar{\top} 0$ is allowed, but not $x \uparrow 0 \wedge x \top 0$. QED.

Example 9.2 For the first variable $n \to \omega$ in the limit expression $\lim_{n \to \omega} 1/n = 0$, if $n \uparrow \omega \wedge n \top \omega$ appears, then it means that the situation $n = \omega$ will sooner or later occur. However, in mathematics, the convention is that all natural numbers must be finite ordinal numbers, that is, $\forall x(x \in N \to x < \omega)$. So, the moment of $n = \omega$ is not at all allowed to occur. So, for the first variable $n \to \omega$ in the given limit expression, the variable n must approach its limit ω in the fashion of potential infinities. That is, only $n \uparrow \omega \wedge n \bar{\top} \omega$ is permitted, but not $n \uparrow \omega \wedge n \top \omega$. QED.

Definition 9.3 Assume that the limit expression $\lim_{x \to x_0} f(x) = A$ satisfies that $f(x)$ is defined at $x = x_0$, and the following conditions:

(i) $x \uparrow x_0 \wedge x \top x_0$, $f(x) \uparrow A \wedge f(x) \top A$. That is, both x and $f(x)$ approach their individual limits in the fashion of actual infinities; and
(ii) $f(x) = A$ iff $x = x_0$.

Then the limit expression $\lim_{x\to x_0} f(x) = A$ is said to be both "definable" and "realizable", or "realizable" for short. Otherwise (for example, $x \uparrow x_0 \wedge x \top x_0 \vdash B, \neg B$, etc.) this limit expression is said to be "definable" but not "realizable", or "not realizable" for short.

Now, we have the following important conclusion:

Proposition 9.19 Each definable limit expression is either realizable or not realizable.

Let us now look at some examples of realizable limit expressions.

Example 9.3 Lobachevsky's function is continuous defined on $[0, +\infty]$ and is shown in Figure 9.6, where the length x is called the parallel distance and $\angle\alpha$ the parallel angle satisfying that the parallel angle α decreases as x gets greater, and increases as x gets smaller. Therefore, the parallel angle $\alpha = \pi(x)$ is a function of the parallel distance and satisfies the following limit expression:

$$\lim_{x\to 0} \pi(x) = \frac{\pi}{2}, \quad \lim_{x\to\infty} \pi(x) = 0.$$

In these limit expressions, $\pi(\mathrm{x})$ is defined at $x =$ 0 and satisfies

(i) $x \uparrow 0 \wedge x \top 0, \pi(x) \uparrow \frac{\pi}{2} \wedge \pi(x) \top \frac{\pi}{2}$; and
(ii) $\pi(x) = \frac{\pi}{2}$ iff $x = 0$.

So, from the definition, the limit expression $\lim_{x\to 0} \pi(x) = \pi/2$ is realizable. QED.

Example 9.4 Assume that we have the limit expression $\lim_{x\to 0} ax = a \lim_{x\to 0} x = a \cdot 0 = 0$. Here, the function $f(x) = x$ is defined at $x = 0$ and satisfies

(i) $x \uparrow 0 \wedge x \top 0, f(x) \uparrow 0 \wedge f(x) \top 0$; and
(ii) $x = f(x) = 0$ iff $x = 0$.

So, the given limit expression is realizable. QED.

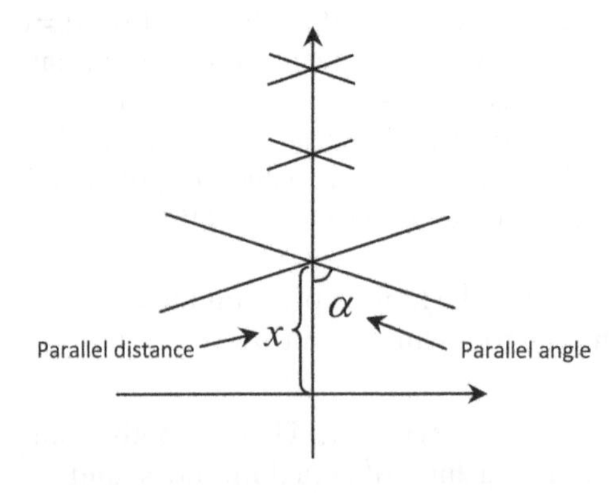

Figure 9.6 The Lobachevsky's function.

Let us conclude this subsection with two examples of definable but not realizable limit expressions.

Example 9.5 In terms of the limit expression $\lim_{n\to\infty} n = \omega$, since in mathematics, it has been clearly written that $\forall x(x \in N \to x < \omega)$, if we assume $n \uparrow \omega \wedge n \top \omega$, then we will have to face the result $n = \omega$ that contradicts the convention that $n < \omega$. That is, we must have $n \uparrow \omega \wedge n \overline{\top} \omega$. So, the given limit expression is not realizable. QED.

Example 9.6 For the limit expression $\lim_{\Delta x\to 0, \Delta y\to 0} \Delta y/\Delta x = dy/dx = A$, since it is known that $0/0$ is not defined mathematically, if we assume $\Delta x \uparrow 0 \wedge \Delta x \top 0$ and $\Delta y \uparrow 0 \wedge \Delta y \top 0$, we would make $\lim_{\Delta x\to 0, \Delta y\to 0} \Delta y/\Delta x$ meaningless. So, this limit expression is not realizable. QED.

Also, the limit expressions $\lim_{x\to 0} 1/x = \omega$ and $\lim_{n\to\omega} 1/n = 0$ in Examples 9.1 and 9.2 are not realizable, either.

9.5.4 New Berkeley paradox in the foundation of mathematical analysis

Let us now recheck how the instantaneous speed of a free falling object at $t = t_0$ is computed in the current theory of limits. As is known, the solution to this problem is obtained by the following expression of limits:

$$
\begin{aligned}
V|_{t=t_0} &= \lim_{\Delta t\to 0} \frac{\Delta S}{\Delta t} = \lim_{\Delta t\to 0} \left(gt_0 + \frac{1}{2} g\Delta t \right) \\
&= gt_0 + \frac{1}{2} g \lim_{\Delta t\to 0} \Delta t \\
&= gt_0 + \frac{1}{2} g \cdot 0 = gt_0 + 0 = gt_0
\end{aligned}
\tag{9.12}
$$

It should be pointed out that under the fixed thinking logic of the current theory of limits, that is, in terms of a variable a and its limit b, the only relationship between a and is $a \uparrow b$ without considering or refusing to consider either $a \top b$ or $a \overline{\top} b$, the previous limit expression in equ. (9.12) for computing the instantaneous speed at $t = t_0$ indeed does not contain an arguable problem, and the theory of limits also seems plausible. However, since now we have introduced in the theory of limits the concept of a variable approaching its limit in the fashion of either potential infinities or actual infinities, that is, $a \uparrow b \wedge a \top b$ and $a \uparrow b \wedge a \overline{\top} b$, potential new abnormalities will appear. To this end, let us discuss as follows.

The previous limit in equ. (9.12) consists of the following two central expressions:

(1) $\lim_{\Delta t\to 0} \frac{\Delta S}{\Delta t} = gt_0$; and

(2) $\lim_{\Delta t\to 0} \Delta t = 0$.

The first variables $\Delta t \to 0$ in both of these limit expressions are the same. However, from Example 9.6 above, it follows that the limit expression (1) is not realizable. So, we must have $\Delta t \uparrow 0 \wedge \Delta t \not\top 0$. That is, the variable Δt must approach its limit 0 in the fashion of potential infinities. Also, from Example 9.4 above, it follows that the limit expression (2) is realizable. So, we must have $\Delta t \uparrow 0 \wedge \Delta t \top 0$. That is, the variable Δt must approach its limit 0 in the fashion of actual infinities.

At this juncture, we have to question naturally: For the same process of solving the same problem and for the same variable $\Delta t \to 0$, how can both $\Delta t \top 0$ and $\Delta t \not\top 0$ be allowed at the same time? This entire process of computing the instantaneous speed of the free falling object seems to suggest that in order to make $\lim_{\Delta t \to 0} \frac{\Delta S}{\Delta t} = gt_0$ meaningful, we took the assumption $\Delta t \uparrow 0 \wedge \Delta t \top 0$. That is, we allowed the variable Δt to approach its limit 0 in the fashion of potential infinities. On the other hand, to obtain $\frac{1}{2}g \lim_{\Delta t \to 0} \Delta t = \frac{1}{2}g \cdot 0$, we changed our mind and allowed $\Delta t \uparrow 0 \wedge \Delta t \top 0$. That is, we allowed the variable Δt to approach its limit 0 in the fashion of actual infinities. This kind of deduction and reasoning is hard for people to accept satisfactorily.

To resolve this dissatisfaction, can we suggest to permit Δt to approach its limit 0 in the fashion of actual infinities? It is impossible, since it would suggest $\Delta t \top 0$. That would make the equation $\lim_{\Delta t \to 0} \frac{\Delta S}{\Delta t} = gt_0$ be equal to the meaningless $\frac{0}{0}$. Since it is so, can we unify the situation by allowing Δt approach its limit 0 in the fashion of potential infinities? This end is also impossible, since under the assumption that $\Delta t \uparrow 0 \wedge \Delta t \not\top 0$, it must cause the limit expression $\lim_{\Delta t \to 0} \Delta t = 0$ to be not realizable. However, from Example 9.4 above, it can be shown that the limit expression $\lim_{\Delta t \to 0} \Delta t = 0$ is realizable. Also, Proposition 9.19 implies that each limit expression is either realizable or not realizable. That is, there does not exist such a limit expression that is both realizable and not realizable. So, of course, the expression $\lim_{\Delta t \to 0} \Delta t = 0$ should not be an exception. In short, after we introduce the concept of approaching limits in the fashion of either potential or actual infinities, various explanations for the variable Δt to approach its limit 0 employed in the process of calculating the instantaneous speed of a free falling object at $t = t_0$ can no longer be plausible.

Our discussion above indicates that as long as in the theory of limits, the concepts of approaching limits in the fashion of either actual or potential infinities are introduced and distinguished, then the shadow of the Berkeley paradox can still be seen.

The notes in Subsection 9.5.2 above have clearly implied that both calculus and the theory of limits have already allowed the thinking logic of both kinds of infinities. Therefore, it should be seen as a logical train of thought for us to explore and to discuss problems existing in the theory of limits using the thinking logic and method of reasoning of allowing both kinds of infinities. Conversely, if there existed such a theoretical system, in which on one hand, for its own existence and development, the thinking logic of allowing both kinds of infinities were employed, and on the other hand, for the purpose of covering up existing contradictions, the thinking logic of allowing both kinds of infinities were disallowed in the analysis of problems, it would be reasonable for anyone to avoid such a theory and to devote his time and effort to study something else.

9.6 INCONSISTENCY OF THE NATURAL NUMBER SYSTEM

The purpose of this section, which is based on (Lin (guest editor), 2008, pp. 482–488), is to employ a third method show that the system of natural numbers is inconsistent.

Without directly employing the concepts of potential and actual infinities, we show in this section that the concept of the set $N = \{x \mid n(x)\}$, where $n(x) =_{def}$ "x is a natural number", of all natural numbers is a self-contradicting, incorrect concept.

9.6.1 Notes and abbreviations

Let us first illustrate the concepts of "the number of natural numbers" and "the magnitude of a natural number." In general, when a concept is defined, one has to make use of some more basic and more general concepts than the one that is to be introduced. Reasoning in this way, it can be seen that there are surely some concepts that can no longer be defined on the basis of other more general and more basic concepts than they are. So, these primary concepts have to be introduced by mutually describing each other, through listing, illustrating and/or describing through using specific examples.

As for the concept of the magnitude of a natural number, it can be understood as follows: For each natural number, it always has a name. For example, for natural number "3", it is read as "three", or its name is "three". Now, for any chosen natural number, its name we use as the magnitude or the value of this natural number. As for the concept of the number or the count of all natural numbers, we have to understand it this way: When faced with a class of objects or elements, we can always count how many objects or elements there are in the class. And, the result of the count can be expressed using some kind of numeral or quantity. It is like when we inform others, we might conveniently say that there are 3 peaches, 8 water melons, and 1,500 hydrogen atoms over here. The only difference for now is that we are looking at nothing else but the set of all natural numbers $N = \{x \mid n(x)\}$, which each and every natural number belongs to. We now can count to "100 natural numbers", "1000 natural numbers, "infinite many natural numbers," etc.

In this section, we use the symbol $kc(n)$, count, to represent the "count of natural numbers" after we have finished counting the first n elements in the natural number sequence λ; the symbol $nv(n)$, numerical value, for the magnitude of the nth natural number in the sequenceλ, the symbol $[\inf]$ for the counting result of "infinitely many," etc. That is, we have introduced the following abbreviations:

$kc(n) =_{def}$ the count of natural numbers after finished counting the first n elements in the sequence λ;
$nv(n) =_{def}$ the magnitude of the nth natural number in the sequence λ;
$[fin] =_{def}$ "finite";
$[\inf] =_{def}$ "infinite or infinitely many";
$[\neg fin] =_{def}$ "not finite".

As a matter of fact, the symbol $kc(n)$ is like the concept of cardinality of a finite or infinite set in set theory, while $nv(n)$ is analogous to that of ordinality of a finite or infinite set in set theory. However, in this section, we will not use these concepts of cardinality and ordinality. For example, the cardinality of the set N of all natural numbers is written as $\overline{\overline{N}} = \aleph_0$, and the ordinality of the sequence λ of all natural numbers is denoted by $\overline{N} = \omega$. Since in this section, we will consider various concepts inside the set N or the sequence λ, a closed system, there will be no chance for us to sense the existence of the whole of N or λ. That is why we have to introduce such abbreviations as $kc(n)$ and $nv(n)$, and why we avoid using symbols like $\aleph_0$ and ω in this section.

9.6.2 The proof of inconsistency of *N*

Now, we are ready to prove the following theorem:

Theorem 9.5: The set $N = \{x \mid n(x)\}$, where $n(x) =_{def}$ "x is a natural number", consisting exactly of all natural numbers, is a self-contradicting non-set. (See Section 9.1, for another proof of a similar result.)

Proof. Let us order the set $N = \{x \mid n(x)\}$ according to the elements' magnitudes starting from the smallest to form the following sequence of natural numbers:

$$\lambda : \{1, 2, 3, \ldots, n, \ldots, \ldots\}.$$

We now conduct the following operation within the inside of the sequence λ with the rule that

Rule 9.1. Starting from 1, count the natural numbers in λ one by one following the given order in λ.

What is found is that every time we pause briefly, our current count (that is, $kc(n)$) of natural numbers always equals the magnitude (that is, $nv(n)$) of that last number just counted. For example, when finished counting the first 5 natural numbers, the value of this last natural number is exactly 5. When finished counting the 1,000th natural number, the magnitude of this last natural number is exactly 1,000. If we don't feel the trouble, we can also apply mathematical induction to prove that under Rule 9.1 above, the identity of the "count" equal to the "value" holds true for every natural number we just finished counting. That is to say, in terms of the symbols $kc(n)$ and $nv(n)$ we introduced in the previous section and the related interpretations, we can prove using mathematical induction that

$$\forall nv(n)(nv(n) = kc(n)).$$

Now, let us look at the detailed argument.

The basic step: When $nv(n) = 1$, we surely have $kc(n) = 1$.

The induction step: Assume that if $nv(n) = m$, the count $kc(n) = m$. Now, let the value $nv(n) = m + 1$. Since the natural number with the value $nv(n) = m + 1$ is exactly that natural number that is ordered right after the natural number of value $nv(n) = m + 1$, in the sequence λ, when we count, starting from natural number 1 along the order in λ and finish right at the natural number with value $nv(n) = m + 1$, we arrive at exactly the natural number which equals adding 1 to the base $kc(n) = m$. That is, at the moment we must have $kc(n) = m + 1$.

Therefore, from mathematical induction, it follows that in terms of the natural number's value $nv(n)$, the equation $kc(n) = nv(n)$ always holds true. That is, we have the following important conclusion:

Proposition 9.20 $\forall nv(n)(nv(n) = kc(n))$.

On the other hand, we have the following well-known theorem:

Proposition 9.21 Each natural number is a finite number. That is, the value of each natural number is finite. In symbols, we have

$$\forall n(n \in N \to nv(n)[fin]).$$

So, from Proposition 9.21, it follows that

$$\forall nv(n)(nv(n)[fin]).$$

From the axiom of equal-value substitutions and the important conclusion in Proposition 9.20 above, it follows that

$$\forall kc(n)(kc(n)[fin]).$$

Since we do our reasoning in the framework of non-constructive mathematics, we can touch on all the natural numbers in the set N or the sequence λ. Or in other words, each of these elements can be counted once. Then, the previous well-known Proposition 9.21, $\forall n(n \in N \to nv(n)[fin])$, tells us that even if you can count every element in the set N or the sequence λ of natural numbers without missing anything, there will not appear the situation of either $nv(n)[\neg fin]$ or $nv(n)[\inf]$. This end implies that the validity and scientificality of the previous conclusion, $\forall nv(n)(nv(n)[fin])$, are a certainty in the framework of non-constructive mathematics.

Secondly, the important conclusion in Proposition 9.20 was proved by using mathematical induction. And, this induction can be applied only on the values of finite natural numbers. So, it is also valid for us to apply Proposition 9.20, which was proven by using mathematical induction, on $nv(n)[fin]$. That is, based on the axiom of equal-value substitutions and Proposition 9.20, the following result must hold true: $\forall kc(n)(kc(n)[fin])$. Therefore, we have

$$\neg\exists kc(n)(kc(n)[\inf]) \quad \text{or} \quad \neg\exists kc(n)(kc(n)[\neg fin]) \tag{9.13}$$

However, on the other hand, we also have the following well-known results.

Proposition 9.22 The natural number set $N = \{x \mid n(x)\}$ or the natural number sequence λ contains infinitely many, that is, either [inf] or [$\neg fin$], natural numbers.

Proposition 9.23 The natural number set $N=\{x \mid n(x)\}$ or the natural number sequence λ contains infinitely many, that is [inf] or $[\neg fin]$, even numbers.

Proposition 9.24 The natural number set $N=\{x \mid n(x)\}$ or the natural number sequence λ contains infinitely many, that is [inf] or $[\neg fin]$, odd numbers.

Proposition 9.25 The natural number set $N=\{x \mid n(x)\}$ or the natural number sequence λ contains infinitely many, that is [inf] or $[\neg fin]$, prime numbers.

Now, Propositions 9.22–9.25 imply the following conclusion: $\exists kc(n)(kc(n)[\text{inf}])$. In fact, if we assume that the opposite holds true: $\neg\exists kc(n)(kc(n)[\text{inf}])$, then we must have $\forall kc(n)(kc(n)[fin])$. This end contradicts with any of Propositions 9.22–9.25 above. So, the assumption by contradiction that $\neg\exists kc(n)(kc(n)[\text{inf}])$ cannot holds true. That is, we have

$$\exists kc(n)(kc(n)[\text{inf}]) \text{ or } \exists kc(n)(kc(n)[\neg fin]) \tag{9.14}$$

Now, we notice that equs. (9.13) and (9.14) contradict each other. Therefore, the set $N=\{x \mid n(x)\}$ of all natural numbers is a self-contradicting non-set. Or in other words, the concept of the natural number set $N=\{x \mid n(x)\}$ is a self-contradicting, incorrect concept. It is just like what Leibniz once pointed out: The phrase of the count of all whole numbers is a self contradictory phrase. It should be abandoned (Kline, 2005).

9.6.3 Discussion and explanations

Both $nv(n)$ and $kc(n)$, studied in Subsection 9.6.2, are variables. And in terms of the variables themselves, both are $[\neg fin]$, since $nv(n)$ can increase without a bound and $kc(n)$ can always reach the next counting level without a ceiling. Therefore, the symbols $nv(n)$ and $kc(n)$ appearing in some of the formulas in the previous subsections are not about the variables themselves, instead they represent all the individuals touched upon by the variables from the respective universes of discourse.

For example, in terms of $nv(n)$ in the formula $\neg\exists nv(n)(nv(n)[\neg fin])$, it represents the statement that under the well-known result $\forall n(n \in N \to nv(n)[fin])$, even if you search through each and every natural number in N or λ without missing any element, you will not find such an element satisfying $nv(n)[\neg fin]$. In terms of the variable $kc(n)$ in either the formula $\exists kc(n)(kc(n)=1{,}000)$ or the formula $\exists kc(n)(kc(n)[\neg fin])$, it stands for the statement that there exists such a $kc(n)$ that equals 1,000, or there exists such a $kc(n)$ which is [inf]. For example, if we count and exhaust all even numbers in N or λ and then stop our counting, then according to the well-known theorem: In N or λ, there are [inf] even numbers, we know that there exists such a $kc(n)$ that equals [inf], or at least there is a $[\neg fin]$ $kc(n)$.

In the process of arguing about the inconsistency of the natural number system in the previous section, even though on the surface the concepts of potential and actual infinite were not employed directly, deep down in the intension of the proof, it is still about the problem that potential infinities cannot be mixed up with actual

$$\lambda: \quad \{ \quad 1, \qquad 2, \cdots\cdots\cdots\cdots, n, \cdots, \cdots\cdots\cdots\}$$
$$\updownarrow \qquad \updownarrow \qquad \updownarrow$$
$$\lambda kc(n): \{kc(n)=1, kc(n)=2, \cdots, kc(n)=n, \cdots, \cdots\}$$
$$\updownarrow \qquad \updownarrow \qquad \updownarrow$$
$$\lambda nv(n): \{nv(n)=1, nv(n)=2, \cdots, nv(n)=n, \cdots, \cdots\}.$$

Figure 9.7 One-to-one correspondences between λ, λnv(n), and λkc(n).

infinities. To this end, let us look at the natural number sequence λ, the value sequence $\lambda nv(n)$ of natural numbers, and the count sequence $\lambda kc(n)$ of natural numbers. Within the framework of non-constructive mathematics, these three sequences are not only one-to-one corresponding to each other, but also all perfect actual infinite, countable sequences, as shown in Figure 9.7.

In particular, let us first analyze $\lambda nv(n)$. Because $nv(n)$ can increase indefinitely, there appears to be such a possibility to head to [inf]. However, under the well-known theorem: $\forall n(n \in N \rightarrow nv(n)[fin])$, in the process for $nv(n)$ to increase without any bound, it is strictly required to satisfy $nv(n)[fin]$. From the discussion in Section 9.3, it follows that there is absolutely no way that $\lambda nv(n)$ can be a perfect (or completed) actually infinite sequence. Instead, it is a standard potentially infinite, spring sequence:

$$\lambda nv(n) : \{nv(n) = 1, nv(n) = 2, \ldots, \overrightarrow{nv(n) = n}).$$

Secondly, let us analyze $\lambda kc(n)$. Because the value $kc(n)$ can become greater without any limitation, it also points to such a possibility of leading to [inf]. Now, from the well-known theorem: "N or λ contains [inf] natural numbers," it follows that in the process for $kc(n)$ to increase without any bound, it has to reach [inf]. Therefore, we are talking about a standard Cantor-Zermelo actually infinite sequence:

$$\lambda kc(n) : \{kc(n) = 1, kc(n) = 2, \ldots, kc(n) = n, \ldots, \ldots\}.$$

That is, for the same system of natural numbers, when it is seen from the angle of $nv(n)$, it is a potentially infinite set $N = \{x \mid n(x)) = \{1, 2, \ldots, \vec{n})$. When the system of natural numbers is seen from the angle of $kc(n)$, it is now an actually infinite, rigid set under the meaning of Cantor-Zermelo $N = \{x \mid n(x)\} = \{1, 2, \ldots, n, \ldots\}$. According to Section 8.3, it is known that in the system of modern mathematics and its theoretical foundations, it is not only that potential infinities are different of actual infinities ($poi \neq aci$), but also satisfies the law of excluded middle $\vdash A \vee \neg A$. Therefore, there is no doubt that an inconsistency in the system of natural numbers exists.

In the proof of such an inconsistency as presented in this section, the potential infiniteness of $\lambda nv(n)$ is expressed in the form of either

$$\forall nv(n)(nv(n)[fin]) \quad \text{or} \quad \neg\exists nv(n)(nv(n)[\neg fin]),$$

and the actual infiniteness of $\lambda kc(n)$ is written as either

$$\neg\forall kc(n)(kc(n)[fin]) \quad \text{or} \quad \exists kc(n)(kc(n)[\text{inf}]).$$

Then, from $\forall nv(n)(nv(n) = kc(n))$ and the axiom of equal-value substitutions, we transformed the implicitly written potential and actual infinities as the final form of a direct contradiction, revealing the existing inconsistency of the natural number system.

There are two plans to resolve the contradiction. One is to overrule the validity of the well-known theorem: "N or λ contains [inf] many elements." By doing so, the unrestrained increase of $kc(n)$ will have to satisfy $kc(n)[fin]$ forever. That is, $\lambda kc(n)$ is now also a potentially infinite spring sequence:

$$\lambda kc(n) : \{kc(n) = 1, kc(n) = 2, \ldots, \overrightarrow{kc(n) = n}).$$

If we employ this plan to resolve our contradiction, we will no longer have (under the meaning of Cantor-Zermelo) the actually infinite set of natural numbers $N = \{x \mid n(x)\} = \{1, 2, \ldots, n, \ldots\}$. Instead, we will only have the potentially infinite, spring set of natural numbers $N = \{x \mid n(x)\} = \{1, 2, \ldots, \vec{n}\}$.

Another plan to resolve our contradiction is to deny the validity of the well-known theorem: $\forall n(n \in N \rightarrow nv(n)[fin])$, and accept, as a fact, that in $\lambda nv(n)$, there is such a natural number n that $nv(n)[fin]$. By doing so, $\lambda nv(n)$ becomes a completed (perfect) actually infinite sequence:

$$\lambda nv(n) : \{nv(n) = 1, nv(n) = 2, \ldots, nv(n) = n, \ldots, nv(n)[fin]\}.$$

However, this outcome disagrees with the following notation of the natural number sequence widely used in modern mathematics and its theoretical foundation:

$$\lambda : \{1, 2, \ldots, n, \ldots\}\omega.$$

In short, no matter which plan is adopted to resolve our contradiction, we have to deal with the consequence that the modern axiomatic set theory has to be modified.

At this juncture, let us notice that if we accept that within N or λ, we have $\exists nv(n)(nv(n)[\text{ inf }])$, then we will have to face the disagreement with the well-recognized notation, intension, and structure for natural number set $N = \{x \mid n(x)\}$ or natural number sequence

$$\lambda : \{1, 2, \ldots, n, \ldots\}\omega,$$

in the system of modern mathematics. And, a class of unseen entities, such as

$$\lambda nv(n) : \{nv(n) = 1, nv(n) = 2, \ldots, nv(n) = n, \ldots, nv(n)[\text{inf}]\},$$

will appear in the study of modern mathematics. That is why in the framework of modern mathematics, it is absolutely not allowed to have formulas like $\exists nv(n)(nv(n)[\text{inf}])$. On the other hand, if we recognize $\exists kc(n)(kc(n)[\text{inf}])$ within N or λ, our consequent work will agree well with the framework of modern mathematics, without causing any damage to the existing intension, structure, and notations of the natural number system as shown in the following diagram:

$$\lambda_1 : \{\underbrace{1, 2, \ldots, n, \ldots, \ldots}_{\exists kc(n)(kc(n)[\text{inf}])}\}$$

$$\lambda_2 : \{\underbrace{2, 4, \ldots, 2n, \ldots, \ldots}_{\exists kc(n)(kc(n)[\text{inf}])}, \underbrace{1, 3, \ldots, 2n + 1, \ldots, \ldots}_{\exists kc(n)(kc(n)[\text{inf}])}\}$$

That is, in this case, from appearance to intension and structure, there is no need for any slightest modification or change to the natural number system. Therefore, it is undoubtedly logical to establish such formulas as $\exists kc(n)(kc(n)[\text{inf}])$ in the framework of modern mathematics.

Part 4

Next stage of mathematics as a systemic field of thought

Chapter 10

Calculus without limit

Due to the importance of calculus, each year thousands of people from around the world study this theory. However, what is shown earlier in this book indicates that there is a theoretical need to rebuild calculus without using the concept of limits. That is exactly the goal of this chapter.

Historically, when Leibniz introduced the concept of derivatives and established the relationship between the tangent line of a curve and the problem of finding the maximum and minimum of a function, an entire class of extremely difficult problems of the past was resolved by using a straightforward method. That of course represented an important breakthrough in science and made mathematicians very excited. However, logically this method suffered from a flaw. In the process of deriving the method, a variable h is first assumed to be not zero; and as soon as it becomes clear that it no longer matters whether it is zero or not, the variable is immediately taken to be zero. For more details, please consult Chapter 6 in this book.

Facing the logical hole, Newton provided the following way to get around the problem: Don't let h become 0 at once; instead, let h become an "infinitesimal". Then what is an infinitesimal? According to Newton, it stands for the final state before a quantity becomes zero; it is not zero; but its absolute value is smaller than any positive number. Because h is not zero, it can be used as a denominator and can be cancelled in division; and because its absolute value is smaller than any positive number, it can be ignored in the final expression.

Since then the problem of what an infinitesimal is had troubled many of the first class mathematicians for more than 100 years. If it is a number, then before it becomes 0, it is still a number. Then, what is the final state whose absolute value is smaller than any positive number? If it is not a number, how can we manipulate it just like a number? Only when Karl Weierstrass established his limit theory in the $\varepsilon - \delta$ language, this problem was considered resolved satisfactorily, where the introduction of limits successfully replaced "equal to zero" by "infinitely approaching 0".

In his life time, Newton created four major works in calculus, where *Treatises on Species and Magnitude of Curvilinear Figures* was the last completed in 1693 (Newton, 1974). However, it was published in 1704, which was the first among his important works in calculus (Li, 2007). What was the reason for the publication of the other three earlier works to be greatly delayed? And why was *Treatises on Species and Magnitude of Curvilinear Figures* published 11 years after it was initially finished? It was because Newton was not satisfied with his calculus, while hoping to remove "the infinitely small quantities" from the theory. In *Treatises on Species and Magnitude of Curvilinear Figures* he gave up the concept of infinitesimals. Instead he employed the concept of initial and finishing ratios. That was the thought of "limits" he proposed in his 1687's monograph *Mathematical Principles of Natural Philosophy*. He said that strictly speaking, the finishing ratio of the diminishing quantities is not that of the last quantities; instead it stands for the limit to which the ratio of these quantities approaches when they decrease infinitely. Of course, looking back into the history, the concept of limits and that of infinitesimals are essentially the same. Newton knew he did not explain the concept of limits well. That was why 17 years after the *Mathematical Principles* was in print, he published *Treatises on Species and Magnitude of Curvilinear Figures* on the basis of the thinking logic of limits. Even so, he referred limits to as initial and finishing ratios. This historical fact indicates that for many years Newton attempted to remove the operation of infinites from calculus.

In his publication *Treatise on Fluxions* of 1742, Maclaurin (1742) attempted to establish a rigorous calculus by using the method of Taylor series, while believing that most problems addressed in calculus could be resolved without relying on the concept of limits. As we know today, that approach cannot really avoid the very concept of limits.

Because his students felt it difficult to comprehend the concepts of infinitesimals and infinities, Joseph-Louis Lagrange later also tried to develop calculus without making use of the concepts of infinitesimals and limits. To this end, he finished and published a monograph in 1797, entitled "*Théorie des fonctions analytiques*", where he attempted to derive the principles of calculus without any involvement of the infinitely small or vanishing quantities, of limits or fluxions by reducing everything to the algebraic analysis of finite quantities. He believed that the ordinary operations of algebra suffice to resolve problems in the theory of curves. Lagrange tried to exclude all of such concepts as differentials, infinitesimals, limits, etc., entirely out of calculus. He first proved Taylor expansion using algebraic means, then defined derivatives as the coefficient of h in the Taylor expansion of the function $f(x+h)$. He believed that by doing so, he could successfully overcome the theoretical difficulty of limits. However, in terms of the problem of convergence of infinite series, he still could not avoid the concept of limits.

For more than 200 years from the time of Lagrange to our modern day, the thought of establishing calculus without involving infinitesimals and limits has always been around. For example, (Sparks, 2004) titled his book using this thought, although he did not succeed in developing a calculus without limits. However, beside these authors, most scholars are convinced that without the concepts of limits or infinitesimals the theory of calculus cannot be erected. In the ocean of calculus and related publications, such conviction is absolutely the majority.

In the past 300 plus years, the question – without limits or infinitesimals can calculus be developed? – has been puzzling the community of mathematicians. What is shown in this chapter will firmly address this question with an answer YES. And, it can be seen additionally that if at the time when calculus was initially invented Newton or Leibniz knew about the concept of uniform inequalities, then the second crisis in the foundations of mathematics would not have existed; and the history of mathematics would be majorly altered. If Cauchy and Weierstrass realized this approach, then the most difficult topics of calculus related theories would have been simplified.

With this new approach of developing calculus, one does not need to study the theory of limits, even not the theory of real numbers. He could start his learning of calculus right with the definition of derivatives. Additionally, all the relevant reasoning is shorted and simplified. For example, let us look at the fact that if the derivative is positive, then the function is increasing. In most of the current textbooks, this theorem is derived on the basis of Lagrange's mean value theorem, where Lagrange mean value theorem comes as a consequence of Rolle's Theorem, while in the proof of Rolle's Theorem, one needs to apply the following fact: Any continuous function, which is defined on a closed interval, reaches its maximum. And the proof of this last proposition involves the theory of real numbers, properties of limits, and is based on the concept of continuous functions. In other words, the proof of such a most commonly used, elementary tool of calculus is so tedious that it requires the reader to walk around circles several times.

By making use uniform inequalities, within half an hour, the fact that if the derivative is positive, then the function is increasing, can be readily proven. That is an independent discovery of (Lin, Q. and Wu, 2002) and (Livshits, 2004). What is presented below for this result is another simplified argument from (Zhang, 2010).

Other than the pedagogical value, as mentioned above, what is presented in this chapter also possesses its scholarly value. For instance, the proofs of the fundamental theorems of calculus are established on the theory of real numbers. A natural question is how much they are dependent of the properties of real numbers. This question has been open for a long time without any definite answer. And now, based on what is presented in this chapter, it is known that such fundamental theorems of calculus as those listed below do not rely on the continuity of real numbers.

- Positive derivative implies an increasing function;
- Taylor formula;
- The fundamental theorem of calculus;

Modern physics recognizes that when the unit of measurement is sufficiently small to a certain magnitude, the properties of both time and space become those of quantum. That understanding cannot be achieved by letting the time interval to approach zero as what has been traditionally done in calculus. That is to say, mathematical models, developed on the theory of real numbers and the concept of limits, are merely approximate descriptions of the realistic physical world. If the new calculus, as presented in this chapter, is applied and uniform inequalities or estimation inequalities are employed as mathematical models, some of this weakness of calculus can be overcome so that the resultant mathematical theory will be more able to describe and represent the worldly reality.

Computers can only compute rational numbers. However, in the traditional calculus, the theory and formulas, which are underneath all computations, are established using the theory of real numbers. So a natural question is: Can we develop specifically for computers a methodology of computation without basing it on the theory of real numbers? To this end, what is presented in this chapter provides such necessary mathematical models and the theoretical foundation for the use of computers.

The theoretical foundation of the traditional calculus is quite complicated with tedious concepts and indirect reasoning. It has been difficult to apply such a theory in the investigation of computer reasoning or computer proofs. What is presented in this chapter provide a simplified and straightforward theoretical framework and a shortened reasoning path, which can be more readily materialized in computer reasoning.

Even though the fundamental theory of this chapter is established without using the theory of real numbers and the concept of limits, it does not exclude this theory and concept. Because the theory of limits has been considered as one essence of mathematics, this theory and the computation of limits have occupied an important position in calculus, and the theory of real numbers is crucial for the proofs of existences, these relevant materials will be arranged accordingly in the new theory and presented in a way that they can be comprehended more easily.

This chapter is organized as follows. Section 10.1 looks at the fundamental problems calculus attempted to address historically. Section 10.2 focuses on the introduction of derivatives by using the concept of estimation inequalities. Section 10.3 establishes the fundamental theorem of calculus and Taylor series expansion of differentiable functions. Section 10.4 investigates the system of real numbers and various existences without using the concept of limits. Section 10.5 provides a different interpretation for the concept of the traditional limits so that this new approach to calculus and the traditional approach of calculus are bridged together. In the final Section 10.6, the concept of Riemann integrals and its connection with that of integral systems are established.

This chapter is mainly based on (Zhang, 2010), including all the notations. If interested in learning more of the relevant details, please consult directly with (Zhang, 2010).

10.1 AN OVERVIEW

Calculus was officially established in the 17th century. At that time, the development in industry and technology raised many questions for mathematics, some of which were continued inquires of the past while others were challenges of the new age.

Investigations of these questions gave birth to calculus. All these relevant questions can be classified into four types:

Type 1: Find the instantaneous speed of a moving object;
Type 2: Compute the tangent line at any chosen point on a curve;
Type 3: Calculate the maximum and minimum values of a function; and
Type 4: Measure the length of a curve, the area of a region, the volume of an enclosed space, and locate the center of mass, etc.

The first two types of problems are about understanding regional properties from the overall attributes of the object or matter of concern. The second two types of problems are the opposite: Derive the overall attributes based on regional properties. So, a natural question arises here: What is the most basic and most fundamental relationship between parts and the whole?

10.1.1 Fundamental relationship between parts and whole

If a runner finished 110 meter hurdles in 12.88 seconds, then his average speed was $110/12.88 \approx 8.54$ m/s, which stands for an overall attribute of the running during the entire period of 12.88 seconds. However, it is imaginable that during this period, at any chosen moment, the runner's speed should be different from one moment to another. So, the question is: What is the connection between the instantaneous speed and this average speed 8.54 m/s?

To address this problem, let us treat the runner as a particle moving along a straight line. So, his movement can be written as a function $S = S(t)$, representing the distance the particle has traveled at time t. If $V = V(t)$ stands for the instantaneous speed of the particle at time t, then the distance traveled from moment u to moment v is $S(v) - S(u)$ so that the average speed during the time interval $[u, v]$ is

$$\frac{S(v) - S(u)}{v - u}$$

Now, there are p and $q \in [u, v]$ such that

$$V(p) \leq \frac{S(v) - S(u)}{v - u} \leq V(q) \tag{10.1}$$

To verify the validity of equ. (10.1), let us look at the motion equation of a free falling object $S(t) = 4.9t^2$ m. Then the average speed over a time interval $[u, v]$ is given as follows:

$$\frac{S(v) - S(u)}{v - u} = 4.9(u + v) \tag{10.2}$$

From the fact $u < v$, it follows that

$$9.8u \leq \frac{S(v) - S(u)}{v - u} \leq 9.8v \tag{10.3}$$

Therefore, by comparing to equ. (10.1), we derive the conclusion $V(t) = 9.8t$, which is the same conclusion as we know in physics.

This reasoning leads to the following conclusion: the overall value (the average speed) is in between two certain local values.

10.1.2 The tangent problem

Secondly, let us look at the problem of tangent lines. If the previous motion equation $S = S(t)$ is seen as a general function $y = f(x)$, then the average speed becomes the average rate of change $\frac{f(v) - f(u)}{v - u}$ of the function over the interval $[u, v]$; and the instantaneous speed is now the local rate of change of the function at point x, denoted $k(x)$. Therefore, there are p and $q \in [u, v]$ such that

$$k(p) \leq \frac{f(v) - f(u)}{v - u} \leq k(q) \tag{10.4}$$

which is essentially the same as equ. (10.1). Now, the question is can we employ this inequality to compute the slope of the curve $y = f(x)$ at any given point x?

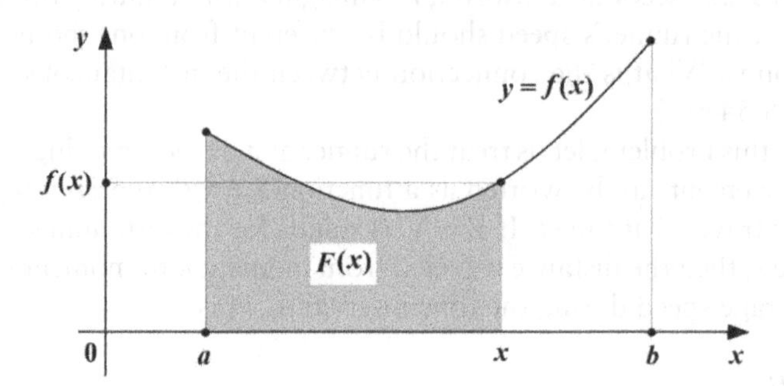

Figure 10.1 The area of a curly top region.

To this end, let us assume $y=f(x)=x^2$. Then for any interval $[u, v]$, we have

$$2u \leq \frac{f(v)-f(u)}{v-u} \leq 2v \tag{10.5}$$

From comparing equs. (10.4) and (10.5) we conclude that the slope of the tangent line at point x is $k(x)=2x$, which turned out to be the correct answer.

10.1.3 Monotonicity of functions

Thirdly, let us look at the problem of finding the maximum and minimum values. To this end, the observation that the overall average is in between two local values once again comes to secure. Equ. (10.4) indicates that if $k(x)>0$ for any $x \in [a, b]$, then for any two points $u, v \in [a, b]$, we always have $f(u)<f(v)$. That is $f(x)$ is increasing on $[a, b]$. Conversely, if $k(x)<0$ for any $x \in [a, b]$, then for any two points $u, v \in [a, b]$, we always have $f(u)>f(v)$. That is $f(x)$ is decreasing on $[a, b]$. Now, we know how to determine where the function $f(x)$ is increasing and where it is decreasing on the interval $[a, b]$. So, the problem of locating maximum and minimum values of the function is resolved.

10.1.4 Area under parabola

Fourthly, let us look at the problem of finding the area under the arch of a parabola. This problem represents one of the ancient quests of man. It was successfully resolved by using the method of exhaustion. In particular, let us consider the area between the curve $y=f(x)$ and the x-axis over the interval $[a, b]$, Figure 10.1. Define $F(x)$ to be the area under the curve $y=f(x)$ from point a to x. Then the desired area is equal to $F(b)-F(a)$. If we replace this curly top area by an equal-area rectangle of length $v-u$, then the height of the rectangle can be seen as the average altitude of the curly top region. It should be in between two values $f(p)$ and $f(q)$ of the function $f(x)$, as long as $f(x)$ is not a constant function, for $p, q \in [u, v] \subseteq [a, b]$. Symbolically, we have

$$f(p) < \frac{F(u)-F(v)}{u-v} < f(q) \tag{10.6}$$

Because equ. (10.6) is essentially the same as equs. (10.1) and (10.4), equ. (10.5) implies that when in equ. (10.6) $f(x)=2x$, the corresponding $F(x)= x^2$. Therefore, when the curly top of the given region is $f(x)=2x$, the enclosed area should be $F(b)-F(a)=b^2-a^2$, which is indeed the answer!

To summarize what is presented above, it can be seen that the mathematical models respectively developed to resolve the four types of problems have been unified together. However, the problems that need to be addressed include: 1) How can the approach be made rigorous? 2) How can one actually compute $f(x)$ if given $F(x)$, or $F(x)$ if given $f(x)$?

10.2 DERIVATIVES

The purpose of this section is to develop the concept of derivatives without employing limits. Instead, the concepts of A and B functions and estimation inequalities are introduced as the necessary bases.

10.2.1 Differences and ratio of differences

Assume the given function $f(x)$ is defined on the set R of all real numbers. In order to study the change of $f(x)$, let us consider the difference $f(v)-f(u)$ of the function values at two different points u and v. without particular explanation, we always assume $u \neq v$. By the step of the difference we mean $h=u-v$. And $\frac{f(v)-f(u)}{v-u}$ is known as the difference ratio of $f(x)$ at u and v. Generally, when $u<v$, this ratio stands for the average rate of change of $f(x)$ over the interval $[u,v]$.

Example 10.1 Study the monotonicity of the function $f(x)=\frac{x^2}{2}+\frac{1}{x}$ by using ratio of differences.

Solution. Assume $u<v$. Then the ratio of differences of $f(x)$ on $[u,v]$ is

$$\frac{f(v)-f(u)}{v-u}=\frac{u+v}{2}-\frac{1}{uv}$$

Because the function $f(x)$ is not meaningful at $x=0$, in the following our discussion will be divided into two cases: $(-\infty,0)$ and $(0,+\infty)$. When $u<v<0$,

$$\frac{f(v)-f(u)}{v-u}=\frac{u+v}{2}-\frac{1}{uv}<0$$

Therefore, the function $f(x)$ is decreasing on the interval $(-\infty,0)$.

When $0<u<v$, if $u<v\leq 1$, then we have

$$\frac{u+v}{2}<1<\frac{1}{uv}$$

so that the following holds true:

$$\frac{f(v)-f(u)}{v-u}<0$$

This end implies that the function $f(x)$ is decreasing on the interval $(0, 1]$. If $1 \leq u < v$, then we have

$$\frac{u+v}{2} > 1 > \frac{1}{uv}$$

so that

$$\frac{f(v) - f(u)}{v - u} > 0$$

Therefore, the function $f(x)$ is increasing on the interval $[1, +\infty)$. QED.

Theorem 10.1: Assume that $f(x)$ is defined at three points $u < a < v$. Let

$$\min\left\{\frac{f(a) - f(u)}{a - u}, \frac{f(v) - f(a)}{v - a}\right\} \quad \text{and} \quad \max\left\{\frac{f(a) - f(u)}{a - u}, \frac{f(v) - f(a)}{v - a}\right\} \tag{10.7}$$

then

$$m \leq \frac{f(v) - f(u)}{v - u} \leq M \tag{10.8}$$

where the equality holds iff $m = M$.

Proof. Without loss of generality, let us assume

$$m = \frac{f(a) - f(u)}{a - u} < \frac{f(v) - f(a)}{v - a} = M$$

Therefore, we have

$$\frac{f(v) - f(u)}{v - u} = \frac{1}{v - u}[(v - a)M + (a - u)m]$$

Hence from $m < M$ it follows that

$$m < \frac{1}{v - u}[(v - a)M + (a - u)m] < M$$

which shows the needed result. QED.

10.2.2 A and B functions

Assume that both functions $F(x)$ and $f(x)$ are defined on $\mathbb{R}$. If for any two points $u < v$, there are $p, q \in [u, v]$ such that

$$f(p) \leq \frac{F(u) - F(v)}{u - v} \leq f(q) \tag{10.9}$$

then $F(x)$ is known as an A function of $f(x)$ over $\mathbb{R}$, and $f(x)$ a B function of $F(x)$; and the previous inequality is known as an estimation inequality.

Theorem 10.2: Based on the properties of a B function, the attributes of an A function can be deducted as follows:

- If the B function is constantly 0, then the A function is a constant;
- If the B function is a constant $\neq 0$, then the A function is linear;
- If the B function is positive over an interval I, then the A function is increasing over I;
- If the B function is negative over an interval I, then the A function ins decreasing over I.

The proofs are straightforward and are omitted. QED.

Example 10.2 Show that the function $g(x)=3x^2$ is a B function of $f(x)=x^3$.

Proof. We compute as follows:

$$\frac{f(v)-f(u)}{v-u}=\frac{v^3-u^3}{v-u}=u^2+uv+v^2$$

So, when $uv>0, u^2+uv+v^2$ is located between $g(u)=3u^2$ and $g(v)=3v^2$. That indicates that on both $(-\infty,0]$ and $[0,+\infty)$ $g(x)$ is a B function of $f(x)$. Then by applying the following result, it follows that $g(x)$ is a B function of $f(x)$ over the interval $(-\infty,+\infty)$. QED.

Theorem 10.3: If $g(x)$ is a B function of $f(x)$ over intervals I and J, where I and J share some common points, then $g(x)$ is a B function of $f(x)$ over the combined interval $K=I\cup J$.

Proof. It suffices to show that for any subinterval $[u,v]$ of K, there are $p,q\in[u,v]$ such that

$$g(p)\leq\frac{f(u)-f(v)}{u-v}\leq g(q) \tag{10.10}$$

In fact, if $[u,v]$ is a subinterval of either I or J, this conclusion surely holds true. Otherwise, there is common point $a\in(u,v)$ such that $[u,a]\subseteq I$ and $[a,v]\subseteq J$. Now, Theorem 10.1 implies that $\frac{f(v)-f(u)}{v-u}$ is between $\frac{f(v)-f(a)}{v-a}$ and $\frac{f(a)-f(u)}{a-u}$. Therefore, the desired conclusion follows. QED.

Example 10.3 Show that the function $g(x)=\cos x$ is a B function of $f(x)=\sin x$ over the interval $(-\infty,+\infty)$.

Proof. We notice that when $0<x<\frac{\pi}{2}$, it is true that $\sin x<x<\tan x$. So, we have $\cos x<\frac{\sin x}{x}<1$. Hence for any integer k, and any $u<v$ from $\left[k\pi-\frac{\pi}{2},k\pi+\frac{\pi}{2}\right]$, we have

$$\frac{\sin v-\sin u}{v-u}=\frac{\sin\frac{v-u}{2}\cdot\cos\frac{v+u}{2}}{\frac{v-u}{2}}=\lambda\cos\frac{v+u}{2}$$

where λ is located between $\cos\frac{v+u}{2}$ and 1. So, we know that $\frac{\sin v-\sin u}{v-u}$ is between $\cos\frac{v-u}{2}\cos\frac{v+u}{2}$ and $\cos\frac{v+u}{2}$. And because

$$\cos\frac{v-u}{2}\cos\frac{v+u}{2}=\frac{\cos u+\cos v}{2}$$

it follows that on the interval $\left[k\pi-\frac{\pi}{2},k\pi+\frac{\pi}{2}\right]$ $\cos x$ is a B function of $\sin x$ so that on $(-\infty,+\infty)$ $\cos x$ is also a B function of $\sin x$. QED.

Theorem 10.4: Regarding the elementary operations, B functions satisfy the following:

(i) The function $g(x)=0$ is a B function of $f(x)=C$;
(ii) The function $g(x)=k$ is a B function of $f(x)=kx+b$;
(iii) If $g(x)$ is a B function of $f(x)$, then $kg(x)$ is a B function of $kf(x)+cx$;
(iv) If $g(x)$ is a B function of $g(x)$, then $kg(kx+c)$ is a B function of $f(kx+c)$.

Proof. Both conclusions (i) and (ii) are obvious. Now, let us look at conclusion (iii). Because $g(x)$ is a B function of $f(x)$, there are $p,q\in[u\cdot v]$ such that

$$g(p)\le\frac{f(u)-f(v)}{u-v}\le g(q) \tag{10.11}$$

That is equivalent to

$$g(p)\le\frac{\left[kf(u)+c\right]-\left[kf(v)+c\right]}{k(u-v)}\le g(q)$$

This end implies that $kg(x)$ is a B function of $kf(x)+cx$.

For conclusion (iv), let $u=kx+c$ and $v=ky+c$. Then we have

$$g(p)\le\frac{f(kx+c)-f(ky+c)}{k(x-y)}\le g(q)$$

Evidently there are r,s in $[x,y]$ or $[y,x]$ such that $p=kr+c$ and $q=ks+c$. That implies that $kg(kx+c)$ is a B function of $f(kx+c)$. QED.

10.2.3 Estimation inequalities

What is shown above indicates that each B function is an estimate of the quotient of differences of an A function, leading to an estimation inequality. To this end, let us look at a few examples of how estimation inequalities can be applied for different purposes.

Example 10.4 Estimate the numerical value of $\sqrt{10}$ by using a B function of $f(x)=\sqrt{x}$.

Solution. First of all, let us find a B function of $f(x)=\sqrt{x}$ on the interval $(0,+\infty)$. Because for any u and v satisfying $0<u<v$ we have

$$\frac{1}{2\sqrt{v}} \leq \frac{f(v)-f(u)}{v-u} = \frac{\sqrt{v}-\sqrt{u}}{v-u} = \frac{1}{\sqrt{v}+\sqrt{u}} \leq \frac{1}{2\sqrt{u}}$$

it follows that $g(x)=\frac{1}{2\sqrt{x}}$ is a B function of $f(x)=\sqrt{x}$. Now, if we let $u=9$ and $v=10$, then the previous inequality becomes

$$\frac{1}{2\sqrt{10}} \leq \frac{\sqrt{10}-\sqrt{9}}{10-9} \leq \frac{1}{2\sqrt{9}}$$

which can be simplified into

$$\frac{1}{2\sqrt{10}} - \frac{1}{6} \leq \sqrt{10} - 3 - \frac{1}{6} \leq 0$$

That implies that $\sqrt{10} \approx 3.167$. QED.

Example 10.5 Find the maximum and minimum values of the cubic function $f(x)=x^3-3x^2-9x+2$ on the interval $[-2, 5]$.

Solution. It can be shown that $g(x)=3x^2-6x-9$ is a B function of the given function $f(x)$. So for any closed interval $[u,v]$ there are $p,q \in [u,v]$ such that $g(p) \leq \frac{f(v)-f(u)}{v-u} \leq g(q)$. So, based on the signs of $g(x)$ it can be determined where $f(x)$ is increasing or decreasing.

In particular, by factoring $g(x)$, we have $g(x)=3x^2-6x-9=3(x+1)(x-3)$ so that when $x \in [-2,-1], g(x)$ is non-negative so that the quotient of differences of $f(x)$ is positive and $f(x)$ is increasing; when $x \in [-1,3], g(x)$ is non-positive so that the quotient of differences of $f(x)$ is negative and $f(x)$ is decreasing; and when $x \in [3,5]$, $g(x)$ is non-negative so that the quotient of differences of $f(x)$ is positive and $f(x)$ is increasing.

That means that at $x=-1$ $f(x)$ reaches its maximum $f(-1)=7$; at $x=3$ $f(x)$ reaches its minimum $f(3)=-25$. Considering the function values at the endpoints of the interval $f(-2)=0$ and $f(5)=7$, it follows that the function $f(x)$ reaches its maximum value 7 at two points $x = -1$ and $x = 5$, and reaches its minimum value -25 at $x = 3$. QED.

Based on the discussion above, we only know that p, q are points in $[u, v]$ without any specific knowledge as for where they are located. If $f(x)$ fluctuates on $[u, v]$ severely, the effect of the estimate will be difficult to control. To this end, when x changes to $x + h$, the function value $f(x)$ changes to $f(x + h)$. So, the degree of fluctuation can be described by $|f(x+h) - f(x)|$. If for $t > 0$ there is a positive monotonic function $\alpha(t)$ satisfying

$$|f(x+h) - f(x)| < \alpha(|h|) \tag{10.12}$$

then $\alpha(t)$ is referred to as a control function of $f(x)$; and equ. (10.12) the control inequality of $f(x)$.

Theorem 10.5: Let $g(x)$ be a B function of $f(x)$ over the interval I. If there is a non-negative, non-decreasing function $\alpha(h)$ defined on $(0, A)$, where A is sufficiently large, making the following inequality meaningful, such that for any two points x and $x + h$ of I, the following holds:

$$|g(x+h) - g(x)| < \alpha(|h|) \tag{10.13}$$

then for any $u < v$ and $s \in [u, v]$ the following holds:

$$\left|\frac{f(v) - f(u)}{v - u} - g(s)\right| < \alpha(v - u) \tag{10.14}$$

Proof. Let $D = \frac{f(v) - f(u)}{v - u}$. From the given condition and the definition of B functions, it follows that there are $p, q \in [u, v]$ such that $g(p) \le D \le g(q)$. So, equ. (10.13) implies that

$$-\alpha(|p - s|) \le g(p) - g(q) \le D - g(s) \le g(p) - g(s) \le \alpha(|q - s|) \tag{10.15}$$

Because $p, q, s \in [u, v]$, it follows that $|p - s| \le v - u$ and $|q - s| \le v - u$. Because $\alpha(h)$ is increasing, it follows that

$$\alpha(|p - s|) \le \alpha(v - u), \quad \alpha(|q - s|) \le \alpha(v - u) \tag{10.16}$$

Now, equ. (10.14) follows from equs. (10.15) and (10.16). QED.

Theorem 10.6: Let $g(x)$ and $f(x)$ be defined on the interval I. If there is a non-negative, non-decreasing function $\alpha(h)$, defined on $(0, A)$ for sufficiently large A making the following inequality meaningful, satisfying for any u and $v = u + h$,

$$|f(v) - f(u) - g(u)h| \le |h|\alpha(|h|) \tag{10.17}$$

then for any u and $v=u+h$ in I, the following holds true:

$$|g(u+h)-g(u)| \leq 2\alpha(|h|). \tag{10.18}$$

Proof. Equ. (10.17) can be rewritten as follows:

$$-|h|\cdot\alpha(|h|) \leq f(v)-f(u)-g(u)h \leq |h|\cdot\alpha(|h|)$$

By exchanging the positions of u and v we have the following:

$$-|h|\cdot\alpha(|h|) \leq f(u)-f(v)+g(v)h \leq |h|\cdot\alpha(|h|)$$

Now the desired result follows from simplying adding these two inequalities together. QED

10.2.4 Derivatives

For $A>0$, a function $d(h)$, defined on $(0,A)$, is known as a d function, provided it is non-negative, non-decreasing, and without any positive lower bound. If necessary, we can expand the domain of such a d function $d(h)$ to $(-A,A)$ such that $d(-h)=d(h)$ and $d(0)=0$. Here, the most important property of the d function $d(h)$ is that it can take values smaller than any positive number. For example, $f(h)=h$ and $g(h)=\sqrt{h}$ are some of the simplest d functions. Gemerally, when $k>0, d(h)=h^k$ is also a d function. In fact, the sum, product, and composite of d functions are still d functions.

Definition 10.1 Let $f(x)$ and $g(x)$ be functions defined on an interval I. If there is a d function $d(h)$ such that for any $x, x+h\in I$, the following holds

$$\left|\frac{f(x+h)-f(x)}{h}-g(x)\right| \leq d(h) \tag{10.19}$$

then the function $y=f(x)$ is known as uniformly differentiable on I; if $f(x)$ is uniformly differentiable on any closed subinterval of I, then it is known as continuously differentiable on I. In either case, $g(x)$ is the derivative or derivative function of $f(x)$, denoted

$$f'(x)=g(x) \quad \text{or} \quad y'=g(x) \quad \text{or} \quad \frac{dy}{dx}=g(x)$$

Definition 10.2 Let $g(x)$ be a function defined on a closed interval I. If there is a d function $d(h)$ such that for any $x, x+h\in I$, the following holds true:

$$|g(x+h)-g(x)| \leq d(h) \tag{10.20}$$

then the function $y=g(x)$ is known as uniformly continuous on the interval I. If $g(x)$ is uniformly continuous on any closed subinterval of I, then $g(x)$ is known as continuous on I.

Example 10.6 From the definition and $|\sin(x+h)-\sin x|\leq|h|$, it follows that the function $f(x)=\sin x$ is uniformly continuous on the interval $(-\infty,+\infty)$.

From $|\sqrt{x+h}-\sqrt{x}|\leq\sqrt{h}$, it follows that the function $f(x)=\sqrt{x}$ is uniformly continuous on the interval $[0,+\infty)$.

It can be readily shown that the function $g(x)=\frac{1}{x}$ is not uniformly continuous on the interval $(0,1]$. QED.

By combining the previous definitions and Theorems 10.5 and 10.6, we immediately have

Theorem 10.7: Assume that $g(x)$ is a B function of $f(x)$ over the interval I. If $g(x)$ is continuous on I, then $f(x)$ is differentiable on I such that $f'(x)=g(x)$. If $g(x)$ is uniformly continuous on the closed interval I, then $f(x)$ is uniformly differentiable such that $f'(x)=g(x)$. QED.

Theorem 10.8: If $f(x)$ is differentiable on the interval I such that $f'(x)=g(x)$, then $g(x)$ is continuous on I. If $f(x)$ is uniformly differentiable on the closed interval I such that $f'(x)=g(x)$, then $g(x)$ is uniformly continuous on the closed interval I. QED.

Theorem 10.9: Assume that $f(x)$ is continuously differentiable on the interval I such that $f'(x)=g(x)$. Then $g(x)$ is a B function of $f(x)$.

Proof. From the definition of differentiability and the given assumption, it follows that for any chosen subinterval $[u,v]$ there is a d function $d(h)$ such that for any $x,x+h\in[u,v]$, the following holds true:

$$\left|\frac{f(x+h)-f(x)}{h}-g(x)\right|\leq d(h) \tag{10.21}$$

Next, we divide the argument into two cases.

Case 1. If the ratio of differences of $f(x)$ is a constant c on $[u,v]$, then for any $u+h\in[u,v]$, equ. (10.21) implies $|c-g(u)|\leq d(h)$. That leads to $g(u)=c$ so that $g(x)$ is a B function of $f(x)$.

Case 2. If the ratio of differences of $f(x)$ is not a constant on $[u,v]$, there must be $[r,s]\subset[u,v]$ and $D>0$ such that

$$\frac{f(s)-f(r)}{s-r}+D=\frac{f(v)-f(u)}{v-u} \tag{10.22}$$

Because $d(h)$ does not have any positive lower bound, there is $w > 0$ such that $d(w) < D$. Take a natural number $n > \frac{s-r}{w}$ and partition $[r, s]$ into n subintervals with the length $h = \frac{s-r}{n} < w$. Then there must be one subinterval $[p, p+h]$ such that the following inequality holds true:

$$\frac{f(p+h) - f(p)}{h} \leq \frac{f(s) - f(r)}{s - r} \tag{10.23}$$

So, equ. (10.21) and the non-decreasing property of $d(h)$ imply that

$$g(p) - \frac{f(p+h) - f(p)}{h} \leq d(h) < d(w) < D \tag{10.24}$$

By combining equs. (10.22)–(10.24) we have

$$g(p) < \frac{f(p+h) - f(p)}{h} + D \leq \frac{f(s) - f(r)}{s - r} + D = \frac{f(v) - f(u)}{v - u} \tag{10.25}$$

To show for some $q \in [u, v]$, $\frac{f(v)-f(u)}{v-u} \leq g(q)$, let $G(x) = -f(x)$. Then from the definition, it follows that $G'(x) = -g(x)$. So, the previous argument implies that there is such a $q \in [u, v]$ such that

$$-g(q) < \frac{G(v) - G(u)}{v - u} = \frac{-(f(v) - f(u))}{v - u}$$

That is equivalent to $\frac{f(v)-f(u)}{v-u} \leq g(q)$. QED.

What the previous three theorems jointly tell is that $g(x)$ is the derivative of $f(x)$ iff $g(x)$ is a continuous B function of $f(x)$.

Example 10.7 Find the derivative of the function $f(x) = x^n$, where n is a positive integer, and point out the d function used in the estimate of the error.

Solution. Based on the given function, we have:

$$f(x+h) - f(x) = (x+h)^n - x^n = nx^{n-1}h + \sum_{k-2}^{n} C_n^k x^{n-k} h^k$$

So when $x \in [a, b]$, we have

$$\left| \frac{f(x+h) - f(x)}{h} - nx^{n-1} \right| = \left| \sum_{k=2}^{n} C_n^k x^{n-k} h^{k-1} \right| \leq 2^n(|a| + |b|)^{n-2}|h|$$

Therefore, from the definition of derivatives, it follows that $f'(x) = (x^n)' = nx^{n-1}$ and the d function is $d(h) = 2^n(|a| + |b|)^{n-2}|h|$. QED.

10.3 INTEGRALS

Assume that $f(x)$ is a function defined on an interval I. If there is a function S of two independent variables u and v, for $u, v \in I$, such that

(i) The additive property: For any $u, v, w \in I$,

$$S(u,v) + S(v,w) = S(u,w)$$

(ii) The mean-value property: For any $u, v \in I$, satisfying $u < v$, there must be two points $p, q \in [u,v]$ such that

$$f(p)(v-u) \le S(u,v) \le f(q)(v-u)$$

then the function S is known as an integral system of $f(x)$ on I. If the integral system S exists uniquely over an interval $[u,v] \subseteq I$, then the function $f(x)$ is said to be integrable on $[u,v]$ and S is known as the definite integral of $f(x)$, denoted symbolically as follows:

$$S(u,v) = \int_u^v f(x)dx$$

Theorem 10.10: If $S(u,v)$ is an integral system, then the following hold true:

(i) $S(u,u) = 0$; and
(ii) $S(u,v) = -S(v,u)$. QED.

Theorem 10.11: Each constant function $f(x) = c$, defined on interval I, has the following unique integral system: $S(u,v) = c(v-u)$.

Proof. From the mean-value property, it follows that $c(v-u) = f(p)(v-u) \le S(u.v) \le f(q)(v-u) = c(v-u)$. QED.

Theorem 10.12: Assume that $S(u,v)$ is an integral system of $f(x)$ on an interval I and c a point in I. Let $F(x) = S(c,x)$. Then $f(x)$ is a B function of $F(x)$ on I. Conversely, if $f(x)$ is a B function of $F(x)$ on I, define $S(u,v) = F(v) - F(u)$. Then $S(u,v)$ is an integral system of $f(x)$ on I.

Proof. Let $S(u,v)$ be an integral system of $f(x)$ on I and $F(x) = S(c,x)$. Then from the additive property, it follows that

$$S(u,v) = S(c,v) - S(c,u) = F(v) - F(u)$$

So, from the mean-value property, it follows that $f(x)$ is a B function of $F(x)$ on I.

On the other hand, if $f(x)$ is a B function of $F(x)$ on I, let $S(u,v) = F(v) - F(u)$. Then from $S(u,v) + S(v,w) = F(v) - F(u) + F(w) - F(v) = F(w) - F(u) = S(u,w)$, it follows that $S(u, v)$ satisfies the additive property. And from the definition of B functions, it follows that $S(u, v)$ satisfies the mean-value property. QED.

10.3.1 The fundamental theorem of calculus

The initial establishment of the following result signaled the birth of calculus over 300 years ago:

Theorem 10.13: (The Fundamental Theorem of Calculus). Assume that a given function $F(x)$ is differentiable on an interval I, satisfying that $F'(x) = f(x)$. Then $S(u, v) = F(v) - F(u)$ is the unique integral system of $f(x)$ on I so that

$$\int_u^v f(x)dx = F(v) - F(u)$$

Proof. From Theorem 10.9, it follows that $F'(x) = f(x)$ is a B function of $F(x)$. From Theorem 10.12 it follows that $S(u, v) = F(v) - F(u)$ is the unique integral system of $f(x)$ on I.

For the rest of the argument, it suffices to show that $S(u, v)$ is a unique integral system of $f(x)$ on I. To this end, assume that $R(u, v)$ is another integral system of $f(x)$ on I. Let us choose a point $c \in I$. Let $G(x) = R(c, x)$. Then Theorem 10.12 implies that $f(x)$ is also a B function of $G(x)$ on I. From Theorem 10.8, it follows that $f(x)$ is continuous on I so that Theorem 10.7 implies that $G(x)$ is differentiable on I and $G'(x) = f(x) = F'(x)$.

So, we conclude that $(G(x) - F(x))' = 0$. Therefore, on the interval I the function $F(u) - G(u) = F(v) - G(v)$ is a constant. That leads to $R(u, v) = G(v) - G(u) = F(v) - F(u) = S(u, v)$, indicating that $S(u, v)$ is the unique integral system of $f(x)$ on I. From the definition of the integral sign, it follows that the symbol of the definite integral is warranted so that from $S(u, v) = F(v) - F(u)$ the desired equality follows. QED.

The previous theorem guarantees the uniqueness of the integral system of $f(x)$ when it is known that the function $f(x)$ is the derivative of a function $F(x)$. So a natural question arises: If we do not know whether a function $f(x)$ has an original function or not, how can we determine whether its corresponding integral system is unique or not? The following result partially addresses this problem.

Theorem 10.14: Assume that $f(x)$ is continuous on an interval I and both $S(u, v)$ and $R(u, v)$ are integral systems of $f(x)$ on I. Then $S(u, v) = R(u, v)$.

Proof. We show this result by using proof by contradiction. Assume that the result does not hold true. Then there are $u, v \in I$ such that $u < v$ and $|S(u, v) - R(u, v)| = a > 0$. Partition the closed interval $[u, v]$ into n pieces with the dividing points $u = x_0 < x_1 < \cdots < x_n = v$. Denote $H = v - u$. From the fact that $f(x)$

is continuous on the interval $[u, v]$, it follows that there is a d function $d(h)$ such that $|f(x+h)-f(x)| \leq d(h)$. So when $x \in [x_{k-1}, x_k]$, we have the following

$$f(x_k) - d\left(\frac{H}{n}\right) \leq f(x) \leq f(x_k) + d\left(\frac{H}{n}\right)$$

From the mean-value property of integral systems, it follows that

$$\left[f(x_k) - d\left(\frac{H}{n}\right)\right]\frac{H}{n} \leq S(x_{k-1}, x_k) \leq \left[f(x_k) + d\left(\frac{H}{n}\right)\right]\frac{H}{n}$$

By summing up k from 1 to n and by denoting $F = f(x_1) + \cdots + f(x_n)$, we obtain

$$\frac{FH}{n} - d\left(\frac{H}{n}\right)H \leq S(u, v) \leq \frac{FH}{n} + d\left(\frac{H}{n}\right)H$$

Similarly, we can obtain

$$\frac{FH}{n} - d\left(\frac{H}{n}\right)H \leq R(u, v) \leq \frac{FH}{n} + d\left(\frac{H}{n}\right)H$$

Therefore,

$$0 < |S(u, v) - R(u, v)| = a \leq 2Hd\left(\frac{H}{n}\right)$$

which contradicts the property of the d function d. QED.

Similarly, the following result holds true.

Theorem 10.15: Assume that $f(x)$ is monotonic on the interval I. Then the integral system of $f(x)$ on the interval I is unique.

Proof. Once again, let us prove this result using proof by contradiction. Without loss of generality, let us assume that the function $f(x)$ is monotonically non-decreasing.

If the conclusion of this theorem is not true so that the function $f(x)$ has two integral systems $S(u, v)$ and $R(u, v)$, then there are $u, v \in I$ such that $|S(u, v) - R(u, v)| = a > 0$. Let us partition the interval $[u, v]$ into n subintervals: $u = x_0 < x_1 < \cdots < x_n = v$. Denote $H = v - u$. From the assumption that $f(x)$ is non-decreasing, it follows that when $x \in [x_{k-1}, x_k]$, we have the following

$$f(x_{k-1}) \leq f(x) \leq f(x_k)$$

So, the mean-value property of integral systems implies that

$$f(x_k) \cdot \frac{H}{n} \leq S(x_{k-1}, x_k) \leq f(x_k) \cdot \frac{H}{n}$$

By summing k from 1 to n and by denoting $F=f(x_1)+\cdots+f(x_n)$, we obtain

$$\frac{[F+f(u)-f(v)]H}{n} \leq S(u,v) \leq \frac{FH}{n}$$

Similarly, we can obtain

$$\frac{[F+f(u)-f(v)]H}{n} \leq R(u,v) \leq \frac{FH}{n}$$

Therefore, we have

$$0 < |S(u,v)-R(u,v)| = a \leq \frac{[f(v)-f(u)]H}{n}$$

which implies that

$$n \leq \frac{[f(v)-f(u)]H}{a}$$

which contradicts the arbitrariness of n. QED.

10.3.2 Some applications of definite integrals

First, let us consider the problem of finding the area enclosed by a general curve L. To this end, assume the curve L is given in the following polar coordinate system: $r=\phi(\theta), \theta\in[\alpha,\beta]$. For any closed subinterval $[u,v]\subseteq[\alpha,\beta]$, let $S(u,v)$ denote the area of the region enclosed by the curve L and the rays $\theta=u$ and $\theta=v$. Then $S(u,v)$ surely satisfies the additive property. Additionally, the radius of the section of such a circle with also the central angle $v-u$ that has the same area as this curly edge region, see Figure 10.2, is given by

$$R=\sqrt{\frac{2S(u,v)}{v-u}}$$

So, there are evidently $p,q\in[u,v]$ such that $\varphi(p)\leq R\leq\varphi(q)$. So we obtain

$$\frac{1}{2}\varphi^2(p)(v-u) \leq S(u,v) \leq \frac{1}{2}\varphi^2(q)(v-u)$$

Therefore, $S(u,v)$ is an integral system of $f(\theta)=\frac{1}{2}\varphi^2(\theta)$ on $[\alpha,\beta]$. If $r=\varphi(\theta)$ is either continuous or monotonic on the interval $[\alpha,\beta]$, then its integral system is unique so that

$$S(\alpha,\beta)=\frac{1}{2}\int_{\alpha}^{\beta}\varphi^2(\theta)d\theta$$

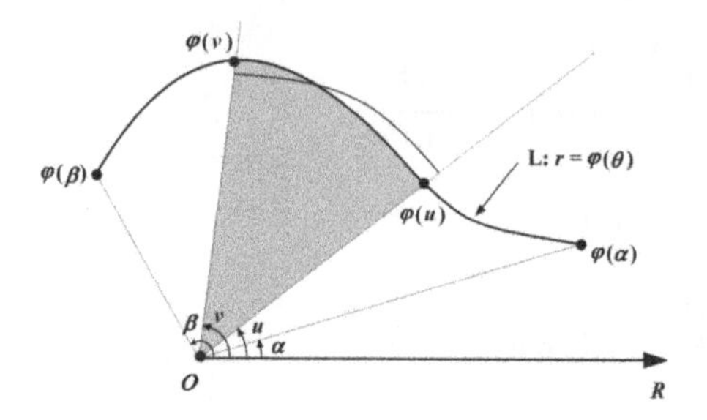

Figure 10.2 The area of the enclosed region by L and $\theta = u$ and $\theta = v$.

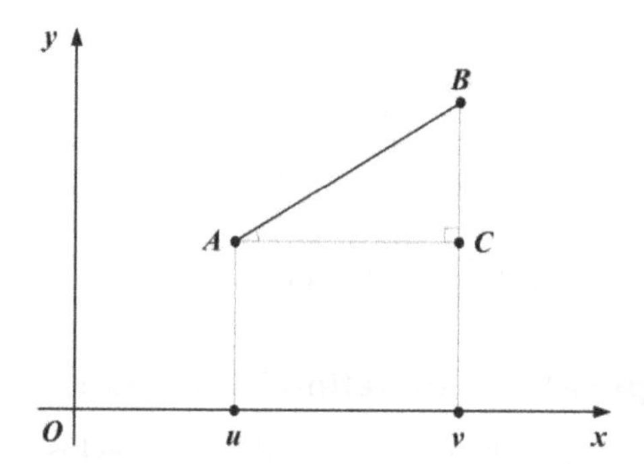

Figure 10.3 Relationship between line length and slope.

That is exactly the same conventional formula developed for the computation of the area of any curly edge fan-like region expressed in the polar coordinate system.

Secondly, let us look at the problem of computing the arc length of a planar curve.

Theorem 10.16: If $f(x)$ and $g(x)$ are piecewise continuous or monotonic on an interval I, then the sum $y = f(x) + g(x)$ has a unique integral system on I. QED

Assume that a function $f(x)$ is defined on the closed interval $[a, b]$, where $[u, v] \subseteq [a, b]$. Let us denote the arc length of the curve $y = f(x)$ over the interval $[u, v]$ as $S(u, v)$. Then $S(u, v)$ evidently satisfies the additive property.

If we can further illustrate the fact that $S(u, v)$ is an integral system of a certain function $g(x)$ on $[a, b]$, then we will develop a method of computing the arc length of the given curve. The key for now is to find the function $g(x)$.

If the curve $y = f(x)$ is a line segment of slope k, then from Pythagorean Theorem it follows that $S(u, v) = (v - u)\sqrt{1 + k^2}$. That is to say, the greater the absolute value of the slope is, the longer the line segment is, see Figure 10.3, where $k = \tan \angle BAC = \frac{BC}{AC}$, $AB = S(u, v) = \sqrt{AC^2 + BC^2} = \sqrt{AC^2 + (k \cdot AC)^2} = AC\sqrt{1 + k^2} = (v - u)\sqrt{1 + k^2}$.

Furthermore, if the curve $y=f(x)$ consists of a series of folding line segments such that for $x \in [u,v]$, the line segment that is a part of the curve $y=f(x)$ has slope $k(x)$, then there must be $p,q \in [u,v]$ such that

$$\sqrt{1+k^2(p)}(v-u) \leq S(u,v) \leq \sqrt{1+k^2(q)}(v-u)$$

which implies that $S(u,v)$ should be an integral system of the function $g(x)=\sqrt{1+k^2(x)}$ on the interval $[a,b]$.

Next let us generalize the previous zigzagged line curve to the situation of a general curve $y=f(x)$, while noticing that the slope of the curve $y=f(x)$ at any point $x \in [u,v]$ is $k(x)=f'(x)$, it is reasonable to see that $S(u,v)$ is an integral system of the function $g(x)=\sqrt{1+(f'(x))^2}$ over $[a,b]$. If the curve $y=f(x)$ is piecewise continuously differentiable, then this integral system is unique. Therefore, we have

$$S(u,v)=\int_u^v \sqrt{1+(f'(x))^2}dx$$

from which we conclude that the formula for computing the arc length of the curve $y=f(x)$ over the interval $[a,b]$ is

$$s=\int_a^b \sqrt{1+(f'(x))^2}dx$$

10.3.3 Taylor series

Through using the expression of definite integrals, the Fundamental Theorem of Calculus reveals the relationship between a function $F(x)$ and its derivative $F'(x)$:

$$\int_u^v F'(x)dx = F(v)-F(u)$$

In the rest of this subsection, let us see what we can gain by look at this formula from the right-hand side to the left-hand side. In particular, let the variable u be fixed, while letting $u=a$ and $v-u=h$. That is, $v=u+h$. Then the previous formula can be rewritten as follows:

$$F(a+h)=F(a)+\int_a^{a+h} F'(x)dx \tag{10.26}$$

which provides an expression that can be employed to compute $F(a+h)$ using $F(a)$ and the definite integral of $F'(x)$.

By applying the substitution $x = a + t$ in the definite integral in equ. (10.26), we obtain

$$F(a+h) = F(a) + \int_0^t F'(a+t)dt \tag{10.27}$$

For the same reasoning, if $F'(x)$ is continuously differentiable, we will have

$$F'(a+h) = F'(a) + \int_0^t F''(a+t)dt \tag{10.28}$$

Substituting equ. (10.28) into equ. (10.27) produces

$$\begin{aligned} F(a+h) &= F(a) + \int_0^h \left[F'(a) + \int_0^t F''(a+t_1)dt_1 \right] dt \\ &= F(a) + \int_0^h F'(a)dt + \int_0^h \int_0^t F''(a+t_1)dt_1 dt \\ &= F(a) + F'(a)h + \int_0^h \int_0^t F''(a+t_1)dt_1 dt \end{aligned} \tag{10.29}$$

If $F''(x)$ is continuously differentiable, its derivative is known as the third order derivative of $F(x)$, denoted $F'''(x)$ or $F^{(3)}(x)$. Speaking generally, the nth derivative of $F(x)$ can be defined using mathematical induction as the derivative of the $(n-1)$th derivative of $F(x)$, denoted $F^{(n)}(x)$. In other words, we have defined $F^{(0)}(x) = F(x)$ and for any natural number n,

$$(F^{(n-1)}(x))' = F^{(n)}(x) \tag{10.30}$$

So, if we apply equ. (10.27) to the function $F''(x)$, we have

$$F''(a+t_1) = F''(a) + \int_0^{t_1} F^{(3)}(a+t_2)dt_2 \tag{10.31}$$

And in general we have

$$F^{(n)}(a+t_{n-1}) = F^{(n)}(\mathrm{a}) + \int_0^{t_{n+1}} F^{(n+1)}(a+t_n)dt_n \tag{10.32}$$

If we substitute equ. (10.31) into the last part of equ. (10.29), then we obtain

$$
\begin{aligned}
F(a+h) &= F(a) + F'(a)h + \int_0^h \int_0^t \left[F''(a) + \int_0^{t_1} F^{(3)}(a+t_2)dt_2 \right] dt_1 dt \\
&= F(a) + F'(a)h + \int_0^h \int_0^t F''(a)dt_1 dt + \int_0^h \int_0^t \int_0^{t_1} F^{(3)}(a+t_2)dt_2 dt_1 dt \\
&= F(a) + F'(a)h + \frac{F''(a)h^2}{2} + \int_0^h \int_0^t \int_0^{t_1} F^{(3)}(a+t_2)dt_2 dt_1 dt \qquad (10.33)
\end{aligned}
$$

In the process of computation, we used

$$
\begin{aligned}
\int_0^h \int_0^t F''(a)dt_1 dt &= \int_0^h \left[F''(a)t_1 \Big|_0^t \right] dt = \int_0^h F''(a)t dt \\
&= \frac{F''(a)t^2}{2} \Bigg|_0^h = \frac{F''(a)h^2}{2} \qquad (10.34)
\end{aligned}
$$

So, if we take $n=3$ in equ. (10.32) and substitute the outcome into equ. (10.33), we obtain

$$
F(a+h) = F(a) + F'(a)h + \frac{F''(a)h^2}{2 \cdot 1} + \frac{F'''(a)h^3}{3 \cdot 2 \cdot 1} + \int_0^h \int_0^t \int_0^{t_1} \int_0^{t_2} F^{(4)}(a+t_3)dt_3 dt_2 dt_1 dt \qquad (10.35)
$$

By either repeating this process or by using mathematical induction, it is not difficult for us to obtain the following result.

Theorem 10.17: If a function $F(x)$ is $(n+1)$ times continuously differentiable on an interval I, and both a and $a+h$ are two points in I, then the following holds true:

$$
F(a+h) = \sum_{k=0}^{n} \frac{F^{(k)}(a)h^k}{k!} + \int_0^h \int_0^t \int_0^{t_1} \cdots \int_0^{t_{n-1}} F^{(n+1)}(a+t_n)dt_n \ldots dt_1 dt \qquad (10.36)
$$

If we let $a+h=x$, then $h=x-a$ and equ. (10.36) becomes

$$
F(x) = \sum_{k=0}^{n} \frac{F^{(k)}(a)(x-a)^k}{k!} + \int_0^h \int_0^t \int_0^{t_1} \cdots \int_0^{t_{n-1}} F^{(n+1)}(a+t_n)dt_n \ldots dt_1 dt \qquad (10.37)
$$

which is known as the nth order Taylor expansion of $F(x)$ at the point $x = a$ or simply Taylor formula. The sum on the right-hand side is known as the nth order Taylor polynomial of the function $F(x)$ at the point $x=a$, which is generally written as follows:

$$T_n(x,F) = \sum_{k=0}^{n} \frac{F^{(k)}(a)(x-a)^k}{k!} \tag{10.38}$$

And the difference between the function $F(x)$ and this nth order Taylor polynomial is known as the remainder of the function's nth order Taylor expansion, denoted

$$R_n(x,F) = F(x) - T_n(x,F) \tag{10.39}$$

When this remainder is written as

$$R_n(x,F) = \int_0^h \int_0^t \int_0^{t_1} \cdots \int_0^{t_{n-1}} F^{(n+1)}(a+t_n)dt_n \dots dt_1 dt \tag{10.40}$$

it is known as the integral representation of the remainder of Taylor expansion. When no confusion will be caused, we simply write $R_n(x)$ and $T_n(x)$ for $R_n(x,F)$ and $T_n(x,F)$, respectively.

When $a=0$ in the previous analysis, the Taylor expansion of $F(x)$ at $x=0$ is also known as Maclaurin expansion of the function $F(x)$.

If when $x \in [a, a+h]$ we have $\left|F^{(n+1)}(x)\right| \leq M$, then it is ready to estimate that

$$|R_n(x,F)| \leq \frac{M|x-a|^{n+1}}{(n+1)!} \tag{10.41}$$

When n is sufficient large or when $|x-a|$ is relative small, the value of $|R_n(x,F)|$ could be very small so that Taylor formula provides an effective way to compute the function value of $F(x)$ using only the four arithmetic operations.

At this juncture, a natural question is: Because Taylor polynomial only involves the function and its derivatives, can we estimate the remainder based on merely the properties of the derivatives? To address this problem, we need the following:

Theorem 10.18: Assume that both $F(x)$ and $G(x)$ are continuously differentiable on the interval $[a,b]$ with $f(x)$ and $g(x)$ being respectively the derivatives of $F(x)$ and $G(x)$. If for any $x \in [a,b]$ the inequality $f(x) \leq g(x)$ holds true, then for all $x \in [a,b]$, the following is true:

$$F(x) - F(a) \leq G(x) - G(a) \tag{10.42}$$

Proof. Let $H(x) = F(x) - G(x)$. Then for any $x \in [a,b]$, we have

$$H'(x) = (F(x) - G(x))' = f(x) - g(x) \leq 0$$

That is, the function $H(x)$ is monotonically non-increasing so that $H(a) \geq H(x)$. That is, equ. (10.42) holds true. QED.

Theorem 10.19: Assume that $H(x)$ is $(n+1)$ times continuously differentiable on the interval $[a, b]$, and satisfies

(i) When $k = 0, 1, 2, \ldots, n, H^{(k)}(a) = 0$; and
(ii) On the interval, $m \leq H^{(n+1)}(x) \leq M$,

then for any $x \in [a, b]$, the following is true:

$$\frac{m(x-a)^{n+1}}{(n+1)!} \leq H(x) \leq \frac{M(x-a)^{n+1}}{(n+1)!} \tag{10.43}$$

Proof. Let us first use mathematical induction on $k = 1, 2, \ldots, n+1$ to show

$$\frac{m(x-a)^k}{k!} \leq H^{(n+1-k)}(x) \leq \frac{M(x-a)^k}{k!} \tag{10.44}$$

In fact, when $k = 1$, we have $m \leq H^{(n+1)}(x) \leq M$. From Theorem 10.18 it follows that

$$m(x-a) \leq H^{(n)}(x) \leq M(x-a) \tag{10.45}$$

Assume that when $k < n+1$ equ. (10.44) holds true. So from Theorem 10.18 it follows that for the situation of $k+1$, we have

$$\frac{m(x-a)^{k+1}}{(k+1)!} \leq H^{(n-k)}(x) \leq \frac{M(x-a)^{k+1}}{(k+1)!} \tag{10.46}$$

Specifically, when $k = n$, we obtain the desired equ. (10.43). QED.

Theorem 10.20: Assume that $F(x)$ is $(n+1)$ times continuously differentiable on the interval $[a, b]$ satisfying that for any $x \in [a, b]$, $|F^{(n)}(x)| \leq M$. Then for any points c and $x \in [a, b]$, the following Taylor expansion holds true:

$$F(x) = F(c) + F'(c)(x-c) + \frac{F^{(2)}(c)}{2!}(x-c)^2 + \cdots + \frac{F^{(n-1)}(c)}{(n-1)!}(x-c)^{n-1} + R_n(x) \tag{10.47}$$

and

$$|R_n(x)| = |F(x) - T_n(x, c)| \leq \frac{M|x-c|^{n+1}}{(n+1)!} \tag{10.48}$$

Proof. Let $H(x) = F(x) - T_n(x, c)$. It can be readily checked that $H(x)$ satisfies the conditions of Theorem 10.19. Therefore, when $x \in [c, b]$, equ. (10.48) holds true. And

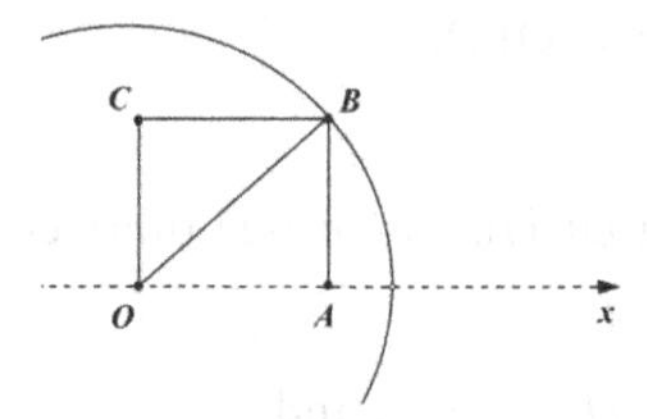

Figure 10.4 An arc passes through a hole in the rational number line.

when $x \in [a, c]$, take $u = -x$ and $G(u) = F(-u)$. Apply the previously established result on $G(u)$ over the interval $[-c, -a]$, and then by substituting G back to F leads to the desired conclusion. QED.

10.4 REAL NUMBER SYSTEM AND EXISTENCES

Although we have established the fundamental concepts and theory of calculus in the previous sections, to prove various existences needed for the further development, we will have to develop the most important properties of real numbers.

Theoretically, the set of real numbers consists of rational and irrational numbers, where we know clearly what rational numbers are and what properties they have. However, when field of rational numbers is expanded to become real numbers, what new properties beyond those of rational numbers do real numbers acquire?

The major difference between the system of real numbers and that of rational numbers is the continuity of real numbers, which is not shared by the system of rational numbers. In particular, the line of rational numbers has many "holes". For example, as indicated in Figure 10.4, all the rational numbers make up the dotted x-axis OX. If we draw the square $OABC$ of side length 1 and a circle centered at point O and passing through point B, then this circular arc will not interact the OX axis. It is because the radius of this circular arc is $\sqrt{2}$. From this fact that the circular arc goes through the number line OX of rational numbers without hitting any point on the line, it follows that the densely distributed rational numbers over the line contain gaps.

When we fill up all these gaps with irrational numbers, we obtain the system of real numbers. That is why the field of real numbers is considered continuous.

10.4.1 Characteristics of real number system

In this subsection, we see how we can employ of the language of mathematics in general and Dedekind cuts in particular to describe the continuity of real numbers.

Dedekind Axiom: Partition the set of all real numbers into two non-empty sets A and B. If for any $x \in A$ and $y \in B$ it is always true that $x < y$, then there is either the maximum number in A or the minimum number in B, exactly one and only one of which holds true.

Assume that S is a set of real numbers and M a real number. If for any $x \in S$ we have $x \leq M$, then the number M is known as an upper bound of S. If for any $x \in S$ we have $x \geq M$, then the number M is known as a lower bound of S. The minimum upper bound of S is known as the supremum, denoted sup$S$, and the maximum lower bound of S the infimum, denoted inf$S$.

Corresponding to this concept, one should be aware that although a non-empty set of rational numbers may have upper (respectively, lower) bounds, the set may not have a supremum (respectively, infimum). For example, the set of all rational numbers whose squares are less than $\sqrt{2}$, does not have a supremum. But for real numbers, we have:

Theorem 10.21: Each non-empty set with upper bounds of real numbers must have a supremum; and each non-empty set with lower bounds of real numbers must have an infimum

Proof. First we show that each non-empty real-number set S with upper bounds must have a supremum.

To this end, let B be the set of all upper bounds of S and A the rest of the real numbers. Then for any $x \in A$ and $y \in B$ we always have $x < y$. From Dedekind Axiom it follows that it must be either that A contains a maximum element or that B contains a minimum number.

If $a \in A$, then a is not an upper bound of S so that there is $c \in S$ such that $a < c$. So, none of the real numbers in the interval (a, c) is an upper bound of S. That implies that A does not have a maximum number. Therefore, B must contain the minimum element, which is the supremum of S.

Secondly, assume that a non-empty real-number set U has a lower bound m. Define $V = \{-u : u \in U\}$. Then, V has a lower bound $-m$. By applying what we have obtained above, it follows that V has its supremum b. Then $-b$ is evidently the infimum of U. QED.

Theorem 10.22: If there is an infinite sequence of closed intervals $\Delta_n = [a_n, b_n]$ such that $\Delta_n \subseteq \Delta_{n+1}$ and the set $\{|b_n - a_n|\}_{n=1}^{\infty}$ of real numbers does not have any positive lower bound, then these intervals have one unique common point.

Proof. From the assumption it follows that each b_k is an upper bound a of the set $\{a_n\}_{n=1}^{\infty}$. So, the number a is not greater than any b_k and has to be a common point of all the given intervals. On the other hand, similarly there is a common point b such that $|b - a|$ is the infimum of $\{|b_n - a_n|\}_{n=1}^{\infty}$. From the assumption that the set $\{|b_n - a_n|\}_{n=1}^{\infty}$ does not have any positive lower bound, it follows that $|b - a| = 0$. That is, $a = b$. That implies that the common point is unique. QED.

10.4.2 Existence of inverse functions

Based on the intuition beyond continuous functions, we have the following result.

Theorem 10.23: (Intermediate value theorem of continuous functions). Assume that a function $f(x)$ is continuous on the closed interval $[a,b]$ such that $f(a)\cdot f(b)<0$. Then there is $c\in[a,b]$ such that $f(c)=0$.

Proof. Without loss of generality, let us assume that $f(a)<0$ and $f(b)>0$. Define

$$A=\{x\in[a,b]:f(x)<0\}$$

Then A is not empty and has upper bounds. Let $c=\sup A$. Then in any neighborhood of the point c the function outputs of $f(x)$ must contain both positive and negative values at the same time. That implies that $f(c)=0$. QED.

Theorem 10.24: (Existence of Inverse Functions). Assume that a function $F(x)$ is continuous and strictly monotonic on the interval $[a,b]$. Let $A=\min\{F(a),F(b)\}$ and $B=\max\{F(a),F(b)\}$. Then there is a unique strictly monotonic function $G(x)$ defined on $[A,B]$ with range $[a,b]$ such that for any $x\in[A,B]$, $F(G(x))=x$; and for any $x\in[a,b]$, $G(F(x))=x$.

Proof. According to the given conditions, when $F(x)$ is increasing, let $G(A)=a$ and $G(B)=b$. When $F(x)$ is decreasing, let $G(A)=b$ and $G(B)=a$. And for $x\in[A,B]$, from Theorem 10.23 it follows that there is $c\in[a,b]$ such that $F(c)=x$. Define $G(x)=c$. Therefore, a function $G:[A,B]\to(a,b)$ is defined satisfying $F(G(x))=F(c)=x$.

To show the uniqueness of this function G, let us assume that there is another function $H(x)$ defined on $[A,B]$ such that $F(H(x))=x$. Then for any $u\in(A,B)$, we have the following

$$F(G(u))=u=F(H(u))$$

The assumed strict monotonicity of the function $F(x)$ implies that $G(u)=H(u)$. That proves the needed uniqueness. QED.

In the previous theorem, the function $G(x)$ is known as the inverse function of $F(x)$ on the interval $[a,b]$.

10.4.3 Existence of definite integrals

In Theorems 10.14–10.16, we have seen that under certain conditions, such as continuity or monotonicity, the existence of integral systems is unique. However, what is left unsettled is how the existence is warranted in the first place. With the properties of real numbers developed, we are now ready to attack this problem.

Assume that $f(x)$ is defined on an interval I. For any subinterval $[u,v]\subseteq I$, let $f(x)$ be bounded on $[u,v]$. If there is a piecewise linear function $g(x)(\le f(x))$ defined on $[u,v]$, then this function is referred to as a lower bound function of $f(x)$. And if there is a piecewise linear function $g(x)(\ge f(x))$ defined on $[u,v]$, then this function is referred to as an upper bound function of $f(x)$. Both upper and lower functions are obviously integrable on $[u,v]$.

Let

$$A = \left\{ \int_u^v g(x)dx : g(x) \text{ is an upper function of } f(x) \right\}$$

then A is a non-empty set of real numbers with lower bounds. Denote its infimum as $H(u,v)$. Similarly, define

$$B = \left\{ \int_u^v g(x)dx : g(x) \text{ is a lower function of } f(x) \right\}$$

then B is a non-empty set of real numbers with upper bounds. Denote its supremum as $L(u,v)$. Then, it is clear that $L(u,v) \leq H(u,v)$. It can be shown that if $L(u,v) = H(u,v)$, let $S(u,v) = L(u,v) = H(u,v)$, then $S(u,v)$ is the unique integral system of the function $f(x)$ on the interval I. Here all the details are omitted. QED.

Corresponding to mathematical induction over the set of natural numbers, the following is known as the continuous mathematical induction over the set of real numbers:

Mathematical Induction over Real Numbers: Assume that P_x is a proposition involving real number x. If the following steps (i) and (ii) hold true,

(i) The initial step: There is a certain real number x_0 such that for any real $x < x_0, P_x$ holds true;
(ii) The inductive Step: If for any $x < y, P_x$ holds true, then there is $z > y$ such that P_x holds true for any $x < z$,

then for any real number x, P_x holds true.

Theorem 10.25: (Zhang, 2010, p. 125). The continuous mathematical induction over real numbers is equivalent to Dedekind Axiom.

Proof. First, let us derive the continuous mathematical induction over real numbers from Dedekind Axiom. To this end, let us use the proof of contradiction.

If the continuous mathematical induction is not true, then there is a proposition P_x involving real number x such that both statements (i) and (ii) hold true, while there is still a real number u such that P_u is not true.

Define two sets A and B as follows:

$$A = \{y \in \mathbb{R} : \forall x \in (-\infty, y)\ P_x \text{ holds true}\}$$

and

$$B = \mathbb{R} - A$$

where $\mathbb{R}$ is the set of all real numbers Then both sets A and B are non-empty; and each element in A is smaller than each element in B. Let $a = \sup A$. Then $a \in A$ or $a \in B$, but not both.

For any $x \in (-\infty, a)$, because $y_0 = \frac{x+a}{2} < a$, $y_0 \in A$. From the fact that $x \in (-\infty, y_0)$ and the definition of the set A, it follows that P_x holds true. Hence for any $x < a$, P_x holds true. From statement (ii) in the continuous mathematical induction, it follows that there is $y > a$ such that P_x holds true for any $x < y$. So, this y belongs to A. That contradicts the definition of $a = \sup A$. Therefore, the assumption of the proof by contradiction is not true. In other words, the continuous mathematical induction holds true.

Next, let us derive Dedekind Axiom from the continuous mathematical induction.

If the set $\mathbb{R}$ of all real numbers is divided into two nonempty sets A and B such that for each $x \in A$ and $y \in B, x < y$, what needs to be shown is that either A has the maximum element or B has the minimum element.

Again, let us use proof by contradiction. If A does not have the maximum element and at the same time B does not the minimum element, let us define a proposition P_x as follows:

$P_x =$ "x belongs to A"

Because A is not empty, take one element $x_0 \in A$. Hence for any $x \in (-\infty, x_0), x$ must belong to A so that P_x holds true. This let us use as the initial step of our continuous mathematical induction.

Assume that for any $x \in (-\infty, y), P_x$ holds true. Because B does not have a minimum element, y belongs to A. And because A does not have a maximum element, there is $z > y$ such that $z \in A$. So, for any $x \in (-\infty, z), P_x$ holds true. This completes the inductive reasoning.

From the continuous mathematical induction, it follows that for any real number x, P_x holds true. That is, every real number belongs to A, which contradicts the assumption that B is not empty. QED.

As applications of this continuous mathematical induction, let us use it to prove a few important properties of real numbers.

Theorem 10.26: Assume that $\{\Delta_s\}_{s \in I}$ is a set of open intervals, where I is an index set. If $\cup_{s \in I} \Delta_s = [a, b]$, for some closed interval $[a, b]$, then there must be a finite subset $J \subseteq I$ such that $\cup_{s \in J} \Delta_s = [a, b]$.

Proof. Define the following proposition:

$P_s = \exists$ finite $J \subseteq I$ $(\cup_{s \in J} \Delta_s = (-\infty, x] \cap [a, b])$

Then we have the following:

(i) When $x < a, P_x$ holds true.
(ii) Assume that for each $x < y, P_x$ holds true.

If $y < a$ or $y > b$, evidently there is $z > y$ such that P_x holds true for any $x < z$. If $y \in [a, b]$, then there is an open interval (c, d) such that $y \in [c, d]$. In this case, take

$x_1 \in (c, y)$. Then from the inductive assumption, it follows that $(-\infty, x_1] \cap [a, b]$ has a finite cover.

For any $x \in [y, d)$, because

$$(-\infty, x] \cap [a, b] = ((-\infty, x_1] \cap [a, b]) \cup ([x_1, x] \cap [a, b])$$

it follows that $(-\infty, x] \cap [a, b]$ has a finite cover. Let $z = d$. Then for any $x < z$, P_x holds true.

From continuous mathematical induction, it follows that for every real number x, P_x holds true. Taking $x > b$ leads to the conclusion that there is a finite subset $J \subseteq I$ such that $\cup_{S \in J} \Delta_s = [a, b]$. QED.

Theorem 10.27: If a sequence of closed intervals $\{[a_n, b_n]\}_{n=1}^{\infty}$ satisfies $a_n \leq a_{n+1} \leq b_{n+1} \leq b_n$, for any natural number n, then there must be a real number c such that $c \in [a_n, b_n]$, for each n. Additionally, if the sequence $\{|a_n - b_n|\}_{n=1}^{\infty}$ does not have any positive lower bound, then such real number c must be unique.

Proof. First, let us prove the existence by using the proof by contradiction. Assume that there is no real number c that belongs to each closed interval $[a_n, b_n]$. Define the following proposition:

$P_x =$ "x is not an upper bound of $\{a_n\}_{n=1}^{\infty}$"

Then the following hold true:

(i) Assume $a = a_1$. Then when $x < a$, P_x holds true; and
(ii) If for each $x < y$, P_x holds true, then it is impossible for y to be an upper bound of $\{a_n\}_{n=1}^{\infty}$; otherwise it belongs to each closed interval $[a_n, b_n]$, which does not agree with the assumption of the proof by contradiction. Hence there is k such that $a_k > y$. Let $z = a_k$, then for each $x < z$, P_x holds true.

From the continuous mathematical induction, it follows that no real number x can be an upper bound of $\{a_n\}_{n=1}^{\infty}$. That contradicts the fact that b_1 is an upper bound of $\{a_n\}_{n=1}^{\infty}$. This end implies that there is real number c that belongs to each closed interval $[a_n, b_n]$.

Next, let us prove the uniqueness. If there is d that belongs to each $[a_n, b_n]$, then $b_n - a_n \geq |c - d|$. Because $\{|b_n - a_n|\}_{n=1}^{\infty}$ does not have any positive lower bound, it means that $|c - d| = 0$. That is $c = d$. QED.

Theorem 10.28: If an infinite set M of real numbers is contained the closed interval $[a, b]$, then there must be a point in $[a, b]$ that is a limit point of M.

Proof. We show this statement by using proof by contradiction. Assume that no point in the closed interval $[a, b]$ is a limit point of M. Let us introduce the following proposition:

$P_x =$ Within $(-\infty, x]$ there are finite many elements of M.

Then we have the following:

(i) If $x_0 = a$, then for each $x < x_0, P_x$ holds true; and
(ii) Assume that there is such y that for each $x < y, P_x$ holds true.

Because y is not a limit point of M, there is an open interval (α, β), within which there are only finite many elements of M. Take $u \in (\alpha, y)$. From the inductive assumption it follows that the interval $(-\infty, u]$ contains only finite many elements of M so that the interval $(-\infty, \beta)$ contains only finite many elements of M. Let $z = \beta > y$, then for each $x < z, P_x$ holds true.

So, continuous mathematical induction implies that for each real number x, the interval $(-\infty, x]$ contains only finite many elements of M. However, when $x = b$, the previous conclusion indicates that M contains only finite many elements, which contradicts the initial assumption of M. That implies that the assumption of the proof by contradiction is incorrect. QED.

10.5 SEQUENCES, SERIES, AND CONTINUITY

Historically, the idea of summing up infinitely many numbers has puzzled the mankind thousands of years. Only after the theory of limits was established, such puzzles are revolved once for all. The purpose of this section is to see how we can successfully avoid the puzzles related to summing up infinitely many numbers and how we can investigate the continuity of functions when the concept of limits is not directly employed. At the same time, this section helps to bridge the new approach, as presented in this chapter, to the traditional approach of calculus.

In particular, Subsection 10.5.1 looks at the limits of sequences and infinite series; Subsection 10.5.2 develops the concept of limits of functions. Then point-wise continuity and differentiability are addressed in Subsections 10.5.3 and 10.5.4.

10.5.1 Limits of sequences and infinite series

For our purposes, any sequence $\{D_n\}_{n=1}^{\infty}$ of non-negative numbers, that are monotonically non-decreasing and unbounded is referred to as a D sequence. And the sequence $\{d_n\}_{n=1}^{\infty}$, where $d_n = \frac{1}{D_n}$ when $D_n \neq 0$, is known as a d sequence. In other words, each d sequence consists of non-negative numbers that are non-increasing and without positive lower bound.

The following some of the most commonly used examples of D sequences:

$$\begin{aligned}
\{n\}_{n=1}^{\infty} &= \{1, 2, 3, \ldots, n, \ldots\} \\
\{n^2\}_{n=1}^{\infty} &= \{1^2, 2^2, 3^2, \ldots, n^2, \ldots\} \\
\{\sqrt{n}\}_{n=1}^{\infty} &= \{\sqrt{1}, \sqrt{2}, \sqrt{3}, \ldots, \sqrt{n}, \ldots\} \\
\{\ln(n)\}_{n=1}^{\infty} &= \{\ln(1), \ln(2), \ln(3), \ldots, \ln(n), \ldots\} \\
\{2^n\}_{n=1}^{\infty} &= \{2^1, 2^2, 2^3, \ldots, 2^n, \ldots\} \\
\{n!\}_{n=1}^{\infty} &= \{1!, 2!, 3!, \ldots, n!, \ldots\}
\end{aligned}$$

Definition 10.3 Assume that $\{a_n\}_{n=1}^{\infty}$ is a sequence of numbers. If there is a D sequence $\{D_n\}_{n=1}^{\infty}$ such that $|a_n| \geq D_n$ except a few finite number of terms, $n = 1, 2, \ldots$, then $\{a_n\}_{n=1}^{\infty}$ is referred to as a sequence of infinity or simply infinity. In this case, we say that a_n approaches infinity and write symbolically as either $a_n \to \infty$ or $\lim_{n\to\infty} a_n = \infty$.

If the $\{a_n\}_{n=1}^{\infty}$ satisfies $a_n \geq D_n$ except a few finite number of terms, $n = 1, 2, \ldots$, then $\{a_n\}_{n=1}^{\infty}$ is referred to as a sequence of positive infinity or simply positive infinity so that we say that a_n approaches positive infinity and write symbolically as either $a_n \to +\infty$ or $\lim_{n\to\infty} a_n = +\infty$. Similarly, we have the concept of a_n approaching negative infinity and write symbolically as either $a_n \to -\infty$ or $\lim_{n\to\infty} a_n = -\infty$.

Definition 10.4 Assume that $\{a_n\}_{n=1}^{\infty}$ is a sequence of numbers. If there is a d sequence $\left\{d_n\right\}_{n=1}^{\infty}$ such that $|a_n| \leq d_n$ except a few finite number of terms, $n = 1, 2, \ldots$, then $\{a_n\}_{n=1}^{\infty}$ is referred to as a sequence of infinitesimal or simply infinitesimal. In this case, we say that a_n approaches 0 and write symbolically as either $a_n \to 0$ or $\lim_{n\to\infty} a_n = 0$.

Definition 10.5 Assume that $\{a_n\}_{n=1}^{\infty}$ is a sequence of numbers. If there is a number a such that $\{a_n - a\}_{n=1}^{\infty}$ is a sequence of infinitesimal, then the number a is known as the limit of the sequence $\{a_n\}_{n=1}^{\infty}$. In this case, we say that a_n approaches a and write symbolically as either $a_n \to a$ or $\lim_{n\to\infty} a_n = a$.

This concept of limits of sequences is different of the traditional one introduced in the $\varepsilon - N$ language. In particular, the new concept here of limits of sequences does not involve potential infinite except that the actual existence of the infinite sequence is assumed, while the traditional concept is established using the language of potential infinite on top of the assumption of an actual infinite. Beyond this difference, the following examples show that this new approach is quite effective in terms of learning of beginners.

Example 10.8 Show $\lim_{n\to\infty} \sqrt[n]{n} = 1$.

Proof. Denote $a_n = \sqrt[n]{n} - 1$. It suffices to show $a_n \to 0$.

Let us raise both sides of the equation $\sqrt[n]{n} = a_n + 1$ to the nth power, respectively. From $a_n \geq 0$ it follows that

$$n = (a_n + 1)^n > 1 + na_n + \frac{n(n-1)}{2}a_n^2 > \frac{n(n-1)}{2}a_n^2$$

So, when $n > 1$, we have

$$|a_n| \leq \frac{2}{\sqrt{n-1}}$$

Because $\left\{\frac{2}{\sqrt{n-1}}\right\}_{n=2}^{\infty}$ is a d sequence, it follows that $a_n \to 0$. QED.

Example 10.9 Show that $\lim_{n\to\infty}(1+\frac{1}{n})^n = e$.

Proof. Because $\ln(v) - \ln(u)$ is equal to the area under the curve $y = \frac{1}{x}$ over the interval $[u, v]$, we have

$$\frac{1}{v} < \frac{\ln(v) - \ln(u)}{v - u} < \frac{1}{u}$$

By taking $u = 1$ and $v = 1 + \frac{1}{n}$, this inequality becomes

$$\frac{n}{n+1} < n\ln(1 + \frac{1}{n}) < 1$$

Therefore, from the monotonicity property of exponential functions, it follows that

$$e^{\frac{n}{n+1}} < (1 + \frac{1}{n})^n < e$$

which implies

$$\left| e - (1 + \frac{1}{n})^n \right| < e - e^{\frac{n}{n+1}} = e^{\frac{n}{n+1}}\left(e^{\frac{1}{n+1}} - 1\right) < 4(e^{\frac{1}{n+1}} - 1)$$

Denote $\beta_n = e^{\frac{1}{n+1}} - 1$. Then $e^{\frac{1}{n+1}} = 1 + \beta_n$. Therefore, we have

$$4 > e = (1 + \beta_n)^{n+1} > 1 + (n+1)\beta_n$$

which implies that $\beta_n < \frac{3}{n+1}$. That shows that $e - (1 + \frac{1}{n})^n$ is a sequence of infinitesimal. Hence $(1 + \frac{1}{n})^n$ approaches e. QED.

Example 10.10 Given

$$\ln(1 + u) = \int_1^{1+u} \frac{dx}{x} \tag{10.49}$$

compute the limit of the sequence $\{S_n\}_{n=1}^{\infty}$, where u is a fixed positive number and

$$S_n = \sum_{k=0}^{n-1} \frac{u}{n + ku}$$

Solution. Let us divide the definite integral of the right-hand side of equ. (10.49) into n pieces as estimate each piece as follows:

$$\frac{u}{n+(k+1)u} = \frac{u}{n}\left(1+\frac{(k+1)u}{n}\right)^{-1}$$
$$< \int_{1+\frac{ku}{n}}^{1+\frac{(k+1)u}{n}} \frac{1}{x}dx$$
$$< \frac{u}{n}\left(1+\frac{ku}{n}\right)^{-1}$$
$$= \frac{u}{n+ku}$$

Summing up this inequality for k from 0 to n − 1 produces

$$\sum_{k=1}^{n}\frac{u}{n+ku} < \ln(1+u) = \int_{1}^{1+u}\frac{1}{x}dx < \sum_{k=0}^{n-1}\frac{u}{n+ku}$$

So, we obtain

$$\left|\ln(1+u) - \sum_{k=0}^{n-1}\frac{u}{n+ku}\right| < \frac{u}{n} + \frac{u}{n(1+u)} < \frac{1}{n}(u+1)$$

This end proves that $S_n \to \ln(1+u)$. QED.

Based on the previous examples, a natural question arises: Under what conditions does a sequence have a limit? The following result provides a fundamental condition for a sequence to have limit.

Theorem 10.29: Each monotonically non-decreasing and bounded sequence must have limit.

Proof. Assume that $\{a_n\}_{n=1}^{\infty}$ is a monotonically non-decreasing sequence with bound. Denote $a = \sup\{a_n\}_{n=1}^{\infty}$. Then $\{|a-a_n|\}_{n=1}^{\infty}$ is a monotonically non-increasing sequence of non-negative numbers. If the sequence $\{|a-a_n|\}_{n=1}^{\infty}$ has lower bound m, then $a_n \le a-m$, for each natural number n, so that we have $a \le a-m$ and $m \le 0$. That is $\{|a-a_n|\}_{n=1}^{\infty}$ is a d sequence without any positive lower bound. Therefore, from definitions of limits, it follows that $a_n \to a$. QED.

Theorem 10.30: If three sequences $\{a_n\}_{n=1}^{\infty}$, $\{b_n\}_{n=1}^{\infty}$, and $\{c_n\}_{n=1}^{\infty}$ satisfy $a_n \le b_n \le c_n$, then if $\lim_{n\to\infty} a_n = \lim_{n\to\infty} c_n = a$, then $\lim_{n\to\infty} b_n = a$. QED.

Next, let us turn our attention to the study of infinite series. Given a series

$$\sum_{n=1}^{\infty} u_n = u_1 + u_2 + u_3 + \cdots + u_n + \cdots$$

denote the partial sum of the first n terms by $S_n = \sum_{k=1}^{n} u_k = u_1 + u_2 + u_3 + \cdots + u_n$. The sequence $\{S_n\}_{n=1}^{\infty}$ is referred to as that of partial sums of the given series.

Definition 10.6 If the partial sums sequence $\{S_n\}_{n=1}^{\infty}$ of the series $\sum_{n=1}^{\infty} u_n$ approaches its limit S, then the series $\sum_{n=1}^{\infty} u_n$ is said to be convergent and write

$$\sum_{n=1}^{\infty} u_n = S$$

If the sequence $\{S_n\}_{n=1}^{\infty}$ does not have any limit, then the series $\sum_{n=1}^{\infty} u_n$ is said to be divergent so that it does not have a sum.

Because the problem of whether a given series is convergent or not is equivalent to that of whether or not a particular sequence has limit, according to what we have established earlier, we can obtain the corresponding results for the convergence of series.

In particular, if each term of a series is positive, then the series is known as one of positive terms. In this case, the partial sums constitute an increasing sequence. Therefore, Theorem 10.29 can be rewritten as follows:

Theorem 10.31: For each positive term series, if its partial sums sequence is bounded, then the series must be convergent. QED.

Theorem 10.32: If the series $\sum_{n=1}^{\infty} u_n$ is convergent, then $\{u_n\}_{n=1}^{\infty}$ is a sequence of infinitesimal. That is, $u_n \to 0$. QED.

Theorem 10.33: A given series $\sum_{n=1}^{\infty} u_n$ is convergent, if and only if

$$\lim_{n\to+\infty} \sup\left\{\left|\sum_{k=n}^{m} u_k\right| : m > n\right\} = 0.\text{QED.}$$

Equivalently, Theorem 10.33 can be written as follows: A given series $\sum_{n=1}^{\infty} u_n$ is convergent, if and only if there is a d sequence $\{d_n\}_{n=1}^{\infty}$ such that when $m > n$, the following holds true:

$$\left|\sum_{k=n}^{m} u_k\right| < d_n.$$

10.5.2 Limits of functions

It has been a long-held belief that only by using the processes of limits one can establish the fundamental concepts of calculus. Although we have shown that this belief is not correct, the concept of limits is still very important.

Each non-decreasing, non-negative, and unbounded function $D(x)$, defined on $[a,+\infty)$, is known as a D function defined on the neighborhood of $+\infty$. Each non-increasing and non-negative function $d(x)$ that does not have a positive lower bound, defined on $[a,+\infty)$, is known as a d function defined on the neighborhood of $+\infty$.

Definition 10.7 Assume that $f(x)$ is a function defined on $[a,+\infty)$. If there is a D function $D(x)$ of the neighborhood of $+\infty$ such that

$$|f(x)| \geq D(x) \tag{10.50}$$

on some $[b,+\infty)$, then $f(x)$ is said to approach infinity as x approaches the positive infinity or $f(x)$ is an infinity as x approaches the positive infinity, denoted

$$\lim_{x\to+\infty} f(x)=\infty \quad \text{or} \quad f(x)\to\infty, \ \text{when}\ x\to+\infty.$$

If equ. (10.50) is replaced by $f(x)\geq D(x)$, then we say that the function $f(x)$ approaches the positive infinity as x approaches the positive infinity, denoted $\lim_{x\to+\infty} f(x)=+\infty$ or $f(x)\to\infty$, when $x\to+\infty$. If equ. (10.50) is replaced by $-f(x)\geq D(x)$, then we say that the function $f(x)$ approaches the negative infinity as x approaches the positive infinity, denoted $\lim_{x\to+\infty} f(x)=-\infty$ or $f(x)\to\infty$, when $x\to-\infty$.

Definition 10.8 Assume that a function $f(x)$ is defined on the interval $[a,+\infty)$. If there is a d function $d(x)$ of the neighborhood $+\infty$ such that

$$|f(x)| \leq d(x) \tag{10.51}$$

on some $[b,+\infty)$, then $f(x)$ is said to approach 0 or to be an infinitesimal as x approaches the positive infinity, denoted

$$\lim_{x\to+\infty} f(x)=0 \quad \text{or} \quad f(x)\to 0, \ \text{when}\ x\to+\infty; \quad \text{or} \quad f(x)=o(1)(x\to+\infty).$$

Definition 10.9 Assume that a function $f(x)$ is defined on the interval $[a,+\infty)$. If there is a real number a such that

$$\lim_{x\to+\infty} [f(x)-a]=0$$

then $f(x)$ is said to approach a as x approaches the positive infinity, denoted

$$\lim_{x\to+\infty} f(x)=a \quad \text{or} \quad f(x)\to a, \ \text{when}\ x\to+\infty; \quad \text{or} \quad f(x)=a+o(1)(x\to+\infty).$$

Definition 10.10 Assume that a function $f(x)$ is defined on the interval $(-\infty, a]$. Then the expression

$$\lim_{x\to-\infty} f(x) = A \quad \text{or} \quad f(x) \to A(x \to -\infty)$$

is equivalent to

$$\lim_{x\to+\infty} f(-x) = A \quad \text{or} \quad f(-x) \to A(x \to +\infty)$$

where A stands for either ∞, or $+\infty$, or $-\infty$, or a real number a.

Definition 10.11 Assume that a function $f(x)$ is defined on $(-\infty, a] \cup [b, +\infty)$. Then the expression

$$\lim_{x\to\infty} f(x) = A \quad \text{or} \quad f(x) \to A(x \to \infty)$$

is equivalent to

$$\lim_{x\to+\infty} f(x) = A \quad \text{or} \quad f(x) \to A(|x| \to +\infty)$$

where A stands for either ∞, or $+\infty$, or $-\infty$, or a real number a.

Definition 10.12 Assume that a function $f(x)$ is defined on $(x_0, x_0 + H]$, where $H > 0$. If there is a real number a and a d function $d(h)$ defined on $(0, H]$ such that for any $x \in (x_0, x_0 + H]$, the following holds true

$$|f(x) - a| \leq d(x - x_0)$$

then $f(x)$ is said to have right-hand side limit a at $x = x_0$, denoted

$$\lim_{x\to x_0^+} f(x) = a \quad \text{or} \quad f(x) \to a(x \to x_0^+)$$

which is also written as $f(x) = a + o(1)$ in a right-hand side neighborhood of $x = x_0$. If $a = 0$, we say that $f(x)$ is an infinitesimal in a right-hand side neighborhood of $x = x_0$.

Similarly, if a function $f(x)$ is defined on $(x_0, x_0 + H]$, where $H > 0$, such that there is a d function $d(h)$ defined on $(0, H]$ satisfying that for any $x \in (x_0, x_0 + H]$, the following holds true

$$f(x) \geq \frac{1}{d(x - x_0)}$$

then $f(x)$ is said to be the positive infinity in a right-hand side neighborhood of $x=x_0$, denoted

$$\lim_{x\to x_0^+} f(x)=+\infty \quad \text{or} \quad f(x)\to+\infty(x\to x_0^+)$$

Similarly, we can establish the concepts of $\lim_{x\to x_0^+} f(x)=-\infty$ and $\lim_{x\to x_0^+} f(x)=\infty$. As for the concept of left-hand side limits, we have

Definition 10.13 Assume that a function $f(x)$ is defined on $[x_0-H,x_0)$, where $H>0$. If there is a real number a and a d function $d(h)$ defined on $(0,H]$ such that for any $x\in[x_0-H,x_0)$, the following holds true

$$|f(x)-a|\le d(x_0-x)$$

then $f(x)$ is said to have left-hand side limit a at $x=x_0$, denoted

$$\lim_{x\to x_0^-} f(x)=a \quad \text{or} \quad f(x)\to a(x\to x_0^-)$$

which is also written as $f(x)=a+o(1)$ in a left-hand side neighborhood of $x=x_0$. If $a=0$, we say that $f(x)$ is an infinitesimal in a left-hand side neighborhood of $x=x_0$.

Definition 10.14 Assume that a function $f(x)$ has both side limits that are the same and equal to a, then $f(x)$ is said to have limit a at $x=x_0$ or to approach a when x approaches x_0, denoted

$$\lim_{x\to x_0} f(x)=a \quad \text{or} \quad f(x)\to a\ (x\to x_0)$$

or $f(x)$ is said to be $f(x)=a+o(1)$ in the neighborhood of $x=x_0$.

Similarly, we can establish the concept of $f(x)$ approaching infinity at $x=x_0$.

Theorem 10.34: If a function $f(x)$ is uniformly continuous on $[a,b]$ and $x_0\in(a,b)$, then the following hold true:

$$\lim_{x\to x_0} f(x)=f(x_0),\quad \lim_{x\to a^+} f(x)=f(a),\quad \lim_{x\to b^-} f(x)=f(b).$$

Proof. According to the definition of uniform continuity, there is a d function $d(h)$ such that for any two points $x_0, x_0+h\in[a,b]$, the following holds true:

$$|f(x_0+h)-f(x_0)|\le d(h)$$

By fixing the point x_0, this inequality can be rewritten as follows:

$$|f(x) - f(x_0)| \leq d(|x - x_0|)$$

which is exactly the definition for $\lim_{x\to x_0} f(x) = f(x_0)$. The other two equations can be shown similarly. The details are omitted. QED.

Theorem 10.35: If a function $f(x)$ is continuously differentiable on $[a,b]$ and $x_0 \in (a,b)$, then the following holds true:

$$\lim_{x\to x_0} \frac{f(x) - f(x_0)}{x - x_0} = f'(x_0)$$

And at the end points the following hold true:

$$\lim_{x\to a^+} \frac{f(x) - f(a)}{x - a} = f'(a), \quad \lim_{x\to b^-} \frac{f(x) - f(b)}{x - b} = f'(b)$$

Proof. According to the relevant definitions, there is a d function $d(h)$ such that for any two points $x_0, x_0 + h$ in $[a,b]$ the following holds true:

$$\left| \frac{f(x_0 + h) - f(x_0)}{h} - f'(x_0) \right| \leq d(h)$$

Let x_0 be fixed and rewrite this inequality as follows:

$$\left| \frac{f(x) - f(x_0)}{x - x_0} - f'(x_0) \right| \leq d(|x - x_0|)$$

Therefore, we obtain $\lim_{x\to x_0} \frac{f(x) - f(x_0)}{x - x_0} = f'(x_0)$. Similar arguments can be used to prove the other two equations. All the details are omitted. QED.

10.5.3 Pointwise continuity

Up to this point, the continuity of functions has been investigated as a property on an interval. In contrast, the traditional textbook of calculus introduces continuity at a particular point first.

Assume that a function $f(x)$ is defined in a neighborhood of a point x_0. If

$$\lim_{x\to x_0} f(x) = f(x_0)$$

then the function $f(x)$ is said to be continuous at the point x_0. If the function $f(x)$ is continuous at every point in an open interval (a,b), then the function is said to be continuous on the interval (a,b).

Equivalently, the concept for a function $y = f(x)$ to be continuous at a point x_0 can be restated as follows: Assume that the function $y = f(x)$ is defined in a neighborhood

Δ of a point x_0. If there is a d function $d(h)$ such that for any point $x \in \Delta$ the following holds true:

$$|f(x) - f(x_0)| \leq d(|x - x_0|)$$

then the function $y = f(x)$ is said to be continuous at the point x_0.

For the pointwise continuity, we evidently have the following properties:

Theorem 10.36: If $f(x)$ is continuous at a point x_0, then $f(x)$ is bounded within a neighborhood of the point x_0. QED.

Theorem 10.37: If $f(x)$ is continuous at a point x_0 and $A < f(x_0) < B$, for some constants A and B, then there is a neighborhood $U(x_0)$ of x_0 such that for every $x \in U(x_0)$, $A < f(x) < B$.

Proof. From the definition of continuous functions, it follows that there are a closed interval $[a, b]$ such that $x_0 \in (a, b)$ and a d function $d(h)$ such that for any $x \in [a, b]$, the following holds true:

$$|f(x) - f(x_0)| \leq d(x - x_0)$$

Let $m = \min\{f(x_0) - A, B - f(x_0)\}$, then $m > 0$. If we want to have $A < f(x) < B$, we only need to have $|f(x) - f(x_0)| < m$. From the definition of a d function it follows that there is a positive number z such that for any $|h| < z$, it is always true that $d(h) < m$. Therefore, let $U(x_0) = (x_0 - z, x_0 + z)$. Then for each $x \in U(x_0)$, we have

$$|f(x) - f(x_0)| \leq d(x - x_0) < d(z) < m$$

That concludes the proof. QED.

Corresponding to the concepts of one-sided limits, we can establish the concepts of left- and right-handed continuity for functions. In particular, we have

Definition 10.15 Assume that a function $f(x)$ is defined in a left neighborhood of x_0. If

$$\lim_{x \to x_0^-} f(x) = f(x_0)$$

then $f(x)$ is said to be continuous from the left at x_0. If the function $f(x)$ is defined in a right neighborhood of x_0 and

$$\lim_{x \to x_0^+} f(x) = f(x_0)$$

then $f(x)$ is said to be continuous from the right at x_0.

If the function $f(x)$ is continuous at each point in (a,b) and continuous from the left at $x=b$ and from the right at $x=a$, then we say that the function $f(x)$ is continuous on $[a,b]$. Generally, if the function $f(x)$ is continuous at each internal point of an interval I and one-sided continuous at each end point of I, then we say that the function $f(x)$ is continuous on the entire interval I.

Pointwise continuity of functions is a local attribute. Next, let us see how we can develop some overall properties based on such local attributes.

Theorem 10.38: If a function $f(x)$ is continuous on a closed interval $[a,b]$, then $f(x)$ reaches its maximum and minimum values on $[a,b]$.

Proof. First let us see why $f(x)$ reaches its maximum. To this end, let us prove by using contradiction. Assume that $f(x)$ does not reaches its maximum on $[a,b]$. Then we introduce the following proposition:

$P_x =$ "There is a point u such that $f(u) > f(x), \forall x \in (-\infty, x]$

(i) For $x=a$, P_x holds true; otherwise $f(a)$ would be the maximum value.
(ii) Assume that there is a certain y such that for any $x<y$, P_x holds true. From the fact that $f(y)$ is not the maximum value it follows that there is v such that $f(v)>f(y)$. From Theorem 10.37, it follows that there is $\delta>0$ such that for any $u\in(y-\delta,y+\delta), f(u)<f(v)$ holds true. Let $z=y+\delta$.

From continuous mathematical induction, it follows that there is w such that $f(w)>f(x)$, for any $x\in(-\infty,y-0.5\delta]$.

Let $f(p)=\max\{f(w),f(v)\}$. Then $f(p)>f(x)$, for any $x\in(-\infty,y+\delta]$. That is, when $x<z$, P_x holds true. So, continuous mathematical induction implies that P_x holds true for each x. In particular, if $x=b$, this end indicates that there is u such that $f(u)>f(x)$, for any $x\in[a,b]$, a contradiction. That implies that $f(x)$ reaches its maximum value on $[a,b]$.

Secondly, let us see why $f(x)$ reaches its minimum. In fact, if we consider the function $-f(x)$. Then it is continuous on the closed interval $[a,b]$. So, from the previous argument, it follows that $-f(x)$ reaches its maximum value at a certain point $x_0\in[a,b]$. That means that $f(x)$ reaches its minimum value at x_0. QED.

Theorem 10.39: If a function $f(x)$ is continuous on a closed interval $[a,b]$ and $f(a)\neq f(b)$, then for any C that lies in between $f(a)$ and $f(b)$, there is at least one point $u\in[a,b]$ such that $f(u)=C$.

Proof. For any $x\in[a,b]$, define $F(x)=f(x)-C$; when $x<a$, define $F(x)=f(a)$; and when $x>b$, define $F(x)=f(b)$. Then $F(x)$ is a continuous function satisfying $F(a)\cdot F(b)<0$. Without loss of generality, let us assume $F(a)<0$ and $F(b)>0$. It suffices to show that there is one point $u\in(a,b)$ such that $F(u)=0$.

Let us prove by contradiction. Assume that for any $u \in [a, b]$, we have $F(u) \neq 0$. Introduce the following proposition:

$P_x =$ "F is forever negative on the interval $(-\infty, x]$"

Then we have the following:

(i) For any $x < a, P_x$ holds true.
(ii) Assume that there is such a y such that when $x < y, P_x$ holds true. That is for each $x < y, F(x) < 0$.

According to the assumption of our proof by contradiction, we know that $F(y) \neq 0$. From the assumption that $F(y)$ is continuous at y and Theorem 10.37, it follows that there is $\delta > 0$ such that $F(x)$ has the same sign as $F(y)$, for each $x \in (y - \delta, y + \delta)$. From $F(y - \frac{\delta}{2}) < 0$, it follows that $F(x) < 0$, for each $x \in (y - \delta, y + \delta)$.

Let $z = y + \delta$. Then when $x < z, P_x$ holds true. From continuous mathematical induction, it follows that P_x holds true, for each x. So when we take $x > b$, we obtain that $F(b) < 0$, a contradiction. QED.

To help our investigation of the relationship between the concepts of pointwise continuity and uniform continuity, let us first look at the concept of amplitudes of functions.

Let $f(x)$ be a bounded function defined on an interval I, and h a positive constant. Denote

$$\omega(h) = \sup\{|f(u) - f(v)| : |u - v| \leq h\}$$

which is referred to as the amplitude of $f(x)$ of step length h over the interval I, and simply known as the amplitude function of $f(x)$. When h is greater than the length of the interval I, $\omega(h)$ is equal to the difference between $\sup\{f(x) : x \in I\}$ and $\inf\{f(x) : x \in I\}$, known as the amplitude of $f(x)$ on I, denoted $\omega(f, I)$.

Evidently, $\omega(h)$ is non-negative and non-decreasing. Denote $\omega_0 = \inf\{\omega(h) : h > 0\} = \lim_{h \to 0} \omega(h)$, known as the maximum local amplitude of the function $f(x)$ on the interval I. If $\omega_0 = 0$, that is, $\omega(h)$ does not have any positive lower bound, then $\omega(h)$ is a d function. In this case, according to the definition, $f(x)$ is uniformly continuous on I.

The concept of pointwsie continuity can also be expressed by using amplitudes. In particular, consider the amplitude $\omega(f, [x_0, x_0 + \delta))$ of $f(x)$ of the half open interval $[x_0, x_0 + \delta)$. When δ decreases, this amplitude of $f(x)$ evidently will not be increasing. Therefore, when $\delta \to 0$, $\omega(f, [x_0, x_0 + \delta))$ has limit, denoted

$$\omega(f, x_0^+) = \lim_{\delta \to 0} \omega(f, [x_0, x_0 + \delta))$$

known as the right-amplitude of $f(x)$ at the point x_0. Similarly, denote

$$\omega(f, x_0^-) = \lim_{\delta \to 0} \omega(f, (x_0 - \delta, x_0])$$

known as the left-amplitude of $f(x)$ at the point x_0. And when $f(x)$ is defined on $(x_0-\delta, x_0+\delta)$, denote

$$\omega(f, x_0) = \lim_{\delta\to 0} \omega(f, (x_0-\delta, x_0+\delta))$$

known as the amplitude of $f(x)$ at the point x_0.

From the concept of amplitudes, it follows that $f(x)$ is continuous at a point x_0 if and only if the amplitude of $f(x)$ at the point x_0 is 0; $f(x)$ is left continuous at x_0 if and only if the left amplitude of $f(x)$ at the point x_0 is 0; and $f(x)$ is right continuous at x_0 if and only if the right amplitude of $f(x)$ at the point x_0 is 0.

Theorem 10.40: If a function $f(x)$ is continuous on a closed interval $[a, b]$, then $f(x)$ is uniformly continuous on $[a, b]$.

Proof. We will show this result by using proof by contradiction. Assume that the infimum ω_0 of the amplitude function $\omega(h)$ of $f(x)$ on $[a, b]$ is positive. In the following we will employ continuous mathematical induction to derive a contradiction.

First, let us define a new function $F(x)$ as follows:

$$F(x) = \begin{cases} f(a), & \text{if } x < a \\ f(x), & \text{if } x \in [a, b] \\ f(b), & \text{if } x > b \end{cases}$$

We denote the maximum local amplitude of $F(x)$ on $(-\infty, x]$ as w_x, and introduce the following proposition:

$$P_x = \text{“}2w_x < \omega_0\text{”}$$

Then we have the following:

(i) For $x < a$, it is obvious that $2w_x = 0 < \omega_0$. That is, P_x holds true.
(ii) Assume that there is y such that when $x < y$, P_x holds true.

From the continuity of $F(x)$ at y, it follows that there is an open interval (α, β) $(\ni y)$ such that

$$\sup\{F(x) : x \in (\alpha, \beta)\} - \inf\{F(x) : x \in (\alpha, \beta)\} < 0.5\omega_0$$

Take $u \in (\alpha, y)$. From the inductive assumption, it follows that $2w_u < \omega_0$. Therefore, we have $2w_\beta < \omega_0$. So, for any $x < z = \beta$, P_x holds true.

So, continuous mathematical induction implies that for every x, P_x holds true. In particular, when $x = b$, because $w_b = \omega_0$, we obtain $2\omega_0 < \omega_0$, a contradiction. QED.

10.5.4 Pointwise differentiability

From Theorem 10.35 we know that when a function $f(x)$ is continuously differentiable on $[a,b]$ and $x_0 \in (a,b)$, then we have:

$$\lim_{x\to x_0}\frac{f(x)-f(x_0)}{x-x_0}=f'(x_0)$$

and

$$\lim_{x\to a^+}\frac{f(x)-f(a)}{x-a}=f'(a),\quad \lim_{x\to b^-}\frac{f(x)-f(b)}{x-b}=f'(b)$$

That is, we can use the concept of limits to deepen our understanding of derivatives, and the knowledge of derivative to help us with the rich intension of limits. In terms of the concept of pointwise differentiability, we have

Definition 10.16 Assume that a function $y=f(x)$ is defined in a neighborhood $U(x_0)$ of the point x_0. If

$$\lim_{h\to 0}\frac{f(x_0+h)-f(x_0)}{h}=A$$

then the function $f(x)$ is said to be differentiable at x_0, and the limit value A is known as the derivative of $f(x)$ at x_0, denoted $f'(x_0)$ or $\left.\frac{dy}{dx}\right|_{x=x_0}$. If this limit does not exist, then the function $f(x)$ is said to be not differentiable at x_0.

Theorem 10.41: If a function $f(x)$ is differentiable at a point x_0, then the function $f(x)$ is continuous at the point x_0.

Proof. Because $f(x)$ is differentiable at a point x_0, from the definition of limits it follows that

$$\left|\frac{f(x_0+h)-f(x_0)}{h}-f'(x_0)\right|\leq d(h)$$

for some d function $d(h)$.

By using the notation of $o(1)$, this inequality can be rewritten as follows:

$$\frac{f(x_0+h)-f(x_0)}{h}-f'(x_0)=o(1)$$

which is equivalent to

$$f(x_0+h)-f(x_0)=f'(x_0)h+o(1)h$$

So, when $h\to 0$, the right-hand side of this equation approaches 0, leading to the conclusion that $f(x_0+h)-f(x_0)=o(1)$. That is, $\lim_{h\to 0} f(x_0+h)=f(x_0)$. QED.

Similar to the concept of one-sided continuity, we can develop the concept of one-sided derivatives. In particular, assume that a function $y = f(x)$ is defined in a right neighborhood $[x_0, x_0 + \delta)$ of the point x_0. If the following right-sided limit exists

$$\lim_{h \to 0^+} \frac{f(x_0 + h) - f(x_0)}{h} = A$$

then A is known as the right derivative of $f(x)$, denoted $f'_+(x_0)$. Similarly, we can define the left derivative as follows:

$$f'_-(x_0) = \lim_{h \to 0^-} \frac{f(x_0 + h) - f(x_0)}{h}$$

If $f(x)$ is differentiable at each point in an interval I, where at any end point the differentiability is simply one-sided, then we say that $f(x)$ is differentiable on I. Then the following result is evident.

Theorem 10.42: If $f(x)$ is uniformly differentiable on the interval $[a, b]$, then $f(x)$ is differentiable at each point in $[a, b]$ and the derivatives under both definitions are the same. QED.

Theorem 10.43: Assume that $f(x)$ is differentiable at each point in an interval I. Then $f'(x) = g(x)$ is a B function of $f(x)$ on I. That is, for any $u, v \in I$, there are points $p, q \in [u, v]$ such that

$$g(p) \le \frac{f(v) - f(u)}{v - u} \le g(q)$$

Proof. We first show there is $p \in [u, v]$ satisfying the left-hand side of the desired inequality.

To this end, denote $\Delta_1 = [a, b]$ and $k = \frac{f(v) - f(u)}{v - u}$ the average rate of change of $f(x)$ on Δ_1. Assume that we have constructed Δ_n and use its mid-point to divide it into two subintervals. From Theorem 10.1, it follows that on one of these two subintervals the average rate of change of $f(x)$ is not greater than that on Δ_n. Denote this particular subinterval as Δ_{n+1}. That is, for each natural number n, a subinterval Δ_n of $[a, b]$ can be constructed so that

$$\Delta_1 \supseteq \Delta_2 \supseteq \cdots \supseteq \Delta_n \supseteq \Delta_{n+1} \supseteq \cdots$$

where the average rate of change of $f(x)$ on Δ_n is no more than k and the length $|\Delta_n| \to 0$ as $n \to \infty$. So, Theorem 10.27 implies that there is a unique $p \in \Delta_n$, for each n.

To show that $f'(p) = g(p)$ satisfies the desired inequality, let us use proof by contradiction so that we have

$$g(p) > k = \frac{f(v) - f(u)}{v - u}$$

then the definition of pointwise derivatives implies that

$$\lim_{h\to 0}\frac{f(p+h)-f(p)}{h}=g(p)$$

So, there is $\delta>0$ such that when $0<|h|<\delta$, we have

$$\frac{f(p+h)-f(p)}{h}>0$$

Therefore, when the length $|\Delta_n|$ is smaller than δ, from $p\in\Delta_n$ and the previous inequality it follows that the average rate of change of $f(x)$ on Δ_n is greater than the overall average k. That contradicts how each Δ_n is constructed. That proves that $f'(p)=g(p)$ satisfies the desired inequality in the theorem.

Secondly, we show that there is $q\in[u,v]$ satisfying the right-hand side of the desired inequality. To this end, consider the function $J(x)=-f(x)$. Then $J'(x)=-g(x)$. According to what we have just shown above, there is $q\in[u,v]$ such that

$$-g(q)\le\frac{J(v)-J(u)}{v-u}=\frac{f(u)-f(v)}{v-u}$$

By multiplying (-1) to this inequality, we obtain the right-hand side of the inequality in the theorem. QED.

In terms of mean-value theorems, we have the following:

Theorem 10.44: (Rolle's Theorem). If a function $f(x)$ satisfies

(i) It is continuous on the closed interval $[a,b]$;
(ii) It is pointwisely differentiable at each point of the open interval (a,b); and
(iii) $f(a)=f(b)$,

then there is $c\in(a,b)$ such that $f'(x)=0$.

Proof. Because $f(x)$ is continuous on $[a,b], f(x)$ must reach its maximum and minimum values on $[a,b]$. If these two values are the same, then $f(x)$ equals a constant on $[a,b]$ so that $f'(x)=0$ for any $x\in[a,b]$. If these two values are different, then either the maximum or the minimum value has to occur at one point $c\in(a,b)$ so that $f'(c)=0$. QED.

Theorem 10.45: (Lagrange mean-value theorem). If a function $f(x)$ satisfies

(i) It is continuous on the closed interval $[a,b]$;
(ii) It is pointwisely differentiable at each point of the open interval (a,b); and

then there is $c\in(a,b)$ such that

$$f'(x)=\frac{f(b)-f(a)}{b-a}$$

Proof. Let us construct the following auxiliary function

$$F(x) = f(x) - \frac{f(b) - f(a)}{b - a} \cdot x$$

Then it can be checked that $F(x)$ satisfies all the conditions in Rolle's Theorem above. So, there is $c \in (a, b)$ such that $F'(x) = 0$. That is, $f'(x) - \frac{f(b) - f(a)}{b - a} = 0$. QED.

10.6 RIEMANN INTEGRALS AND INTEGRABILITY

One of the natural questions at this juncture is how the concept of integrals developed in this chapter is related to that of Riemann integrals. In this section we will visit the basics of Riemann integrals and investigate the relationship between Riemann integrability and the uniqueness of integral systems developed earlier in this chapter.

10.6.1 The concept of Riemann integrals

Based on the arguments for Theorems 10.15 and 10.16, the following concept of Riemann definite integrals becomes natural. In particular, let $f(x)$ be a function defined on the interval $[a, b]$. Partition this interval $[a, b]$ into n subintervals by inserting $n - 1$ dividing points as follows:

$$a = x_0 < x_1 < x_2 < \cdots < x_{n-1} < x_n = b$$

Denote $\Delta_k = [x_{k-1}, x_k]$; let its length be $\Delta x_k = x_k - x_{k-1}$, and choose a point $\xi_k \in [x_{k-1}, x_k]$, for $k = 1, 2, \ldots, n$. Then the following sum

$$\sum_{k=1}^{n} f(\xi_k) \Delta x_k$$

is known as the Riemann sum or integral sum of $f(x)$ on $[a, b]$, corresponding to the partition $T = \{\Delta_k\}_{k=1}^{n}$ and the choice of points $\{\xi_k\}_{k=1}^{n}$.

Let $\lambda(T) = \max_{1 \leq k \leq n} \{\Delta x_k\}$. If no matter which partitioned T of the interval $[a, b]$ is employed and no matter how the points $\xi_k \in [x_{k-1}, x_k]$, $k = 1, 2, \ldots, n$, are chosen, as long as $\lambda(T) \to 0$, the limit J of the previous Riemann sum exists, then $f(x)$ is said to be (Riemann) integrable on $[a, b]$. The limit value

$$J = \lim_{\lambda(T) \to 0} \sum_{k=1}^{n} f(\xi_k) \Delta x_k \tag{10.52}$$

is known as the (Riemann) integral of $f(x)$ over $[a, b]$, denoted symbolically

$$\int_a^b f(x) dx$$

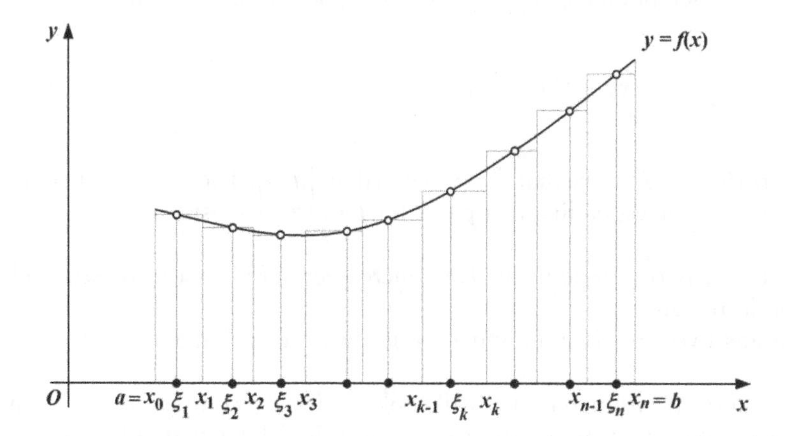

Figure 10.5 Riemann sum.

Earlier we have studied the concepts of limits of sequences and functions. However, Riemann sum is neither a sequence nor a function, how could we understand the process of limit in equ. (10.52)?

According to what has been developed in this chapter, equ. (10.52) can be simply comprehended as follows: There is a d function $d(x)$ such that for any partition $T=\{\Delta_k\}_{k=1}^n$ of the interval $[a,b]$ and any choice of the points $\{\xi_k\}_{k=1}^n$ that corresponds to the partition T, the following always holds true:

$$\left|\sum_{k=1}^{n} f(\xi_k)\Delta x_k - J\right| \le d(\lambda(T))$$

Figure 10.5 depicts the geometry of Riemann sum, where the curve $y=f(x)$ over the interval $[a,b]$, a partition T of $[a,b]$, and a choice of the points $\{\xi_k\}_{k=1}^n$ are given. The area of each small rectangular strip corresponds to a term in the sum. The total sum of all these rectangular areas is the Riemann sum. Intuitively speaking, as long as the partition is sufficiently dense, no matter how the points $\{\xi_k\}_{k=1}^n$ are chosen, the Riemann sum should be very close to the true area of the curly top enclosed region.

From the definition it can be readily seen that if $f(x)$ is Riemann integrable on $[a,b]$, then $f(x)$ must be bounded on $[a,b]$. So, in the following, we only consider the case where $f(x)$ is bounded on $[a,b]$.

Assume that $T=\{\Delta_k\}_{k=1}^n$ is an arbitrary partition of $[a,b]$. Because where $f(x)$ is bounded, let us defined

$$M_k = \sup_{x\in\Delta_k} f(x), \quad m = \inf_{x\in\Delta_k} f(x)$$

for $k=1,2,\ldots,n$. Define respectively the following upper and lower sums of $f(x)$ with respect to the partition T:

$$S(T) = \sum_{k=1}^{n} M_k\Delta x_k, \quad s(T) = \sum_{k=1}^{n} m_k\Delta x_k$$

So, for any chosen points $\{\xi_k\}_{k=1}^n$, where $\xi_k \in \Delta_k, k=1,2,\ldots,n$, we have

$$s(T) \leq \sum_{k=1}^{n} f(\xi_k)\Delta x_k \leq S(T)$$

Theorem 10.46: Assume that $f(x)$ is defined on $[a,b]$. Then the upper sum $S(T)$ and lower sum $s(T)$ corresponding to a partition T of $[a,b]$ satisfy

(i) When the partition points in T are increased, $S(T)$ is not increased while $s(T)$ is not decreased.
(ii) For any two partitions T_1 and T_2, it is always true that $S(T_1) \geq s(T_2)$.

Proof. Because (i) is obvious, we will look at (ii). By combining the partitions T_1 and T_2, we produce a new partition T. From (i) it follows that $S(T_1) \geq S(T) \geq s(T) \geq s(T_2)$. QED.

Assume that $f(x)$ is defined on $[a,b]$. Then $S = \sup\{S(T) : T$ is a partition of $[a,b]\}$ is referred to as the upper integral of $f(x)$ on $[a,b]$, while $s = \inf\{s(T) : T$ is a partition of $[a,b]\}$ the lower integral of $f(x)$ on $[a,b]$.

Theorem 10.47: Assume that $f(x)$ is defined on $[a,b]$. Then if T stands for a partition of $[a,b]$, and $\lambda(T) = \max_{1 \leq k \leq n}\{\Delta x_k\}$, then $\lim_{\lambda(T)\to 0} S(T) = S$ and $\lim_{\lambda(T)\to 0} s(T) = s$. QED.

Based on the preparation above, we are already to establish the following:

Theorem 10.48: Assume that $f(x)$ is bounded on $[a,b]$. Then the following conditions are equivalent:

(i) The upper and lower integrals of $f(x)$ on $[a,b]$ are equal;
(ii) The function $f(x)$ is Riemann integrable on $[a,b]$; and
(iii) For any $\varepsilon > 0$, there is a partition T of $[a,b]$ such that $S(T) - s(T) < \varepsilon$. QED.

Theorem 10.49: If $f(x)$ is Riemann integrable on $[a,b]$, then there is a point $x_0 \in (a,b)$ such that $f(x)$ is continuous at x_0.

Proof. From Theorem 10.48 (iii), it follows that for any $\varepsilon > 0$, there is a partition T of $[a,b]$ such that

$$S(T) - s(T) = \sum_{k=1}^{n}(M_k - m_k)\Delta x_k < (b-a)\varepsilon$$

So, there is a $k \leq n$ such that $M_k - m_k < \varepsilon$. That is, on the subinterval $\Delta_k = [x_{k-1}, x_k]$, the amplitude of $f(x)$ is smaller than ε. Denote $D_1 = \Delta_k = [x_{k-1}, x_k] = [a_1, b_1]$. Because $f(x)$ is Riemann integrable on D_1, there is an interval $D_2 = [a_2, b_2] \subseteq D_1$ such that the amplitude of $f(x)$ on D_2 is smaller than $\frac{\varepsilon}{2}$.

Reasoning similarly, for each natural number n, we can construct an interval $D_n = [a_n, b_n]$ such that

$$a_1 < a_2 < \cdots < a_n < a_{n+1} < \cdots < b_{n+1} < b_n < \cdots < b_2 < b_1$$

such that the amplitude of $f(x)$ on D_n is smaller than $\frac{\varepsilon}{n}$. Let $c = \sup\{a_n : n = 1, 2, \ldots\}$. Then c belongs to each D_n so that the amplitude of $f(x)$ at c is 0. Therefore, $f(x)$ is continuous at c. QED.

10.6.2 Riemann integrability and uniqueness of integral systems

Based on the concept of Riemann integrals studied in the previous subsection, we have the following result that establishes the relationship between Riemann integrals and integral systems investigated earlier.

Theorem 10.50: If a function $f(x)$ is Riemann integrable on $[a, b]$, then the existence of integral system of $f(x)$ on $[a, b]$ must be unique.

Proof. Because $f(x)$ is Riemann integrable on $[a, b]$, it is integrable on any subinterval $[u, v]$ of $[a, b]$. From Theorem 10.48 it follows that the upper and lower integrals of $f(x)$ on $[a, b]$ are the same.

Assume that $S(u, v)$ is an integral system of $f(x)$ on $[a, b]$. Then for each subinterval $\Delta_k = [x_{k-1}, x_k]$ of any partition T of $[u, v]$, based on the mean-value property of integral systems we have

$$m_k(x_k - x_{k-1}) \leq S(x_k, x_{k-1}) \leq M_k(x_k - x_{k-1})$$

By making use of the additive property of integral systems and by summing up the previous inequalities with respect to k, we obtain $s(T) \leq S(u, v) \leq S(T)$. Because the upper and lower integrals of $f(x)$ on $[u, v]$ are the same, the value of $S(u, v)$ is uniquely determined. QED.

As for the existence of integral systems, we have

Theorem 10.51: Assume that a function $f(x)$ is bounded on the interval $[a, b]$ and $[u, v] \subseteq [a, b]$. Let $H(u, v)$ and $L(u, v)$ be respectively the upper and lower integral of $f(x)$ on $[u, v]$. Take $\lambda \in [0, 1]$. Define

$$S(u, v) = \lambda H(u, v) + (1 - \lambda)L(u, v)$$

along with the convention that $S(u, v) = -S(u, v)$. Then when $H(a, b) = L(a, b)$, $S(u, v)$ is an integral system of $f(x)$ on $[a, b]$.

Proof. The particularly defined $S(u, v)$ evidently satisfies the additive property. To show the mean-value property, it suffices to prove the existence of $p, q \in [u, v]$ such that

$$f(p) \leq \frac{S(u, v)}{v - u} \leq f(q) \tag{10.53}$$

For convenience, denote $S(u,v)=k(v-u), M=\sup\{f(x):x\in[u,v]\}$, and $m=\inf\{f(x):x\in[u,v]\}$.

If $M=m$, because $\lambda\in[0,1]$, we have $m(v-u)\le L(u,v)\le S(u,v)\le H(u,v)\le M(v-u)$. Hence $m\le k=\frac{S(u,v)}{v-u}\le M$. So, $m=k=M$. The desired conclusion follows naturally.

If $M>m$ and $k\in(m,M)$, the conclusion is also obvious. It is because if $f(x)<k$ holds true on $[u,v]$, then from the definition of supremum it follows that $M\le k$. That contradicts the assumption that $k\in(m,M)$. Therefore, there must be $q\in[u,v]$ such that $k\le f(q)$. Similarly, there is $p\in[u,v]$ such that $f(q)\le k$.

What is left for the proof consists of two cases: $k\ge M>m$ and $k\le k<M$. In the following we will focus on the first case. The second case can be shown similarly and is omitted.

When $k\ge M>m$, it suffices to show there is $q\in[u,v]$ such that $f(q)\ge k$.

Because the upper integral is the supremum of the upper sums, for any partition $T=\{\Delta_i\}_{i=1}^n$ the following holds true:

$$S(T)=\sum_{i=1}^{n}M_i\Delta_i\ge H(u,v)\ge\lambda H(u,v)+(\lambda-1)L(u,v)=k(v-u)$$

Because $M\ge M_i$, for each $i=1,2,\ldots,n$, we have

$$M(v-u)=\sum_{i=1}^{n}M\Delta_i\ge\sum_{i=1}^{n}M_i\Delta_i\ge k(v-u)$$

Combined with the assumption that $k\ge M$, it follows that $M=k$ so that both $\ge$ signs in the previous equation all becomes $=$. That leads to

$$H(u,v)=\lambda H(u,v)+(\lambda-1)L(u,v)=k(v-u)$$

and

$$\sum_{i=1}^{n}(M-M_i)\Delta_i=0$$

Therefore, the condition $\lambda\in[0,1]$ implies that $H(u,v)=L(u,v)$. From Theorem 10.49, it follows that $f(x)$ is continuous at $x=q$ on $[u,v]$. The previous equation implies that $M=M_i$, for each $i=1,2,\ldots,n$. That is, the supremum of $f(x)$ on any subinterval $[x_{k-1},x_k]\subseteq[u,v]$ is the same as $M=k$ so that $f(q)=k$. QED.

Theorem 10.52: Assume that $f(x)$ is bounded on $[a,b]$. Then $f(x)$ is Riemann integrable on $[a,b]$, if and only if $f(x)$ has a unique integral system $S(u,v)$ on $[a,b]$. In this case, the following holds true:

$$S(u,v)=\int_{u}^{v}f(x)dx$$

Proof. Sufficiency. If $f(x)$ is Riemann integrable on $[a,b]$, then the function $f(x)$ is Riemann integrable on any subinterval $[u,v]\subseteq[a,b]$. Let $R(u,v)$ be the Riemann integral of $f(x)$ on $[u,v]$. From Theorem 10.48 it follows that the upper integral $H(u,v)$, the lower integral $L(u,v)$, and the Riemann integral $R(u,v)$ of $f(x)$ on $[u,v]$ are the same. From Theorem 10.51, it follows that from $H(a,b)=L(a,b)$ we obtain $H(a,b)$ is an integral system of $f(x)$ on $[a,b]$. Now from Theorem 10.50 it follows that the existence of the integral system of $f(x)$ on $[a,b]$ is unique.

Necessity. If $f(x)$ is not Riemann integrable on $[a,b]$, then from Theorem 10.48 it follows that $H(a,b)>L(a,b)$. From Theorem 10.51, it follows that when $\lambda\in(0,1)$, $H(u,v)=H(u,v)+(1-)L(u,v)$ is an integral system of $f(x)$ on $[a,b]$. So, for different values of λ, a different integral system is produced. That is, the integral system of $f(x)$ on $[a,b]$ is not unique. QED.

Chapter 11

New look at some of the historically important paradoxes

One consequence of the previous discussions on actual and potential infinities is that when a necessary layer structure is spotted and clearly described, then the relevant differences can be more adequately discovered. In particular, if potential infinite is seen as an ongoing progressive process and actual infinite as looking back at a finished process, then separations between these two concepts of infinite can be clearly made. Such a metaphor of actual and potential infinite can lead to the following discussion on how to get rid of some other paradoxes, known as either simple or composite contradictory paradoxes, as soon as conceptual criterion and judgmental criterion are distinguished from each other.

A proposition is known as a paradox, provided that it contradicts the common sense knowledge. Speaking more clearly, a paradox stands for such a proposition that with a straightforward reasoning one always reaches an unacceptable conclusion, while for a long period of time no plausible explanation for why that is the case has been derived. Historically, there have been many paradoxes that have troubled the mankind for more than two thousand years. The Berkeley's paradox of the 16th hundred has been known as the cause of a major crisis in mathematics and motivated the community of mathematicians to rebuild the foundation of calculus in particular and the foundation of mathematics in general, for details, please consult Chapter 6 and Section 9.5 in this book. And at the start of the 20th century, Russell's paradox led to the third crisis in the foundations of mathematics, for more details, please consult Chapter 6 in this book. After then, a great number of new paradoxes have been constructed. Although efforts have been made to get rid of the paradoxes in the past one hundred years, as of this writing this third crisis has not been truly resolved. As a matter of fact, it is shown by (Lin (guest editor), 2008a) that a more wide spread problem is discovered in the entire spectrum of mathematics, for more details, please consult Part 3 of this book.

This chapter is organized as follows. In Section 11.1, we will establish the symbolism and the ground necessary for us to practically distinguish conceptual criterion and judgmental criterion. In Section 11.2, we will show how self-referential paradoxes are in fact contradictory fallacies. Based on the results in this section, we look at paradoxes of composite contradictory type in Section 11.3, while proving that they also represent self-contradictory fallacies. Based on these investigations, Section 11.4 looks at the structural characteristics of contradictory paradoxes. Section 11.5 focuses on self-referential propositions by showing that each self-substituting proposition violates the law of identity, and that self-substituting propositions might also violate the

Table 11.1 Examples of negation pairs.

I	Truth	Right	Finite	The same	Can do
¬I	False	Wrong	Infinite	Not the same	Cannot do

law of contradiction. This chapter is based on (Wen, 2005; 2006) and the presentations of Wen Bangyan delivered at the 2002 International Congress of Mathematicians in Beijing.

11.1 THE NECESSARY SYMBOLISM

The system of symbols introduced in this section will be beneficially useful for the rest of this chapter in order to produce the desired conclusions.

11.1.1 Proposition and its negation

Let I stand for a chosen object (proposition) and ¬I the negation of the object (proposition) I. For instance, Table 10-1 provides a list of examples of how I and ¬I are as related with each other, where soon as we specify the meaning of I we will be able to compute the negation ¬I.

In this section, we use the notation $I == \neg I$ to stand for a proposition that is self-contradictory, while $I == \neg I \leftrightarrow \emptyset$ represents the judgmental conclusion that the self-contradictory proposition is a fallacy, which is meant to be a commonly recognized incorrect proposition. The same symbolic expression $I == \neg I \leftrightarrow \emptyset$ can also be used to represent the law of contradiction of the classical logic: Each object (proposition) can only be either positively or negatively confirmed but not both at the same time.

11.1.2 Two layers of thinking

In the current textbooks of logic, the expression $(I \cap \neg I)$ is employed to represent the law of contradiction, where the symbol $\cap$ stands for the operation of taking "and" or "intersection" or conjunction of two propositions. If we consider the layer structure in our description and judgment, then the correct interpretation of the expression $(I \cap \neg I)$ should be the proposition constructed by taking the conjunction of two propositions of mutual negation of each other. And at the layer of judgment, the law of contradiction says that $(I \cap \neg I) \leftrightarrow \emptyset$, which says equivalently that the truth value of the combined proposition $(I \cap \neg I)$ is false. That is, the proposition of interest is either true or false, but not both. It represents the judgmental evaluation about the truth value of the combined statement $(I \cap \neg I)$. For example, $3+4$ stands for an arithmetic problem, while $3+4=7$ represents an equation of computational scheme. They belong to two different layers of thinking. In the rest of this work, we will clearly separate these two layers.

For example, $(I \cap \neg I)$ represents the conjunction of the proposition that "my sword can cut through any shield" and that "my sword cannot cut through a particular shield", two negating propositions. That is, $(I \cap \neg I)$ is a proposition. On the other hand,

$(I \cap \neg I) \leftrightarrow \emptyset$ says that it is impossible for my sword to cut through any shield and not to break a particular shield, which is a judgment of the truth value of the proposition $(I \cap \neg I)$. Here, the symbol $\emptyset$ means "empty" or "non-existent". The significance of the law of contradiction is to avoid the appearance of propositions of the nature of $(I \cap \neg I)$. Therefore, we can also write this law as $I \neq \neg I$, and $I == \neg I$ as a contradictory proposition with $(I == \neg I) \leftrightarrow \emptyset$ being the judgment that the contradictory proposition is a fallacy. Here, both of the symbols $==$ and $\leftrightarrow$ stand for "equivalent".

In terms of the law of excluded middle, it says that when a concept or proposition is negated, the two entities are not only negating of each other but also complement each other so that these two mutually negating parts constitute a whole. For instance, true and not true (false), correct and incorrect (wrong), have and not have, etc., are examples of division into only two parts. That is the law of excluded middle as summarized by Aristotle in logic: It has to be either "yes" or "no" and nothing else.

11.1.3 Judgmental statements

In logic, the symbol $(I \cup \neg I)$ is commonly used to represent the law of excluded middle. Once again, considering the layer structure in human thinking, let us interpret $(I \cup \neg I)$ as the "union" or disjunction of the proposition of concern and its negation, and $(I \cup \neg I) \leftrightarrow \Omega$ as the judgmental statement of the law of excluded middle, where Ω stands for the whole of concern. That is, the whole consists of only two possibilities of I and $\neg I$. Additionally, this law of excluded middle can also be written as $I == \neg(\neg I)$, indicating that if not "no", it must be "yes" with only two possibilities.

When we either classify a property or make a judgment, if the conclusion is definite, then the conclusion must satisfy both laws of contradiction and excluded middle. In this case, we say that the classification or the judgment satisfies the property of right-wrong (or yes-no) distinction. In other words, the classification or the judgment has to take one and only one of the "right" and "wrong" (or yes and no) scenarios. This concept of right-wrong distinction can be written as follows:

$$(I \neq \neg I) \wedge (I == \neg(\neg I)).$$

In the following when we discuss elements and sets, as in the current literature we assume that the membership relationship is yes-no distinctive. That is, an arbitrary element x either belongs to a given set X or does not, and one and only one of these two possibilities holds true. Symbolically, we have

$$(a \in I) \leftrightarrow (a \notin \neg I) \quad \text{and} \quad (a \notin I) \leftrightarrow (a \in \neg I).$$

In the following, we will see how our symbolism can be successfully employed to show that a whole class of paradoxes is contradictory fallacies.

11.2 SELF-REFERENTIAL PARADOXES AND CONTRADICTORY FALLACIES

A paradox is seen as simply contradictory, if the given proposition both confirms and rejects a same object or matter.

11.2.1 Plato-Socrates' paradox

Example 11.1 (Plato-Socrates' Paradox (Kretzmann and Stump, 1988, p. 349)). Plato says: "What Socrates says below is false." Socrates says: "What Plato says above is true." Now, the question is whether or not the proposition said by Plato is true or false?

If what Plato said is true, then it follows that what Socrates said is false. So, what Socrates said implies that what Plato said is false, which contradicts the initial assumption. So, what Plato said cannot be true. Now, let us assume the opposite, that what Plato said is false. Then it follows that what Socrates said is true, which in turn implies that what Plato said is true. This conclusion once again indicates that the initial assumption is incorrect.

So, based on the previous analysis, what Plato said can be neither true nor false. That conclusion is really unacceptable. Since no one provided any convincing explanation for this situation in the past over two thousand years, it becomes a paradox. However, by using the symbolism introduced above, we have the following result.

Theorem 11.1: The Plato-Socrates' paradox is a contradictory fallacy.

Proof. Let P represent what Plato said, S what Socrates said, $\in$ "the content is", $\leftrightarrow$ equivalence, I true, and $\neg$I false. Then, what Plato said implies that

$$P \in I \leftrightarrow S \in \neg I \tag{11.1}$$

What Socrates said implies that

$$S \in I \leftrightarrow P \in I \tag{11.2}$$

Therefore, by applying the meaning of $\leftrightarrow$ equs. (11.1) and (11.2) lead to

$$S \in I \leftrightarrow S \in \neg I \tag{11.3}$$

If we let $A == (S \in I)$, then we have $\neg A = (S \notin I) == (S \in \neg I)$. That is, equ. (11.3) implies that $A == \neg A$. Now, the judgment on the proposition $A == \neg A$ is that $(A == \neg A) \leftrightarrow \emptyset$, indicating a contradictory fallacy. Therefore, the Plato-Socrates' paradox is a self-contradictory fallacy. QED.

11.2.2 The liar's paradox

Example 11.2 (The liar's paradox). "This statement is a lie." It represents one of the versions of an ancient paradox of over two thousand years old.

If we assume that this sentence is a lie, then saying it is a lie becomes a truth, which contradicts the initial assumption. On the other hand, if we assume that this sentence is true, then it really should be a lie according to what the sentence says. That once again contradicts the initial assumption. Because neither of the assumptions can hold true without encountering difficulty and for a long time no acceptable explanation could be established, it has been known as a paradox.

Also, according to the previous analysis, instead of calling it a paradox, one can reasonably conclude that this statement is neither true nor false, or "this statement is not true." To this end, if this sentence is neither true nor false, then it must be not true. Because that is what this sentence states, it means that the sentence must be true. That is, another paradox arises. In fact, historically many scholars, such as Alfred Tarski (1901–1983), Arthur Prior (1914–1969), Saul Kripke (1940–), Jon Barwise (1942–2000) and John Etchemendy (1952–), Chris Langan (1952–), and others, had looked at this paradox and provided their varied explanations, although not totally convincingly.

Along with the advances of the computer technology, people had high hopes of using computers to resolve this paradox (Chaitin, 1999). However, what comes out of a computer is a non-terminating string of "true, false, true, false, ..." or a string of "tight, wrong, right, wrong, ..." As long as the computer is not broken or interrupted, it will forever continue with the generation of the output string.

Theorem 11.2: The liar's paradox is a self-contradictory fallacy.

Proof. Let A stand for "This Statement", I true, ¬I false, ∈ "be", and ↔ equivalent. Then according to the analysis above we have

$$A \in I \leftrightarrow A \in \neg I \tag{11.4}$$

Let $B == (A \in I)$. Then we have

$$\neg B == (A \notin I) == (A \in \neg I) \tag{11.5}$$

From equs. (11.4) and (11.5), it follows that

$$B == \neg B \leftrightarrow \emptyset$$

which implies that $B == \neg B$ is a self-contradictory fallacy. QED.

11.2.3 The barber's paradox

Example 11.3 (The barber's paradox). Bertrand Russell in 1918 constructed the following barber's paradox: Suppose that a barber who lives in a village shaves all those and only those men of the same village who do not shave themselves. Now the question is whether or not the barber should shave himself.

If the barber shaves himself, then according to the rule he should not shave for himself. On the other hand, if the barber does not shave himself, then the rule says that he would have to shave himself. So, no matter what, the barber does not know what to do. And for nearly one hundred years, no convincing explanation to this imaged scenario has been established. That is why it is known as a paradox.

Theorem 11.3: The barber's paradox is a contradictory fallacy.

Proof. Let us use ordered pairs (x, y) to represent the relationship of shaving, where the first coordinate x stands for the actor and the second coordinate y the person being acted upon. Assume that N_i represents the ith man of the village, while N_0 the barber, when $i = 0$. Therefore, the ordered pair $A == (N_i, N_i)$ means the relationship that the man N_i shaves himself, $B == (\neg N_i, N_i)$ that the man N_i does not shave himself, and $L == (N_0, N_i)$ that the barber shaves the man N_i.

Based on the notations, it follows that the rule of the barber's shop is $L == B \neq A$, which means that the barber shaves those and only those men who do not shave themselves. Hence, we have

$$N_0 == \neg N_i \neq N_i. \tag{11.6}$$

That is, the rule of the shop is only relevant to the actor without involving the person of being acted upon.

Let $i = 0$. Then equ. (11.6) implies that $N_0 == \neg N_0$. In terms of the consequent judgment we have

$$N_0 == \neg N_0 \leftrightarrow \emptyset$$

which implies that in terms of the barber himself, the rule of the shop is a self-contradictory fallacy. QED.

Note: In the previous proof, the purpose of introducing the second person in each ordered pair is for the convenience of dealing with the general relation. In fact, the argument can be greatly simplified as follows:

Let N stand for the case that the barber shaves himself. Then, $\neg$N will represent the case that the barber does not shave himself. Now, if he does not shave himself, then the rule of the shop mandates $\neg N == N$. If he does shave himself, then the rule of the shop requires that $N == \neg N$. That is, no matter what is the case, we have

$$N == \neg N \leftrightarrow \neg N == N \leftrightarrow \emptyset$$

Therefore, the rule of the shop is a self-contradictory fallacy to the barber. QED.

11.3 PARADOXES OF COMPOSITE CONTRADICTORY TYPE

Because in the liar's paradox and barber's paradox, as discussed in the previous section, the difficulties occur on a certain single entity, it is why they are seen as paradoxes of simple contradictory type. In this section, we will look at some paradoxes in which difficulties occur with collections of objects. So, we will refer these paradoxes to as the composite contradictory type. When seen at the conceptual level, the simple type can be seen as element type, while the composite type as set type.

11.3.1 Russell's paradox

Example 11.4 (Russell's paradox (van Heijenoort, 1967)). Bertrand Russell in 1901 showed that Georg Cantor's naive set theory leads to a contradiction. The same paradox had been discovered a year earlier by Ernst Zermelo but he did not publish the idea, which remained known only to Hilbert, Husserl and other members of the University of Göttingen.

In June of 1902, Russell informed logician Gottlob Frege through a postal mail that he discovered the following paradox, which is later named after Russell. It is this paradox that caused the third crisis in the foundations of mathematics to appear.

Divide the totality of all sets into two classes A and B so that A contains all such sets each of which contains itself as an element and B all such sets each of which is not an element of itself. Symbolically, we have

$$A = \{S|S \in S\} \quad \text{and} \quad B = \{S|S \notin S\} \tag{11.7}$$

For instance, the set of all sets is an element in A and the set of all concepts also belongs to A because set is also a concept, while the set of all people is known as mankind and belongs to B, where mankind is not an individual person. The set of all oranges, apples, ... constitutes the class of fruits. However, this class of fruits is not an orange, not an apple, ...

Let $C = \{B|B \notin B\}$ be the set of all sets B such that B is not an element of itself B. Now, the question is which class of A and B should C belong to?

If $C \in A$, then from the definition of A, it follows that C must be an element of itself. That is, $C \in C$. However, from the definition of C, it follows that C is not an element of itself. That is $C \notin C$. That is surely a contradiction. On the other hand, if $C \in B$, then from the definition of B, it follows that C does not belong to C, $C \notin C$, as an element. However, all sets in C are in B so that C must be an element in itself, $C \in C$, which once again is a contradiction. To summarize, no matter whether $C \in A$ or $C \in B$, we have $(C \in C) \leftrightarrow (C \notin C)$. Because of this unexplainable contradiction, this is known as a paradox.

Theorem 11.4: Russell's paradox is a self-contradictory fallacy.

Proof. Let $B = \{s|s \notin s\}$ and $\neg B = \{s|s \in s\}$, where s stands for a set. Then $(B \in B) \leftrightarrow (B \notin B)$ and $(B \notin B) \leftrightarrow (B \in B)$. Define $I == (B \in B)$. Then $\neg I == (B \in \neg B) == (B \notin B)$. Consequently, we have $I == \neg I \leftrightarrow \emptyset$. That is, Russell's paradox is a self-contradictory fallacy. QED.

11.3.2 The catalogue paradox

Example 11.5 (Catalogue paradox (Gonseth, 1936)). The Swiss mathematician Ferdinand Gonseth used the following story to propose a paradox, which is later known as the catalogue or Gonseth's paradox.

A librarian of the long-standing Library of Alexandria was editing the table of contents of the works of Aristotle and his followers. However, he experienced the following difficulty: He classified all the catalogs into two portions A and B, where each catalog in A lists itself as an entry, while each catalog in B does not list itself as an entry. After the librarian finished the overall catalog C consisting of all the items in B, he started to wonder whether this overall catalog C should belong to A class or B class.

To answer this question, he reasoned as follows: If this overall catalog C belongs to class A, then according to the criterion of classification, C should list itself. However, as the overall catalog of the items in class B, C cannot list items from class A into the overall catalog of class B. On the other hand, if the overall catalog C belongs to class B, then according to the criterion of classification, since C is the overall catalog of the items in class B, all items in B including C should be listed in the overall catalog C. To summarize, no matter whether the overall catalog C belongs to class A or class B, one is trapped in the situation that the overall catalog C of items in class B lists itself if and only if C does not list itself, a paradoxical scenario known also as the catalogue paradox.

In a simplified version, this paradox is stated as follows: Consider a library which compiles a bibliographic catalog of all (and only those) catalogs which do not list themselves. Then does the library's catalog list itself?

Theorem 11.5: The catalogue paradox is a self-contradictory fallacy.

Proof. Let $B = \{s | s \notin s\}$ and $\neg B = \{s | s \in s\}$, where s stands for a catalog of works and B and ¬B two overall catalogs. Then $(B \in B) \leftrightarrow (B \notin B)$ and $(B \notin B) \leftrightarrow (B \in B)$. Define $I == (B \in B)$. Then $\neg I == (B \in \neg B) == (B \notin B)$. Consequently, we have $I == \neg I \leftrightarrow \emptyset$. That is, the catalogue paradox is a self-contradictory fallacy. QED.

11.3.3 Richard's paradox

Example 11.6 (Richard's Paradox (Richard, 1905)). This paradox is a semantic antinomy in set theory and natural language. The original statement of it has a relation to Cantor's diagonal argument on the uncountability of the set of real numbers.

It is true that certain expressions in the English language unambiguously define real numbers, while other expressions in English do not. For example, "The real number whose integer part is 18 and whose nth decimal place is 1 if n is even and 2 if n is odd" defines the real number 18.21212121..., while the phrase "London is a city" does not define a real number.

Because there is an infinite list of English phrases, where each phrase is of finite length that varies in the list, each of which unambiguously defines a real number. Let us arrange this list first by length and then by dictionary order so that the ordering is canonical. Then, this list defines an infinite list of the corresponding real numbers: $r_1, r_2, \ldots, r_n, \ldots$

Now let us construct a new real number r as follows. The integer part of r is 0, the nth decimal place of r is 1 if the nth decimal place of r_n is not 1, and the nth decimal place of r is 2 if the nth decimal place of r_n is 1. This sentence is an expression in English which unambiguously defines a real number r. Thus r must be one of the numbers r_n. However, r was constructed so that it cannot equal any of the r_n. This is the paradoxical contradiction.

For our purpose here, let us consider a variation of the paradox that uses integers instead of real-numbers, while preserving the self-referential character of the original. Once again, let us consider the English language with which the arithmetical properties of integers are defined. For example, "the first natural number" defines the property of being the first natural number 1. "Not divisible by any natural number other than 1 and itself" defines the property of being a prime number.

Since the list of all such possible definitions is itself infinite, it is readily seen that each individual definition is composed of a finite number of words, and therefore also a finite number of characters. Since this is true, we can order all the definitions, first by length of word and then by dictionary order. Now, we map these definitions to the set of integers: $a_1, a_2, \ldots, a_n, \ldots$, such that the definition with the smallest number of characters and alphabetical order will correspond to the number 1, the next definition in the series will correspond to 2, and so on.

Because each definition is associated with a unique integer, it is possible that the integer assigned to a definition by chance *fits* that definition. For example, the number of letters in the definition equals the integer. If, for example, the 43 letters long (ignoring the spaces) description "not divisible by any integer other than 1 and itself" were assigned to the number 43, then this would be true. Since 43 is itself not divisible by any integer other than 1 and itself, then the number of this definition has the property of the definition itself. However, such match may not always be the case. If the definition of "the first natural number" were assigned to the number 4, then the number of the definition would not have the property of the definition itself. This latter example will be termed as having the property of being *Richardian*. Thus, a number is Richardian, if the definition corresponding to that number is a property that the number itself does not have. Speaking formally, a natural number n is Richardian, provided that the natural number n does not have the property designated by the defining expression with which n is correlated in the serially ordered set of definitions.

Because the property of being Richardian is itself a numerical property of integers, it surely belongs in the list of all definitions of properties. Therefore, the property of being Richardian is assigned with some integer, n. Now, the question is: Is n Richardian?

If we suppose that n is Richardian, then it is only possible if n does not have the property designated by the defining expression which n is correlated with. In other words, that means that n is not Richardian, contradicting the initial assumption. However, if we suppose that n is not Richardian, then it does have the defining property which it corresponds to. This, by definition, means that it is Richardian, again contrary to assumption. Thus, the statement "n is Richardian" cannot consistently be designated as either true or false.

Because this contradictory situation cannot be explained plausibly, it has been seen as a paradox.

Theorem 11.6: Richard's Paradox is a contradictory fallacy.

Proof. Let the set of all Richardian numbers be $a_N = \{a_n | n \notin a_n\}$, where the symbol $n \notin a_n$ means that the definition corresponding to the number a_n is a property that the number a_n itself does not have, and N the integer that corresponds to the property of all Richardian numbers. Then the set of all non-Richardian numbers is $\neg a_N = \{a_n | n \in a_n\}$.

Now, from the previous definitions, it follows that $N \in a_N \leftrightarrow N \notin a_N$ and $N \notin a_N \leftrightarrow N \in a_N$. That is, no matter what, we have $I == \neg I \leftrightarrow \emptyset$, where $I = (N \in a_N)$. That is, Richard's paradox is a contradictory fallacy. QED.

11.4 STRUCTURAL CHARACTERISTICS OF CONTRADICTORY PARADOXES

In this section, we will analyze the general structural characteristics of both simple and composite paradoxes in order to explore the root reasons for the contradictory fallacies to appear. By doing so, we will be able to develop ways to prevent these paradoxical situations from occurring.

11.4.1 Characteristics of simple contradictory paradoxes

Let us first look at the structural characteristics of simple contradictory paradoxes.

Characteristic # 11.1: Division between "yes" and "no" (and "right" and "wrong") is clearly given and then apply both "yes" and "no" on the same entity.

For instance, for the barber's shaving, the criterion for "yes" and "no" is unambiguously given. If the criterion is applied to different entities, there will appear no difficulty. The fallacy immediately occurs as soon as the criterion of "yes" and "nor" is imposed on the same entity, say the barber.

Characteristic # 11.2: After establishing a clear division between "yes" and "no", a self-criticism is made for negative proposition while confusing different layers of language.

"This statement is false" is a negative proposition. It first definitely divides what is "right" and what is "wrong". "This statement" could be right and could also be wrong. In order to make such a judgment, some external criteria will be needed, because the given sentence is not sufficient enough to make such a call. That is to say, judging whether the given sentence is right or wrong belongs to a different hierarchical layer of thinking. As soon as the layers that "this statement" and "this statement is right or wrong" belong to individually are confused with each other, a fallacy appears. In particular, let us look at the following analysis.

Let A be "this statement", I and $\neg I$ respectively right and wrong, and B the sentence that "this statement is wrong": $B = (A \in \neg I)$. When a judgment is made about B, there are two possibilities in equs. (11.8) and (11.9):

$$B \in I \leftrightarrow (A \in \neg I) \in I \leftrightarrow A \notin I \tag{11.8}$$

where because A and B belong to two different layers, there is no contradiction appearing here.

$$B \in \neg I \leftrightarrow (A \in \neg I) \in \neg I \leftrightarrow A \in I \tag{11.9}$$

where as long as both A and B belong to their respective distinctive layers, ther is no contradiction appearing here either.

However, if one assumes that both A and B belong to the same layer of language, that is, $A == B$, then equs. (11.8) and (11.9) respectively imply the following contradictions: $B \in I \leftrightarrow B \notin I$ and $B \in \neg I \leftrightarrow B \in I$.

Note: If the proposition is given positively instead of negatively, then no contradictory fallacy will be resulted. In particular, let A stand for "this statement", while B for "this statement is right". That is, $B = (A \in I)$. Then we have

$$B \in I \leftrightarrow (A \in I) \in I \leftrightarrow A \in I \tag{11.10}$$

So, even if both A and B belong to the same language layer so that $A == B$, there is still no contradiction appearing here. At the same time, we have

$$B \in \neg I \leftrightarrow (A \in I) \in \neg I \leftrightarrow A \in \neg I \tag{11.11}$$

Once again, even if $A == B$ are from the same language layer, no contradiction occurs.

Characteristic # 11.3: A sufficient and necessary condition for a paradox of the contradictory type to appear.

Based on the discussions above, it can be seen that a necessary condition for a paradox of the contradictory type to appear is negative definition or negative proposition, and a sufficient condition is apply affirmation and negation on a same entity. That is the general structural characteristics of paradoxes of simple contradictory type. Self-judging a negative proposition and confusing the proposition with the outcome of the judgment are the exact moment when affirmation and negation are applied on the same entity, causing a contradictory fallacy to appear.

11.4.2 Characteristics of composite contradictory paradoxes

Next, let us look at the reasoning and structural characteristics of composite contradictory paradoxes.

In terms of the reasoning applied in paradoxes of the composite contradictory type, let us first look at the assumptions.

1 Elements are dichotomically definite. In set theory, it is assumed that the properties an element satisfies are definite. If we write a as an element that satisfies a particular property Φ, and we use $\neg a$ to mean an element that does not satisfy the said property Φ, then any chosen element either satisfies the property Φ or does not, and it cannot be falling in between. Symbolically, we have: $(a \neq \neg a) \wedge (a == \neg(\neg a))$.
2 The membership relation of elements. For a given property Φ, let A be the set of all elements a that satisfy the property Φ. In this case, we say that such an element a belongs to the set A, written as $a \in A$. Each element $\neg a$ that does not satisfy the property Φ is not a member of the set A. All these elements are located outside of the set A. The totality of these elements is written as $\neg A$.

3 Sets are dichotomically definite. Both sets A and ¬A are defined unambiguously on the basis of the dichotomy of their elements. Therefore, we have: $x \notin \mathrm{A} \leftrightarrow x \in \neg\mathrm{A}$. That is, if an element does not belong to A, it has to belong to ¬A. Symbolically, we have $(\mathrm{A} \neq \neg\mathrm{A}) \wedge (\mathrm{A} == \neg(\neg\mathrm{A}))$.

In terms of properties like $\mathrm{A} \in \mathrm{A}$ or $\mathrm{A} \notin \mathrm{A}$ they are specifically related to the relationship of self-containment. If A stands for the set of all such sets each of that belongs to itself as an element and $\neg A$ the set of all those sets each of which does not belong to itself, then the situation is quite special. In this case, if A is defined affirmatively, that is, $A = \{S|S \in S\}$, then $\neg A = \{S|S \notin S\}$. In this circumstance, we have

$$A \in A \leftrightarrow A \in A \quad \text{or} \quad A \in \neg A \leftrightarrow A \notin A \tag{11.12}$$

where no contradiction appears. On the other hand, if A is defined negatively, that is, $A = \{S|S \notin S\}$, then $\neg A = \{S|S \in S\}$ so that the following hold true:

$$A \in A \leftrightarrow A \notin A \quad \text{or} \quad A \in \neg A \leftrightarrow A \in A \tag{11.13}$$

which always represents an inconsistency, because within the discourse of properties only one of the following can hold true: the set of all sets each of which is an element of itself and the set of all those sets each of which is not an element of itself. Here, if A is defined to be the former with definite affirmation, then there will appear no contradiction. However, if A is defined with a negation, then it will both affirm and negate this set A, leading to a contradiction without any doubt.

When a set A is not the same as the set A that is defined by the way of negation as described above, we have

$$\mathrm{A} \in \mathrm{A} \leftrightarrow \mathrm{A} \notin \mathrm{A} \quad \text{or} \quad \mathrm{A} \in \neg A \leftrightarrow \mathrm{A} \in \mathrm{A} \tag{11.14}$$

In this case, although A is defined negatively, no contradiction will appear. In particular, let us look at thc catalogue paradox as example. When we complete cataloging all items in class B except the overall catalog, if there is no requirement regarding this "overall catalog", we then have finished the task without the need to list the overall catalog. It is exactly because of the need for the overall catalog to include every class B item that this requirement cannot be satisfied, leading to the appearance of a contradiction.

Summarizing the discussion above, we establish the following structural characteristics of paradoxes of the composite contradictory type.

Characteristic # 11.4: There is such a set that does not belong to itself; and there is such a set that is a member of itself. That is, there is such a set that does not satisfy the properties of elements.

Characteristic # 11.5: There is a totality that contains all such sets each of that is not an element of itself; and there is such a totality that contains all such sets each of that belongs to itself as an element. That is, there is such a totality A that is made up of all the sets each of that does not satisfy all the properties of elements.

Characteristic # 11.6: When the totality A, as defined in Characteristic # 11.5, needs to be classified against the property of being an element of self, the consequence is that no matter what, a contradiction is developed. Here, if this totality A is an element of itself, it will then not satisfy all the properties of elements; if it is not an element of itself, it will then possess all the properties of elements.

Characteristic # 11.7: A necessary condition for a contradiction to appear is to unambiguously define the property for a set to belong to self by using the form of negation. A sufficient condition for a contradiction to appear is that the totality of all the previously defined sets constitutes a set and then it is classified by using the property for a set to belong to self.

11.5 SELF-REFERENTIAL PROPOSITIONS

The discussions in the previous sections lead us to consider self-referential propositions, because such propositions might be paradoxical. This section is based on (Wen, 2006). For other related discussions please consult this reference and the references listed there.

11.5.1 Evaluation of self-referential propositions

Without loss of generality, a self-referential proposition, as commonly talked about in logic, can be defined as a statement that involves the statement itself. For example, "this proposition is not true", "what is said here is correct", "this sentence is written in English", "what I say is a lie", etc., are self-referential propositions. Based on the form of judgment, a self-referential proposition can be classified as either affirmative or negative. For instance, "what is said here is correct" and "this sentence is written in English" are affirmative propositions, while "this proposition is not true" and "what I say is a lie" are examples of negative propositions.

If a self-referential proposition provides a judgment about itself and one needs to make a true-false or right-wrong judgment on this proposition, then such a process of decision making is referred to as a judging (and evaluating) the self-referential proposition. To this end, we need to be cautious that

1 To evaluate and judge a proposition, one has to know exactly what he is evaluating and have a corresponding set of evaluation criteria.
2 When making the "true-false" and "right-wrong" judgment about such propositions as "this statement is not true (or true)", it will be difficult or impossible to draw a definite conclusion.

Establishing a judgmental conclusion regarding a self-referential proposition is the same as evaluating any proposition of other kind. One firstly has to be clear about what content is to be evaluated and what the relevant criteria are. For instance, if the given proposition is that "the sun rises in the east", then it is a true statement. If the proposition is that "the sun rises in the west", then it is false. In such cases, the criteria of the judgments are the facts that the earth spins in the direction from west to east.

If the proposition is "this spear pokes through that shield and at the same time it does not poke through that shield", then it is an incorrect statement. If this sentence is stated as "this spear either pokes through that shield or does not poke through that shield", then it represents a correct statement. It is because the logical criterion is that of the two possible judgmental outcomes of right and wrong one and only one will be and can be the conclusion.

When such propositions as "this statement is not true" are evaluated for a definite right-wrong or true-false judgmental conclusion, the only possible answer is that "this statement" could be either true or not true; the definite conclusion cannot be drawn without further information. It is because "this statement" does not contain any particular content so that we cannot find the relevant criteria for making the evaluation; therefore, no judgmental conclusion can be derived. Similarly, for the proposition that "this statement is true" we cannot derive any judgmental conclusion, either.

11.5.2 Self-substituting propositions

A self-referential proposition is referred to as self-substituting, provided the proposition uses itself to substitute for the contents it tries to state. Symbolically, a self-substituting proposition U is denoted as follows:

$$U = (U \in I) \quad \text{or} \quad U = (U \notin I) \tag{11.15}$$

For instance, the content of "this statement" is "this statement is written in English", and "this sentence" implies "this sentence is incorrect" are examples of self-substituting propositions.

Based on their contents, self-substituting propositions are classified into two types: factual and yes-no judgmental. For instance, if the content of "this sentence" is "this sentence is not written in Chinese", then this proposition belongs to the factual type; if "this proposition" means "this proposition is true", then it represents a yes-no judgmental type.

In terms of their forms of judgment, self-substituting propositions are classified into affirmative and negative types. For instance, if "this proposition" means "this proposition is not true", then it belongs to the negative type; if the content of "this sentence" is that "this sentence is written in English", then it represents an affirmative type.

Language layers exist within self-substituting and self-referential propositions. "This proposition" represents one layer, even though no content of "this proposition" is given here. "This proposition is not true (or true)" provides a judgment about "this proposition". Even though the content of "this proposition" is still not known, one point is clear here: "this proposition" and "this proposition is not true (or true)" respectively belong to their different language layers. "'This proposition is not true (or true)' is true (or not true)" stands for a judgment of "this position is not true (or true)". Once again, even though the content of "this proposition" is still not known, what is clear is the increased number of layers. Such propositions as that "this statement" means "this statement is not true" are exactly the self-substituting propositions that we investigate here by first separating them from the general self-referential propositions.

Theorem 11.7: Each self-substituting proposition violates the law of identity.

Proof. The law of identity lays out the rule of definite thinking: Within a determined discourse of discussion, any object, such as a concept or proposition, can only represent the same object of thinking instead of multiple different objects; otherwise, it would lead to confusion.

If the logic of thinking is written by using symbols, then the law of identity can be explained with symbolism. One of the commonly accepted expressions is A = A, while the following also reflects the content of this law of identity:

$$(C = A) \wedge (C = B) \rightarrow (A = B) \tag{11.16}$$

which can be interpreted as follows: Two different symbols A and B can be employed to represent the same object C of thinking; in this case, the meanings of A and B are identical. At the same time, this law of identity can also be written as follows:

$$(C = A) \wedge (C = B) \rightarrow \neg(A \neq B) \tag{11.17}$$

which can be interpreted as follows: If the same symbol C represents two objects A and B of thinking, then the conclusion that A and B are different is erroneous.

Within the self-substituting proposition $U = (U \notin I)$, although U and $(U \notin I)$ are two different objects of thinking, they are treated as substitutable with one another. That evidently confuses the objects of thinking from two different layers so that it violates the law of identity. So, it has to be incorrect. Here, the symbol "=" means that the expressions on each side can be replaced by each other (Wen, 2005). QED.

11.5.3 Violation of the law of identity

Regarding Theorem 11.7, some disagreement might appear. For instance, we can look at the following sentence: "The statement on the screen is false." That naturally provides the following self-substituting proposition: "The statement on the screen" is that "the statement on the screen is false". Because the sentence is uniquely determined, how can it violate the law of identity?

To address this inquiry, let us think through the details by asking and answering the following questions:

Q: Can "the sentence on the screen" mean "the sentence on the screen"?
A: Of course, it can. There is no doubt that A = A.
Q: If "the sentence on the screen" means "the sentence on the screen", and at the same time, it represents "the sentence on the screen is false", will then such different assignments cause confusion?
Q: Is "the sentence on the screen" a proposition?
A: Yes, it is a proposition, because "the sentence on the screen is false" is a proposition. And since "the sentence on the screen" means this proposition, "the sentence on the screen" is consequently a proposition.

Note: "The sentence on the screen" might not be a proposition, because one has to point to it that might lead to contradictory answers.

11.5.4 Violation of the law of contradiction

Theorem 11.8: Self-substituting propositions might also violate the law of contradiction.

Proof. When substitution is used, the negative self-substituting proposition $U = (U \notin I)$ of the yes-no judgmental type will experience contradictions. Symbolically, we have the following:

$$U \notin I \leftrightarrow (U \notin I) \notin I \leftrightarrow U \in I \tag{11.18}$$

$$U \in I \leftrightarrow (U \notin I) \in I \leftrightarrow U \notin I \tag{11.19}$$

where U is substituted by $(U \notin I)$. Either of the previous lines represents a contradiction. So, the substitutions are erroneous; and the contradictions arise from the self-substituting proposition itself.

When substitution is applied to the affirmative self-substituting proposition $U = (U \in I)$ of the yes-no judgmental type, although no contradiction will be produced, no judgmental conclusion can be made. To this end, we can also use the following symbolism:

$$U \in I \leftrightarrow (U \in I) \in I \leftrightarrow U \in I \tag{11.20}$$

And, although no contradiction is led to, one still cannot conclude that $(U \in I)$ is correct, either. In particular, we have

$$U \notin I \leftrightarrow (U \in I) \notin I \leftrightarrow U \notin I \tag{11.21}$$

That is, either $U \in I$ or $U \notin I$ can hold true. And, in equs. (11.20) and (11.21), they represent assumptions. Only when the specific content of U is known and this content is compared to a set of criteria, one can derive the judgmental evaluation regarding both $U \in I$ and $U \notin I$.

When substitution is applied to self-substituting propositions of the factual type, there will be no practical significance in terms of logic reasoning. For such propositions, because their criteria of evaluation are physical facts, instead of possibly producing confusion or erroneous conclusions, the operation of substitution does not in general provide much help with formal reasoning. In order not to make this presentation too long, all the details are omitted. QED.

Chapter 12

An attempt to rebuild the system of mathematics

As the last chapter in this book, in the following we will see how we can rebuild the system of mathematics where only potential infinities are allowed.

This chapter is organized as follows. Section 12.1 lays out a detailed plan for such a desired system by formulating the formal basis of the necessary philosophical, logical, and set theoretic foundations. Section 12.2 addresses several open problems about the infinity between predicates and infinite sets. In the final section, 12.3, we consider the intension and structure of actually infinite, rigid sets.

12.1 MATHEMATICAL SYSTEM OF POTENTIAL INFINITE

The co-existence of potential and actual infinities in modern mathematics and its theoretical foundation are studied in Section 7.4. It is shown that not only the whole system of modern mathematics but also the subsystems that directly deal with infinities have permitted the co-existence of these two kinds of infinities. In Section 8.3, some mutually contradicting implicit rules of thinking, widely employed in the system of modern mathematics, are discussed.

Specifically, some axioms of logic and some non-logical axioms of modern mathematics and its theoretical foundation have implicitly carry out the rule of thinking that potential infinite is the same as actual infinite. And, at the same time, some other axioms of either logic or not logic have implicitly employed the rule of thinking that potential infinite is different of actual infinite. Therefore, the results presented in Sections 7.4 and 8.3 tell us that both the method of deduction of allowing both kinds of infinites and the rule of thinking that potential infinite is different of actual infinite are innate to the system of modern mathematics. They are not only domestic to the system, not artificially introduced into the system, but also provide us with the feasible bases for anyone to apply this method of deduction and the rule of thinking to solve problems in the framework of modern mathematics.

In Sections 9.1 to 9.3, from different angles and using various methods, it is shown that infinite set $A = \{x | P(x)\}$ of various kinds, studied in modern mathematics, represents a self-contradicting incorrect concept. Applying the method of analysis of equating both kinds of infinites, as being innate to the theory of limits, Section 9.5 discussed some implicit problems existing in that theory. It is found that the shadow of the Berkeley paradox does not truly disappear from the theory of limit. So, we face directly with two problems that urgently need to be resolved. One of the problems is how to select an appropriate theoretical foundation for modern mathematics and the theory of computer science. The other problem is, under what interpretation can modern mathematics and the theory of computer science be kept in their entirety?

In this section, we will construct the mathematical system of potential infinities in an effort to address the two afore-mentioned problems. In order not to make each subsection too long, the entire effort is divided into four parts based on their relevant contents: "Preparation," "Formal System of Logical Basis," "Meta theory of Logical Basis," and "Set Theoretical Foundation," with the study of the consistency problem included. As shown in Subsection 12.1.1.2 below, the said mathematical system of potential infinities is completely different of the mathematical system constructed on the basis of intuitionism.

This section is based on (Lin (guest editor), 2008, pp. 489–525).

12.1.1 Preparation

In this subsection we will look at the division principle of the background world and relevant comments on the effort of constructing the mathematics of potential infinite.

12.1.1.1 The division principle of the background world

Let us now introduce the following set of symbols in order to represent the finite and the infinite background worlds, respectively.

1 $\Omega_F(R;x)$ stands for a finite background world;
2 $\Omega_P(R;x)$ stands for a potentially infinite world; and
3 $\Omega_A(R;x)$ stands for an actually infinite world,

where the subscript *F* means finite, *P* potentially infinite, and *A* actually infinite, the variables *R* and *x* represent respectively the collection of all predicates and the subject of study in the background world.

From the angle of history, since the believers of actual infinite and those of potential infinite have been constantly debating about which side is correct by mutually negating each other, the common belief is that these two kinds of infinites cannot co-exist side by side. Hence, these scholars insist on the law of dividing the background world into two parts. That is, believers of actual infinite see the background world as either finite or actually infinite. On the contrary, those who believe in potential infinite recognize the background world as either finite or potentially infinite. If we use symbols, these two negating beliefs can be written as follows:

1 The law of dividing the background world into two for actual infinite believers: $\vdash \Omega_F \vee \Omega_A$;
2 The law of dividing the background world into two for potential infinite believers: $\vdash \Omega_F \vee \Omega_P$.

If we refer to the first law above in short as law (I) of dividing into two and the second as law (II), then we have

Law (I) of dividing into two: $\vdash \Omega_F \vee \Omega_A$;
Law (II) of dividing into two: $\vdash \Omega_F \vee \Omega_P$.

In the light of epistemology, there is a well-recognized universal principle: For each existing matter or concept, there must exist a background world for the matter or concept. Conversely, any matter or concept without its background world must not have its foundation. And, any matter or concept without its foundation will not exist or develop for long.

Now, because it has been a long time since the time when the two concepts of infinite appeared, established, and debated upon, and the areas of learning these concepts have touched are so wide-ranging, we have strong reasons to believe that each of these two concepts of infinite has its own strong and solid background world. No matter whether this kind of background world is a reflection of the objective world or a reasonable extension of human intelligence, whether it is a certain kind of combination of some realistic matters and events or a philosophical exploration, its existence is reasonable and undoubted.

On the other hand, in this world, anything reasonable can exist and last a long time and must have its background world. Conversely, each matter or concept with its own background world, which experiences long-lasting growth and development, must possess its own reasons and rationality. In terms of the concepts of potential and actual infinite, even though they are different, they have existed and been developed for thousands of years. So, they must possess their own individual, respective rationalities. Here, one point everyone would agree upon is: Whatever is reasonable, we should provide it with space to exist and to evolve and should not deny it in its entirety. Not only this, all reasonable matters in this world can tolerate each other so that they can co-exist and evolve together in the same space. Therefore, for the two concepts of infinite, they should look for their own ways to exist and to mature without completely denying the righteousness of the other; and they should not try to engrave designs for their own existence and development by rejecting or overtalking the background world of the other kind of infinite. To this end, we advocate the co-existence of these two different kinds of infinites. That is, we favor the establishment of such a law that divides the finite or infinite background world into three portions. In particular, we would unconditionally recognize our background world as either finite, or potentially infinite, or actually infinite. In symbols, this law can be written as follows:

$$\vdash \Omega_F \vee \Omega_P \vee \Omega_A.$$

In this subsection, this law of dividing the background world into three finite or infinite portions is abbreviated as the law of dividing into three. That is, we have:

The law of dividing into three: $\vdash \Omega_F \vee \Omega_P \vee \Omega_A$.

12.1.1.2 On constructing mathematics of potential infinite

In the previous subsection, we discussed the laws of dividing the background world into two or three finite or infinite portions. According to the essential intensions of these laws, if the system of a theory is established in the spirit of the law of dividing into three, then this system has to allow potential and actual infinites. If the system is constructed in the spirit of either law (I) or law (II) of dividing into two, that system will not allow both kinds of infinites.

In terms of the development history of mathematics, the constructive system of mathematics of the intuitionistic school should have been established in the spirit of law (II) of dividing into two, because in the system of constructive mathematics, the purpose of completely excluding actual infinities is perfectly accomplished. It is only under the rigorous requirement that "existence must mean constructible (within a finite number of steps)", that too many reasonable contents have to be lost under this overreaching restriction so that the majority of the mathematicians do not like this approach. This end explains why the development of this system of mathematics has been very limited. On the other hand, from Cantor's naïve set theory to Zermelo's modern axiomatic set theory, actual infinities are strongly emphasized. On the surface, these theories seem to be established in the spirit of carrying out law (I) of dividing into two. However, as a matter of fact, these theories did not completely exclude potential infinities out of their systems. For detailed discussions about this end, please consult Section 7.4. Also, from the discussions in Sections 8.3 and Chapter 9, we know that there exist some deeply rooted contradictions in the logic and the set theoretic foundation of modern mathematics. Therefore, the desperate questions, which have to be addressed, include: How can an appropriate theoretical foundation be chosen for modern mathematics and computer science? Under what method of interpretation can modern mathematics and computer science be kept as completely as possible?

These are some of the basic problems we need to face and to resolve. In this section, we will construct the mathematical system of potential infinite as our effort to address the afore-mentioned problems. The said mathematical system of potential infinite will be denoted as PIMS, in which we will still employ the classical two-value logical calculus as the tool of deduction. Of course, some modifications will be introduced accordingly. And, in the PIMS, the non-logical axioms about existence and construction of sets will be established on the basis of the non-logical axioms of the modern axiomatic set theory. And, each of these new axioms will be refined and interpreted one by one.

In a certain sense, the PIMS can be seen as a modified version of the modern axiomatic set theory. There is no doubt that the PIMS will be constructed in the spirit of carrying out law (II) of dividing into two. So, in the PIMS, the concept of potential infinite will be completely and totally carried out without allowing any actual infinite. Because of this, some scholars might not help but ask, if the PIMS is constructed in this manner, will it be a copy of the constructive mathematics of the intuitionism?

To this question, our answer is: No, there will be a difference between the two systems of mathematics! In particular,

1 The logical tool corresponding to the constructive mathematics of the intuitionism is the intuitionistic logic, while the tool of reasoning for the PIMS is the classical two-value logical calculus with some modifications added;
2 The constructive mathematics of the intuitionism does not start with a set theory of any kind. Instead, it begins with the natural number theory of the dual intuition of Brouwer's objects; and
3 The constructive analysis of the intuitionism is founded on Brouwer's spread continuum, while the analysis of the PIMS is developed on the basis of spring power sets of the spring natural number system.

To summarize, both mathematical systems have their own perspective starting points and their own schemes of construction so that they possess different degrees of richness in their individual intensions.

To conclude this subsection, we also like to point out the fact that in the historical spectrum of mathematics, there has not been a plausible mathematical system that clearly is constructed in the spirit of the law of dividing into three. As a matter of fact, constructing such a mathematical system that allows both kinds of infinites will be a difficult project, which should be one goal for us all to collectively pursue and materialize with all of our might!

12.1.2 Formal basis of logical systems

In this subsection, we continue to resolve the following two problems that badly need an answer by establishing the logical foundation for the mathematical system of potential infinite: (1) How can an appropriate theoretical foundation be chosen for modern mathematics and computer science? And, (2) under what interpretations can modern mathematics and the theory of computer science be kept as completely as possible?

12.1.2.1 Introduction

Continued the previous subsection, let us use the abbreviation PIMS to denote the mathematical system of potential infinite, which we will construct in the rest of this section. As is pointed earlier, the tool of logical reasoning in the PIMS will still be the classical two-value logical calculus with appropriate modifications added. Since most of our readers would be quite familiar with the system's framework of the classical two-value logical calculus, in this subsection, when we provide the logical calculus system of the PIMS, a modified version of the two-value logic, other than the locations where modifications are made, and appropriate illustrations are given, all other relevant materials constitute the regulations only for composite formulas and the storage of the symbols of the formal language of the natural deduction system. We will list all the familiar rules of deduction without any wordy explanation. As a matter of fact, the so-called appropriate modification to the classical two-value logical calculus system is nothing more than adding a listing quantifier and relevant conventions on how to interpret the quantifier to the logic of predicates.

12.1.2.2 Deduction system of PIMS's propositional logic

First, let us look at the basic symbols of the formal language L_{pia}.

1 Propositional connectives: $\neg, \rightarrow, \wedge, \vee, \leftrightarrow$.
2 Propositional words: $p_1, q_1, r_1, \ldots, p_i, q_i, r_i$, where $i = 1, 2, \ldots, \vec{n}$.
3 Technical signs: (,).

We often use lower case letters p, q, r to represent arbitrary propositional words, and capital letters A, B, C arbitrary composite formulas. Here, we should notice that the notation $\{1, 2, \ldots, \vec{n})$ stands for the potentially infinite spring set $N = \{x|n(x))$ as introduced in Section 9.3. In the PIMS system, each similar situation, or any set $\{a_1, a_2, \ldots, a_{\vec{n}}\}$ or sequence $a_1, a_2, \ldots, a_{\vec{n}}$ with the set $\{1, 2, \ldots, \vec{n}\}$ as the range of the subscripts will be understood similarly. That is, they are either a potentially infinite spring set or a potentially infinite spring sequence, named in short as potential sequence.

Secondly, let us look at the formation rules of composition formulas (Wff) in L_{pia}.

1 Each independent propositional word is Wff of L_{pia}.
2 A is a Wff of $L_{pia} \Rightarrow (\neg A)$ is a Wff of L_{pia}.
3 A, B are Wff of $L_{pia} \Rightarrow (A \to B)$, $(A \wedge B)$, $(A \vee B)$, $(A \leftrightarrow B)$ are all Wff of L_{pia}.
4 Inductive definition: An L_{pia}'s formula A is a Wff iff A is generated by the Wff formation rules in L_{pia}.

We will use symbols Σ, Γ, Δ, etc., to stand for sets of formulas. That is, they represent the metavariables of formula sets in the P^{PIN} system.

Thirdly, let us look at the deduction rules of P^{PIN} that correspond to L_{pia}.

(Ref) $A \vdash A$.
(τ) $\Gamma \vdash \Delta, \Delta \vdash A \Rightarrow \Gamma \vdash A$.
$(\neg)$ $\Gamma, \neg A \vdash B, \neg B \Rightarrow \Gamma \vdash A$.
$(\to_-)$ $A \to B, A \vdash B$.
$(\to_+)$ $\Gamma, A \vdash B \Rightarrow \Gamma \vdash A \to B$.
$(\wedge_-)$ $A \wedge B \vdash A, B$.
$(\wedge_+)$ $A, B \vdash A \wedge B$.
$(\vee_-)$ $A \vdash C, B \vdash C \Rightarrow A \vee B \vdash C$.
$(\vee_+)$ $A \vdash A \vee B, B \vee A$.
$(\leftrightarrow_-)$ $A \leftrightarrow B, A \vdash B; A \leftrightarrow B, B \vdash A$.
$(\leftrightarrow_+)$ $\Gamma, A \vdash B, \Gamma, B \vdash A \Rightarrow \Gamma \vdash A \leftrightarrow B$.

For the sake of convenience, we will write the formula set $\Sigma = \{A_0, A_1, \ldots, A_{n-1}, A_{\vec{n}})$ directly as the following form of a formula sequence: $A_0, A_1, \ldots, A_{n-1}, A_{\vec{n}}$. Of course, since Σ is a set, the ordering of the formulas in the sequence can be arbitrary and has nothing to do with the order the formulas are listed. Therefore, $\{A_0, A_1, \ldots, A_{n-1}, A_{\vec{n}}) \vdash B$, $\Sigma \cup \{A\} \vdash B$, $\Sigma \cup \Sigma' \vdash B$ can respectively be written as: $A_0, A_1, \ldots, A_{n-1}, A_{\vec{n}} \vdash B$, Σ, $A \vdash B$, and $\Sigma, \Sigma' \vdash B$. And, we denote the sequence $\Sigma \vdash A_1, \ldots, \Sigma \vdash A_n$ by $\Sigma \vdash A_1, \ldots, A_n$.

Definition 12.1 Formula A can be deducted formally from a formula set Σ in the propositional logic P^{PIN}, denoted symbolically $\Sigma \vdash A$, iff $\Sigma \vdash A$ can be generated by the rules of deduction in finite many steps.

If $\Sigma \vdash A$, then the formula A is referred to as Σ's grammatical successor.

Note: The symbol $\vdash$ represents the relationship of deductibility and is not a symbol in the formal language. And, $\Sigma \vdash A$ is not a formula in the formal language. Instead, it stands for such a metalanguage symbol that expresses the relationship between the formula set Σ and the formula A.

Definition 12.2 If a formula A is deductible in the $\emptyset$ form, then the formula A is said to be provable. Each formula sequence, formally deductible from $\emptyset$ to A, is called a proof of the formula A. If a formula A is provable, then the formula A is seen as a theorem in the system P^{PIN}, denoted symbolically $\vdash A$.

Theorem 12.1: If $\Sigma \vdash A$, then there exists a finite set Σ^*, $\Sigma^* \subseteq \Sigma(\Sigma^* \overrightarrow{\subseteq} \Sigma)$, such that $\Sigma^* \vdash A$. QED.

Here, we introduced another symbol $\overrightarrow{\subseteq}$ to describe the relationship between a set Σ and its subset Σ^*. When Σ is a finite set, the relationship between the set Σ and its subset Σ^* is written symbolically as $\Sigma^* \subseteq \Sigma$. When Σ is a potentially infinite spring set, the relationship between the set Σ and its subset Σ^* is denoted by $\Sigma^* \overrightarrow{\subseteq} \Sigma$. For a more detailed discussion on this symbolic expression, please consult Subsection 12.1.4 below.

Fourthly, let us look at the semantics of P^{PIN}.

Definition 12.3 A truth–value assignment v is a function with such a domain that contains every set of formulas and with $\{0, 1\}$ as its range, satisfying the following conditions:

(1) $v(\neg A) = 1$ iff $v(A) = 0$.
(2) $v(A \rightarrow B) = 1$ iff $v(A) = 0$ or $v(B) = 1$.
(3) $v(A \vee B) = 1$ iff $v(A) = 1$ or $v(B) = 1$.
(4) $v(A \wedge B) = 1$ iff $v(A) = 1$ and $v(B) = 1$.
(5) $v(A \leftrightarrow B) = 1$ iff "if $v(A) = 1$, then $v(B) = 1$" and "if $v(B) = 1$, then $v(A) = 1$".

Definition 12.4 Assume that Σ is a set of formulas and A a formula. If there exists an assignment v such that $v(A) = 1$, then the formula A is called satisfiable. And, $v(\Sigma) = 1$, iff for any formula C, if $C \in \Sigma$, then $v(C) = 1$. If there exists an assignment v such that $v(\Sigma) = 1$, then the formula set Σ is said to be satisfiable. If an assignment v satisfies a formula A or a formula set Σ, then v is also referred to as an FPIN model of the formula A or the formula set Σ.

Definition 12.5 Let Σ be a set of formulas and A a given formula. The formula A is a grammatical successor of Σ, denoted $\Sigma \vDash A$, iff for any assignment v, if $v(\Sigma) = 1$, then $v(A) = 1$.

Definition 12.6 A propositional logical formula A is called effective, iff for any assignment v, it is always true that $v(A) = 1$. A propositional logic formula A is termed as to a contradiction, iff for any assignment v, it is always true that $v(A) = 0$.

Sometimes, a propositional logically effective formula is also termed to as tautology, denoted $\vDash A$. In the P^{PIN} system, the following results can be shown:

Theorem 12.2: (The reliability theorem). (1) If $\Sigma \vdash A$, then $\Sigma \vDash A$; (2) If $\vdash A$, then $\vDash A$. QED.

Theorem 12.3: (The completeness theorem). (1) If $\Sigma \vDash A$, then $\Sigma \vdash A$; (2) If $\vDash A$, then $\vdash A$. QED.

That is, the deduction system P^{PIN} possesses the properties of reliability and the completeness.

12.1.2.3 Deduction system of PIMS's predicate logic

First, let us look at the basic notations of the formal language L_{pie}. In the storage of symbols of L_{pie}, let it first keep all the basic symbols from L_{pia}, that is, the five propositional connectives, $\neg$, $\rightarrow$, $\wedge$, $\vee$, $\leftrightarrow$, the propositional words, $p_1, q_1, r_1, \ldots, p_i, q_i, r_i$, where $i = 1, 2, \ldots, \vec{n}$, and the technical symbols (,), etc. Beyond these symbols, we need to add the following additional classes of basic symbols:

(1) The quantifiers: $\forall$, $\exists$, E.
(2) Individual words: $a_1, a_2, \ldots, a_{\vec{n}}$.
(3) Predicates: $F_1, G_1, H_1, \ldots, F_{\vec{n}}, G_{\vec{n}}, H_{\vec{n}}$.
(4) Restricted variables: $x_1, y_1, z_1, \ldots, x_{\vec{n}}, y_{\vec{n}}, z_{\vec{n}}$; and
(5) Technical symbols: , , (,).

Evidently, the storage of symbols of L_{pie} is a true expansion of that of L_{pia}. We will often use lower case letters a, b, c to represent arbitrary individual words, and capital letters F, G, H for arbitrary predicates, and the lower case letters x, y, z for arbitrary restricted variables.

The name of the quantifier $\forall$ is the universal quantifier, which is interpreted and read as "for all". The formal symbol $\forall$ is from the first letter of the English word "All". The name of E is the listing quantifier, which is interpreted and read as "for each" or "for any". The formal symbol E is from the first letter of the English word "every".

In the logical calculus of the PIMS, the universal quantifier $\forall$ cannot be interpreted or read as "for each" or "for any", while the listing quantifier E cannot be interpreted or read as "for all". These two quantifiers are strictly separated, and this end is a modification about the interpretations of quantifiers in the classical two-value logical calculus.

In Section 7.3, we have clearly defined the concept of a "listing procedure." In terms of a non-finite listing procedure, if this non-terminating listing procedure did not proceed to its end, that is, the listing procedure does not exhaust all its possibilities, then what the procedure faces must be a potential infinite of a forever present progressive tense (ongoing). However, if the listing procedure is completed, that is, the procedure has exhausted all its possibilities, then what the procedure faces is an actual infinite of a perfect tense (done). Therefore, the transition from listing to being exhausted is a transition from a progressive tense (ongoing) to a perfect tense (done), and also one from potential infinite to actual infinite.

From the discussion in Example 7.7 on listing procedures in Section 7.3 and the convention on the interpretations of the quantifiers in the PIMS's logical calculus, it follows that the interpretation "for all" of the universal quantifier $\forall$ indicates a perfect tense (done), while the interpretation "for each" or "for any" of the listing quantifier E implies a (present) progressive tense (ongoing).

On the other hand, since the PIMS is constructed in the spirit of law (II) of dividing into two, in the PIMS, one faces either something finite or potentially infinite, and the concept of actual infinite is absolutely not allowed. Each potential infinite must be a present progressive tense (ongoing) and whatever is finite means the listing has to be exhausted or finished in a finite number of steps. Hence, in a finite background world, the perfect tense (done) is the most essential. So, based on our discussion above, we can conclude that since the universal quantifier $\forall$ is interpreted as "for all" and what faces $\forall$ is a perfect tense (done), the universal quantifier $\forall$ in the PIMS should belong to a finite background world Ω_F instead of a potentially infinite background world Ω_P. Conversely, since the listing quantifier E is interpreted as "for each" or "for any" and is faced with a present progressive tense (ongoing), the listing quantifier E in the PIMS must belong to the potentially infinite background world instead of the finite background world Ω_F. As discussed above, the conclusions can be graphically shown as below:

$$\text{PIMS}\begin{cases} E \in \Omega_p \\ \forall \in \Omega_F \end{cases}$$

The name of $\exists$ is the existence quantifier, which is interpreted as "there is (are)" or "has (have)." The formal symbol $\exists$ is from the first letter of the English word "exist".

Secondly, let us look at the formation rules of composite formulas (Wff) in the formal language L_{pie}.

1 Each single propositional word is an L_{pie} Wff.
2 If F_n is a predicate in n variable, and $a_1,\ldots,a_n$ are n arbitrary individual words, then $F_n(a_1,\ldots,a_n)$ is an L_{pie} Wff.
3 A is an L_{pie} Wff $\Rightarrow (\neg A)$ is an L_{pie} Wff.
4 A, B are L_{pie} Wff $\Rightarrow (A \to B)$, $(A \wedge B)$, $(A \vee B)$, $(A \leftrightarrow B)$ are all L_{pie} Wffs.
5 $G(a)$ is an L_{pie} Wff, where a appears without x appearing, $\Rightarrow \forall x G(x)$, $ExG(x)$, $\exists x G(x)$ are all L_{pie} Wffs.
6 Inductive definition: An L_{pie} formula A is L_{pie} Wff, iff A is generated by using the Wff formation rules in L_{pie}.

We will use the symbols Σ, Γ, Δ, etc., to represent any set of formulas. That is, they are the meta-variables for formula sets in the P^{PIN} system.

Thirdly, let us look at the deduction rules corresponding to L_{pie} in F^{PIN}. In the set of deduction rules corresponding to L_{pie} in F^{PIN}, let it first keep all the deduction rules in P^{PIN} corresponding to L_{pia}. These rules are: (Ref), (τ), $(\neg)$, $(\to_-)$, $(\to_+)$, $(\wedge_-)$, $(\wedge_+)$, $(\vee_-)$, $(\vee_+)$, $(\leftrightarrow_-)$, $(\leftrightarrow_+)$. Next, let us add the following deduction rules:

$(\forall_-)$ $\forall x A(x) \vdash A(a)$.
(E_-) $ExA(x) \vdash A(a)$.
(E_+) $\Gamma \vdash A(a)$, a is not in Γ, $\Rightarrow \Gamma \vdash \exists x A(x)$
$(\exists_-)$ $A(a) \vdash B$, a does not appear in B, $\Rightarrow \exists x A(x) \vdash B$.
$(\exists_+)$ $A(a) \vdash \exists x A(x)$, where $A(x)$ is obtained by replacing each a's appearance in $A(a)$ by x.

The deduction rule $(\forall_-)$ is named the universal quantifier's cancellation law. In the traditional two-value logical calculus, this law reflects the following thinking of reasoning applied in an argument: When all elements in the universe of discourse of a discipline satisfy a certain property, one can then conclude that for any chosen element from this universe of discourse, this element must satisfy this property. Now, in the PIMS, since we have $\forall \in \Omega_F$, this $(\forall_-)$ law reflects such a thinking logic that if all the elements in the finite universe of discourse of a discipline satisfy a certain property, we can then conclude that for each chosen element in this finite universe of discourse, this element must satisfy the said property.

In terms of a research object x from the potentially infinite (spring infinite set) or actually infinite (rigid infinite set) universe of discourse of a certain discipline, from the fact that all x satisfy a property P, it can be drawn that each object x satisfies the property P. Furthermore, one can also draw the conclusion that for any chosen object a, selected from the universe of discourse without any restriction, it satisfies the property P. Using symbols, we have

$$\forall xP(x) \vdash ExP(x) \vdash P(a),$$

where both $ExP(x)$ and $P(a)$ not only are listing procedures, but also belong to the same potentially infinite present progressive tense (ongoing). Therefore, we have

$$ExP(x) \vdash P(a).$$

However, $\forall xP(x)$ is an actually infinite perfect tense (done). When faced with an infinite universe of discourse, it is not feasible to derive actually infinite (*aci*) perfect tense (done) from a potentially infinite present progressive tense (ongoing). It is because each perfect tense (done) grows out of a present progressive tense (ongoing), and each potential infinite (*poi*) is in fact an initial section of an actual infinite (*aci*), as said (Robinson, 1964). So, any thought of the following kinds is impossible to hold:

$$P(a) \vdash \forall xP(x) \quad \text{or} \quad ExP(x) \vdash \forall xP(x).$$

That is, only the following kinds of thinking are reasonable and feasible:

$$\forall xP(x) \vdash P(a) \quad \text{or} \quad \forall xP(x) \vdash ExP(x).$$

What we like to say is that in terms of the deduction tools for allowing both kinds of infinites, if we introduce the following law for the universal quantifier in the two-value logical calculus,

$$(\forall_+)\ \Gamma \vdash A(a),\ a \text{ does not appear in } \Gamma,\ \Rightarrow \Gamma \vdash \forall xA(x),$$

it is not hard to see that a kind of incorrect rule of thinking is carried on implicitly. So, in the PIMS, and even in the future construction of a new mathematical system of actual infinite, such a law as $(\forall_+)$ has to be abandoned. On the other hand, in terms of finite rigid sets, $(\forall_+)$ can still be applied. However, in order to avoid confusion, we will permanently avoid $(\forall_+)$ in the logical calculus of the PIMS. Since, in terms of finite rigid universe of discourse, we will have to employ the universal quantifier $\forall$, we will keep the right to use the deduction rule $(\forall_-)$.

The deduction rule (E_-) is referred to as the listing quantifier's cancellation law. Since in the PIMS, we have $E \in \Omega_P$, (E_-) reflects the following thinking of deduction: If every element in a potentially infinite universe of discourse of a certain discipline satisfies a property, then it can be concluded that for an arbitrarily chosen element in this potentially infinite universe of discourse, this selected element must satisfy the property.

The deduction rule (E_{+}) is referred to as the listing quantifier's introduction law. Since in the PIMS, we have $E \in \Omega_P$, what (E_{+}) reflects is the following thinking of deduction in PIMS: In a potentially infinite universe of discourse of a discipline, if we select an element a under no restriction and can always prove that under a certain condition the element a satisfies a property A, then we can conclude that under the same condition, each element in the potentially infinite universe of discourse satisfies the property A. Here, it is very important to assume that the selection of the element a is under no restriction, especially, its selection cannot be done under the condition that the element satisfies the property A.

The deduction rule $(\exists_{-})$ is referred to as the existence quantifier's cancellation law. Since the PIMS is constructed in the spirit of law (II) of dividing into two, that is, $\vdash \Omega_F \vee \Omega_P$, in the PIMS, what $(\exists_{-})$ reflects is such a thinking of deduction: For an arbitrarily chosen element a from a finite or potentially infinite spring universe of discourse of a discipline, as long as a satisfies a property A, then we can derive conclusion B, then it is affirmed that under the condition that there is such an element in the finite or potentially infinite spring universe of discourse that satisfies the property A, we will be able to derive the conclusion B. What also needs to be pointed out is that the selection of the element a is done without any restriction. Especially, it is not restricted by having to derive the conclusion B.

The deduction rule $(\exists_{+})$ is referred to as the existence quantifier's introduction law. In the PIMS, if it is known that an element a in a universe of discourse, which is either a finite or a potentially infinite spring universe, satisfies a property A, then we can conclude that there indeed exist (at least one) elements in this finite or potentially infinite spring universe of discourse that satisfy the property A.

For the sake of convenience, we write the formula set $\Sigma = \{A_0, A_1, \ldots, A_{n-1}, A_{\vec{n}})$ directly in the form of a sequence of formulas: $A_0, A_1, \ldots, A_{n-1}, A_{\vec{n}}$. Of course, since Σ is a set, the order of the formulas, in which they appear in the sequence, does not really matter. In this case, $\{A_0, A_1, \ldots, A_{n-1}, A_{\vec{n}}) \vdash B$, $\Sigma \cup \{A\} \vdash B$, $\Sigma \cup \Sigma' \vdash B$ can be respectively written as: $A_0, A_1, \ldots, A_{n-1}, A_{\vec{n}} \vdash B$, $\Sigma, A \vdash B$, and $\Sigma, \Sigma' \vdash B$.

We will write the sequence $\Sigma \vdash A_1, \ldots, \Sigma \vdash A_n$ briefly as $\Sigma \vdash A_1, \ldots, A_n$.

Definition 12.6 A formula A is formally deductible from a set Σ of formulas in the predicate logic F^{PIN}, denoted symbolically $\Sigma \vdash A$, iff $\Sigma \vdash A$ can be generated using deduction rules in finite many steps.

If $\Sigma \vdash A$, then formula A is called a grammatical successor of Σ.

Definition 12.7 If a formula A can be derived in the $\emptyset$ format, then the formula A is called provable. Each sequence of formally derivable formulas from $\emptyset$ to A is called a proof of the formula A. If a formula A is provable, then A is called a theorem in the F^{PIN} system, denoted $\vdash A$.

Theorem 12.4: If $\Sigma \vdash A$, then there exists a finite set Σ^* satisfying that $\Sigma^* \subseteq \Sigma$ $(\Sigma^* \overrightarrow{\subseteq} \Sigma)$ and $\Sigma^* \vdash A$. QED.

Fourthly, let us look at the semantics of F^{PIN}.

Definition 12.8 (Model) *A* model **M** of the first order language PIMS is an ordered triple

$$<M, \{R_i^{\mathbf{M}}\}_{i\in I}, \{a_k^{\mathbf{M}}\}_{k\in K}>,$$

which is made up of three parts:

(1) A non-empty finite or potentially infinite spring set M, called the universe of the model **M**;
(2) $\mathbf{M}(R_i) \subseteq M^n$ $(\mathbf{M}(R_i) \overrightarrow{\subseteq} M^n)$, for each n-ary relation symbol R_i in the PIMS;
(3) $\mathbf{M}(a_k) \in M$ $(\mathbf{M}(a_k) \overrightarrow{\in} M)$, for each constant element symbol in the PIMS.

Here, we have employed the relationship symbol between a set Σ and its element x. When Σ is a finite set, the membership relation between the set Σ and its element x is written as $x \in \Sigma$. When Σ is a potentially infinite set, the membership relationship is denoted as $x \overrightarrow{\in} \Sigma$. For more detailed discussions on these symbols, please consult Subsection 12.1.4 below.

Definition 12.9 Let **M** be a model. Then, the model $\mathbf{M}(a_i/m_j)$ stands for

$$\mathbf{M}(a_i/m_j)(w) = \begin{cases} \mathbf{M}(w), & \text{if } w \neq a_i \\ m_j, & \text{if } w = a_i \end{cases}$$

Definition 12.10 (Basic semantics of formulas) Assume that **M** is a model and *A* a formula. If **M** satisfies the following conditions, then **M** is called an F^{PIN} model.

(1) $\mathbf{M}(p_i) \in \{0, 1\}$;
(2) $\mathbf{M}(F_i^n(a_1, \ldots, a_n)) = 1$, iff$\langle \mathbf{M}(a_1), \ldots, \mathbf{M}(a_n)\rangle \in \mathbf{M}(F_i^n)$ $(\langle \mathbf{M}(a_1), \ldots, \mathbf{M}(a_n)\rangle \overrightarrow{\in} \mathbf{M}(F_i^n))$;
(3) $\mathbf{M}(\neg A) = 1$, iff $\mathbf{M}(A) = 0$;
(4) $\mathbf{M}(A \to B) = 1$, iff $\mathbf{M}(A) - 0$ or $\mathbf{M}(B) = 1$;
(5) $\mathbf{M}(A \vee B) = 1$, iff $\mathbf{M}(A) = 1$ or $\mathbf{M}(B) = 1$;
(6) $\mathbf{M}(A \wedge B) = 1$, iff $\mathbf{M}(A) = 1$ and $\mathbf{M}(B) = 1$;
(7) $\mathbf{M}(A \leftrightarrow B) = 1$, iff "$\mathbf{M}(A) = 1$ and $\mathbf{M}(B) = 1$" or "$\mathbf{M}(A) = 0$ and $\mathbf{M}(B) = 0$";
(8) If $\mathbf{M}(\forall x A(x)) = 1$, then for each $m \in M(m \overrightarrow{\in} M)$, $\mathbf{M}(a_i/m)(A(a_i)) = 1$;
(9) $\mathbf{M}(\exists x A(x)) = 1$, iff there is $m \in M(m \overrightarrow{\in} M)$ such that $\mathbf{M}(a_i/m)(A(a_i)) = 1$;
(10) $\mathbf{M}(ExA(x)) = 1$, iff for each $m \in M(m \overrightarrow{\in} M)$, $\mathbf{M}(a_i/m)(A(a_i)) = 1$.

Theorem 12.5: Assume that **M** is an F^{PIN} model and *A* an arbitrary formula. Then it is true that $\mathbf{M}(A) \in \{1, 0\}$. QED.

Definition 12.11 (Satisfiability) Assume that *A* is a formula and Σ a set of formulas. The formula *A* is said to be satisfiable, iff there exists an F^{PIN} model M satisfying that $M(A) = 1$. The set of formulas Σ is said to be satisfiable, iff there exists an F^{PIN} model M such that for any B, if $B \in \Sigma(B \overrightarrow{\in} \Sigma)$, then $M(B) = 1$, denoted in short as $M(\Sigma) = 1$.

Definition 12.12 (Effectiveness) Let A be a formula. A is said to be effective, iff for any F^{PIN} model M, $M(A)=1$ holds true.

Definition 12.13 Assume that A is a formula and Σ a set of formulas. A is a grammatical successor of Σ, denoted $\Sigma \models A$, iff for any F^{PIN} model M, $M(\Sigma)=1$ implies $M(A)=1$.

Evidently, $\emptyset \models A$ iff A is an effective formula.

Theorem 12.6: (Ref) $A \models A$;

(τ) $\Gamma \models \Delta,\ \Delta \models A \Rightarrow \Gamma \models A$;
$(\neg)$ $\Gamma, \neg A \models B, \neg B \Rightarrow \Gamma \models A$;
$(\rightarrow_-)$ $A \rightarrow B, A \models B$;
$(\rightarrow_+)$ $\Gamma, A \models B \Rightarrow \Gamma \models A \rightarrow B$;
$(\wedge_-)$ $A \wedge B \models A, B$;
$(\wedge_+)$ $A, B \models A \wedge B$;
$(\vee_-)$ $A \models C, B \models C \Rightarrow A \vee B \models C$;
$(\vee_+)$ $A \models A \vee B, B \vee A$;
$(\leftrightarrow_-)$ $A \leftrightarrow B, A \models B$;
$A \leftrightarrow B, B \models A$;
$(\leftrightarrow_+)$ $\Gamma, A \models B, \Gamma, B \models A \Rightarrow \Gamma \models A \leftrightarrow B$;
$(\forall_-)$ $\forall x A(x) \models A(a)$;
(E_-) $ExA(x) \models A(a)$;
(E_+) $\Gamma \models A(a)$, a does not appear in Γ, $\Rightarrow \Gamma \models ExA(x)$;
$(\exists_-)$ $A(a) \models B$, a does not appear in B, $\Rightarrow \exists x A(x) \models B$;
$(\exists_+)$ $A(a) \models \exists x A(x)$, $A(x)$ is obtained by replacing each a appearing in $A(a)$ by x.

Proof. We will only show the parts involving quantifiers.

(1) $\forall x A(x) \models A(a)$.

Assume that $\forall x A(x) \models A(a)$ does not hold true. Then, there exists an F^{PIN} model M such that

(i) $\mathsf{M}(\forall x A(x))=1$

and

(ii) $\mathsf{M}(A(a))=0$.

From (i), it follows that

(iii) for any $m \in M(m \overrightarrow{\in} M)$, $\mathsf{M}(a_i/m)(A(a_i))=1$.

So, for $M(a) \in M(M(a) \overrightarrow{\in} M)$, we have $\mathsf{M}(a_i/\mathsf{M}(a))(A(a_i))=1$. That is,

(iv) $\mathsf{M}(a_i/\mathsf{M}(a))(a_i) \in \mathsf{M}(a_i/\mathsf{M}(a))(A)$
$(\mathsf{M}(a_i/\mathsf{M}(a))(a_i) \overrightarrow{\in} \mathsf{M}(a_i/\mathsf{M}(a))(A))$.

However, $\mathsf{M}(a_i/\mathsf{M}(a))(a_i)=\mathsf{M}(a)$ and $\mathsf{M}(a_i/\mathsf{M}(a))(A)=\mathsf{M}(A)$. So, we have

(v) $\mathsf{M}(a)\in\mathsf{M}(A)(\mathsf{M}(a)\overrightarrow{\in}\mathsf{M}(A))$.

Therefore, we have

(vi) $\mathsf{M}(A(a))=1$.

Since (ii) and (vi) contradict with each other, our assumption does not hold true. So, we have $\forall xA(x)\vDash A(a)$.

(2) $ExA(x)\vDash A(a)$.

The proof is similar to that of part (1) and is omitted.

(3) $\Sigma\vDash A(a)$, a does not appear in Σ, $\Rightarrow\Sigma\vDash ExA(x)$.

Assume that $\Sigma\vDash A(a)$, where a does not appear in Σ, holds true, while $\Sigma\vDash ExA(x)$ does not. Then, there exists an F^{PIN} model M such that

(i) $\mathsf{M}(\Sigma)=1$

and

(ii) $\mathsf{M}(ExA(x))=1$.

From (ii), it follows that

(iii) there is $m\in M(m\overrightarrow{\in}M)$ such that $\mathsf{M}(a/m)(A(a))=0$.

Since a does not appear in Σ, both (i) and (iii) imply

(iv) $\mathsf{M}(a/m)(\Sigma)=1$.

Since $\Sigma\vDash A(a)$ and a does not appear in Σ, we have

(v) $\mathsf{M}(a/m)(A(a))=1$.

Now, (iii) and (v) contradict with each other. So, our initial assumption does not hold true. Therefore, $\Sigma\vDash A(a)$, where a does not appear in Σ, $\Rightarrow\Sigma\vDash ExA(x)$.

About the proof on effectiveness of the deduction rules of the existence quantifier, the details are the same as those of the classical logic. So, the proof is omitted. QED.

Theorem 12.7: In F^{PIN}, the following hold true:

(1) $\vDash\forall xP(x)\rightarrow P(a)$;
(2) $\vDash\forall xP(x)\rightarrow eVxP(x)$;
(3) $\nvDash ExP(x)\rightarrow\forall xP(x)$. QED.

12.1.3 A meta theory of logical basis

This subsection continues our effort to resolve the following two problems, which urgently need an answer: (1) How can an appropriate theoretical foundation be chosen for modern mathematics and computer science? And, (2) under what interpretations can modern mathematics and the theory of computer science be kept as completely as possible? In particular, in this subsection, we will establish a meta theory of the logical foundation for the mathematical system of potential infinite, while proving the relevant results on the reliability and the completeness of the logical system.

12.1.3.1 Reliability of F^{PIN}

Based on all the concepts and notations introduced in Subsection 12.1.2, we can now show the following result on the reliability of the formal logical system F^{PIN} of the mathematical system of potential infinite.

Theorem 12.8: (Reliability Theorem of F^{PIN})

(1) If $\Sigma \vdash A$, then $\Sigma \vDash A$.
(2) If $\vdash A$, then $\vDash A$.

Proof. Let us apply the induction on the deduction length N of $\Sigma \vdash A$, where the set Σ is excluded in the calculation of the degree of deduction.

Basic Step. When N = 1, the deduction in this case can be done by directly applying the rules (Ref), $(\rightarrow_-)$, $(\wedge_-)$, $(\wedge_+)$, $(\vee_+)$, $(\leftrightarrow_-)$, $(\forall_-)$, (E_-) and $(\exists_+)$. So, we have $\Sigma \vDash A$.

Inductive Step: Assume that when N = k, the theorem holds true. Now, let N = k + 1. In this case, the deduction could be directly done by using the rules (Ref), $(\rightarrow_-)$, $(\wedge_-)$, $(\wedge_+)$, $(\vee_+)$, $(\leftrightarrow_-)$, $(\forall_-)$, (E_-), and $(\exists_+)$. So, we have $\Sigma \vDash A$.

If the deduction is done by following other rules, then the inductive assumption and Theorem 12.4 above imply that $\Sigma \vDash A$.

Now, statement (2) is a special case of (1). So, this ends the proof. QED.

Definition 12.14 (Harmonicity) *A* set Σ of formulas is harmonic, iff there does not exist such a formula *A* that $\Sigma \vdash A$ and $\Sigma \vdash \neg A$.

Evidently, harmonicity is only a concept of pure semantics.

Definition 12.15 A set Σ of formulas is potentially maximum harmonic, iff Σ satisfies.

(1) Σ is harmonic; and
(2) for any $A \notin \Sigma$ $(A \overrightarrow{\notin} \Sigma)$, $\Sigma \cup \{A\}$ is not harmonic.

Theorem 12.9: Assume that Σ is a potentially maximum harmonic set of formulas. For any formula *A*, $\Sigma \vdash A$ iff $A \in \Sigma$ $(A \overrightarrow{\in} \Sigma)$.

Proof. Assume that $A \in \Sigma$ $(A \overrightarrow{\in} \Sigma)$, then it is obvious that $\Sigma \vdash A$.

Now, assume that $\Sigma \vdash A$. If $A \notin \Sigma$ $(A \overrightarrow{\notin} \Sigma)$, then from the assumption that Σ is potentially maximum harmonic, we must have that $\Sigma \cup \{A\}$ is not harmonic. So, there exists a formula B such that $\Sigma \cup \{A\} \vdash B$ and $\Sigma \cup \{A\} \vdash \neg B$. So, we have $\Sigma \vdash \neg A$. This end contradicts the harmonicity of Σ. So, we must conclude $A \in \Sigma$ $(A \overrightarrow{\in} \Sigma)$. QED.

Theorem 12.10: Assume that Σ is a potentially maximum harmonic set. For any formula A, $\Sigma \vdash \neg A$ iff $\Sigma \nvdash A$. QED.

Theorem 12.11: Any harmonic set of formulas can be expanded to a potentially maximum harmonic set.

Proof. Let Σ be a harmonic set of formulas. And, let

$$[*] : A_1, A_2, \ldots, A_{\vec{n}},$$

be a potentially infinite spring sequence of all formulas in the PIMS. We now define a potentially infinite spring sequence of formula sets: $\Sigma_0, \Sigma_1, \Sigma_2, \ldots, \Sigma_{\vec{n}}$, as follows:

(1) $\Sigma_0 = \Sigma$;

(2) $$\Sigma_{n+1} = \begin{cases} \Sigma_n \cup \{A_{n+1}\}, & \text{if } \Sigma_n \cup \{A_{n+1}\} \text{ is harmonic} \\ \Sigma_n, & \text{otherwise} \end{cases}$$

So, we have

(3) $\Sigma_n \subseteq \Sigma_{n+1}$ $(\Sigma_n \overrightarrow{\subseteq} \Sigma_{n+1})$,

(4) For each $n (\overrightarrow{\in} \omega)$, Σ_n is harmonic.

Define $\Sigma^* = \bigcup_{n \overrightarrow{\in} \vec{n}} \Sigma_n$. Then, we have

(5) $\Sigma \subseteq \Sigma^*$ $(\Sigma \overrightarrow{\subseteq} \Sigma^*)$, and

(6) Σ^* is the potentially maximum harmonic set.

For statement (6), let us see the proof below. First, let us check the harmonicity of Σ^*. By contradiction, assume that Σ^* is not harmonic. Then there exists a formula B such that $\Sigma^* \vdash B$ and $\Sigma^* \vdash \neg B$. According to Theorem 12.4 above, there are finite many formulas in Σ^*: $B_1, \ldots, B_k, B_{k+1}, \ldots, B_{k+l}$, satisfying that

(7) $B_1, \ldots, B_k \vdash B$,

(8) $B_{k+1}, \ldots, B_{k+l} \vdash \neg B$.

Therefore, $\{B_1, \ldots, B_k, \ldots, B_{k+l}\}$ is not harmonic. Assume

$$B_i \in \Sigma_{m_i} (B_i \overrightarrow{\in} \Sigma_{m_i}),\ (1 \le i \le k + l, m_i \overrightarrow{\in} \omega),$$

and let

$$m = max(m_1, \ldots, m_k, \ldots, m_{k+l}).$$

Then statement (3) above indicates that $\{B_1, \ldots, B_k, \ldots, B_{k+l}\} \subseteq \Sigma_m$ ($\{B_1, \ldots, B_k, \ldots, B_{k+l}\} \overrightarrow{\subseteq} \Sigma_m$). Therefore, Σ_m is not harmonic. This end contradicts the statement (4) above. So, Σ^* is harmonic.

Secondly, we show Σ^* is potentially maximum. Assume that $C \notin \Sigma^* (C \overrightarrow{\notin} \Sigma^*)$ is a formula. Then, the definition of Σ^* implies that for any $n(\overrightarrow{\in} \omega)$, $C \notin \Sigma_n (C \overrightarrow{\notin} \Sigma_n)$. Assume that in sequence [*], $C = A_{j+1}$. From the definition of Σ_{j+1}, it follows that $\Sigma_j \cup \{C\}$ is not harmonic. Since if $\Sigma_j \cup \{C\}$ is harmonic, then $\Sigma_{j+1} = \Sigma_j \cup \{C\}$, and so $C \in \Sigma_{j+1} (C \overrightarrow{\in} \Sigma_{j+1})$. This end contradicts the assumption that for any $n(\overrightarrow{\in} \omega)$, $C \notin \Sigma_n$ $(C \overrightarrow{\notin} \Sigma_n)$. Hence since $\Sigma_j \subseteq \Sigma^* (\Sigma_j \overrightarrow{\subseteq} \Sigma^*)$, $\Sigma^* \cup \{C\}$ is not harmonic. That is, we have shown that Σ^* is potentially maximum. QED.

12.1.3.2 Potentially maximum harmonicity in PIMS$^+$

Definition 12.16 The first order language PIMS$^+$ is made up of the entire PIMS and a sequence of additional symbols for constants:

$$u'_1, u'_2, \ldots, u'_{\vec{n}}.$$

We will use u', v', and w' to represent arbitrarily chosen constant symbols from this list.

Definition 12.17 Let Σ be a set of formulas from the PIMS. The set Σ is said to have the property of existence, iff for each existence formula $\exists x A(x)$, if $\exists x A(x) \in \Sigma$ $(\exists x A(x) \overrightarrow{\in} \Sigma)$, then there exists u' such that $A(u') \in \Sigma (A(u') \overrightarrow{\in} \Sigma)$.

Theorem 12.12: Assume that Σ is a set of formulas in the PIMS. If Σ is harmonic, then Σ can be enlarged to the potentially maximum harmonic set Σ^* in the PIMS$^+$ such that Σ^* satisfies the property of existence.

Proof. Since in the PIMS$^+$, each set of formulas is either a potential infinite spring set or a potential sequence, each infinite subset of the given set of formulas, constructed by using an existence formula, must also be a potentially infinite spring set or a potential sequence. Let

$$\exists x A_1(x), \exists x A_2(x), \exists x A_3(x), \ldots, \exists x A_{\vec{n}}(x) \tag{12.1}$$

be an arbitrary ordering of the elements in a chosen subset. Now, we define a potentially infinite sequence of formula sets Σ_n in the PIMS$^+$ as follows:

$$\Sigma_0, \Sigma_1, \Sigma_2, \ldots, \Sigma_{\vec{n}},$$

where $\Sigma_0 = \Sigma$.

Take the first existence formula $\exists x A_1(x)$ in equ. (12.1). Since the length of $\exists x A_1(x)$ is finite, we can always find a certain u' such that u' does not appear in $\exists x A_1(x)$. Since $\Sigma_0 = \Sigma$, u' does not appear in Σ_0, either. Let

$$\Sigma_1 = \Sigma_0 \cup \{\exists x A_1(x) \to A_1(u')\}.$$

Assume that we have defined $\Sigma_0, \Sigma_1, \ldots, \Sigma_n$. Let us now take the existence formula $\exists x A_{n+1}(x)$ in equ. (12.1). Since $\{u'_1, u'_2, \ldots, u'_{\vec{n}}\}$ is a potentially infinite set, and the number of new constants used in $\exists x A_i(x)$ in each $\Sigma_i (1 \le i \le n)$, and in $A_i(w')$ are finite, we can always find a certain v' such that v' does not appear in $\exists x A_{n+1}(x)$ and in Σ_n. Now, define

$$\Sigma_{n+1} = \Sigma_n \cup \{\exists x A_{n+1}(x) \to A_{n+1}(v')\}.$$

Evidently, we have

$$\text{For each } n(n \overrightarrow{\in} \omega), \Sigma_n \subseteq \Sigma_{n+1}; \tag{12.2}$$

$$\text{For each } n(n \overrightarrow{\in} \omega), \Sigma_n \text{ is harmonic.} \tag{12.3}$$

This statement in equ. (12.3) can be shown by using the induction. In particular, we have the following details:

The basic step: Σ_0 is harmonic.

The inductive step: Assume that Σ_n is harmonic. If Σ_{n+1} is not harmonic, then we have

$$\begin{aligned}
&\Sigma_n \vdash \neg(\exists x A_{n+1}(x) \to A_{n+1}(v')),\\
&\Sigma_n \vdash \exists x A_{n+1}(x) \wedge \neg A_{n+1}(v'),\\
&\Sigma_n \vdash Ey(\exists x A_{n+1}(x) \wedge \neg A_{n+1}(y)),\\
&\Sigma_n \vdash Ey(\exists x A_{n+1}(x) \wedge \neg A_{n+1}(y)) \to (\exists x A_{n+1}(x) \wedge Ey\neg A_{n+1}(y)),\\
&\Sigma_n \vdash \exists x A_{n+1}(x) \wedge Ey\neg A_{n+1}(y),\\
&\Sigma_n \vdash \exists x A_{n+1}(x) \wedge Ex\neg A_{n+1}(x),\\
&\Sigma_n \vdash \exists x A_{n+1}(x) \wedge \neg\exists x A_{n+1}(x).
\end{aligned}$$

This end contradicts the inductive assumption that Σ_n is harmonic. Therefore, Σ_{n+1} is harmonic.

Define $\Sigma' = \bigcup_{n \overrightarrow{\in} \vec{n}} \Sigma_n$. Then, Σ' is harmonic. In fact, if Σ' is not harmonic, there then exists a formula B such that $\Sigma' \vdash B$ and $\Sigma' \vdash \neg B$. From Theorem 12.4 above, it follows that there are finite many formulas $B_1, \ldots, B_k, B_{k+1}, \ldots, B_{k+l}$ in Σ' such that

$$\begin{aligned}
&B_1, \ldots, B_k \vdash B, \quad \text{and}\\
&B_{k+1}, \ldots, B_{k+l} \vdash \neg B.
\end{aligned}$$

That is, $\{B_1, \ldots, B_k, \ldots, B_{k+l}\}$ is not harmonic. Assume that

$$B_i \in \Sigma_{m_i},\ \ 1 \le i \le k+l,\ \ m_i \overrightarrow{\in} \omega.$$

Define $m = max\{m_1, \ldots, m_k, \ldots, m_{k+l}\}$. From the statement in equ. (12.2) above, it follows that

$$\{B_1, \ldots, B_k, \ldots, B_{k+l}\} \subseteq \Sigma_m(\{B_1, \ldots, B_k, \ldots, B_{k+l}\} \overrightarrow{\subseteq} \Sigma_m).$$

So, Σ_m is not harmonic. This end contradicts the statement (3) above. Therefore, Σ' must be harmonic.

Now, Theorem 12.11 implies that Σ' can be enlarged to the potentially maximum harmonic set $\Sigma^* \overrightarrow{\subseteq} Form(L^{FO+})$.

As the end of this proof, let us now prove that Σ^* satisfies the property of existence.

For any existence formula $\exists x A(x) \in \Sigma^*$ in the PIMS, assume that this formula $\exists x A(x)$ is the kth entry $\exists x A_k(x)$ in the sequence in equ. (12.1) above. So, there is u' such that

$$\exists x A_k(x) \to A_k(u') \in \Sigma_k(\exists x A_k(x) \to A_k(u') \overrightarrow{\in} \Sigma_k).$$

So, we have

$$\exists x A_k(x) \to A_k(u') \in \Sigma^*,$$
$$\Sigma^* \vdash \exists x A_k(x) \to A_k(u'),$$
$$\Sigma^* \vdash \exists x A_k(x),$$
$$\Sigma^* \vdash A_k(u'),$$
$$A_k(u') \overrightarrow{\in} \Sigma^*.$$

That is, Σ^* satisfies the property of existence. QED.

12.1.3.3 Properties of potentially maximum harmonic sets

Theorem 12.13: Assume that Σ is a potentially maximum harmonic set, satisfying the property of existence. Let $M = \{a | a.$ is a constant element appearing in $\Sigma\}$. Then,

(1) $A \in \Sigma(A \overrightarrow{\in} \Sigma)$ iff $\neg A \notin \Sigma(\neg A \overrightarrow{\notin} \Sigma)$;
(2) $(A \to B) \in \Sigma((A \to B) \overrightarrow{\in} \Sigma)$ iff $A \notin \Sigma(A \overrightarrow{\notin} \Sigma)$ or $B \in \Sigma(B \overrightarrow{\in} \Sigma)$;
(3) $(A \vee B) \in \Sigma((A \vee B) \overrightarrow{\in} \Sigma)$ iff $A \in \Sigma(A \overrightarrow{\in} \Sigma)$ or $B \in \Sigma(B \overrightarrow{\in} \Sigma)$;
(4) $(A \wedge B) \in \Sigma((A \wedge B) \overrightarrow{\in} \Sigma)$ iff $A \in \Sigma(A \overrightarrow{\in} \Sigma)$ and $B \in \Sigma(B \overrightarrow{\in} \Sigma)$;
(5) $(A \leftrightarrow B) \in \Sigma((A \leftrightarrow B) \overrightarrow{\in} \Sigma)$ iff "$A \in \Sigma(A \overrightarrow{\in} \Sigma)$ and $B \in \Sigma(B \overrightarrow{\in} \Sigma)$" or "$A \notin \Sigma (A \overrightarrow{\notin} \Sigma)$ and $B \notin \Sigma(B \overrightarrow{\notin} \Sigma)$";
(6) If $\forall x A(x) \in \Sigma(\forall x A(x) \overrightarrow{\in} \Sigma)$, then for each $a \in M(a \overrightarrow{\in} M)$, $A(a) \in \Sigma(A(a) \overrightarrow{\in} \Sigma)$;
(7) $\exists x A(x) \in \Sigma(\exists x A(x) \overrightarrow{\in} \Sigma)$, iff there is $a \in M(a \overrightarrow{\in} M)$, such that $A(a) \in \Sigma\ (A(a) \overrightarrow{\in} \Sigma)$;
(8) $ExA(x) \in \Sigma(ExA(x) \overrightarrow{\in} \Sigma)$, iff for each $a \in M(a \overrightarrow{\in} M)$, $A(a) \in \Sigma(A(a) \overrightarrow{\in} \Sigma)$.

Proof. (1) First, we prove $\neg A \in \Sigma\ (\neg A \overrightarrow{\in}\ \Sigma) \Rightarrow A \notin \Sigma\ (A \overrightarrow{\notin}\ \Sigma)$. Assume that $\neg A \in \Sigma$ $(\neg A \overrightarrow{\in}\ \Sigma)$. If $A \in \Sigma\ (A \overrightarrow{\in}\ \Sigma)$, then we must have $\Sigma \vdash A$ and $\Sigma \vdash \neg A$. So, Σ is not harmonic. This conclusion contradicts the assumption that Σ is a potentially maximum harmonic set. Therefore, $A \notin \Sigma\ (A \overrightarrow{\notin}\ \Sigma)$.

Secondly, we prove $A \notin \Sigma\ (A \overrightarrow{\notin}\ \Sigma) \Rightarrow \neg A \in \Sigma\ (\neg A \overrightarrow{\in}\ \Sigma)$. Assume that $A \notin \Sigma$ $(A \overrightarrow{\notin}\ \Sigma)$. If $\neg A \notin \Sigma\ (\neg A \overrightarrow{\notin}\ \Sigma)$, then both $\Sigma \cup \{A\}$ and $\Sigma \cup \{\neg A\}$ are not harmonic. So, the proof of Theorem 12.9 implies that $\Sigma \vdash \neg A$ and $\Sigma \vdash \neg\neg A$. That is, Σ is not harmonic. This end contradicts the assumption that Σ is a potentially maximum harmonic set. Therefore, $\neg A \in \Sigma\ (\neg A \overrightarrow{\in}\ \Sigma)$.

(2) If $A \to B \in \Sigma\ (A \to B \overrightarrow{\in}\ \Sigma)$ and $A \in \Sigma\ (A \overrightarrow{\in}\ \Sigma)$, then we have $\Sigma \vdash A \to B$ and $\Sigma \vdash A$. So, $\Sigma \vdash B$. From Theorem 12.9, it follows that $B \in \Sigma\ (B \overrightarrow{\in}\ \Sigma)$.

If $A \to B \notin \Sigma\ (A \to B \overrightarrow{\notin}\ \Sigma)$, then from the statement in equ. (12.1) above, it follows that $\neg(A \to B) \in \Sigma\ (\neg(A \to B) \overrightarrow{\in}\ \Sigma)$. So, from Theorem 12.9, we have $\Sigma \vdash \neg(A \to B)$. So, $\Sigma \vdash A$ and $\Sigma \vdash \neg B$. Based on Theorem 12.9, we have $A \in \Sigma\ (A \overrightarrow{\in}\ \Sigma)$ and $\neg B \in \Sigma$ $(\neg B \overrightarrow{\in}\ \Sigma)$. Therefore, the statement in equ. (12.1) above implies that $A \in \Sigma\ (A \overrightarrow{\in}\ \Sigma)$ and $B \notin \Sigma\ (B \overrightarrow{\notin}\ \Sigma)$. That is, we have that it is not if $A \in \Sigma\ (A \overrightarrow{\in}\ \Sigma)$ then $B \in \Sigma$ $(B \overrightarrow{\in}\ \Sigma)$.

(3) Assume that $(A \vee B) \in \Sigma\ ((A \vee B) \overrightarrow{\in}\ \Sigma)$, $A \notin \Sigma\ (A \overrightarrow{\notin}\ \Sigma)$, and $B \notin \Sigma\ (B \overrightarrow{\notin}\ \Sigma)$. Then, since $A \notin \Sigma\ (A \overrightarrow{\notin}\ \Sigma)$, the statement in equ. (12.1) above implies that $\neg A \in \Sigma$ $(\neg A \overrightarrow{\in}\ \Sigma)$. So, Theorem 12.9 implies that $\Sigma \vdash \neg A$. For the same reason we can obtain $\Sigma \vdash \neg B$. Therefore, $\Sigma \vdash \neg A \wedge \neg B$ and $\Sigma \vdash \neg(A \vee B)$. Also, from Theorem 12.9, it follows that $\neg(A \vee B) \in \Sigma\ (\neg(A \vee B) \overrightarrow{\in}\ \Sigma)$. From $(A \vee B) \in \Sigma\ ((A \vee B) \overrightarrow{\in}\ \Sigma)$, it follows that Σ is not harmonic.

This end contradicts the assumption that Σ is a potentially maximum harmonic set. Hence, our assumption does not hold true, and the following hold true:

If $(A \vee B) \in \Sigma((A \vee B) \overrightarrow{\in}\ \Sigma)$, then $A \in \Sigma(A \overrightarrow{\in}\ \Sigma)$ or $B \in \Sigma(B \overrightarrow{\in}\ \Sigma)$.

Assume that $(A \vee B) \notin \Sigma\ ((A \vee B) \overrightarrow{\notin}\ \Sigma)$, but either $A \in \Sigma\ (A \overrightarrow{\in}\ \Sigma)$ or $B \in \Sigma\ (B \overrightarrow{\in}\ \Sigma)$. Then, since $(A \vee B) \notin \Sigma\ ((A \vee B) \overrightarrow{\notin}\ \Sigma)$, the statement in equ. (12.1) above implies that $\neg(A \vee B) \in \Sigma\ (\neg(A \vee B) \overrightarrow{\in}\ \Sigma)$. So, Theorem 12.9 implies that $\Sigma \vdash \neg(A \vee B)$. And in turn, we have $\Sigma \vdash \neg A \wedge \neg B$ and then $\Sigma \vdash \neg A$ and $\Sigma \vdash \neg B$. Similarly, Theorem 12.9 implies $\neg A \in \Sigma\ (\neg A \overrightarrow{\in}\ \Sigma)$ and $\neg B \in \Sigma\ (\neg B \overrightarrow{\in}\ \Sigma)$. Now, the statement in equ. (12.1) above implies that $A \notin \Sigma\ (A \overrightarrow{\notin}\ \Sigma)$ or $B \notin \Sigma\ (B \overrightarrow{\notin}\ \Sigma)$. This end contradicts that $A \in \Sigma$ $(A \overrightarrow{\in}\ \Sigma)$ or $B \in \Sigma\ (B \overrightarrow{\in}\ \Sigma)$.

Therefore, if $A \in \Sigma\ (A \overrightarrow{\in}\ \Sigma)$ or $B \in \Sigma\ (B \overrightarrow{\in}\ \Sigma)$, then $(A \vee B) \in \Sigma\ ((A \vee B) \overrightarrow{\in}\ \Sigma)$.

(4) Assume that $(A \wedge B) \in \Sigma\ ((A \wedge B) \overrightarrow{\in}\ \Sigma)$. Then we have $\Sigma \vdash A \wedge B$ and further: $\Sigma \vdash A$ and $\Sigma \vdash B$. Therefore, Theorem 12.9 implies: $A \in \Sigma\ (A \overrightarrow{\in}\ \Sigma)$ and $B \in \Sigma$ $(B \overrightarrow{\in}\ \Sigma)$.

Assume that $A \in \Sigma\ (A \overrightarrow{\in}\ \Sigma)$ and $B \in \Sigma\ (B \overrightarrow{\in}\ \Sigma)$. Then we have $\Sigma \vdash A$ and $\Sigma \vdash B$. Furthermore, we have $\Sigma \vdash A \wedge B$. So, Theorem 12.9 implies $(A \wedge B) \in \Sigma\ ((A \wedge B) \overrightarrow{\in}\ \Sigma)$.

(5) Let us prove $(\Rightarrow)$ first. Assume that $(A \leftrightarrow B) \in \Sigma\ ((A \leftrightarrow B) \overrightarrow{\in}\ \Sigma)$ and that it is not true that "$A \in \Sigma\ (A \overrightarrow{\in}\ \Sigma)$ and $B \in \Sigma\ (B \overrightarrow{\in}\ \Sigma)$." Then, it follows that $A \notin \Sigma\ (A \overrightarrow{\notin}\ \Sigma)$ or $B \notin \Sigma\ (B \overrightarrow{\in}\ \Sigma)$.

Assume $A \notin \Sigma$ ($A \overrightarrow{\notin} \Sigma$). From the statement in equ. (12.1) above, it follows that $\neg A \in \Sigma$ ($\neg A \overrightarrow{\in} \Sigma$). So, $\Sigma \vdash \neg A$. From $(A \leftrightarrow B) \in \Sigma$ ($(A \leftrightarrow B) \overrightarrow{\in} \Sigma$), it follows that $\Sigma \vdash A \leftrightarrow B$. Therefore, we have $\Sigma \vdash \neg B$, and furthermore, $\neg B \in \Sigma$ ($\neg B \overrightarrow{\in} \Sigma$). From Theorem 12.9, it follows that $B \notin \Sigma$ ($B \overrightarrow{\notin} \Sigma$). So, we obtain $A \notin \Sigma$ ($A \overrightarrow{\notin} \Sigma$) and $B \notin \Sigma$ ($B \overrightarrow{\notin} \Sigma$).

Assume that $B \notin \Sigma$ ($B \overrightarrow{\notin} \Sigma$). Similarly, we can obtain: $A \notin \Sigma$ ($A \overrightarrow{\notin} \Sigma$) and $B \notin \Sigma$ ($B \overrightarrow{\notin} \Sigma$).

So, it is always true that $A \notin \Sigma$ ($A \overrightarrow{\notin} \Sigma$) and $B \notin \Sigma$ ($B \overrightarrow{\notin} \Sigma$).

Summarizing what we have obtained, we have shown: If $(A \leftrightarrow B) \in \Sigma$ ($(A \leftrightarrow B) \overrightarrow{\in} \Sigma$), then either "$A \in \Sigma$ ($A \overrightarrow{\in} \Sigma$) and $B \in \Sigma$ ($B \overrightarrow{\in} \Sigma$)" or "$A \notin \Sigma$ ($A \overrightarrow{\notin} \Sigma$) and $B \notin \Sigma$ ($B \overrightarrow{\notin} \Sigma$)".

Next, let us prove ($\Leftarrow$). Assume either that "$A \in \Sigma$ ($A \overrightarrow{\in} \Sigma$) and $B \in \Sigma$ ($B \overrightarrow{\in} \Sigma$)" or that "$A \notin \Sigma$ ($A \overrightarrow{\notin} \Sigma$) and $B \notin \Sigma$ ($B \overrightarrow{\notin} \Sigma$)". Then, if "$A \in \Sigma$ ($A \overrightarrow{\in} \Sigma$) and $B \in \Sigma$ ($B \overrightarrow{\in} \Sigma$)" holds true, then $\Sigma \vdash A$ and $\Sigma \vdash B$ and further $\Sigma \vdash A \wedge B$. Since $A \wedge B \vdash A \leftrightarrow B$, we have $\Sigma \vdash A \leftrightarrow B$. So, Theorem 12.9 implies that $(A \leftrightarrow B) \in \Sigma$ ($(A \leftrightarrow B) \overrightarrow{\in} \Sigma$).

If "$A \notin \Sigma$ ($A \overrightarrow{\notin} \Sigma$) and $B \notin \Sigma$ ($B \overrightarrow{\notin} \Sigma$)" holds true, then we have $\neg A \in \Sigma$ ($\neg A \overrightarrow{\in} \Sigma$) and $\neg B \in \Sigma$ ($\neg B \overrightarrow{\in} \Sigma$). So, $\Sigma \vdash \neg A$ and $\Sigma \vdash \neg B$, and further, $\Sigma \vdash \neg A \wedge \neg B$. Since $\neg A \wedge \neg B \vdash A \leftrightarrow B$, we obtain $\Sigma \vdash A \leftrightarrow B$. Theorem 12.9 implies that $(A \leftrightarrow B) \in \Sigma$ ($(A \leftrightarrow B) \overrightarrow{\in} \Sigma$).

Therefore, no matter which situation holds true, we always have $(A \leftrightarrow B) \in \Sigma$ ($(A \leftrightarrow B) \overrightarrow{\in} \Sigma$).

(6) Assume that $\forall x A(x) \in \Sigma$ ($\forall x A(x) \overrightarrow{\in} \Sigma$), but there exists a constant element $a \in M$($a \overrightarrow{\in} M$) such that $A(a) \notin \Sigma$ ($A(a) \overrightarrow{\notin} \Sigma$).

Since $\forall x A(x) \in \Sigma$ ($\forall x A(x) \overrightarrow{\in} \Sigma$), $\Sigma \vdash \forall x A(x)$. Since $\forall x A(x) \vdash A(a)$, $\Sigma \vdash A(a)$. Based on Theorem 12.9, it follows that $A(a) \in \Sigma$ ($A(a) \overrightarrow{\in} \Sigma$). This end contradicts that $A(a) \notin \Sigma$ ($A(a) \overrightarrow{\notin} \Sigma$). Therefore, our assumption does not hold true. Therefore, if $\forall x A(x) \in \Sigma$ ($\forall x A(x) \overrightarrow{\in} \Sigma$), for each $a \in M$($a \overrightarrow{\in} M$), we have $A(a) \in \Sigma$ ($A(a) \overrightarrow{\in} \Sigma$).

(7) Assume that $\exists x A(x) \in \Sigma$ ($\exists x A(x) \overrightarrow{\in} \Sigma$). Since Σ is a potentially maximum harmonic set, satisfying the property of existence, there exists $a \in M$($a \overrightarrow{\in} M$) such that $A(a) \in \Sigma$ ($A(a) \overrightarrow{\in} \Sigma$).

Assume that there exists $a \in M$($a \overrightarrow{\in} M$) such that $A(a) \in \Sigma$ ($A(a) \overrightarrow{\in} \Sigma$). Theorem 12.9 implies that $\Sigma \vdash A(a)$. Since $A(a) \vdash \exists x A(x)$, we have $\Sigma \vdash \exists x A(x)$. Similarly, Theorem 12.9 implies that $\exists x A(x) \in \Sigma$ ($\exists x A(x) \overrightarrow{\in} \Sigma$).

(8) Assume that $ExA(x) \in \Sigma$ ($\exists x A(x) \overrightarrow{\in} \Sigma$). Theorem 12.9 implies that $\Sigma \vdash ExA(x)$. Since for each $a \in M$($a \overrightarrow{\in} M$) we have $ExA(x) \vdash A(a)$, we obtain $\Sigma \vdash A(a)$. Similarly, Theorem 12.9 implies that for each $a \in M$ ($a \overrightarrow{\in} M$), $A(a) \in \Sigma$ ($A(a) \overrightarrow{\in} \Sigma$).

Assume that $ExA(x) \notin \Sigma$ ($ExA(x) \overrightarrow{\notin} \Sigma$). The statement in equ. (12.1) above implies that $\neg ExA(x) \in \Sigma$ ($\neg ExA(x) \overrightarrow{\in} \Sigma$). Therefore, we have $\Sigma \vdash \neg ExA(x)$. Since $\neg ExA(x) \vdash \exists x \neg A(x)$, we get $\Sigma \vdash \exists x \neg A(x)$. Therefore, $\exists x \neg A(x) \in \Sigma$ ($\exists x \neg A(x) \overrightarrow{\in} \Sigma$). Since Σ is a potentially maximum harmonic set, satisfying the property of existence, there exists $a \in M$ ($a \overrightarrow{\in} M$) such that $\neg A(a) \in \Sigma$ ($\neg A(a) \overrightarrow{\in} \Sigma$). So, the statement in equ. (12.1) above implies that there exists $a \in M$ ($a \overrightarrow{\in} M$) such that $A(a) \notin \Sigma$ ($A(a) \overrightarrow{\notin} \Sigma$). QED.

12.1.3.4 Model M* *and the completeness of* F^{PIN}

Definition 12.18 The model M*, generated by a potentially maximum harmonic set Σ^*, satisfying the property of existence, consists of the following:

(1) $M^* = \{a^* | a \text{ is a constant element in } \Sigma^*\}$;
(2) For any constant element a, $\mathsf{M}^*(a) = a^* \in M^*(a^* \overrightarrow{\in} M^*)$; for any constant element u', $\mathsf{M}^*(u') = u'^* \in M^*(u'^* \overrightarrow{\in} M^*)$;
(3) For any symbol R of an n-ary relation and arbitrary constant elements $a_1^*, \ldots, a_n^* \in M^*(a_1^*, \ldots, a_n^* \overrightarrow{\in} M^*)$,

$$<a_1^*, \ldots, a_n^*> \in \mathsf{M}^*(R)(<a_1^*, \ldots, a_n^*> \overrightarrow{\in} \mathsf{M}^*(R)) \Leftrightarrow R(a_1, \ldots, a_n) \in \Sigma^*(R(a_1, \ldots, a_n) \overrightarrow{\in} \Sigma^*).$$

(4) For any p_i, $p_i \in \Sigma^*(p_i \overrightarrow{\in} \Sigma^*)$ iff $\mathsf{M}^*(p_i) = 1$.

Theorem 12.14: For an arbitrary constant element c, $\mathsf{M}^*(c) = c^* \in M^*(c^* \overrightarrow{\in} M^*)$. QED.

Theorem 12.15: For any formula A, let $\mathsf{M}^*(A) = 1$ iff $A \in \Sigma^*(A \overrightarrow{\in} \Sigma^*)$. Then M* is an F^{PIN} model.

Proof. Based on Definition 12.8, it is obvious that M* is a model. In the following, we show that M* is an F^{PIN} model.

(1) $\mathsf{M}^*(p_i) \in \{0, 1\}$.

Since Σ^* is a potentially maximum harmonic set, we have either $p_i \in \Sigma^*(p_i \overrightarrow{\in} \Sigma^*)$ or $p_i \notin \Sigma^*(p_i \overrightarrow{\notin} \Sigma^*)$. So, based on Definition 12.18, it is obvious that $\mathsf{M}^*(p_i) \in \{0, 1\}$.

(2) $\mathsf{M}^*(F_i^n(a_1, \ldots, a_n)) = 1$ iff $<\mathsf{M}^*(a_1), \ldots, \mathsf{M}^*(a_n)> \in \mathsf{M}^*(F_i^n)$ $(<\mathsf{M}^*(a_1), \ldots, \mathsf{M}^*(a_n)> \overrightarrow{\in} \mathsf{M}^*(F_i^n))$.
This is because $\mathsf{M}^*(F_i^n(a_1, \ldots, a_n)) = 1$
iff $F_i^n(a_1, \ldots, a_n) \in \Sigma^*(F_i^n(a_1, \ldots, a_n) \overrightarrow{\in} \Sigma^*)$
iff $\langle a_1, \ldots, a_n \rangle \in \mathsf{M}^*(F_i^n)(\langle a_1, \ldots, a_n \rangle \overrightarrow{\in} \mathsf{M}^*(F_i^n))$
iff $\langle \mathsf{M}^*(a_1), \ldots, \mathsf{M}^*(a_n) \rangle \in \mathsf{M}^*(F_i^n)(\langle \mathsf{M}^*(a_1), \ldots, \mathsf{M}^*(a_n) \rangle \overrightarrow{\in} \mathsf{M}^*(F_i^n))$.
(3) $\mathsf{M}^*(\neg A) = 1$ iff $\mathsf{M}^*(A) = 0$.
This is because $\mathsf{M}^*(\neg A) = 1$ iff $\neg A \in \Sigma^*(\neg A \overrightarrow{\in} \Sigma^*)$
iff $A \notin \Sigma^*(A \overrightarrow{\notin} \Sigma^*)$
iff $\mathsf{M}^*(A) = 0$.
(4) $\mathsf{M}^*(A \to B) = 1$ iff $\mathsf{M}^*(A) = 0$ or $\mathsf{M}^*(B) = 1$.
This is because $\mathsf{M}^*(A \to B) = 1$ iff $(A \to B) \in \Sigma^*((A \to B) \overrightarrow{\in} \Sigma^*)$
iff $A \notin \Sigma^*(A \overrightarrow{\notin} \Sigma^*)$ or $B \in \Sigma^*(B \overrightarrow{\in} \Sigma^*)$
iff $\mathsf{M}^*(A) = 0$ or $\mathsf{M}^*(B) = 1$.

(5) $\mathsf{M}^*(A \vee B) = 1$ iff $\mathsf{M}^*(A) = 1$ or $\mathsf{M}^*(B) = 1$.
This is because $\mathsf{M}^*(A \vee B) = 1$ iff $(A \vee B) \in \Sigma^*((A \vee B) \overrightarrow{\in} \Sigma^*)$,
iff $A \in \mathsf{M}^*(A \overrightarrow{\in} \mathsf{M}^*)$ or $B \in \mathsf{M}^*(B \overrightarrow{\in} \mathsf{M}^*)$,
iff $\mathsf{M}^*(A) = 1$ or $\mathsf{M}^*(B) = 1$.

(6) $\mathsf{M}^*(A \wedge B) = 1$ iff $\mathsf{M}^*(A) = 1$ and $\mathsf{M}^*(B) = 1$.
This is because $\mathsf{M}^*(A \wedge B) = 1$ iff $(A \wedge B) \in \Sigma^*((A \wedge B) \overrightarrow{\in} \Sigma^*)$
iff $A \in \Sigma^*(A \overrightarrow{\in} \Sigma^*)$ and $B \in \Sigma^*(B \overrightarrow{\in} \Sigma^*)$
iff $\mathsf{M}^*(A) = 1$ and $\mathsf{M}^*(B) = 1$.

(7) $\mathsf{M}^*(A \leftrightarrow B) = 1$ iff "$\mathsf{M}^*(A) = 1$ and $\mathsf{M}^*(B) = 1$" or "$\mathsf{M}^*(A) = 0$ and $\mathsf{M}^*(B) = 0$."
This is because $\mathsf{M}^*(A \leftrightarrow B) = 1$ iff $(A \leftrightarrow B) \in \Sigma^*((A \leftrightarrow B) \overrightarrow{\in} \Sigma^*)$
iff "$A \in \Sigma^*(A \overrightarrow{\in} \Sigma^*)$ and $B \in \Sigma^*(B \overrightarrow{\in} \Sigma^*)$" or "$A \notin \Sigma^*(A \overrightarrow{\notin} \Sigma^*)$ and $B \notin \Sigma^*$ $(B \overrightarrow{\notin} \Sigma^*)$"
iff "$\mathsf{M}^*(A) = 1$ and $\mathsf{M}^*(B) = 1$" or "$\mathsf{M}^*(A) = 0$ and $\mathsf{M}^*(B) = 0$."

(8) If $\mathsf{M}^*(\forall x A(x)) = 1$, then for each $m \in M^*(m \overrightarrow{\in} M^*)$, $\mathsf{M}^*(a_i/m)(A(a_i)) = 1$.
This is because if $\mathsf{M}^*(\forall x A(x)) = 1$ iff $\forall x A(x) \in \Sigma^*(\forall x A(x) \overrightarrow{\in} \Sigma^*)$,
then for each $a \in M(a \overrightarrow{\in} M)$, $A(a) \in \Sigma^*(A(a) \overrightarrow{\in} \Sigma^*)$,
iff for each $a^* \in M^*(a^* \overrightarrow{\in} M^*)$, $\mathsf{M}^*(A(a)) = 1$
iff for each $a^* \in M^*(a^* \overrightarrow{\in} M^*)$, $\mathsf{M}^*(a) \in \mathsf{M}^*(A)(\mathsf{M}^*(a) \overrightarrow{\in} \mathsf{M}^*(A))$
iff for each $a^* \in M^*(a^* \overrightarrow{\in} M^*)$, $a^* \in \mathsf{M}^*(A)(a^* \overrightarrow{\in} \mathsf{M}^*(A))$
iff for each $m \in M^*(m \overrightarrow{\in} M^*)$, $m \in \mathsf{M}^*(A)(m \overrightarrow{\in} \mathsf{M}^*(A))$
iff for each $m \in M^*(m \overrightarrow{\in} M^*)$, $\mathsf{M}^*(a_i/m)(a_i) \in \mathsf{M}^*(a_i/m)(A)$
$(\mathsf{M}^*(a_i/m)(a_i) \overrightarrow{\in} \mathsf{M}^*(a_i/m)(A))$
iff for each $m \in M^*(m \overrightarrow{\in} M^*)$, $\mathsf{M}^*(a_i/m)(A(a_i)) = 1$.

(9) $\mathsf{M}^*(\exists x A(x)) = 1$ iff there exists $m \in M^*(m \overrightarrow{\in} M^*)$ such that $\mathsf{M}^*(a_i/m)\ (A(a_i)) = 1$.
This is because $\mathsf{M}^*(\exists x A(x)) = 1$ iff $\exists x A(x) \in \Sigma^*(\exists x A(x) \overrightarrow{\in} \Sigma^*)$
iff there exists $a \in M(a \overrightarrow{\in} M)$ such that $A(a) \in \Sigma^*(A(a) \overrightarrow{\in} \Sigma^*)$
iff there exist $a^* \in M^*(a^* \overrightarrow{\in} M^*)$ such that $\mathsf{M}^*(A(a)) = 1$
iff there exists $a^* \in M^*(a^* \overrightarrow{\in} M^*)$ such that $\mathsf{M}^*(a) \in \mathsf{M}^*(A)$
$(\mathsf{M}^*(a) \overrightarrow{\in} \mathsf{M}^*(A))$
iff there exists $a^* \in M^*(a^* \overrightarrow{\in} M^*)$ such that $a^* \in \mathsf{M}^*(A)(a^* \overrightarrow{\in} \mathsf{M}^*(A))$
iff there exists $m \in M^*(m \overrightarrow{\in} M^*)$ such that $m \in \mathsf{M}^*(A)(m \overrightarrow{\in} \mathsf{M}^*(A))$
iff there exists $m \in M^*(m \overrightarrow{\in} M^*)$ such that $\mathsf{M}^*(a_i/m)(a_i) \in \mathsf{M}^*(a_i/m)(A)$
$(\mathsf{M}^*(a_i/m)(a_i) \overrightarrow{\in} \mathsf{M}^*(a_i/m)(A))$
iff there exists $m \in M^*(m \overrightarrow{\in} M^*)$ such that $\mathsf{M}^*(a_i/m)(A(a_i)) = 1$.

(10) $\mathsf{M}^*(ExA(x)) = 1$ iff for each $m \in M^*(m \overrightarrow{\in} M^*)$, $\mathsf{M}^*(a_i/m)(A(a_i)) = 1$.
This is because $\mathsf{M}^*(ExA(x)) = 1$ iff $(ExA(x)) \in \Sigma^*((ExA(x)) \overrightarrow{\in} \Sigma^*)$
iff for each $a \in M(a \overrightarrow{\in} M)$, $A(a) \in \Sigma^*(A(a) \overrightarrow{\in} \Sigma^*)$
iff for each $a^* \in M^*(a^* \overrightarrow{\in} M^*)$, $\mathsf{M}^*(A(a)) = 1$
iff for each $a^* \in M^*(a^* \overrightarrow{\in} M^*)$, $\mathsf{M}^*(a) \in \mathsf{M}^*(A)(\mathsf{M}^*(a) \overrightarrow{\in} \mathsf{M}^*(A))$
iff for each $a^* \in M^*(a^* \overrightarrow{\in} M^*)$, $a^* \in \mathsf{M}^*(A)(a^* \overrightarrow{\in} \mathsf{M}^*(A))$
iff for each $m \in M^*(m \overrightarrow{\in} M^*)$, $m \in \mathsf{M}^*(A)(m \overrightarrow{\in} \mathsf{M}^*(A))$
iff for each $m \in M^*(m \overrightarrow{\in} M^*)$, $\mathsf{M}^*(a_i/m)(a_i) \in \mathsf{M}^*(a_i/m)(A)$
$(\mathsf{M}^*(a_i/m)(a_i) \overrightarrow{\in} \mathsf{M}^*(a_i/m)(A))$
iff for each $m \in M^*(m \overrightarrow{\in} M^*)$, $\mathsf{M}^*(a_i/m)(A(a_i)) = 1$. QED.

Theorem 12.16: Assume that Σ is a set of formulas in the PIMS and A a formula in the PIMS.

(1) If Σ is harmonic, then Σ is satisfiable.
(2) If A is harmonic, then A is satisfiable.

Proof. (1) If Σ is harmonic, from Theorem 12.12, it follows that Σ can be enlarged to the potentially maximum harmonic set Σ^* in the PIMS^+, satisfying the property of existence. From Theorem 12.15, it follows that under the F^{PIN} model M^*, Σ is satisfiable.

(2) is a special case of (1). QED.

Theorem 12.17: (Completeness Theorem of F^{PIN}) Assume that Σ is a set of PIMS formulas and A a PIMS formula. Then,

(1) If $\Sigma \models A$, then $\Sigma \vdash A$; and
(2) If $\emptyset \models A$, then $\emptyset \vdash A$.

Proof. (1) If $\Sigma \nvdash A$, then $\Sigma \cup \{\neg A\}$ is harmonic. So, from Theorem 12.16, it follows that $\Sigma \cup \{\neg A\}$ is satisfiable. That is, there is a model M such that $\mathsf{M}(\Sigma \cup \{\neg A\}) = 1$. That is, there exists such a model M that satisfies $\mathsf{M}(\Sigma) = 1$ and $\mathsf{M}(\neg A) = 1$. Therefore, there exists a model M such that $\mathsf{M}(\Sigma) = 1$ and $\mathsf{M}(A) = 0$. So, $\Sigma \nvDash \mathrm{A}$.

Result (2) is a special case of (1). QED.

12.1.4 The set-theoretic foundation

Continued from the previous section, this subsection lays out the set theoretical foundation for the mathematical system of potential infinite. It represents the non-logical axiomatic part of the mathematical system of potential infinite: the axiomatic set theoretic system. At the end, the problem of consistency of this axiomatic set theory is discussed.

12.1.4.1 Basics of the PIMS-Se

In Section 9.6, the mathematical system of potential infinite is denoted as PIMS, whose logical axioms, that is, the logical calculus system, were established in both Sections 9.6 and 12.1.2. In this subsection, we will develop the non-logical axioms for the PIMS. They will make up of the axiomatic set theoretical system of the PIMS. For the sake of convenience, the logical calculus system of the PIMS will be written as PIMS-Ca, while the corresponding axiomatic set theoretical system as PIMS-Se.

In the PIMS-Se, the logical tool employed will be the PIMS-Ca. The formal language of the PIMS-Se includes the following symbols:

(1) Symbols for variables: $x_1, x_2, x_3, \ldots, x_{\vec{n}}; x'_1, x'_2, x'_3, \ldots, x'_{\vec{n}}$;
(2) Symbols for predicates: Univariate predicate symbols: *FRig*, *PSpr*;
 Bivariate predicate symbols: $\in$, $\vec{\in}$;
 Other symbols for predicates: $F_1, G_1, H_1, \ldots, F_{\vec{n}}, G_{\vec{n}}, H_{\vec{n}}$;

(3) Symbols for connectives: $\neg$, $\rightarrow$, $\wedge$, $\vee$, $\leftrightarrow$;
(4) Symbols of quantifiers: $\forall$, $\exists$, E;
(5) Symbol for equal words: $=$;
(6) Technical symbols: (,).

For convenience, we will often use x, y, z to represent an arbitrary element in $x_1, x_2, x_3, \ldots, x_{\vec{n}}$, use x', y', z' to represent an arbitrary variable symbol from $x'_1, x'_2, x'_3, \ldots, x'_{\vec{n}}$, and use u, v, w, and these symbols together with subscripts to represent arbitrary variables.

Each formula in the PIMS-Se will be written in one of the capital letters A, B, C, etc. Formulas in the PIMS-Se are defined as follows:

(1) For an arbitrary variable symbol x, $FRig(x)$ is a formula; for an arbitrary variable symbol x', $PSpr(x')$ is a formula; for n variable symbols $u, \ldots, w$, and an n-ary predicate symbol G, $G(u, \ldots, w)$ is a formula;
(2) For any variable symbols u, v, $u = v$ is a formula; for any variable symbols x, y, x', y', $x \in y$, $x' \in y$, $x \vec{\in} y'$, $x' \vec{\in} y'$ are all formulas;
(3) If A, B are formulas, so are $(\neg A)$, $(A \rightarrow B)$, $(A \wedge B)$, $(A \vee B)$, $(A \leftrightarrow B)$;
(4) If $A(u)$ is a formula, contained variable symbol u, and u appears in a formula of the form $u \in y$, then $\forall u A(u)$ is a formula. If $A(u)$ is a formula, containing variable symbol u, then $\exists u A(u)$, $EuA(u)$ are formulas;
(5) All sequences of symbols, obtaining by applying the previous rules a finite number of times, are formulas.

All rules on parentheses and omissions are the same as in the conventional set theory.

Intuitively speaking, (1) there are two classes of variable symbols in the PIMS-Se: (i) variable x for a finite rigid set, denoted $FRig(x)$, and (ii) variable x' for a potentially infinite spring set, denoted $PSpr(x')$. In the following, we will often use English letters $a, b, c, \ldots$, to represent arbitrary, finite, rigid sets, and use $\alpha, \beta, \gamma, \ldots$, to represent arbitrary, potentially infinite spring sets. And, in some Wff, when necessary, we will also sometimes use $\xi, \eta, \zeta, \ldots$, to represent finite rigid sets under certain conditions or potentially infinite spring sets under some other conditions. That is, for the sake of convenience of expression, these last symbols will be employed for flexibility in our communication. Also, we stipulate the interpretation of $FRig(x)$ as that "x is a finite, rigid set," and that of $PSpr(x')$ as that "x' is a potentially infinite, spring set."

(2) There are two constant bivariate predicates in the PIMS-Se: (i) "$\in$", which is interpreted as "belongs to", and (ii) "$\vec{\in}$", which is interpreted as "be included in". As for the scientificality and necessity of introducing the bivariate constant predicate $\vec{\in}$, please consult Subsection 12.1.4.

(3) In terms of the afore-mentioned rules for formula constructions, none of $x \in y'$, $x' \in y'$, $x \vec{\in} y$, $x' \vec{\in} y$, is a formula. That is to say, "belongs to" ($\in$) is only used to describe the relationship between a variable and a finite, rigid set, while "be included in" ($\vec{\in}$) is designed to express the connection between a variable and a potentially finite, spring set.

(4) The universal quantifier $\forall$ is only used on $FRig(a)$, because in the PIMS-Se, only $FRig(a)$ is a perfect tense. Other than this circumstance, the quantifier E will be used.

12.1.4.2 Axioms of the PIMS-Se

Axiom (0): (1) $ExFRig(x)$;

(2) $Ex'PSpr(x')$;

(3) $Eu(FRig(u) \leftrightarrow \neg PSpr(u))$;

(4) $EuEv(FRig(u) \wedge v \subseteq u \rightarrow FRig(v))$;

(5) $EuEv(PSpr(u) \wedge u \overrightarrow{\subseteq} v \rightarrow PSpr(v))$; and

(6) $Eu(FRig(u) \wedge Ev(v \in u \rightarrow G(v)) \rightarrow \forall v(v \in u \rightarrow G(v)))$.

As a matter of facts, lines (4) and (5) in Axiom (0) are not independent of each other. On the bases of other axioms, they can be derived from each other. However, for the sake of convenience, we have listed them together as axioms. This situation exists not only in this system of axioms. In fact, it also appears in other axioms. For example, the axiom of empty set is not independent of other axioms in the ZFC set theory.

Axiom (**I**) (The Axioms of Extensionality):

(1) $EaEb(\forall u(u \in a \leftrightarrow u \in b) \rightarrow a = b)$;

(2) $E\alpha E\beta(Eu(u \overrightarrow{\in} \alpha \leftrightarrow u \overrightarrow{\in} \beta) \rightarrow \alpha = \beta)$.

Definition 12.19

(1) $a \subseteq b =_{df} \forall u(u \in a \rightarrow u \in b)$;

(2) $a \subset b =_{df} a \subseteq b \wedge a \neq b$;

(3) $a \supseteq b =_{df} b \subseteq a$;

(4) $a \supset b =_{df} b \subset a$;

(5) $a \not\subseteq b =_{df} \neg a \subseteq b$;

(6) $a \not\subset b =_{df} \neg a \subset b$;

(7) $a \overrightarrow{\subseteq} \alpha =_{df} \forall u(u \in a \rightarrow u \overrightarrow{\in} \alpha)$;

(8) $a \overrightarrow{\subset} \alpha =_{df} a \overrightarrow{\subseteq} \alpha \wedge a \neq \alpha$;

(9) $\alpha \overleftarrow{\supseteq} a =_{df} a \overrightarrow{\subseteq} \alpha$;

(10) $\alpha \overleftarrow{\supset} a =_{df} a \overrightarrow{\subset} \alpha$;

(11) $a \overrightarrow{\not\subseteq} \alpha =_{df} \neg a \overrightarrow{\subseteq} \alpha$;

(12) $a \overrightarrow{\not\subset} \alpha =_{df} \neg a \overrightarrow{\subset} \alpha$;

(13) $\alpha \overrightarrow{\subseteq} \beta =_{df} Eu(u \overrightarrow{\in} \alpha \rightarrow u \overrightarrow{\in} \beta)$;

(14) $\alpha \overrightarrow{\subset} \beta =_{df} \alpha \overrightarrow{\subseteq} \beta \wedge \alpha \neq \beta$;

(15) $\alpha \overleftarrow{\supseteq} \beta =_{df} \beta \overrightarrow{\subseteq} \alpha$;

(16) $\alpha \overleftarrow{\supset} \beta =_{df} \beta \overrightarrow{\subset} \alpha$;

(17) $\alpha \overrightarrow{\not\subseteq} \beta =_{df} \neg \alpha \overrightarrow{\subseteq} \beta$; and

(18) $\alpha \overrightarrow{\not\subset} \beta =_{df} \neg \alpha \overrightarrow{\subset} \beta$.

The interpretations of these symbols are given as follows: $a \subseteq b$ is read as "*a* is contained in *b*", $a \subset b$ as "*a* is properly contained in *b*", $a \supseteq b$ as "*a* contains *b*", $a \supset b$ as "*a* properly contains *b*", $a \not\subseteq b$ as "*a* is not contained in *b*", $a \not\subset b$ as "*a* is not properly contained in b." $a \overrightarrow{\subseteq} \alpha$ is read as "a is enclosed in α", $a \overrightarrow{\subset} \alpha$ as "a is properly enclosed in α", $\alpha \overleftarrow{\supseteq} a$ as "α encloses a", $\alpha \overleftarrow{\supset} a$ as "α properly encloses a", $a \overrightarrow{\nsubseteq} \alpha$ as "a is not enclosed in α", $a \overrightarrow{\not\subset} \alpha$ as "a is not properly enclosed in α". For all other formulas, their interpretations and readings are similarly done.

At this juncture, we should notice that for $FRig(a)$, $FRig(b)$, and $PSpr(\alpha)$, $PSpr(\beta)$, that is when a, b are finite rigid sets, and α, β are potentially infinite spring sets, it is impossible to appear situations like: $a \subseteq \alpha$, $\alpha \subseteq a$, $\alpha \subseteq \beta$, $a \overrightarrow{\subseteq} b$, $\alpha \overrightarrow{\subseteq} a$, etc. Of course, it is very likely and reasonable to have a situation like $\exists x(x \overrightarrow{\in} \alpha \wedge x \in a)$.

Definition 12.20

(1) $\xi \neq \zeta =_{df} \neg(\xi = \zeta)$;
(2) $x \notin a =_{df} \neg(x \in a)$; and
(3) $u \overrightarrow{\notin} \alpha =_{df} \neg(u \overrightarrow{\in} \alpha)$.

Axiom (II) (The Axiom of the Empty Set). $\exists x \forall v(v \notin x)$.

12.1.4.3 Properties and more axioms

Theorem 12.18: $Eu(\forall v(v \notin u) \rightarrow FRig(u))$.

Proof. Assume that $\forall v(v \notin u)$. Then, it is evident that $Ev(v \in u \rightarrow v \in x_1)$. So, $u \subseteq x_1$. From Axiom (0) (1), it follows that $FRig(x_1)$. Now, Axiom (0) (4) implies that $FRig(u)$. QED.

Theorem 12.19: There is a unique set, which contains no element.

Proof. The property of existence of such a set is guaranteed by the axiom of the empty set. Now, we only need to show the uniqueness of such existence.

Assume that both u_1 and u_2 are sets containing no elements. That is,

$$\forall v(v \notin u_1) \quad \text{and} \quad \forall v(v \notin u_2).$$

Then, Theorem 12.18 implies both $FRig(u_1)$ and $FRig(u_2)$. From $\forall v(v \notin u_1)$, it follows that $Ev(v \in u_1 \rightarrow v \in u_2)$. From Axiom (0) (5), it follows that

$$\forall v(v \in u_1 \rightarrow v \in u_2).$$

From $\forall v(v \notin u_2)$, it follows that $Ev(v \in u_2 \rightarrow v \in u_1)$. Similarly, Axiom (0) (6) implies that

$$\forall v(v \in u_2 \rightarrow v \in u_1).$$

That is, we have $\forall v(v \in u_1 \leftrightarrow v \in u_2)$. Now, the axiom of extensionality (Axiom (I)(1)) implies that $u_1 = u_2$. QED.

Definition 12.21 The set, which contains no element, is denoted $\emptyset$. That is, $\forall v(v \notin \emptyset)$.

Theorem 12.18 indicates that $FRig(\emptyset)$. That is, $\emptyset$ is a finite rigid set.

Axiom (III) (The Axiom of Pairs). $EuEv\exists a\forall w(w \in a \leftrightarrow w = u \vee w = v)$.

Axiom (IV) (The Axiom of Unions: the Elementary Form).

(1) $EaEb\exists c\forall u(u \in c \leftrightarrow u \in a \vee u \in b)$;
(2) $E\alpha E\beta\exists\gamma Eu(u \vec{\in} \gamma \leftrightarrow u \vec{\in} \alpha \vee u \vec{\in} \beta)$;
(3) $EaE\beta\exists\gamma Eu(u \vec{\in} \gamma \leftrightarrow u \in a \vee u \vec{\in} \beta)$.

Axiom (V) (The Axiom of Power Sets).

(1) $Ea\exists b\forall u(u \in b \leftrightarrow u \subseteq a)$;
(2) $E\alpha\exists\beta Eu(u \vec{\in} \beta \leftrightarrow u \vec{\subseteq} \alpha)$.

Axiom (VI) (The Axiom of Subsets).

(1) $Ea\exists b\forall u(u \in b \leftrightarrow u \in a \wedge \_\_\_\_)$;
(2) $E\alpha(\exists b\forall u(u \in b \leftrightarrow u \vec{\in} \alpha \wedge \_\_\_\_) \vee \exists\beta Eu(u \vec{\in} \beta \leftrightarrow u \vec{\in} \alpha \wedge \_\_\_\_))$.

Theorem 12.20: (1) $Ea\exists! b\forall u(u \in b \leftrightarrow u \in a \wedge \_\_\_\_)$;
(2) $E\alpha(\exists! b\forall u(u \in b \leftrightarrow u \vec{\in} \alpha \wedge \_\_\_\_) \vee \exists! \beta Eu(u \vec{\in} \beta \leftrightarrow u \vec{\in} \alpha \wedge \_\_\_\_))$. QED.

Theorem 12.21: (1) For each non-empty set a, there is an unique set c that contains exactly each such element that belongs to a set that belongs to a; (2) For each non-empty set α, there exists a unique set c that contains exactly each such element that belongs to a set that belongs to α; and (3) For each non-empty set α, there is a unique set β that contains exactly each such element that is an element of a set that belongs to α.

Proof. (1) Since a is not empty, we can pick an arbitrary set b among those that belong to a. From Theorem 12.20, it follows that there exists a unique set c such that

$$\begin{aligned} c &= \{u | u \in b \wedge \forall v(v \in a \rightarrow u \in v \vee u \vec{\in} v)\} \\ &= \{u | \forall v(v \in a \rightarrow u \in v \vee u \vec{\in} v)\}. \end{aligned}$$

The arguments for (2) and (3) are similar and are omitted here. QED.

Definition 12.22 (1) Assume that $a \neq \emptyset$, write $\cap a = \{u | \forall v(v \in a \rightarrow u \in v \vee u \vec{\in} v)\}$; (2) Assume that $\alpha \neq \emptyset$, write $\cap \alpha = \{u | \forall v(v \vec{\in} \alpha \rightarrow u \in v \vee u \vec{\in} v)\}$.

Axiom (VII) (The Axiom of Unions: The Advanced Form).

(1) $Ea(\forall u(u \in a \rightarrow FRig(u)) \rightarrow \exists b \forall v(v \in b \leftrightarrow \exists u(u \in a \wedge v \in u)));$
(2) $Ea(\forall u(u \in a \rightarrow PSpr(u)) \rightarrow \exists b Ev(v \overrightarrow{\in} \beta \leftrightarrow \exists u(u \in a \wedge v \overrightarrow{\in} u)));$
(3) $Ea(\exists u(u \in a \wedge FRig(u)) \wedge \exists u(u \in a \wedge PSpr(u)) \rightarrow$
$\exists \beta Ev(v \overrightarrow{\in} \beta \leftrightarrow \exists c(c \in a \wedge v \in c) \vee \exists \gamma(\gamma \in a \wedge v \overrightarrow{\in} \gamma)));$
(4) $E\alpha(Eu(u \overrightarrow{\in} \alpha \rightarrow FRig(u)) \rightarrow \exists \beta Ev(v \overrightarrow{\in} \beta \leftrightarrow \exists u(u \overrightarrow{\in} \alpha \wedge v \in u)));$
(5) $E\alpha(Eu(u \overrightarrow{\in} \alpha \rightarrow PSpr(u)) \rightarrow \exists \beta Ev(v \overrightarrow{\in} \beta \leftrightarrow \exists u(u \overrightarrow{\in} \alpha \wedge v \overrightarrow{\in} u)));$
(6) $E\alpha(\exists u(u \overrightarrow{\in} \alpha \wedge FRig(u)) \wedge \exists u(u \overrightarrow{\in} \alpha \wedge PSpr(u)) \rightarrow$
$\exists \beta Ev(v \overrightarrow{\in} \beta \leftrightarrow \exists c(c \overrightarrow{\in} \alpha \wedge v \in c) \vee \exists \gamma(\gamma \overrightarrow{\in} \alpha \wedge v \overrightarrow{\in} \gamma))).$

Definition 12.23 For a set u, its successor (set) u^+ is defined as

$$u^+ = u \cup \{u\}.$$

Definition 12.24 A set u is an inductive set, denoted $Ind(u)$, iff

(1) $\emptyset \overrightarrow{\in} u$; and
(2) $Ev(v \overrightarrow{\in} u \rightarrow v^+ \overrightarrow{\in} u)$.

Axiom (VIII) (The Axiom of Infinity). $\exists u(\emptyset \overrightarrow{\in} u \wedge Ea(a \overrightarrow{\in} u \rightarrow a^+ \overrightarrow{\in} u))$.

What this axiom implies is that we unconditionally recognize the existence of potentially infinite inductive set which starts from $\emptyset$.

Theorem 12.22: If u is a non-empty inductive set, then so is $\cap u$. QED.

Theorem 12.23: There is a unique inductive set, which is contained in every inductive set. QED.

Definition 12.25 $Eu(Ind(u) \rightarrow \omega \overrightarrow{\subseteq} u)$.

Definition 12.26 $Eu(\omega \overrightarrow{\subseteq} u \leftrightarrow PSpr(u))$.

In the PIMS-Sc, since a is a finite rigid set, its successor a^+ is still a finite rigid set. So, the successor of a^+, $a^{++} = a^+ \cup \{a^+\}$, is also a finite rigid set. Reasoning along this line, we see that $\overbrace{a^{++\cdots+}}^{n \text{ times}}$ is always a finite, rigid set. On the other hand, starting from a, we can continue to construct the successor, the successor of the successor, ..., without any ending in sight. And, this procedure of constructing successor is of course a listing procedure. In the PIMS, since we can entirely exclude actual infinities, any listing procedure in this system cannot be exhausted. That is, in the PIMS, there is not such a possibility of completing an infinite listing procedure. Therefore, the listing procedure of indefinitely constructing successors from a produces a potentially infinite spring set of the following form:

$$\{a, a^+, a^{++}, \ldots, a^{\overbrace{++\ldots++}^{\tilde{n} \text{ times}}}\}$$

In the system PIMS-Se since we have $FRig(\emptyset)$, similar to the discussion above, we can also obtain the following spring set, starting from $\emptyset$:

$$\{\emptyset, \emptyset^+, \emptyset^{++}, \ldots, \emptyset^{\overbrace{++\ldots+}^{\tilde{n} \text{ times}}}\}$$

Now, we assume that we are given a potentially infinite spring set α. Analogous to what is done above, we can define the successor α^+ of α by $\alpha^+ = \alpha \cup \{\alpha\}$. Since α is a potentially infinite spring set, its successor α^+ is also a potentially infinite spring set. To summarize, we should be able to obtain such potentially infinite spring sets as

$$\alpha, \alpha^+, \alpha^{++}, \ldots, \alpha^{\overbrace{++\ldots++}^{n \text{ times}}}, \text{ etc.}$$

Similar to above, starting from α, we can produce the following spring set:

$$\{\alpha, \alpha^+, \alpha^{++}, \ldots, \alpha^{\overbrace{++\ldots+}^{\tilde{n} \text{ times}}}).$$

Definition 12.27 (1) If a potentially infinite spring set α satisfies: (i) $\emptyset \overset{\rightarrow}{\in} \alpha$, and (ii) $Ea(a \overset{\rightarrow}{\in} \alpha \rightarrow a^+ \overset{\rightarrow}{\in} \alpha)$, then α is called a potentially infinite inductive set, starting from $\emptyset$.

(2) If a potentially infinite spring set α satisfies: (i) $a \overset{\rightarrow}{\in} \alpha$, and (ii) $Eb(b \overset{\rightarrow}{\in} \alpha \rightarrow b^+ \overset{\rightarrow}{\in} \alpha)$, then α is called a potentially infinite inductive set, starting from a.

(3) If a potentially infinite spring set α satisfies: (i) $\beta \overset{\rightarrow}{\in} \alpha$, and (ii) $E\gamma(\gamma \overset{\rightarrow}{\in} \alpha \rightarrow \gamma^+ \overset{\rightarrow}{\in} \alpha)$, then α is called a potentially infinite inductive set with initial element β.

In the following, we will specifically name the condition $Ea(a \overset{\rightarrow}{\in} \alpha \rightarrow a^+ \overset{\rightarrow}{\in} \alpha)$ and $E\gamma(\gamma \overset{\rightarrow}{\in} \alpha \rightarrow \gamma^+ \overset{\rightarrow}{\in} \alpha)$ as successive inclusion.

Axiom (IX) (The Axiom of Choice).

(1) $Ea(a \neq \emptyset \wedge \forall u(u \in a \rightarrow u \neq \emptyset) \wedge \forall u \forall v(u \in a \wedge v \in a \wedge u \neq v \rightarrow u \cap v = \emptyset)$
$\rightarrow \exists b \forall w(w \in b \leftrightarrow \exists u(u \in a \wedge (w \in u \vee w \overset{\rightarrow}{\in} u) \wedge u \cap b = \{w\})))$.

(2) $E\alpha(Eu \overset{\rightarrow}{\in} \alpha \rightarrow u \neq \emptyset) \wedge EuEv(u \overset{\rightarrow}{\in} \alpha \wedge v \overset{\rightarrow}{\in} \alpha \wedge u \neq v \rightarrow u \cap v = \emptyset)$
$\rightarrow \exists \beta Ew(w \overset{\rightarrow}{\in} \beta \leftrightarrow \exists u(u \overset{\rightarrow}{\in} \alpha \wedge (w \in u \vee w \overset{\rightarrow}{\in} u) \wedge u \cap \beta = \{w\})))$.

Axiom (X) (The Axiom of Replacement).

(1) $Ea(\forall u \forall v_1 \forall v_2(u \in a \wedge \varphi(u, v_1) \wedge \varphi(u, v_2) \rightarrow v_1 = v_2)$
$\rightarrow \exists b \forall v(v \in b \leftrightarrow \exists u(u \in a \wedge \varphi(u, v))))$,
where b does not appear in $\varphi(u, v)$).

(2) $E\alpha(\forall u \forall v_1 \forall v_2(u \overset{\rightarrow}{\in} \alpha \wedge \varphi(u, v_1) \wedge \varphi(u, v_2) \rightarrow v_1 = v_2)$
$\rightarrow \exists \beta Ev(v \overset{\rightarrow}{\in} \beta \leftrightarrow \exists u(u \overset{\rightarrow}{\in} \alpha \wedge \varphi(u, v))))$,
where b does not appear in $\varphi(u, v)$).

Axiom (XI) (The Axiom of Regularity)

(1) $Ea(a \neq \emptyset \rightarrow \exists u(u \in a \wedge u \cap a = \emptyset))$;
(2) $E\alpha\exists u(u \overrightarrow{\in} \alpha \wedge u \cap \alpha = \emptyset)$.

Definition 12.28 The interpretation A^* of a formula A is generated by the following:

(1) $(FRig(x))^* = x$ is a variable symbol without an apostrophe;
(2) $(PSpr(x'))^* = x'$ is a variable symbol with an apostrophe;
(3) $(G(u, \ldots, w))^* = G(u, \ldots, w)$;
(4) $(u = v)^* = u = v$;
(5) $(x \in y)^* = x \in y$;
(6) $(x' \in y)^* = x' \in y$;
(7) $(x \overrightarrow{\in} y')^* = x \in y'$;
(8) $(x' \overrightarrow{\in} y')^* = x' \in y'$;
(9) $(\neg A)^* = (\neg A^*)$;
(10) $(A \rightarrow B)^* = (A^* \rightarrow B^*)$;
(11) $(A \wedge B)^* = (A^* \wedge B^*)$;
(12) $(A \vee B)^* = (A^* \vee B^*)$;
(13) $(A \leftrightarrow B)^* = (A^* \leftrightarrow B^*)$;
(14) $(\forall u A(u))^* = \forall u(A(u)^*)$;
(15) $(\exists u A(u))^* = \exists u(A(u)^*)$; and
(16) $(EuA(u))^* = \forall u(A(u)^*)$;

Theorem 12.24: If A is a theorem in the PIMS-Se, then A^* is a theorem in the ZFC axiomatic set theory system.

Proof. It suffices to apply induction on the length of proof for the theorem A in the PIMS-Se. All the details are omitted. QED.

Theorem 12.25: If the ZFC axiomatic set theory system is consistent, then the potentially infinite set theory system PIMS-Se is also consistent.

Proof. Assume that the potentially infinite set theory system PIMS-Se is not consistent. Then, there exists a formula A such that both A and $\neg A$ are theorems in the PIMS-Se system. From Theorem 12.24, it follows that A^* and $\neg A^*$ are theorems in the ZFC axiomatic set theory system. That is, the ZFC system is not consistent. QED.

12.1.5 Some comments

As is known, because of the appearance of paradoxes in the naive set theory, modern axiomatic set theory was developed. However, in terms of the progress on the problem of consistency accomplished by the axiomatic set theory, the work can be summarized in two aspects as follows:

(1) For all the historically existing paradoxes of the two-value logic, corresponding interpretations can be obtained within the axiomatic set theory system. That is, none of these known paradoxes can appear within the new theory system.
(2) As of this writing, it has not been shown theoretically that there will not be any new paradoxes in the modern axiomatic set theory system. On the other hand, as of this writing, all existing and known paradoxes of the logic can be classified into three types: (A) Two-value logical paradoxes; (B) Multi-value logical paradoxes, including any n-value logical paradoxes, $3 \leq n < \omega$, and infinite-value logical paradoxes; (C) paradoxes on different concepts of infinities, such as the implicit contradiction between potential infinities = actual infinities and potential infinities ≠ actual infinities, as pointed out in Section 7.4, the shadow of the Berkeley paradox in the theory of limits in Section 9.5, and the inconsistencies, existing in the concepts of countable and uncountable infinite sets, as discussed in Sections 9.1 and 9.2.

At this juncture, we can clearly point out that all the historically known paradoxes of the two-value logic will not reappear in the PIMS. Otherwise, assume that there is a historically known paradox of the two-value logic, which appears in the PIMS. Then, from Theorem 12.25, it follows that this paradox has to appear in the modern axiomatic set theory. This end contradicts the fact that no such two-value logical paradox reappears in the modern set theory system.

Also, since the logical tool, established purposely for the PIMS, is a modified version of the two-value logical calculus, in the framework of the two-value logic, there is no need to interpret any of the finite-valued ($3 \leq \mathrm{n} < \omega$) and the infinite-valued paradoxes. Because no actual infinity is allowed in the PIMS, the problem of whether or not potential infinities equal actual infinities does not exist. Also, the problem of juxtaposing affirmed perfect tense (⊤) and negated perfect tense ($\not\top$) does not exist, either. Therefore, all the paradoxes related to the concepts of infinites, as discussed in Sections 8.3, 9.1, and 9.2, will not appear in the PIMS. That is, the achievement on the problem of consistency made in the PIMS can be seen as no worse than in the modern axiomatic set theory system.

12.2 PROBLEM OF INFINITY BETWEEN PREDICATES AND INFINITE SETS

This section, which is based on (Lin (guest editor), 2008, pp. 526–533), reconsiders the feasibility at both the heights of mathematics and philosophy of the statement that each predicate determines a unique set.

In the naïve and the modern axiomatic set theories, it is a well-known fact that each predicate determines precisely one set. That is to say, for any precisely defined predicate P, there is always $A = \{x|P(x)\}$ or $x \in A \to P(x)$. However, when we are influenced by the thinking logic of allowing both kinds of infinites and compare these two kinds of infinites, and potentially infinite and actually infinite intervals and number sets, it is found that the expressions of these number sets are not completely reasonable.

12.2.1 Expressions for a variable to approach its limit

As is known, we often use [a, b] to stand for a closed interval, where the endpoints a and b are included in the interval. The symbol (a, b) is often employed to denote an open interval with the endpoints a and b excluded from the interval. In order to be more intuitive and more to the point to reflect the fact that the endpoints a and b are excluded from the interval, the symbol for open interval (a, b) should be written as a(,)b. As a matter of fact, in a certain sense, any open interval should not have any endpoints. To summarize, "b]" means that the point b is included, while ")b" for the notation that b is excluded. We stipulate that if a variable x indefinitely approaches its limit b and will eventually reach the limit b (done), then we say that the variable x approaches b in the form of actual infinity. Now, if a variable x approaches indefinitely to its limit b and will never reach the limit b (ongoing), then we will say that the variable x approaches its limit b in the form of potential infinity. Therefore, in terms of the form in which a variable x approaches its limit, it approaches its limit either in the form of potential infinity or actual infinity.

Let us now analyze the situation for a variable x to approach b indefinitely from within the closed interval [a, b] along the x-axis in the positive direction. Since b is a point of the interval, the variable x can not only indefinitely approach its limit b, but also reaches its limit b from within the interval. However, when the variable x approaches its limit b in the open interval a(,)b along the positive direction of the x-axis, even though the variable x can indefinitely approach b and has b as its limit, since b is located outside the open interval a(,)b, the variable x can never, ever reach its limit b within its open interval domain. That is, the variable x can only in the form of potential infinity approach its limit b from within the interval. That is, when the interval is closed, the process for the generating variable x to approach its limit is a perfect tense. When the interval is open, the process for the generating variable x to approach its limit b is a present progressive tense.

That is, the method of expression "b]" represents not only the fact that b is in the interval and so in turn it is reachable, but also the closeness of the interval and a perfect tense. As for the method of expression ")b," on one hand, it reflects the fact that the limit b is located outside the interval so that it cannot be reached. On the other hand, it shows the openness of a present progressive tense. So, it is reasonable to employ "b]", with b included in the inner side of the interval bracket, and ")b", with b located in the outsider side of the interval parenthesis, as methods of representation for actual and potential infinities. They respectively reflect the actual infiniteness and the potential infiniteness. Conversely, if we used the symbols "]b" and "b)" for the purpose of representing either kind of infinity, it does not seem reasonable. It is because the first symbol (with b located outside the bracket) reflects a contradiction with a closed perfect tense (done) and an unreachable outside target (ongoing), while the second symbol (with b located inside the parenthesis) shows the contradiction of an open progression (ongoing) and an internal reachability (done).

In terms of intervals, other than open intervals like a(,)b and closed intervals like [a, b], as discussed above, we of course have such intervals as half open one, such as [a,)b, infinite closed intervals like [a,+∞], infinite half open intervals like [a,)+∞, etc.

In the following, on the basis of the expressions for the two kinds of infinite intervals, we will further our discussion on the representations of number sets of the two kinds of infinites. In Section 9.3, we have talked about the actually infinite, rigid set of natural numbers $N = \{x|n(x)\} = \{1, 2, \ldots, n, \ldots\}\omega$ and the potentially infinite, spring set of natural numbers $\mathrm{N} = \{x|n(x)\} = \{1, 2, \ldots, \vec{n})\omega$. So, the reasonable correspondence between our notation of intervals and that of number sets should be: The bracket of the intervals "]" should correspond to the number set braces "}". In fact, both of these symbols represent closed, perfect tense. And, the parenthesis ")" of the intervals corresponds to the parenthesis ")" of the number set. Both of these symbols, which happen to be the same, stand for openness and progressive tense. What is left now is the symbol +∞ of the intervals, that should correspond to the symbol ω of the number sets. Both of them represent the limit of their own perspective variables.

At this juncture, we notice that in the notation of the actually infinite, rigid set of natural numbers $N = \{x|n(x)\} = \{1, 2, \ldots, n, \ldots\}\omega$, as well-studied in the naïve set theory and the modern axiomatic set theory, the symbol "}ω" corresponds exactly to that "]+∞" of the notation of intervals. However, as discussed above, the notation "]+∞" is not reasonable, since this notation stands for the contradiction between "a closeness and perfect tense" and "an external unreachability." Therefore, in the conventional notation of the number set $\{1, 2, \ldots, n, \ldots\}\omega$, the symbol "}$\omega$" is not a reasonable notation, instead a self-contradiction. This fact reflects from a different angle the inconsistency of the actually infinite, rigid set of all natural numbers, see Section 9.1, 9.3, and 9.6. However, some conventional rules of thinking implicitly employed in both the naive and the modern axiomatic set theories have forced people to accept such unreasonable notation as "}ω". It is because in these set theories, the following two well-known theorems exist:

Proposition 12.1 There are ω many pairwisely different naturally numbers in the set N, consisting of all natural numbers. QED.

Proposition 12.2 Each natural number is a finite ordinal number. So, each natural number is less than ω. In symbols, for each natural number n, $n < \omega$. QED.

So, from Proposition 12.1, it follows that the set N of all natural numbers is a completed, actually infinite rigid set. Hence $N = \{x|n(x)\} = \{1, 2, \ldots, n, \ldots\}$ must close on the right-hand end with a brace "}", which cannot be replaced by a parenthesis ")", as used in the notation of the potentially infinite, spring set of natural numbers $\mathrm{N} = \{x|n(x)) = \{1, 2, \ldots, \vec{n})$. Next, since the transfinite ordinal number ω is not finite, Proposition 12.2 implies that there should be a ω placed outside the ending brace so that one ends up with a combined notation "}ω", an unreasonable notation which represents the contradiction between a closed perfect tense and an external unreachability. That is why in the naïve and the modern axiomatic set theories, it is a must to have appeared such a self-contradictory symbol as "}ω".

To this end, we have once again confirmed the correspondence between the representation of the actually infinite perfect tense for a variable x to approach its limit b from within the closed interval, and the reasonable notation $N=\{1,2,\ldots,n,\ldots,\omega\}$ of the actually infinite, rigid set of natural numbers. Also, we confirmed that corresponding to the representation of the potentially infinite progressive tense for a variable x to approach its limit b from within the open interval a(,)b should be the reasonable notation $\mathrm{N}=\{1,2,\ldots,\vec{n})\omega$ for the potentially infinite spring set of natural numbers. It is exactly because of the transition from the form of actual infinity for a variable to approach its limit to the form of potential infinity for a variable to approach its limit, as seen from the methods of representation of the transition process, that it is a change from a bracket "}" to a parenthesis ")", and a dislocation of the limit point b moved from inside the bracket to the outside of the parenthesis. So, from the previous comparative analysis between the methods of representation of intervals and number sets, it follows that when we take out the symbol ω from the reasonable notation $N=\{1,2,\ldots,n,\ldots,\omega\}$ of the actually infinite, rigid set of natural numbers, and place it outside the right-side brace, the brace "}" must be transformed into a parenthesis ")". At the same time, the method of actual infinity for the variable n to approach its limit ω has been transformed to the method of potential infinity for the variable to approach its limit. Therefore, we can draw the following important result from our comparative analysis between the methods of representation of intervals and number sets:

Proposition 12.3 When we take out ω from the reasonable representation $N=\{1,2,\ldots,n,\ldots,\omega\}$ for the actually infinite, rigid set of natural numbers, the set N must be transformed to the potentially infinite, spring set $\mathrm{N}=\{x|n(x))=\{1,2,\ldots,\vec{n})$ of natural numbers, which can no longer be the actually infinite, rigid set $N=\{x|n(x)\}=\{1,2,\ldots,n\ldots\}$ of natural numbers, as conventionally believed.

This proposition is obtained in the comparative analysis of the methods of representation of intervals and number sets. As a matter of fact, the validity of the conclusion in Proposition 12.3 can also be produced from an analysis of the common laws in epistemology and scientific philosophy. It is because for each natural number n, n is a finite ordinal number. And ω is a transfinite ordinal number. So, both n(finite) and ω (transfinite) represent a pair of contrary opposites. Since each actual infinite must be a perfect tense and a perfect tense stands for a completed transition from the opposite. In particular, for our analysis it represents the completed transition from finiteness (n) to transfiniteness (ω). This completed transition must reflect the formation of $N=\{1,2,\ldots,n,\ldots,\omega\}$. Now, if ω is taken out from N, it means that we will reverse the perfect tense back to an imperfect tense, which is a present progressive tense. So, we have to convert N, which reflects the actually infinite, rigid set N of natural numbers, back to N, which represents the potentially infinite, spring set of natural numbers.

However, according to the traditional convention, the intension of taking out the transfinite number ω from N is completely identical to that of taking out a finite number n from N. So, the commonly accepted practice is that we can order natural numbers from the smallest to the larger and continue this ranking process until reaching ω. So, all natural numbers are ordered, which is of course a perfect tense. After that, we remove ω from the list, leaving all the natural numbers in the list. That is why what is left is still the actually infinite, rigid set of natural numbers. This end is completely opposite and inconsistent of the conclusion in Proposition 12.3. This philosophical analysis is valid, and the comparative analysis between number sets and intervals is extremely intuitive and natural.

Of course, intuitions might not be enough to be used as evidence. What we need is the outcome of correct theoretical reasoning. However, by glancing through the history of science, which system of rational thinking did not contain realistic backgrounds and objective models that were based on intuitive, empirical thinking? The most of the difference between different systems of rational thinking is that the realizations of the base, intuitive, and empirical reasoning are seen at different levels. In particular, sometimes, the rational thinking of a certain level is used as the intuitive model of the rational abstraction of a higher level. Of course, as seen from the angle of reasoning, this comparison and intuitive analysis are indeed very superficial.

In terms of closed and open intervals, the foundation of our intuitive thinking is only about how a variable x moves continually along the positive direction of the x-axis toward b from within the given interval. When the interval is closed, the point b is part of the interval so that the variable x can move straight to and arrive at point b. When the interval is open, the limit point b is located outside the interval. So, the variable x can certainly get near point b indefinitely, but will never arrive at point b. It is the same situation for the actually infinite, rigid set of natural numbers, and the potentially infinite, spring set of natural numbers. Imagine that we walk along the natural numbers from the smallest one to the larger ones. Let us arrive at ω first and then delete the transfinite ω. The actually infinite, rigid set of natural numbers is traditionally constructed this way. As a matter of fact, when we truly rethink about the situation from the angle of rational thinking, we can see that no matter whether the variable x moves toward its limit point b or the natural number variable n towards the limit ω, a transfinite ordinal number, it is an extremely complicated process. And after n has arrived at ω and then looks back to remove ω, it is absolutely not as easy as we imagine. This is also a very complicated matter. All the complications involved in these two imagined acts will be addressed specifically in a future occasion at the height of rational reasoning.

12.2.2 Actually infinite rigidity of natural number set and medium transition

As is known, in the medium logical calculus system and the medium axiomatic set theory system, both of which will be written as the medium system in the rest of this section, the formal symbol ╕ is introduced for the contrary negation. It is interpreted and read as "contrary to". The fuzzy negation ~ is also introduced and interpreted as "partially". So, for any given predicate P, the contrary opposite of P is written as ╕P. For example, let us be given the predicate P: "man". Then, its contrary opposite "woman" should be written as ╕P. Ordinarily, P(x) stands for the fact that the object x completely satisfies the predicate P. So, ~P(x) stands for the object x to partially satisfy P, and ╕P(x) the object x completely satisfies the contrary opposite of P. Besides, in the medium system, for any given contrary opposites P and ╕P, if there is such an object x that partially satisfies P and partially satisfied ╕P, then the object x is called a medium object of the contrary opposites, denoted ~P(x).

Also, in the establishment and development of the medium system, the spirit of the "medium principle" has to be implicitly carried on throughout the work. The so-called medium principle says that unconditionally accept the objective existence of such contrary opposites that have medium object(s). On the other hand, the medium principle does not say that all contrary opposites have medium objects. It only means that some contrary opposites might not have medium objects. As a matter of fact, there is such a commonly known and important principle in epistemology and scientific philosophy that implies: During the mutual transition between two contrary opposites, one has to pass through such a medium state that involves both sides of the opposites. For example, the dawn is the medium state of the transition from the dark night to the bright daylight. Zero (0) is such a neutral number that is both negative and positive. Here, we only list two such examples among many others. That is, this commonly accepted and important principle of epistemology is the philosophical background for the medium principle, introduced in the medium system. Conversely, the concept of medium objects of contrary opposites, as studied of the medium system, is a practical realization in the medium system of the medium states of the epistemology.

Now, let us stand on the medium principle as our footing to analyze the complicated process for a variable x to move to its limit b in an interval and a variable n to travel to its limit ω in the set N of natural numbers. In fact, any finite ordinal number n and the transfinite ordinal ω are a specific model for the pair of contrary opposites: "finite" and "actually infinite". So, when seen from the angle of medium transition, without passing through a medium state, how can one side be transformed into the opposite side? And, what could be the medium objects between the finite and transfinite ordinal numbers?

First, in terms of philosophy, potential infinities are the medium between the contrary opposites of finiteness and actual infiniteness. It is because each potential infinity is not finite, instead it possesses the possibility of materializing an actual infinity. So, it satisfies part of the properties of actual infinities. In the contrary, since each potential infinite is forever a present progressive tense, it will never reach an actual infinity. So it does not completely go beyond the concept of finiteness. And, it also possesses partially the properties of finiteness. Secondly, speaking mathematically, the varying, ending element $\vec{n}$, which can increase indefinitely in the potentially infinite, spring set $\mathrm{N} = \{x|n(x)) = \{1, 2, \ldots, \vec{n})$ of natural numbers, is the medium state between the finite ordinal number n and the transfinite ordinal number ω. Since $\vec{n}$ can increase without any bound, it can satisfy partially the properties of transfinite numbers. However, in the process for $\vec{n}$ to increase indefinitely, it is required to satisfy the property that $\vec{n} < \omega$. So, $\vec{n}$ must also partially satisfy the properties of finite ordinal numbers.

However, in the naïve and modern axiomatic set theories, the thinking mode that each predicate determines exactly one (actually infinite, rigid) set completely simplifies the phenomenon of medium transition between two contrary opposites. Of course, that is an inevitable consequence of the thinking method of the classical two-value logic. At the same time, it is under the influence of this simplified thinking method that the representation $\{x|\ \mathrm{P}(x)\}$ of actually infinite sets, as seen as meaningless by (Robinson, 1964), becomes conventionally accepted and widely used. It is exactly because the thinking of medium transition between opposites cannot be organically implemented that the rationality of creating the actually infinite, rigid set N of natural numbers by "letting the natural number variable n travel to ω and then removing ω" is accepted.

However, when seen from the angle of medium transition, the actual situation is way more complicated. As a matter of fact, in that completed, perfect actually infinite set of natural numbers, the inclusion and exclusion of ω can no longer be seen similarly to the inclusion and exclusion of an ordinary natural number n. On the other hand, in the naïve and modern axiomatic set theories, the principle that each predicate determines exactly one set is mainly reflected in the following rule of thinking: For any given set-building predicate P, the collection of all the objects that satisfy this predicate can be organized into a whole or set, denoted $\{x|\ P(x)\}$. And, as soon as this set is formed, it becomes a ready, independent object for investigation with such a regulation of its identity that this independently existing entity, called a set, consists purely of the objects satisfying the property P. However, the scientific philosophy believes that when the boundary of an object is stipulated, the boundary has been surpassed. That is, when we affirm the stipulation about the content of an entity, we have already denied the rule. So, the absolute purity does not exist.

Therefore, when we collect all the objects with the property P to form a single object for investigation, we have already gone beyond those rules of contents of this individual object. For example, in terms of such a class of objects as natural numbers, it cannot be a simply self-closed system while surpassing its own boundary. That is, when we collect all such objects as natural numbers together to form the sequence λ, where all natural numbers are ordered according to their natural magnitudes, we have already gone beyond the rules about the essence of those natural numbers (finite ordinal numbers) that form the sequence λ. Based on this principle of recognition, as soon as the sequence λ is formed, there must be such quantitative object(s), different of natural numbers, that have been included. However, Cantor himself and other followers did not realize this problem and stayed with the traditional concept of sets $\{x|\ P(x)\}$.

What needs to be reminded here is that what we are talking about is the actually infinite, rigid sets. That is, in order to gather infinitely many entities to form a completed object for investigation, one cannot list the entities one by one as when he constructs a finite set. And, he cannot, either, stay forever in the process of accepting more entities as when he constructs a potentially infinite, spring set. So, he has to complete his task of gathering all the needed entities through introducing rules. However, by using this approach, the existence of the desired set can only be constructed by affirming the rules on contents and at the same time by rejecting these rules. Therefore, the root cause of the one-sidedness existing in modern system of mathematics is the phenomenon of looking at rigid infinite sets as objects of investigations without considering the philosophical issue of how entities are gathered into individual objects.

In a next section, we will further investigate the intension and structure of actually infinite, rigid sets on the basis of the afore-mentioned principle of scientific philosophy and the thinking method of medium transitions.

12.3 INTENSION AND STRUCTURE OF ACTUALLY INFINITE, RIGID SETS

Continuing the discussion in the previous section, this section, which is based on (Lin (guest editor), 2008, pp. 534–542), further studies the problem of infinity existing in between predicates and sets. Consequently, the conventional rule of thinking that each predicate determines a unique set will be modified, and a principle regarding the relationship between predicates and sets will be established. At the end, the structures of actually infinite, rigid sets will be investigated.

12.3.1 Relation between predicates and sets in infinite background world

In order to understand the relationship between predicates and sets in an infinite background world in the framework of medium transitions, let us first start with the specific predicate P: "natural number." Then, we abstract this particular predicate to an arbitrary predicate P and generalize the precise set-building predicates in Cantor's sense to fuzzy predicates. Besides, we will generalize the specific contrary opposites, finite and transfinite ordinal numbers, to the general contrary opposites: P and ╕P.

As is known, in the medium system, the formal definitions of precise and fuzzy predicates are given (Zhu and Xiao, 1988), and it is shown that these two kinds of predicates satisfy the law of excluded middle. (It should be noticed that the medium principle does not claim that each pair of contrary opposites has a medium.) So, for any given predicate P, it is either a precise predicate or a fuzzy predicate. In terms of a precise predicate P, there does not exist such an object x that satisfies ~P(x). And, in terms of a fuzzy predicate P, there must exist such an object x that partially satisfies the predicate P, that is, ~P(x). Figure 12.1 shows the structure of all predicates.

In the medium system, the formal symbol ~ is known as the fuzzy negation, which is interpreted as "partially", and the formal symbol ╕is known as the contrary negation, that is read as "contrary to". So, $P(x)$ means that the object x satisfies or possesses the property P completely, ~P(x) partially, and ╕P satisfies or possesses the property that is contrary to P. Hence, for any given pair of contrary opposites (P, ╕P), if there exists an object x that satisfies $\sim P(x) \wedge P(x)$, then x is called a medium object of (P, ╕P).

At this juncture, there is one point we need to pay our attention to. Even though there are formal definitions of precise and fuzzy predicates in the medium system, for any given predicate P, it is an empirical experience or intuition that helps to determine whether this predicate is precise or fuzzy. For example, for the predicate P = "natural number" (or "finite ordinal number"), the conventional belief is that this predicate is precise. It is because, intuitively speaking, it seems to be impossible for us to find such an object that partially possesses the property of being a "natural number". Or in other words, under the traditional thinking logic of the standard analysis, in terms

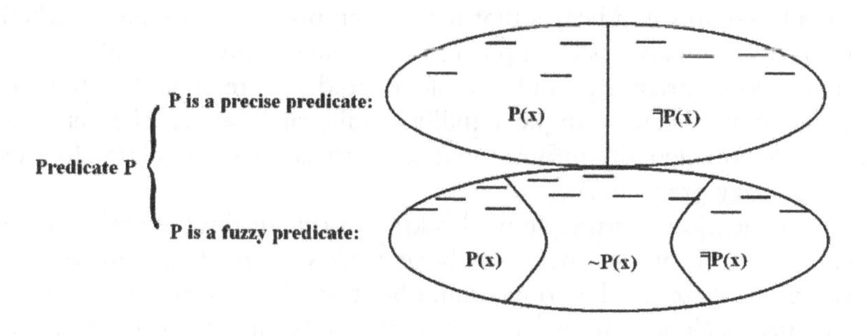

Figure 12.1 Classifications of predicates.

of the predicate P = "natural number", it is a common belief that there is no x such that ~P(x). Now, let us define the predicate P: "handsome man." Immediately, we all recognize that it is a fuzzy predicate, since in real life, there must be such an object x that satisfies ~P(x). All these judgments in fact come from intuition or empirical experience. However, when seen from the angle of rational thinking, for the predicate P of "natural number" or "finite ordinal number", how can we be so sure that there is no such an object x that ~P(x)? As a matter of fact, when our rational thinking continually expands and becomes more abstract, we just might discover such objects x that ~P(x) holds true. For example, in the nonstandard analysis, after the field R of real numbers is enlarged to *R, there are objects x that make ~P(x) true. So, on the level of rational thinking, what is more important is to study the situation of fuzzy predicates. In fact, only when we do not need to consider ~P(x) in solving a real-life problem, we can determine the predicate P as a precise predicate. On the other hand, assume that P is a predicate. Then, P, ~P, and ⌉P can be seen as independent predicates according to the specific circumstances involved. For example, if the predicate P: "man", is not defined, then ⌉P can be seen as an independent predicate P': "woman", and ~P another independent predicate P'': "neutral human being."

As is known, in the naïve and the modern axiomatic set theories, there is such a well-known principle: Each predicate determines a unique set. And, in these set theories, each set-building predicate is precise. That is, fuzzy predicates are not part of the study and not considered for set building. Besides, each infinite set is a perfect actually infinite set. That is, in there, such concept as potential infinities is not talked about. And of course, such entities as potentially infinite, spring sets are nonexistent. However, in our work here, we talk about not only potentially infinite, spring sets, but also actually infinite, rigid sets. Of course, in terms of finite sets, they are all rigid sets. At this juncture, we earnestly point out that any finite set does not belong to our discussion below. That is, all sets considered in the following are either potentially infinite spring sets or actually infinite rigid sets. It is our goal to completely change the previously mentioned well-known principle of set theory and establish a new principle on the relationship between predicates and sets. The intension of this new principle contains the following two points:

(1) Each precise predicate can only determine one potentially infinite spring set.
(2) Each fuzzy predicate cannot only define an actually infinite rigid set, but also a potentially infinite spring set.

What is listed above indicates that for a given predicate, no matter whether it is precise or fuzzy, it can always be applied to determine a potentially infinite spring set. However, no precise predicate can be employed to define an actually infinite rigid set. Only with fuzzy predicates, unique actually infinite rigid sets can be determined. Or in other words, any actually infinite rigid set must be uniquely defined by using the meaning of a fuzzy predicate.

The afore-mentioned principle or this kind of rule of thinking is reasonable. For example, in terms of the statement (1) above, since we already assume that the predicate P is precise, there will be no medium object for the pair of contrary opposites P and ⌉P. Because each actually infinite rigid set stands for a perfect tense, the perfect tense has to emphasize on the transition from one opposite to the other. Therefore, from one of the common laws of epistemology and scientific philosophy, it follows that the mutual transition between the opposites can only been accomplished through the medium. So, for precise predicates without any medium, no such mutual transition can be completed. Therefore, any precise predicate cannot be employed to define an actually infinite rigid set. Since potentially infinite spring sets are forever a present progressive tense, to any potentially infinite process, there does not exist the problem of transition from one opposite side to the other. The only requirement is that those objects, satisfying a predicate P completely, can be collected one by one into the corresponding spring set. So, there is not a single problem about using a precise predicate to determine a potentially infinite spring set. And, in terms of the statement (2) above, if P is a fuzzy predicate, then there must exist such a medium phenomenon for the contrary opposites P and ˥P, where the medium reflects both of the opposites. That is, there exists an object x such that ~P(x). So, a transition from P to ˥P through ~P can be accomplished. That is, using a fuzzy predicate P to determine a completed, perfect actually infinite rigid set can be materialized. On the other hand, we can also employ this fuzzy predicate P to determine a potentially infinite spring set. It is because in this case, we simply do not think about the mutual transition between P and ˥P. For this fuzzy predicate P, it is possible to realize that all those objects satisfying P completely are included one by one into the corresponding spring set. This is a process that needs to be done first and that can be done without any doubt. In short, such a constructive progressive process can be materialized for any predicate in any infinite background world.

In the rest of this section, we will consider two different background conditions in our study of the structure of actually infinite rigid sets. The first condition is to develop our most general discussion without being constrained by the medium system. The second is to evolve our discussion in the framework of the medium system. The essential difference between these two conditions is whether or not precise predicates or fuzzy predicates satisfy the law of excluded middle, because in the medium system, these two kinds of predicates are shown, as a theorem, to satisfy the law of excluded middle. That is, this conclusion is predetermined when the logical and non-logical axioms of the medium system are initially established. And, the medium principle does not claim that each pair of contrary opposites has a medium.

12.3.2 Constructive mode of actually infinite rigid sets without constraints

Let us first develop our most general discussion under no constraints.

Since each actually infinite rigid set must be a perfect tense, one has to complete the transition from the set-building predicate P to ˥P. To do so, the transition process has to go through the medium ~P between P and ˥P. On the other hand, since what we talk about is on the problem of infinite, we also have to complete the transition from a potential infinity to the eventual actual infinity. Since potential infinite and actual infinite are also a pair of contrary opposites, to complete such a transition from one opposite to the other, a medium between the two kinds of infinites has to be gone through.

Let us use б to represent the medium. So, the symbol б also represents the abstract medium state existing in the process of transition from a potentially infinite spring set to an actually infinite rigid set. When seen only from the angle of existence, the elements of a potentially infinite spring set are collected one by one. So, between these elements, they can be seen as discrete. That is, each element always has a left and a right neighbor. Since the opposite of discreteness is continuity, let us use δ to represent the medium state of the contrary opposites: discreteness and continuity. So, the transition from a potentially infinite spring set to an actually infinite rigid set, in terms of the form of existence of sets, faces the transition from the discreteness and continuity. And, to accomplish this transition from one opposite to the other, the transition process has to pass through their medium δ. In short, to construct an actually infinite rigid set on the basis of a potentially infinite spring set, one has to face the transition process of three pairs of contrary opposites: the set-building predicate P and ⌉P, potential infinity and actual infinity, and discreteness and continuity. The transition of each of these pairs is accomplished by passing through their respective media ~P, б and δ. Here, the media ~P, б, and δ are not on the same level, because ~P is relatively special, which is determined by the set-building predicate P. That is, different set-building predicate P leads to different ~P. And both ~P and ⌉P are independent set-building predicates of the same level. However, б and δ are different. They are not affected by any specific set-building predicate. For any set-building predicate, the existence of both б and δ stays the same. Therefore, both б and δ belong to a higher level, the abstract level, than ~P. And, each object, satisfying the set-building predicate, belongs to a lower level than the predicate and the resultant set. To summarize, both б and δ belong to the level of abstraction, ~P the level of the predicate P and the resultant set, and each element or object x the next lower level of all objects.

If we say that under the convention that each predicate determines a unique set (an actually infinite rigid set), the afore-mentioned media ~P, б and δ have disappeared in the naïve and the modern axiomatic set theories due to the applied over-simplification, then for our discussion here of relevant problems, we have to focus on not making the matter too complicated. Otherwise, it would not be advantageous for future work to follow. To accomplish this end, we will only focus on two media: ~P and б, since the third medium δ only deals with the form of existence without any direct connection with the fundamental intension of the matter.

Based on our discussion above, for an arbitrarily given set-building predicate P, the constructive mode of the actually infinite rigid set, corresponding to P, should include the contents the of following aspects: First, the potentially infinite spring set $\alpha = \{x| P(x)\}$ of all objects x satisfying P; secondly, the potentially infinite spring set $\beta = \{x| \sim P(x))$ of all objects satisfying the predicate ~P; thirdly, a certain constant element a satisfying the predicate ⌉P; and fourthly, those objects y that fit the existence of б.

In terms of the fourth aspect above, because the potential and actual infinities and their medium б are not any specific set-building predicate, and do not belong to the level of predicates and resultant sets, they are certain abstract existence of a higher level. As an existence, there should exist objects to fit the existence. Since б is not any specific set-building predicate, those existing objects y that fit б do not lead to a spring or rigid set. The only fact we know is that these y should be included in the actually infinite rigid set gradually constructed from the predicate P. Similar to the notation P(x), which means the object x completely satisfying the predicate

P, we use (y)б to describe the existence of object y that fits б. For those y's that fit б, they should first partially possess the property of actual infinite as a perfect tense. So, they have partially reached the constant element a that satisfies ⌉P. Next, these y should also partially possess the property of potential infinite as a progressive tense. So, they will partially not reach that constant element a that satisfies ⌉P. That is, we can only say that each y has partially been glued to a, where a satisfies the predicate ⌉P.

Now, let us once again look at these y that fit б from the angle of discreteness and continuity. Their form of existence should fit δ. That is, these objects y, which are partially glued to a (⌉P(a)), should be partially discrete and partially continuous to each other. In the real analysis of mathematics, those objects which are spaced out from each other are called discrete, those objects which are not spaced away from each other but are separated by tiny gaps are called dense everywhere, and those which are not spaced away from each other and not separated by gaps are called continuous. So, being dense everywhere can be seen as the medium state δ of the contrary opposite pair: discreteness and continuity. So, those y which are partially glued to a (⌉P(a)) and fit б are analogous to the rational numbers ordered in their natural order which are everywhere densely glued to the constant element a. It is exactly because these y which fit б are everywhere densely glued to the constant element a that when we remove this constant element a from this constructive mode, these y which are glued to a would also be removed. Here, we need to notice that the existence of the constant element a (⌉P(a)) in the actually infinite rigid set signals the perfect completion of the transition from P to ⌉P. So, removing the constant element a from this actually infinite rigid set also indicates the disappearance of the perfect tense and the recovery of the present progressive tense. Or in other words, we have returned to the potential infinity from the actual infinity. That is to say, the actually infinite rigid set has been destroyed and no longer exists. On the other hand, those y which are everywhere densely glued to the constant element a and fit б are also removed together with a. It implies that at the same time when the actually infinite rigid set is destroyed, what is left are those two potentially infinite spring sets determined respectively by P and ~P. And, the union of these two spring sets is still a potentially infinite spring set.

Based on the discussion above, the analytic expression of the actually infinite rigid set A, determined by the predicate P, can be written as:

$$A = \{x, y, a | P(x) \text{ or } \sim P(x) \text{ or } (y)\partial \text{ or } a\},$$

where a satisfies the predicate ⌉P. The constructive mode of the actually infinite rigid set A, determined by the predicate P, can be intuitively shown in Figure 12.2.

The previous analytic expression of the set A can be imagined as a partially raw and partially cooked egg with two yolks. The so-called partially raw and partially cooked means that the egg yolks have not started to solidify while the egg white has been in a partial state of coagulation. The two potentially infinite spring sets determined by P and ⌉P, respectively, correspond to the egg yolks, which have not started to solidify. The egg shell stands for the boundary of the actually infinite rigid set A. Those existing y that fit б and are glued to a (⌉P(a)), together with the part of the constant element

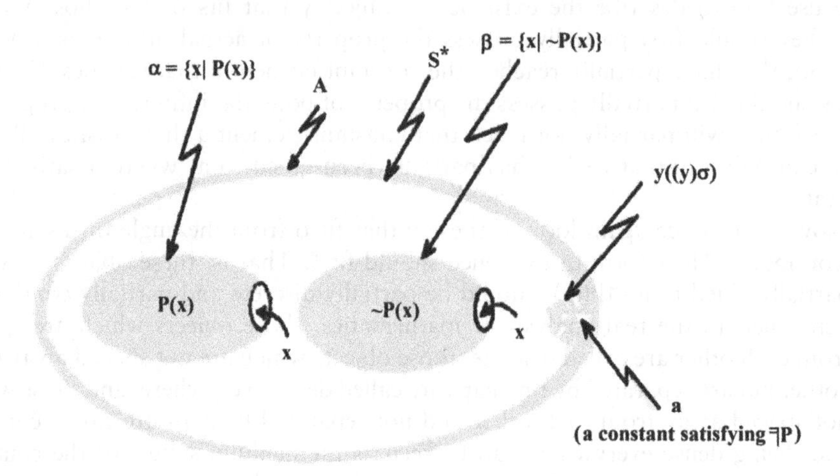

Figure 12.2 The construction of the actually infinite set A.

a, that are glued to y, are fastened in the partially coagulating egg white. The part of a that is glued to y stays inside the egg shell, while being exposed outside the partially coagulating egg white. The partially coagulating egg white can be seen as the medium between the spring and rigid sets, denoted S*. Since on the level of predicates and sets, potential infinities and actual infinities correspond respectively to spring sets and rigid sets, S* should be the corresponding entity of б on the level of sets. When necessary, we can call S* as a quasi-rigid body or an ultra-spring body.

12.3.3 Constructive mode of actually infinite rigid sets with constraints

In this subsection, we study the constructive mode of actually infinite rigid sets under the constraints of the framework of the medium system. The said constraints are to accept the result, shown in the medium system, that "both precise predicates and fuzzy predicate satisfy the law of excluded middle." That is, each given predicate is either precise or fuzzy. Or in other words, there does not exist any medium between the contrary opposites of precise predicates and fuzzy predicates. Because it is so, since each precise predicate determines a spring set and each spring set is a realization of potential infinite on the level of predicates and sets, and rigid sets are determined by fuzzy predicates and are realizations of actual infinite on the ground of predicates and sets, as long as there is no medium between precise predicates and fuzzy predicates, their corresponding spring sets and rigid sets, and potential and actual infinites do not have any medium, either. Therefore, the actually infinite rigid sets we intend to construct will only deal with the following two transitions: One is between a set-building predicate P and ⌉P; and the other between the contrary opposites of discreteness and continuity. These transitions are accomplished respectively through their media ~P and δ of different levels. Now, for any given set-building predicate P, the constructive mode of the actually infinite rigid set corresponding to P should include the contents of the following aspects:

(1) The potentially infinite spring set $\alpha = \{x | P(x)\}$ of all objects satisfying the predicate P;
(2) A certain constant element a satisfying the predicate ˥P; and
(3) Those objects y that satisfy the predicate ~P.

In terms of aspect (3) above, there are two possibilities to consider: The first is to emphasize on those objects y that satisfy the predicate ~P and that constitute a spring set $\beta = \{y | \sim P(y)\}$. Now, the actually infinite rigid set A which we are constructing is made up of two spring sets α and β and a constant element α that satisfies the predicate ˥P. And, the existence of that constant element a in A signals the completion of the transition process from P to ˥P, and the perfect tense of the actually infinite rigid set A.

The second possibility is not to emphasize what spring set the objects that satisfy the predicate ~P constitute, instead it specifically focuses on the characteristic that each object satisfying ~P must partially satisfy the predicate ˥P. So, these objects have already partially reached the constant element a (˥P(a)). Similar to the discussion of the existence of those objects y that fit б under no constraints, if follows that those objects y that satisfy the predicate ~P are everywhere densely glued to the constant element a (˥P(a)). So, the characteristic of the perfect tense of the transition from the predicate P to ˥P directly through ~P is greatly highlighted.

By referencing to Figure 12.2, the intuitive graph of the egg with two yolks drawn for the structure of the actually infinite rigid set A under no constraint, and the relevant discussion, it becomes easier for one to imagine such a similar graph for each of the previous two actually infinite rigid sets A subject to the current constraint. And, similar discussion can be provided. Summarizing what we have done above, we see a central point here: No matter how the actually infinite rigid set A is constructed, the constant element a, satisfying the predicate ˥P, must be included; and as soon as a is take out from A, the set A will definitely be transformed to the potentially infinite spring set A.

To conclude this section, we still need to clarify two points of recognition in terms of the constructive mode of the previous, actually infinite rigid sets.

(1) In terms of the rigid set A, even though we know that all the objects x satisfying P or ~P are included in this actually infinite rigid set A, we still cannot answer the question about the cardinality of the set of all these x satisfying P or ~P. The difficulty here is that as soon as we ask such a question, we have already assumed the validity of the statement that each precise predicate (P or ~P) determines a unique actually infinite rigid set. However, in our discussion here, the rule of thinking is that each precise predicate defines a unique potentially infinite spring set. So, no such an actually infinite rigid set A exists that satisfies one and only one precise predicate.

(2) How can we comprehend and recognize the cardinality of an actually infinite rigid set? According to the modes of construction of actually infinite rigid sets on the basis of potentially infinite spring sets, it should be recognized that the cardinality of an actually infinite rigid set is dependent on those of the base spring sets. That is, if the spring sets α and β are respectively proper subsets of rigid sets A and B, and if the cardinality of α is smaller than that of β, then we can conclude that the cardinality of A is less than that of B. So, the rule of thinking that "the cardinality of an actually

infinite rigid set is dependent on the cardinality of the spring set on which the rigid set is constructed" is reasonable.

As a matter of fact, in the conventional set theory, the concept of cardinality is established on the basis of one-to-one correspondences. And the principle of one-to-one correspondence is only used as a rule of matching. After that, what is left is an arbitrary recursive listing of elements in the sets involved. However, any recursive listing of elements can at most be a potentially infinite process and can at the best be applied to potentially infinite spring sets. There is no foundation for Cantor to arbitrarily employ the one-to-one principle, established on arbitrary recursive listings, on actually infinite rigid sets. In particular, he has no foundation to apply the one-to-one correspondence principle to uncountable sets, because even in the conventional sense of set theory, a recursive listing can at most be applied to countable infinities. Since it is so, how can the one-to-one correspondence principle, established on arbitrary recursive listings, be employed to resolve the problem of cardinality of uncountable actually infinite sets?

Therefore, the capability of applying the one-to-one correspondence principle to determine the cardinalities of various kinds of actually infinite rigid sets must be very limited. Of course, since there does not exist the concept of potentially infinite spring sets in the traditional set theory, the concept of cardinality is defined only for infinite rigid sets. In terms of our potentially infinite spring sets, we could employ such a phrase as "contained quantity" or "containing cardinality" to replace the word "cardinality" in the traditional set theory. Then, we can stipulate that the cardinality of an actually infinite rigid set is determined by the containing cardinality of the base (potentially infinite) spring set. In terms of the containing cardinalities of potentially infinite spring sets, the diagonal method and power sets can be reasonably and sensibly applied to classify and to compare the magnitudes of containing cardinalities.

Afterword: A systemic yoyo model prediction

In the recorded history of mathematics, there have appeared three crises in the foundations of mathematics. The first crisis arose in the fifth century B.C. out of its own discovery of the Pythagorean School. In particular, the Pythagoreans discovered the Pythagorean Theorem regarding right triangles, which led to the proof of the existence of irrational numbers. The existence of these new numbers was a mortal blow to the Pythagorean philosophy that all in the world depend on the whole numbers. The final resolution of this crisis was successfully achieved in about 370 B.C. by the Eudoxus by redefining proportions. What good comes out of this crisis is that the axiomatic method is formulated and adopted in the system of mathematics.

During this time period, notions connected with infinitesimals, limits, and summation processes were considered. Zeno paradoxes were devised (ca. 450 B.C.) to address the question of whether a magnitude is infinitely divisible or it is made up of a very large number of tiny indivisible atomic parts. ((Lin and Ma, 1987a) shows that under ZFC, each system must be finitely divisible). There is evidence that since the Greek time, schools of mathematical reasoning developed their thoughts using both of these possibilities.

The second crisis in the foundation of mathematics appeared at around the end of the eighteenth century when a number of absurdities and contradictions had crept into mathematics. The early calculus was established on the ambiguous and vague concept of infinitesimals, attracting reproaches from various angles, the most representative of which is the Berkeley paradox published in 1734. In 1754, Jean-le Rond d'Alembert observed that a theory of limits was needed to resolve this serious crisis. In 1821, a great stride was made when Cauchy successfully executed d'Alembert's suggestion by developing an acceptable theory of limits. From our discussions in Section 6.3, it can be seen that the way how the Berkeley paradox is avoided by using the modern theory of limits is to assume that actual infinite is the same as potential infinite. This assumption beautifully hides the "fallacy of a shift in hypothesis," as said Berkeley, so that the increment of a variable can be assumed to be zero or non-zero at will, see discussion in Section 9.5 for more details.

The third crisis in the foundations of mathematics appeared at the end of the nineteenth century out of the continued effort of developing the foundation of mathematics onto the level of the naïve set theory, where a set is defined as a collection of objects. Russell's paradox (1902) proved that this definition of sets is the source of a lot of troubles. Soon after that, it was found that throughout the history, similar paradoxes in logic have appeared at various times and the root to all these paradoxes is the application of circular definitions.

To resolve this crisis, several schools of thought emerged. However, each of these schools met with difficulties in their attempt to reconstruct the edifice of the system of mathematics. As a joint effort, several axiomatic set theoretic systems were established, including the ZFC system. To avoid the known paradoxes, for instance, in the ZFC system, the concept of sets is left undefined. Similar treatment was repeatedly employed throughout the history of mathematics when introducing elementary concepts, such as that of points in geometry. Even though such a procedure has been criticized for merely avoiding the paradoxes without explaining any of them and carries no guarantee that other kinds of paradoxes will not crop up in the future, the community of mathematicians has been feeling pretty fortunate that it has been a century without any new paradox publicized.

Since the ultimate goal of reconstructing the naïve set theory is to lay the foundation of the modern system of mathematics on it, the new theory of sets has to be rich enough and powerful enough to handle infinite. As what we have discussed in this book, this is exactly where new contradictions and paradoxes are found and where the Berkeley paradox made its way back.

Now, if we model the history of mathematics mentally as an abstract systemic yoyo, then we can see that the black-hold side sucks in empirical data and puzzles we collect and experience in daily lives. Through our human capability available to us naturally, we process the data with rational thinking (the narrow neck of the yoyo), then we produce (out of the big-bang side) all kinds of concepts, theories, and products that we can physically use in our lives. Now, some concepts, theories, and products become obsolete and disregarded over time, while the rest are recycled back into the black-hole side. These recycled concepts, theories and products, combined with new observations and newly collected empirical data, through the tunnel of rational thinking again, lead to new or renewed concepts, theories, and products, which help to bring the quality of daily lives to another level. That is, we have the yoyo model shown in Figure Af.1 for the general structure of the system of mathematics.

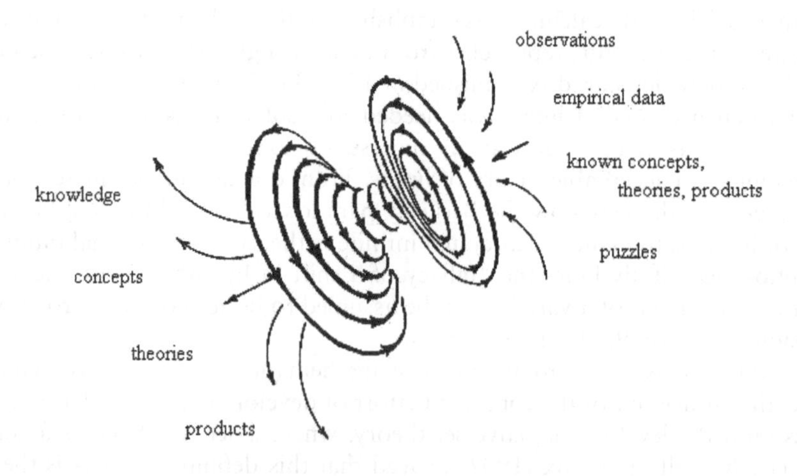

Figure Af.1 The yoyo structure of the system of mathematics.

Now, if we stand at a distance away from this spinning yoyo structure and look at the spinning field from either the black-hole side or the big-bang side, what we see is like a huge whirlpool of data, observations, puzzles, concepts, theories, products, etc. What the dishpan experiment suggests is that this huge whirlpool goes through pattern changes periodically, alternating between uniform and symmetric flows to chaotic currents with local eddies formed.

Naturally, we can identify the chaotic currents with local eddies with one of the crises in the history of mathematics, while equating the uniform and symmetric flow pattern with one of the quiet periods in the history between two consecutive crises. Then, what we can expect is that after the relatively quiet twentieth century, we should be able to see another major crisis of mathematics in the making. As the previous crises, this forthcoming event will also be as devastating, as frustrating as any of the known crises. Its impact will be felt throughout most major corners of the modern system of mathematics. Here, we notice that some of the major concepts and puzzles, such as those of infinitesimals, limits, and summations of numbers, etc., played a role in each of the past crises. So, they can be seen as some of those knowledge, concepts, theories, puzzles, etc., which are sucked into the black-hole of our yoyo structure of the system of mathematics. After being processed by human thinking and reasoning, they are spit out of the big-bang side in their renewed forms.

The discussions in the chapters of Part 3 of this book indicate that the impact of the hidden contradictions existing in modern mathematics, the problem with the self-contradictory non-sets existing in the foundation of modern mathematics, and the return of the Berkeley paradox, should be seen and felt in all major corners of the system of modern mathematics in the years and decades to come. To be clear, what we try to say is that as predicted by the yoyo model, established mentally for the structure of the system of modern mathematics, the fourth crisis in the foundations of mathematics has appeared.

Here, how do we explain the periodic pattern change in the dishpan experiment with our model of mathematics? The reason why we need to address this question is that in the history of mathematics, the time span between the appearance of the first crisis and that of the second crisis is about 23 centuries. The time span between the appearances of the second and the third crisis is about two centuries. And, the time span between the third crisis to the fourth, which it was just declared in 2008, is about one century. With such a drastic decrease in time spans, how can we understand the appearance of the crises as "periodic" pattern changes in the dishpan experiment?

To this end, let us look at the word "periodic" from the amount of mathematical research produced. In particular, we have the following statistics (Eves, 1992). Moritz Cantor's history of mathematics stops at the end of the eighteenth century. This work consists of four volumes of about one thousand pages each. It is conservatively estimated that if the history of mathematics of the nineteenth century should be written with the same details as in Cantor's work, it would require at least fourteen more such volumes. No one has tried to estimate the number of volumes needed for a similar treatment of the history of mathematics of the twentieth century, which is by far the most active era of all.

Here is another piece of statistics (Eves, 1992). Prior to 1700, there were only seventeen periodicals containing articles of mathematical contents. In the eighteenth century, there were 210 such periodicals, and in the nineteenth century, 950 of them.

The number has increased enormously during the twentieth century, by one count done before 1990, to some 2,600.

Here is still another piece of statistics (Eves, 1992). It has been estimated that more than 50% of all known mathematics was created during the past fifth years, and that 50% of all mathematicians who have ever lived are alive today.

It is surely our hope that these statistics well define the word "periodic" for any of our doubtful readers.

As in the past with the previous crises in the foundations of mathematics, one possible resolution for this current crisis is to select an appropriate theoretical foundation for modern mathematics and to choose a right interpretation so that modern mathematics can be kept in its entirety. In particular, we construct the needed mathematical system of potential infinite, which is different to that constructed on the basis of intuitionism, and the mathematical system of actual infinite. When putting these two systems together, we cover the entirety of the current mathematics plus whatever else might be gained. For more details about this end, please consult Part 4 of this book.

To conclude this book, here is our systemic prediction for the future of mathematics:

> Analogous to the historical development of physics, from this point on, mathematics will no longer be seen as a block of truths and absolute truths. And theories of mathematics will be subject to interpretation. From time and again, the truthfulness of each particular mathematical theory will be questioned, doubted and reshaped in order to better reflect the nature.

Part 5

Appendices: A close look at the systemic yoyo model

Appendix A

Theoretical foundation of the Yoyo model

This appendix looks at the theoretical foundation on why such an intuition as the systemic yoyo model of general systems holds for each and every system that is either tangible or imaginable.

In particular, Section A.1 looks at the concept of blown-ups that commonly occur in evolutions of systems. Section A.2 summarizes the mathematical properties of transitional changes that exist along with each evolution. Section A.4 provides a brand new understanding of the quantitative infinity. And Section A.4 focuses on the study of equal quantitative movements and equal quantitative effects.

A.1 BLOWN-UPS: MOMENTS OF EVOLUTIONS

When we study the nature and treat everything we see as a system (Klir, 2001), then one fact we can easily see is that many systems in nature evolve in concert. When one thing changes, many other seemingly unrelated things alter their states of existence accordingly. That is why (OuYang, Chen, and Lin, 2009) proposes to look at the evolution of a system or event of concern as a whole. That is, when developments and changes naturally existing in the natural environment are seen as a whole, we have the concept of whole evolutions. And, in whole evolutions, other than continuities, as well studied in modern mathematics and science, what seems to be more important and more common is discontinuity, with which transitional changes (or blown-ups) occur. These blown-ups reflect not only the singular transitional characteristics of the whole evolutions of nonlinear equations, but also the changes of old structures being replaced by new structures. By borrowing the form of calculus, we can write the concept of blown-ups as follows: For a given (mathematical or symbolic) model, that truthfully describes the physical situation of our concern, if its solution $u = u(t; t_0, u_0)$, where t stands for time and u_0 the initial state of the system, satisfies

$$\lim_{t \to t_0} |u| = +\infty, \tag{A.1}$$

and at the same time moment when $t \to t_0$, the underlying physical system also goes through a transitional change, then the solution $u = u(t; t_0, u_0)$ is called a blown-up solution and the relevant physical movement expresses a blown-up. For nonlinear models in independent variables of time (t) and space (x, x and y, or x, y, and z), the concept of blown-ups are defined similarly, where blow-ups in the model and the underlying physical system can appear in time or in space or in both.

A.2 PROPERTIES OF TRANSITIONAL CHANGES

To help us understand the mathematical characteristics of blown-ups, let us look at the following constant-coefficient equation:

$$\dot{u} = a_0 + a_1 u + \cdots + a_{n-1}u^{n-1} + u^n = F, \tag{A.2}$$

where u is the state variable, and a_0, $a_1, \ldots, a_{n-1}$ are constants. Based on the fundamental theorem of algebra, let us assume that equ. (A.2) can be written as

$$\dot{u} = F = (u - u_1)^{p_1} \cdots (u - u_r)^{p_r}(u^2 + b_1 u + c_1)^{q_1} \cdots (u^2 + b_m u + c_m)^{q_m}, \tag{A.3}$$

where p_i and q_j, $i = 1, 2, \ldots, r$ and $j = 1, 2, \ldots, m$, are positive whole numbers, and $n = \sum_{i=1}^{r} p_i + 2\sum_{j=1}^{m} q_j$, $\Delta = b_j^2 - 4c_j < 0$, $j = 1, 2, \ldots, m$. Without loss of generality, assume that $u_1 \geq u_2 \geq \cdots \geq u_r$, then the blown-up properties of the solution of equ. (A.3) are given in the following theorem.

Theorem A.1: The condition under which the solution of an initial value problem of equ. (A.2) contains blown-ups is given by

1 When u_i, $i = 1, 2, \ldots, r$, does not exist, that is, $F = 0$ does not have any real solution; and
2 If $F = 0$ does have real solutions u_i, $i = 1, 2, \ldots, r$, satisfying $u_1 \geq u_2 \geq \cdots \geq u_r$,

   (a) When n is an even number, if $u > u_1$, then u contains blow-up(s);
   (b) When n is an odd number, no matter whether $u > u_1$ or $u < u_r$, there always exist blown-ups.

A detailed proof of this theorem can be found in (Wu and Lin, 2002, p. 65–66) and is omitted here. And for higher order nonlinear evolution systems, please consult (Lin, 2008b).

A.3 QUANTITATIVE INFINITY

One of the features of the concept of blown-ups is the quantitative infinity ∞, which stands for indeterminacy mathematically. So, a natural question is how to comprehend this mathematical symbol ∞. In applications, this symbol causes instabilities and calculation spills that have stopped each and every working computer.

To address the previous question, let us look at the mapping relation of the Riemann ball, which is well studied in complex functions (Figure A.1). This so-called Riemann ball, a curved or curvature space, illustrates the relationship between the

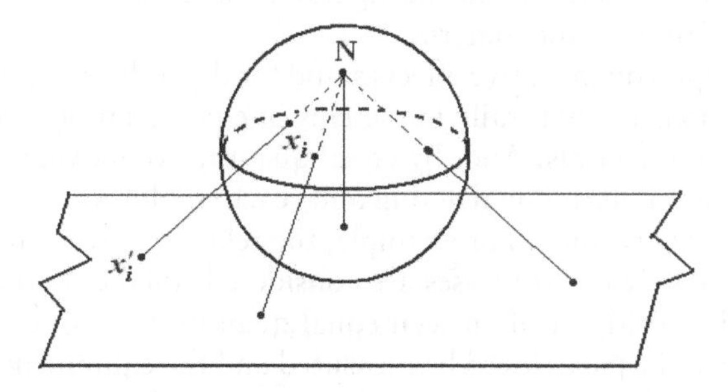

Figure A.1 The Riemann ball – relationship between planar infinity and three-dimensional North Pole.

infinity on the plane and the North Pole N of the ball. Such a mapping relation connects $-\infty$ and $+\infty$ through a blown-up. Or in other words, when a dynamic point x_i travels through the North Pole N on the sphere, the corresponding image x_i' on the plane of the point x_i shows up as a reversal change from -∞ to $+\infty$ through a blown-up. So, treating the planar points $\pm\infty$ as indeterminacy can only be a product of the thinking logic of a narrow or low dimensional observ-control, since, speaking generally, these points stand implicitly for direction changes of one dynamic point on the sphere at the polar point N. Or speaking differently, the phenomenon of directionless, as shown by blown-ups of a lower dimensional space, represents exactly a direction change of movement in a higher dimensional curvature space. Therefore, the concept of blown-ups can specifically represent implicit transformations of spatial dynamics. That is, through blown-ups, problems of indeterminacy of a narrow observ-control in a distorted space are transformed into determinant situations of a more general observ-control system in a curvature space. This discussion shows that the traditional view of singularities as meaningless indeterminacies has not only revealed the obstacles of the thinking logic of the narrow observ-control (in this case, the Euclidean space), but also the careless omissions of spatial properties of dynamic implicit transformations (bridging the Euclidean space to a general curvature space).

Summarizing what has been discussed above in this appendix, we can see that nonlinearity, speaking mathematically, stands (mostly) for singularities in Euclidean spaces, the imaginary plane discussed above. In terms of physics, nonlinearity represents eddy motions, the movements on curvature spaces, the Riemann ball above. Such motions are a problem about structural evolutions, which are a natural consequence of uneven evolutions of materials. So, nonlinearity accidentally describes discontinuous singular evolutionary characteristics of eddy motions (in curvature spaces) from the angle of a special, narrow observ-control system, the Euclidean spaces.

A.4 EQUAL QUANTITATIVE MOVEMENTS AND EFFECTS

Another important concept studied in the blown-up theory is that of equal quantitative effects. Even though this concept was initially proposed in the study of fluid motions, it essentially represents the fundamental and universal characteristics of all movements of materials. What's more important is that this concept reveals the fact that nonlinearity is originated from the figurative structures of materials instead of non-structural quantities of the materials.

The so-called equal quantitative effects stand for the eddy effects with non-uniform vortical vectorities existing naturally in systems of equal quantitative movements due to the unevenness of materials. And, by equal quantitative movements, it is meant to be the movements with quasi-equal acting and reacting objects or under two or more quasi-equal mutual constraints. For example, the relative movements of two or more planets of approximately equal masses are considered equal quantitative movements. In the microcosmic world, an often seen equal quantitative movement is the mutual interference between the particles to be measured and the equipment used to make the measurement. Many phenomena in daily lives can also be considered equal quantitative effects, including such events as wars, politics, economies, chess games, races, plays, etc.

Comparing to the concept of equal quantitative effects, Aristotelian and Newtonian framework of separate objects and forces is about unequal quantitative movements established on the assumption of particles. On the other hand, equal quantitative movements are mainly characterized by the kind of measurement uncertainty that when I observe an object, the object is constrained by me. When an object is observed by another object, the two objects cannot really be separated apart. At this juncture, it can be seen that the Su-Shi Principle of Xuemou Wu's panrelativity theory (1990), Bohr (N. Bohr, 1885–1962) principle and the relativity principle about microcosmic motions, von Neumann's Principle of Program Storage, etc., all fall into the uncertainty model of equal quantitative movements with separate objects and forces.

What's practically important and theoretically significant is that eddy motions are confirmed not only by daily observations of surrounding natural phenomena, but also by laboratory studies from as small as atomic structures to as huge as nebular structures of the universe. At the same time, eddy motions show up in mathematics as nonlinear evolutions. The corresponding linear models can only describe straight-line-like spraying currents and wave motions of the morphological changes of reciprocating currents. What is interesting here is that wave motions and spraying currents are local characteristics of eddy movements. This fact is very well shown by the fact that linearities are special cases of nonlinearities. Please note that we do not mean that linearities are approximations of nonlinearities.

The birth-death exchanges and the non-uniformity of vortical vectorities of eddy evolutions naturally explain where and how quantitative irregularities, complexities and multiplicities of materials' evolutions, when seen from the current narrow observ-control system, come from. Evidently, if the irregularity of eddies comes from the unevenness of materials' internal structures, and if the world is seen at the height of structural evolutions of materials, then the world is simple. And, it is so simple that there are only two forms of motions. One is clockwise rotation, and the other counter clockwise rotation. The vortical vectority in the structures of materials has very intelligently resolved the Tao of Yin and Yang of the "*Book of Changes*" of the eastern mystery (Wilhalm and Baynes, 1967), and has been very practically implemented in the common form of motion of all materials in the universe. That is when the concept of invisible organizations of the blown-up system comes from.

The concept of equal quantitative effects not only possesses a wide range of applications, but also represents an important omission of modern science, developed in the past 300 plus years. Evidently, not only are equal quantitative effects more general than the mechanic system of particles with further reaching significance, but also have they directly pointed to some of the fundamental problems existing in modern science.

In order for us to intuitively see why equal quantitative effects are so difficult for modern science to handle by using the theories established in the past 300 plus years, let us first look at why all materials in the universe are in rotational movements. According to Einstein's uneven space and time, we can assume that all materials have uneven structures. Out of these uneven structures, there naturally exist gradients. With gradients, there will appear forces. Combined with uneven arms of forces, the carrying materials will have to rotate in the form of moments of forces. That is exactly what the ancient Chinese Lao Tzu, (English and Feng, 1972) said: "Under the heaven, there is nothing more than the Tao of images," instead of Newtonian doctrine of particles (under the heaven, there is such a Tao that is not about images but sizeless and volumeless particles). The former stands for an evolution problem of rotational movements under stirring forces. Since structural unevenness is an innate character of materials, that is why it is named second stir, considering that the phrase of first push was used first in history (OuYang, et al., 2000). What needs to be noted is that the phrases of first push and second stir do not mean that the first push is prior to the second stir.

Now, we can imagine that the natural world and/or the universe be composed of entirely with eddy currents, where eddies exist in different sizes and scales and interact with each other. That is, the universe is a huge ocean of eddies, which change and evolve constantly. One of the most important characteristics of spinning fluids, including spinning solids, is the difference between the structural properties of inwardly and outwardly spinning pools and the discontinuity between these pools. Due to the stirs in the form of moments of forces, in the discontinuous zones, there exist sub-eddies and sub-sub-eddies (Figure A.2, where sub-eddies are created naturally by the large eddies *M* and *N*). Their twist-ups (the sub-eddies) contain highly condensed amounts of materials and energies. Or in other words, the traditional frontal lines and surfaces (in meteorology) are not simply expansions of particles without any structure. Instead, they represent twist-up zones concentrated with irregularly structured materials and energies (this is where the so-called small probability events appear and small-probability information is observed and collected so such information (event) should also be called irregular information (and event)). In terms of basic energies,

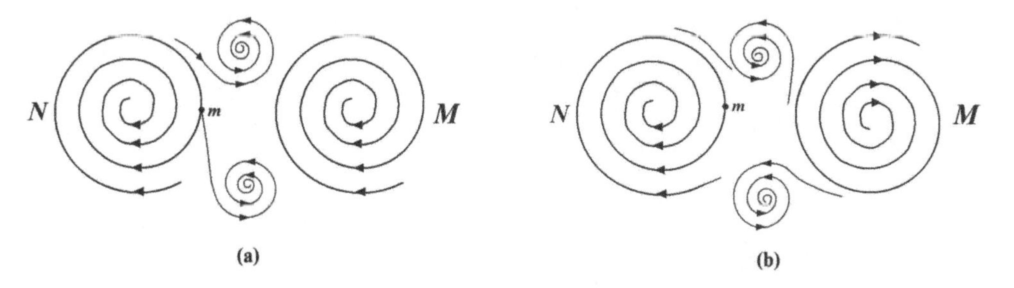

Figure A.2 Appearance of sub-eddies.

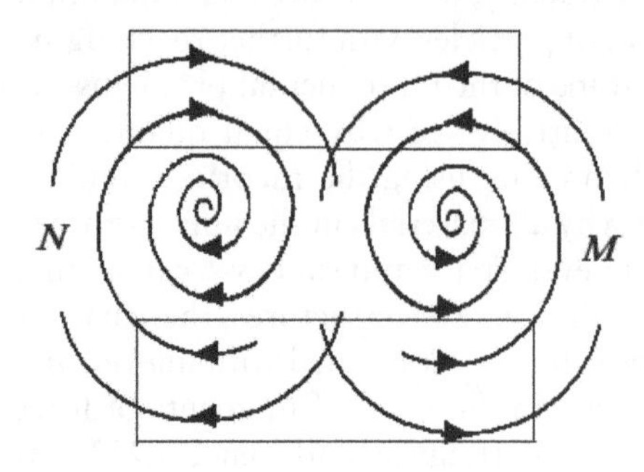

Figure A.3 Structural representation of equal quantitative effects.

these twist-up zones cannot be formed by only the pushes of external forces and cannot be adequately described by using mathematical forms of separate objects and forces. Since evolution is about changes in materials' structures, it cannot be simply represented by different speeds of movements. Instead, it is mainly about transformations of rotations in the form of moments of forces ignited by irregularities. The enclosed areas in Figure A.3 stand for the potential places for equal quantitative effects to appear, where the combined pushing or pulling is small in absolute terms. However, it is generally difficult to predict what will come out of the power struggles. In general, what comes out of the power struggle tends to be drastic and unpredictable by using the theories and methodologies of modern science.

Appendix B

Empirical evidence of the yoyo model

Continuing on what was done in Appendix A, we will in this appendix study several empirical evidences and observations that underline the existence of the yoyo structure behind each and every system, which either tangibly exists or is intellectually imaginable.

In particular, Section B.1 looks at the circulations that exist in fluids. Section B.2 introduced the law of conservation of informational infrastructure. And Section B.3 lists two examples of silent human evaluations of each other or of a situation.

B.1 CIRCULATIONS IN FLUIDS

Based on the previous discussions, it is found that nonlinearity accidentally describes discontinuous singular evolutionary characteristics of eddy motions (in curvature spaces) from the angle of a special, narrow observ-control system, the Euclidean spaces (the imaginary plane). To support this end, let us now look at the Bjerknes' Circulation Theorem (1898).

At the end of the 19th century, V. Bjerkes (1898) (Hess, 1959) discovered the eddy effects due to changes in the density of the media in the movements of the atmosphere and ocean. He consequently established the well-known circulation theorem, which was later named after him. Let us look at this theorem briefly.

By a circulation, it is meant to be a closed contour in a fluid. Mathematically, each circulation Γ is defined as the line integral about the contour of the component of the velocity vector locally tangent to the contour. In symbols, if $\vec{V}$ stands for the speed of a moving fluid, S an arbitrary closed curve, $\delta\vec{r}$ the vector difference of two neighboring points of the curve S (Figure B.1), then a circulation Γ is defined as follows:

$$\Gamma = \oint_S \vec{V}\delta\vec{r}. \tag{B.1}$$

Through some very clever manipulations, we can produce the following well-known Bjerknes' Circulation Theorem:

$$\frac{d\vec{V}}{dt} = \iint_\sigma \nabla\left(\frac{1}{\rho}\right) \times (-\nabla p) \cdot \delta\sigma - 2\Omega\frac{d\sigma}{dt}, \tag{B.2}$$

where $\vec{V}$ stands for the velocity of the circulation, σ the projection area on the equator plane of the area enclosed by the closed curve S, p the atmospheric pressure, ρ the density of the atmosphere, and Ω the earth's rotational angular speed.

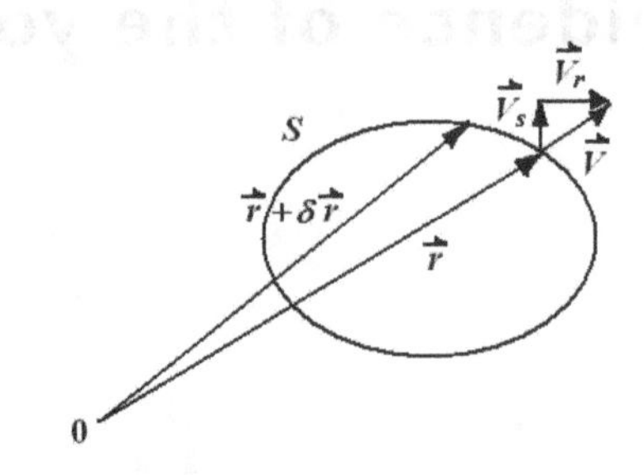

Figure B.1 The definition of a closed circulation.

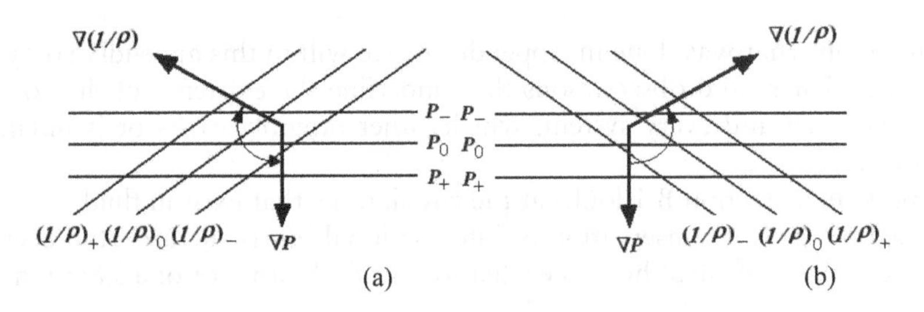

Figure B.2 A diagram for solenoid circulations.

The left-hand side of equ. (B.2) represents the acceleration of the moving fluid, which according to Newton's second law of motion is equivalent to the force acting on the fluid. On the right-hand side, the first term is called a solenoid term in meteorology. It is originated from the interaction of the p- and ρ-planes due to uneven density ρ so that a twisting force is created. Consequently, materials' movements must be rotations with the rotating direction determined by the equal p- and ρ-plane distributions (Figure B.2). The second term in equ. (B.2) comes from the rotation of the earth. In short, when a force is acting on a fluid, a rotation is created.

B.2 INFORMATIONAL INFRASTRUCTURE

Some branches of modern science were made "exact" by introducing various laws of conservation, even though, at the times when they were proposed, there might not have been any "theoretical or mathematical" foundations for these laws. Walking along the similar lines, Lin (1995) developed a theoretical foundation for some laws of conservation, such as the laws of conservation of matter-energy, of fundamental particles, etc., on the basis of general systems theory. By addressing some problems related to the discussions in (Lin, 1995), Lin and Fan (1997) systematically showed that human understanding of nature can be very much limited by our sensing organs, even though our constant attempts do help us get closer to the true state of the nature.

To form the heuristic foundation of the conservation law of informational infrastructure, let us first look at the intuitive understanding of the concept of general systems. From a practical point of view, a system is what is distinguished as a system (Klir, 1985). From a mathematical point of view, a system is defined as follows (Lin, 1987): S is a (general) system, provided that S is an ordered pair (M, R) of sets, where M is the set of objects of the system S and R a set of some relations on the set M. The sets M and R are called the object set and the relation set of the system S, respectively. (For those readers who are sophisticated enough in mathematics, for each relation r in R, it implies that there exists an ordinal number $n = n(r)$, a function of r, such that r is a subset of the Cartesian product of n copies of the set M.) The idea of using an ordered pair of sets to define the general system is to create the convenience of comparing systems. In particular, when two systems S_1 and S_2 are given, by writing each of these systems as an ordered pair (M_i, R_i), $i = 1, 2$, we can make use of the known facts of mathematics to show that $S_1 = S_2$, if, and only if $M_1 = M_2$ and $R_1 = R_2$. When two systems S_1 and S_2 are not equal, then with their ordered pair structures, we can readily investigate their comparisons, such as how and when they are similar, congruent, or one is structurally less than the other, and other properties between systems. For more details about this end, please consult (Lin, 1999).

By combining these two understandings of general systems, we can intuitively see the following: Each thing that can be imagined in human minds is a system according to Klir's definition so that this thing would look the same as that of an ordered pair (M, R) according to Lin's definition. Furthermore, relations in R can be about some information of the system, its spatial structure, its theoretical structure, etc. That is, there should exist a law of conservation that reflects the uniformity of all tangible and imaginable things with respect to:

1 The content of information;
2 Spatial structures;
3 Various forms of movements, etc.

Based on this intuition of (general) systems and by looking at the available data from particle physics, field theory, astronomy, celestial mechanics, geo-physics, and meteorology, it is shown that between the macrocosm and the microcosm, between the electromagnetic interactions of atomic scale and the strong interactions of Quark's scale, between the central celestial body and the circling celestial bodies of celestial systems, between and among the large, medium, small, and micro-scales of the earth atmosphere, there are exist laws of conservation of products of spatial physical quantities. In particular, it can be conjectured that there might be a more general law of conservation in terms of structure, in which the informational infrastructure, including time, space, mass, energy, etc., is approximately equal to a constant. In symbols, this conjecture can be written as follows:

$$AT \times BS \times CM \times DE = a \tag{B.3}$$

or more generally,

$$AT^{\alpha} \times BS^{\beta} \times CM^{\gamma} \times DE^{\varepsilon} = a, \tag{B.4}$$

where α, β, γ, ε, and a are constants, T, S, M, E and A, B, C, D are respectively time, space, mass, energy, and their coefficients. These two formulas can be applied to various conservative systems of the universal, macroscopic, and microscopic levels. The constants α, β, γ, ε and a are determined by the initial state and properties of the natural system of interest.

In equ. (B.3), when two (or one) terms of choice are fixed, the other two (or three) terms will vary inversely. For example, under the conditions of low speed and the macrocosm, all the coefficients A, B, C, and D equal 1. In this case, when two terms are fixed, the other two terms will be inversely proportional. This end satisfies all principles and various laws of conservation in the classical mechanics, including the laws of conservation of mass, momentum, energy, moment of momentum, etc. So, the varieties of mass and energy in this case are reflected mainly in changes in mass density and energy density. In the classical mechanics, when time and mass are fixed, the effect of a force of a fixed magnitude becomes the effect of an awl when the cross section of the force is getting smaller. When the space and mass are kept unchanged, the same force of a fixed magnitude can have an impulsive effect, since the shorter the time the force acts the greater density the energy release will be. When time and energy are kept the same, the size of working space and the mass density are inversely proportional. When the mass is kept fixed, shrinking acting time and working space at the same time can cause the released energy density reaching a very high level.

Under the conditions of relativity theory, that is, under the conditions of high speeds and great amounts of masses, the coefficients in equ. (B.3) are no longer equal to 1, and equ. (B.4) becomes more appropriate, and the constants A, B, C, D, and a and the exponents α, β, γ, and ε satisfy relevant equations in relativity theory. When time and space are fixed, the mass and energy can be transformed back and forth according to the well-known mass-energy relation:

$$E = mc^2.$$

When traveling at a speed close to that of light, the length of a pole will shrink when the pole is traveling in the direction of the pole and any clock in motion will become slower. When the mass is sufficiently great, light and gravitation deflection can be caused. When a celestial system evolves to its old age, gravitation collapse will appear and a black hole will be formed. We can imagine based on equ. (B.3) that when our earth evolves sufficiently long, say a billion or trillion years, the relativity effects would also appear. More specifically speaking, in such a great time measurement, the creep deformation of rocks could increase and solids and fluids would have almost no difference so that solids could be treated as fluids. When a universe shrinks to a single point with the mass density infinitely high, a universe explosion of extremely high energy density could appear in a very short time period. So, a new universe is created!

If the previous law of conservation of informational infrastructure holds true, (all the empirical data seem to suggest so), its theoretical and practical significance is obvious. The hypothesis of the law contains the following facts:

1 Multiplications of relevant physical quantities in either the universal scale or the microscopic scale approximately equal a fixed constant; and
2 Multiplications of either electromagnetic interactions or strong interactions approximately equal a fixed constant.

In the widest domain of human knowledge of our modern time, this law of conservation deeply reveals the structural unification of different layers of the universe so that it might provide a clue for the unification of the four basic forces in physics. This law of conservation can be a new evidence for the big bang theory. It supports the model for the infinite universe with border and the oscillation model for the evolution of the universe, where the universe evolves as follows:

$$\ldots \rightarrow \text{explosion} \rightarrow \text{shrinking} \rightarrow \text{explosion} \rightarrow \text{shrinking} \rightarrow \ldots$$

It also supports the hypothesis that there exist universes "outside of our universe." The truthfulness of this proposed law of conservation is limited to the range of "our universe," with its conservation constant being determined by the structural states of the initial moment of "our universe."

All specific examples analyzed in establishing this law show that to a certain degree, the proposed law of conservation indeed holds true. That is, there indeed exists some kind of uniformity in terms of time, space, mass, and energy among different natural systems of various scales under either macroscopic or microscopic conditions or relativity conditions. Therefore, there might be a need to reconsider some classical theoretical systems so that our understanding about nature can be deepened. For example, under the time and space conditions of the earth's atmosphere, the traditional view in atmospheric dynamics is that since the vertical velocity of each atmospheric huge scale system is much smaller than its horizontal velocity, the vertical velocity is ignored. As a matter of fact (Ren and Nio, 1994), since the atmospheric density difference in the vertical direction is far greater than that in the horizontal direction, and since the gradient force of atmospheric pressure to move the atmospheric system 10 m vertically is equivalent to that of moving the system 200 km horizontally, the vertical velocity should not be ignored. The law of conservation of informational infrastructure, which holds true for all scales used in the earth's atmosphere, might provide conditions for a unified atmospheric dynamics applicable to all atmospheric systems of various scales. As a second example, in the situation of our earth where time and mass do not change, in terms of geological time measurements (sufficiently long time), can we imagine the force, which causes the earth's crust movements? Does it have to be as great as what is believed currently?

As for applications of science and technology, tremendous successes have been made in the macroscopic and microscopic levels, such as shrinking working spatial sectors, shortening the time length for energy releasing, and sacrificing partial masses (say, the usage of nuclear energy). However, the law of conservation of informational infrastructure might very well further the width and depth of applications of science and technology. For example, this law of conservation can provide a theory and thinking logic for us to study the movement evolution of the earth's structure, the source of forces or structural information which leads to successful predictions of major earthquakes, and to find the mechanisms for the formation of torrential rains and for the arrival of earthquakes (Ren, 1996).

Philosophically speaking, the law of conservation of informational infrastructure indicates that in the same way as mass, energy is also a characteristic of physical entities. Time and space can be seen as the forms of existence of physical entities with motion as their basic characteristics. This law of conservation connects time, space, mass, and motion closely into an inseparable whole. So, time, space, mass, and energy can be seen as attributes of physical entities. With this understanding, the concept of mass is generalized and the wholeness of the world is further proved and the thoughts of never diminishing mass and never diminishing universes are evidenced.

B.3 SILENT HUMAN EVALUATIONS

In this subsection, we will look at how the systemic yoyo model is manifested in different areas of life by briefly visiting relevant experimental and clinical evidences. All the omitted details can be found in the relevant references.

Based on the systemic yoyo model, each human being is a 3-dimensional realization of a spinning yoyo structure of a certain dimension higher than three. To illustrate this end, let us consider two simple and easy-to-repeat experiments.

Experiment #1: feel the vibe

Let us imagine we go to a sport event, say a swim meet. Assume that the area of competition contains a pool of the Olympic size and along one long side of the pool there are about 200 seats available for spectators to watch the swim meet. The pool area is enclosed with a roof and walls all around the space.

Now, let us physically enter the pool area. What we find is that as soon as we enter the enclosed area of competition, we immediately fall into a boiling pot of screaming and jumping spectators, cheering for their favorite swimmers competing in the pool. Now, let us pick a seat a distance away from the pool deck anywhere in the seating area. After we settle down in our seat, let us purposelessly pick a voluntary helper standing or walking on the pool deck for whatever reason, either for her beauty or for his strange look or body posture, and stare at him intensively. Here is what will happen next:

Magically enough, before long, our stare will be felt by the person from quite a good distance; she/he will turn around and locate us in no time out of the reasonably sized and boiling audience.

By using the systemic yoyo model, we can provide one explanation for why this happens and how the silent communication takes place. In particular, each side, the person being stared at and us, is a high dimensional spinning yoyo. Even though we are separated by space and possibly by informational noise, the stare of one party on the other has directed that party's spin field of the yoyo structure into the spin field of the yoyo structure of the other party. Even though the latter party initially did not know the forthcoming stare, when her/his spin field is interrupted by the sudden intrusion of another unexpected spin field, the person surely senses the exact direction and location where the intrusion is from. That is the underlying mechanism for the silent communication to be established.

When this experiment is done in a large auditorium where the person being stared at is on the stage, the afore-described phenomenon does not occur. It is because when many spin fields interferes the field of a same person, these interfering fields actually destroy their originally organized flows of materials and energy so that the person who is being stared at can only feel the overwhelming pressure from the entire audience instead of from individual persons.

This easily repeatable experiment in fact has been numerously conducted by some of the high school students in our region. When these students eat out in a restaurant and after they run out of topics to gossip about, they play the game they call "feel the vibe." What they do is to stare as a group at a randomly chosen guest of the restaurant to see how long it takes the guest to feel their stares. As described in the situation of swim meet earlier, the chosen guest can almost always feel the stares immediately and can locate the intruders in no time.

Experiment #2: she does not like me!

In this case, let us look at the situation of human relationship. When an individual A has a good impression about another individual B, magically, individual B also has a similar and almost identical impression about A. When A does not like B and describes B as a dishonest person with various undesirable traits, it has been clinically proven in psychology that what A describes about B is exactly who A is himself (Hendrix, 2001).

Once again, the underlying mechanism for such a quiet and unspoken evaluation of each other is based on the fact that each human being stands for a high dimensional spinning yoyo and its rotational field. Our feelings toward each other are formed through the interactions of our invisible yoyo structures and their spin fields. So, when person A feels good about another person B, it generally means that their underlying yoyo structures possess the same or very similar attributes, such as spinning in the same direction, both being either divergent or convergent at the same time, both having similar intensities of spin, etc. When person A does not like B and lists many undesirable traits B possesses, it fundamentally stands for the situation that the underlying yoyo fields of A and B are fighting against each other in terms of either opposite spinning directions, or different degrees of convergence, or in terms of other attributes. For a more in depth analysis along a line similar to this one, please consult Part 2: The Mind, of this book.

Such quiet and unspoken evaluations of one another can be seen in any working environment. For instance, let us consider a work situation where quality is not and cannot be quantitatively measured, such as a teaching institution in the USA. When one teacher does not perform well in his line of work, he generally uses the concept of quality loudly in day to day settings in order to cover up his own deficiency in quality. When one does not have honesty, he tends to use the term honesty all the time. It is exactly as what Lao Tzu (time unknown, Chapter 1) said over 2,000 years ago: "The one who speaks of integrity all the time does not have integrity."

When we tried to repeat this experiment with local high school students, what is found is that when two students A and B, who used to be very good friends, turned away from each other, we ask A why she does not like B anymore. The answer is exactly what we expect: "Because she does not like me anymore!"

Bibliography

Allen, Bem P. (2005). Personality Theories: Development, Growth, and Diversity (5th Edition). Upper Saddle River, NJ: Pearson Education/Allyn & Bacon.

Aristotle. (1984). The Complete Works. Jonathan Barnes (ed.), Princeton University Press.

Armson, R. (2011). *Growing Wings On The Way: Systems Thinking for Messy Situations.* Devon, UK: Triarchy Press.

Basdevant, J. L., Rich, J., and Spiro, M. (2005). *Fundamentals in Nuclear Physics: From Nuclear Structure to Cosmology*. Springer.

Baum, Eric (2004). *What is Thought.* Boston: MIT Press.

Becker, G. (1974). A theory of social interactions. *Journal of Political Economy*, vol. 82, no. 6, pp. 1063–1093.

Bell, E. T. (1945). *The Development of Mathematics* (2nd edition). McGraw-Hill.

Benacerraf, P. (1962). Tasks, super-tasks, and modern eleatics. *Journal of Philosophy*, vol. LIX, pp. 76–784.

Benditt, Theodore (1998). *Philosophy Then and Now* with eds. Arnold and Graham. Oxford: Blackwell Publishing.

Bergstrom T. (1989). A fresh look at the Rotten Kid Theorem and other household mysteries. *Journal of Political Economy*, vol. 97, no. 5, pp. 1138–1159.

Berlinski, D. (1976). *On Systems Analysis*. Cambridge, MA: MIT Press.

Bernays, E. (1945). *Public Relations*. Boston, MA: Bellman Publishing Company.

Berresford, G. C. (1981). A note on Thompson's lamp 'paradox'. *Analysis*, vol. 41, pp. 1–5.

Berridge, K. C., Robinson, T. E. & Aldridge, J. W. (2009). Dissecting components of reward: 'liking', 'wanting', and learning. *Current Opinion in Pharmacology*, vol. 9, 65–73.

Bhikkhu, Thanissaro (1996). The Wings to Awakening. Dhamma Dana Publications.

Blauberg, I. V., Sadovsky, V. N., and Yudin, E. G. (1977). *Systems Theory, Philosophy and Methodological Problems*. Moscow: Progress Publishers.

Block, Ned (2007). *Consciousness, Function, and Representation: Collected Papers* (Volume 1). Boston: MIT Press.

Blumenthal, L. M. (1940). A paradox, a paradox, a most ingenious paradox. *The American Mathematical Monthly*, vol. 47, pp. 346–353.

Bollobás, B. (ed.) (1986). *Littlewood's Miscellany*. Cambridge, MA: Cambridge University Press. chapt 7, 8.

Bolzano, B. (1950). Paradoxes of the Infinite. Translated by F. Prihonsky, London: Routledge & Kegan Paul.

Bos, H. J. (1981). On the representation of curves in Descartes' Geometrie. *Arch. Hist. Ex. Sc.*, vol. 24, pp. 295–338.

Bottazzini, U. (1986). *The Higher Calculus: A History of Real and Complex Analysis from Euler to Weierstrass*. Berlin: Springer-Verlag.

Brandelik, R. et al. (1979). Evidence for Planar Events in e^+e^- Annihilation at High Energies. *Phys. Lett. B* 86: 243–249.

Branden, Nathaniel (1969). *The Psychology of Self-Esteem*, p. 41, Nash Publishing Corp.
Brouwer, L. E. J. (1983a). Intuitionism and formalism. In: Paul Benacerraf and Hilary Putnam (eds), *Philosophy of Mathematics*, Cambridge University Press.
Brouwer, L. E. J. (1983b). Consciousness, philosophy and mathematics. In: Paul Benacerraf and Hilary Putnam (eds), *Philosophy of Mathematics*, Cambridge University Press.
Brun, Jean (1978). *Socrate* (6th edition). Presses universitaires de France.
Brunet, A., Akerib, V., and Birmes, P. (2007). "Don't throw out the baby with the bathwater (PTSD is not overdiagnosed)". *Can J Psychiatry* 52 (8): 501–2; discussion 503.
Bunge, B. (1977). *Treatise on Basic Philosophy*. Vol. 4: A World of Systems. Holland: Reidel Publishing Company.
Burali-Forti, C. (1967). A question on transfinite numbers. Trans. Jean van Heijenoort, in Jean van Heijenoort (ed.), *From Frege to Godel: A source Book in Mathematical Logic, 1879–1931*. Cambridge, MA: Harvard University Press.
Buss, David (2004). *Evolutionary Psychology: The New Science of the Mind*. Boston, MA: Pearson Education, Inc.
Butler, T. (2006). http://www.suitcaseofdreams.net/Wizard_Mermaid.htm (accessed on March 28, 2012).
Byl, J. (2000). On resolving the Littlewoo–Ross Paradox. *Missouri Journal of Mathematical Sciences*, vol. 12, no. 1, pp. 42–47.
Cajori, F. (1918). Origin of the name "mathematical induction". *The American Mathematical Monthly*, vol. 25, no. 5, pp. 197–201.
Carlyle, Thomas (1841). *On Heroes, Hero-Worship, and the Heroic History*. Boston, MA: Houghton Mifflin.
Cecil A. Gibb (1970). *Leadership (Handbook of Social Psychology)*. Reading, Mass.: Addison-Wesley.
Chaitin, G. J. (1999). *The Unknowable*. New York: Springer-Verlag.
Cajori, F. (1913). History of exponential and logarithmic concepts. *The American Mathematical Monthly*, vol. 20, several issues.
Cajori, F. (1918). Origin of the name "mathematical induction". *The American Mathematical Monthly*, vol. 25, no. 5, pp. 197–201.
Cantor, G. (1955). *Contributions to the founding of the Theory of Transfinite Numbers*. trans. Phili E. B. Jourdain, New York: Dover.
Chaitin, G. J. (1999). *The Unknowable*. New York: Springer.
Chemers, M. M. (2001). Cognitive, social, and emotional intelligence of transformational leadership: Efficacy and Effectiveness. In R. E. Riggio, S. E. Murphy, F. J. Pirozzolo (Eds.), *Multiple Intelligences and Leadership*, pp. 139–160, Mahwah, NJ: Lawrence Erlbaum.
Chen, G. R. (2007). *The Original State of the World: The Theory of Ether Whirltrons*. Hong Kong: Tianma Books Limited.
Cohen, Anthony P. (1994). *Self-Consciousness*, Routledge.
Collins, Steven (1982). *Selfless Persons: Thought and Imagery in Theravada Buddhism*. Cambridge University Press.
Colvin, Geoff (2008). Talent Is Overrated: What Really Separates World-Class Performers from Everybody Else. Portfolio Hardcover.
Cooke, Edward F. (1974). *A Detailed Analysis of the Constitution*. Littlefield Adams & Co.
Corazza, P. (2000). Consistency of $V = HOD$ with the wholeness axiom. *Archives of Mathematical Logic*, vol. 39, pp. 219–226.
Cornacchio, J. V. (1972). Topological concepts in the mathematical theory of general systems. In: *Trends in General Systems Theory*, edited by G. Klir, John Wiley, New York.
Cornford, F. M. (1939). *Plato and Parmenides*. Routledge.
Covey, Stephen R. (1989). *The 7 Habits of Highly Effective People: Powerful Lessons in Personal Change*. New York:Free Press.

Dauben, J. W. (1979). *Georg Cantor: His Mathematics and Philosophy of the Infinite.* Cambridge, Massachusetts: Harvard University Press.

Dauben, J. W. (1988). *Cantor's Mathematics and Philosophy of Infinities* (in Chinese, translated by L. X. Zheng and X. L. Liu). Nanjing: Press of Jiangsu Education.

Davis, P. J. (1964). Numbers. *Scientific American*, vol. 211, (Sept. 1964), pp. 51–59.

De Sua, F. (1956). Consistency and completeness: a resume. *The American Mathematical Monthly*, vol. 63, pp. 295–305.

Dennett, D. (1984). *Elbow Room: The Varieties of Free Will Worth Wanting.* Bradford Books.

Descartes, Rene (Author) (translated by Stephen Voss in 1990) (1649). The Passions of the Soul: An English Translation of Les Passions De L'Ame. Hackett Publishing Company.

Descartes, R. (1984). *The Philosophical Writings.* Trans. John Cottingham, Robert Stoothoff, and Dugald Murdoch, Cambridge University Press.

Edwards, C. H. (1979). The Historical Development of the Calculus. Berlin: Springer-Verlag.

Egan, Kieran (1992). *Imagination in Teaching and Learning.* Chicago: University of Chicago Press.

Einstein, A. (1983). *Complete Collection of Albert Einstein.* Trans. by L. Y. Xu. Beijing: Commercial Press.

Einstein, Albert (1987). *The Collected Papers of Albert Einstein.* Princeton, NJ: Princeton University Press.

Einstein, A. (1997). *Collected Papers of Albert Einstein.* Princeton, NJ: Princeton University Press.

Engelfriet, J. and Gelsema, T. (2004). A new natural structural congruence in the pi-calculus with replication. *Acta Informatica*, vol. 40, pp. 385–430.

English, J. and Feng, G. F. (1972). *Tao De Ching.* New York: Vintage Books.

Euclid (1956). *Elements.* ed. T. L. Heath, New York: Dover.

Evans, Dylan (1996). An Introductory Dictionary Of Lacanian Psychoanalysis. Routledge.

Eves, J. (1992). *An Introduction to the History of Mathematics with Cultural Connections* (6th edition). Fort Wroth, TX: Saunders College Publishing.

Forrest, J., Wen, B. Y., and Panger, Q. (to appear). Actual-potential infinite and some related matters. Submitted for publication.

Frege, G. (1980). *The Foundations of Arithmetic: A Logico-Mathematical Inquiry into the Concept of Number.* Trans. J. L. Austin, Oxford, Blackwell.

Fultz, D., Long, R. R., Owens, G. V., Bohan, W., Kaylor, R., and Weil, J. (1959). *Studies of thermal convection in a rotating cylinder with some implications for large-scale atmospheric motion.* Meteorol. Monographs (American Meteorological Society) vol. 21, no. 4.

Furnari, G. (2010). *Calculus Without Limits.* lulu.com.

Gardner, M. (1976). Mathematical games, in which 'monster' curves force redefinition of the word 'curve'. *Scientific American*, vol. 235, pp. 124–133.

Gardner, R. J., and Wagon, S. (1989). At long last the circle has been squared. *Notices of American Mathematical Society*, vol. 36, pp. 1338–1343.

Gerassi, John (1989). *Jean-Paul Sartre: Hated Conscience of His Century. Volume 1: Protestant or Protester?* Chicago: University of Chicago Press.

Ginet, C. (1983). "In Defense of Incompatibilism. " *Philosophical Studies* 44, pp. 391–400.

Gleick, J. (1987). *Chaos: Making a New Science.* New York: Viking.

Godel, K. (1967). On formally undecidable propositions of principia mathematica and related systems I. Trans. Jean van Heijenoort, in: Jean van Heijenoort (ed.), *From Frege to Godel: A Source Book in Mathematical Logic, 1879–1931* Cambridge, Mass.: Harvard University Press.

Gonseth, F. (1936). La structure du paradoxe des catalogues. In: *Les mathematiqueset la realite: Essai sur la method axiomatique.* Paris: Félix Alcan, pp. 255–257.

Grattan-Guinness, I., and Bornet, G. (1997). *George Boole: Selected Manuscripts on Logic and its Philosophy*, Berlin: Birkhäuser Verlag

Gratzer, G. (1978). *Universal Algebra*. New York: Springer-Verlag.

Greene, J. and Cohen, J. (2004). For the law, neuroscience changes nothing and everything. *Philosophical Transactions of the Royal Society of London B*, 359, 1775–1785.

Greiner, W. (1993). *Quantum Mechanics: An Introduction* (2nd Edition). Berlin: Springer-Verlag.

Griffiths, D. J. (1998). *Introduction to Electrodynamics* (3rd ed.). Upper Saddle River, New Jersey: Prentice Hall.

Griffiths, D. J. (2004). *Introduction to Quantum Mechanics* (2nd edition). Upper Saddle River, New Jersey: Prentice Hall.

Griggs, Edward Howard (1969). *Great Leaders in Human Progress*. Freeport, N.Y: Books for Libraries Press.

Hailperin, T. (2001). Potential infinite models and ontologically neutral logic. *Journal of Philosophical Logic*, vol. 30, pp. 79–96.

Hall, A. D. and Fagen, R. E. (1956). Definitions of systems. *General Systems*, vol. 1, pp. 18–28.

Hare, R. D. (1970). *Psychopathy: Theory and Research*. New York: John Wiley.

Harris, Paul (2000). The Work of the Imagination. New York: Wiley-Blackwell.

Hausdorff, F. (1935). *Mengenlehve*. Watter ac. Hruyler.

Hausdorff, F. (1962). *Set Theory* (2nd edition). New York: Chelsea.

Hegel, G. W. F. (translated by A. V. Miller) (1807). Phenomenology of Spirit (1979 edition). Oxford University Press.

Hegel, G. W. F. (1969). *Science of Logic*. Trans. A. V. Miller, London, Allen & Unwin.

Hemphill, John K. (1949). *Situational Factors in Leadership*. Columbus: Ohio State University Bureau of Educational Research.

Hendrix, H. (2001). *Getting the Love You Want: A Guide for Couples*. New York: Owl Books.

Hergenhahn, B. R., (2005). *An Introduction to the History of Psychology*. Belmont, CA, USA: Thomson Wadsworth.

Hess, S. L. (1959). *Introduction to Theoretical Meteorology*. New York: Holt, Rinehart and Winston.

Hide, R. (1953). Some experiments on thermal convection in a rotating liquid. *Quart. J. Roy. Meteorol. Soc.*, vol. 79, pp. 161.

Hill, Napoleon (1928). *The Law of Success: In Sixteen Lessons* (reprint of 2007). BN Publishing, www.bnpublishing.com.

House, Robert J. (1996). "Path-goal theory of leadership: Lessons, legacy, and a reformulated theory". *Leadership Quarterly*, vol. 7, no. 3, pp. 323–352.

Hoyle, John R. (1995). *Leadership and Futuring: Making Visions Happen*. Thousand Oaks, CA: Corwin Press, Inc.

Hume, David (1740). *A Treatise of Human Nature* (1967 edition). Oxford: Oxford University Press.

Institute of Foreign History of Philosophy, Beijing University (1962) *The Philosophy of Ancent Greeks* (in Chinese) Beijing: Commercial Press

James, William (1984). "What is Emotion". *Mind*, vol. 9, pp. 188–205.

Kahneman, D. and Thaler, R. (2006). Utility maximization and experienced utility. *J. Econ., Perspect*, vol. 20, pp. 221–234.

Kane, Robert (1996). *The Significance of Free Will*. New York: Oxford University Press.

Kanovei, V., and Reeken, M. (2000). Extending standard models of ZFC to models of nonstandard set theory. *Studia and Logica*, vol. 64, pp. 37–59.

Kant, Immanuel (translated by Paul Guyer and Eric Matthews in 2001) (1790). *Critique of the Power of Judgment*. Cambridge University Press.

Kawabata, H. and Zeki, S. (2008). The Neural Correlates of Desire. *PLoS ONE* 3(8): e3027. doi: 10.1371.

Kirkpatrick, Shelley and Locke, Edwin (1991). Leadership: Do traits matter? *Academy of Management Executive*, vol. 5, no. 2, pp. 48–60.

Kleiner, I., and Movshovitz-Hadar, N. (1994). The role of paradoxes in the evolution of mathematics. *The American Mathematical Monthly*, vol. 101, no. 10, pp. 963–974.

Kline, M. (1972). *Mathematical Thought from Ancient to Modern Times*. Oxford: Oxford University Press.

Kline, M. (1979). *Mathematical Thoughts from Ancient to Modern Times* (vol. 1, in Chinese). Shanghai Press of Science and Technology, Shanghai.

Kline, M. (2005). *Mathematics in Western Civilization* (in Chinese, translated by Z. G. Zhang). Shanghai: Fudan University Press.

Klir, G. (1970). *An Approach to General Systems Theory*. Princeton, New Jersey: Van Nostrand.

Klir, G. (1985). *Architecture of Systems Problem Solving*. New York, NY: Plenum Press.

Klir, G. (2001). *Facets of Systems Science*. New York: Springer.

Knowles, Henry P. and Saxberg, Borje O. (1971). *Personality and Leadership Behavior*. Reading, Mass.: Addison-Wesley.

Kofman, Sarah (translated by Catherine Porter) (1998). Socrates: Fictions of a Philosopher. Cornell University Press.

Kouzes, J., and Posner, B. (2007). The Leadership Challenge. CA: Jossey Bass.

Kretzmann, N., and Stump, E. (1988). *The Cambridge Translations of Medieval Philosophical Texts. Volume 1: Logic and the Philosophy of Language*. New York: Cambridge University Press.

Kuhn, Thomas (1996). *The Structure of Scientific Revolutions*. Chicago: University Of Chicago Press.

Landau, L. D., and Lifshitz, E. M. (1971). *Classical Theory of Fields* (3rd ed.). London: Pergamon.

Lao Tzu (time unknown). *Tao Te Ching*. A new translation by Gia-fu Feng and Jane English. New York: Vintage Books, 1972.

Lao Tzu (unknown). *Tao Te Ching: The Classic Book of Integrity and the Way*. An Entirely New Translation Based on the Recently Discovered Ma-Wang-Tui Manuscript. Translated, annotated, and with an Afterword by V. H. Mair, Bantam Books, New York, 1990.

Leapfrogs (1980). *Complex Numbers*. England: E. G. M. Mann & Son.

Leibniz, G. W. (1960–1). *Philosophische Schriften*. Hildesheim.

Lenin V. (1959). *Complete Collection of V. Lenin* (in Chinese). Beijing: People's Press

Li, X. C. (2007). *Founders and Pioneers of Calculus* (3rd edition). Beijing: Press of Higher Education.

Libet, Benjamin (2003). "Can Conscious Experience affect brain Activity?", *Journal of Consciousness Studies*, vol. 10, no. 12, pp. 24–28.

Lilienfeld, D. (1978). *The Rise of Systems Theory*. New York: Wiley.

Lin, Q. (2008). *Free Calculus: A Liberation from Concepts and Proofs*. Singapore, World Scientific.

Lin, Q., and Wu, C. X. (2002). *College Mathematics for Humanity Majors*. Baoding, Hebei: Press of Hebei University.

Lin, Y. (1987). A model of general systems. *Mathematical Modeling: An International Journal*, vol. 9, pp. 95–104.

Lin, Y. (1988). Can the world be studied in the viewpoint of systems? *Mathl. Comput. Modeling*, vol. 11, pp. 738–742.

Lin, Yi (1988a). Tree-like hierarchies of systems. *International Journal of General Systems*, vol. 14, pp. 33–44.

Lin, Yi (1988b). An application of systems analysis in sociology. *Cybernetics and Systems: An International Journal*, vol. 19, pp. 267–278.

Lin, Yi (1990). Periodic linked hierarchies of systems. *Cybernetics and Systems: An International Journal*, vol. 21, pp. 59–77.

Lin, Yi (1991). Similarity between general systems. *Systems Analysis Modelling Simulation*, vol. 8, pp. 607–615.

Lin, Y. (1995). Developing a theoretical foundation for the laws of conservation. *Kybernetes: The International Journal of Systems and Cybernetics*, vol. 24, pp. 52–60.

Lin, Y. (1998). Discontiunity: a weakness of calculus and beginning of a new era. *Kybernetes: The International Journal of Systems and Cybernetics*, vol. 27, pp. 614–618.

Lin, Yi (1998b). Discontiunity: a weakness of calculus and beginning of a new era. *Kybernetes: The International Journal of Systems and Cybernetics*, vol. 27, pp. 614–618.

LinY. (guest editor) (1998). Mystery of nonlinearity and Lorenz's chaos. *Kybernetes: The International Journal of Systems and Cybernetics*, vol. 27, nos. 6/7, pp. 605–854.

Lin, Yi (guest editor) (1998a). Mystery of Nonlinearity and Lorenz's Chaos. A special double issue, *Kybernetes: The International Journal of Cybernetics, Systems and Management Science*, vol. 27, nos. 6–7, pp. 605–854.

Lin, Y. (1999). *General Systems Theory: A Mathematical Approach*. New York: Kluwer Academic Publishers.

Lin, Y. (2007). Systemic yoyo model and applications in Newton's, Kepler's Laws, etc., *Kybernetes: The International Journal of Cybernetics, Systems and Management Science*, vol. 36, no. 3–4, pp. 484–516.

Lin, Yi (2008b). *Systemic Yoyos: Some Impacts of the Second Dimension*. New York: Auerbach Publications, an imprint of Taylor and Francis.

Lin, Yi (guest editor) (2008a). Systematic Studies: The Infinity Problem in Modern Mathematics. *Kybernetes: The International Journal of Cybernetics, Systems and Management Sciences*, vol. 37, no. 3–4, pp. 385–542. chapt 4, 7, 9.

Lin, Y. (guest editor) (2010). Research Studies: Systemic Yoyos and Their Applications. *Kybernetes: The International Journal of Systems, Cybernetics, and Management Science*, vol. 39, no. 2, pp. 174–378.

Lin, Y. and Fan, T. H. (1997). The fundamental structure of general systems and its relation to knowability of the physical world. *Kybernetes: The International Journal of Systems and Cybernetics*, vol. 26, pp. 275–285.

Lin, Y., and Forrest, B. (to appear). Natural causes for differences between civilizations: a systemic point of view. Submitted for publication.

Lin, Y. and Forrest, B. (2010a). The state of a civilization. *Kybernetes: The International Journal of Cybernetics, Systems and Management Science*, vol. 39, no. 2, pp. 343–356.

Lin, Y. and Forrest, B. (2010b). Interaction between civilizations. *Kybernetes: The International Journal of Cybernetics, Systems and Management Science*, vol. 39, no. 2, pp. 367–378.

Lin, Y. and Forrest, B. (2010c). *Systemic Structure behind Human Organizations: From Civilizations to Individuals*. New York: Springer.

Lin, Y., and Forrest, B. (2011). *Systemic Structure behind Human Organizations: From Civilizations to Individuals*. New York: CRC Press, an imprint of Taylor and Francis.

Lin, Y. and Forrest, B. (to appear 1). Nature, human, and the phenomenon of self-awareness. *Kybernetes: The International Journal of Cybernetics, Systems and Management Science*, in press.

Lin, Y. and Forrest, B. (to appear 2). The mechanism behind imagination, conscience, & free will. *Kybernetes: The International Journal of Cybernetics, Systems and Management Science*, in press.

Lin, Y. and Forrest, B. (to appear 3). Structures of human character and thought. *Kybernetes: The International Journal of Cybernetics, Systems and Management Science*, in press.

Lin, Y. and Forrest, B. (to appear 4). Structures of human desire and enthusiasm. *Kybernetes: The International Journal of Cybernetics, Systems and Management Science*, in press.
Lin Y. and Forrest, D. (2008A). Economic yoyos and Becker's rotten kid theorem. *Kybernetes: The International Journal of Systems, Cybernetics and Management Science*, vol. 37, no. 2, pp. 297–314
Lin Y. and Forrest, D. (2008B). Economic yoyos and never-perfect value system. *Kybernetes: The International Journal of Systems, Cybernetics and Management Science*, vol. 37, no. 1, pp. 149–165.
Lin Y. and Forrest, D. (2008C). Economic yoyos, parasites and child labor. *Kybernetes: The International Journal of Systems, Cybernetics and Management Science*, vol. 37, no. 6, pp. 757–767
Lin, Y., Hu, Q. P., and Li, D. (1997). Some unsolved problems in general systems theory (I). *Cybernetics and Systems: An International Journal*, vol. 28, pp. 287–303.
Lin, Yi and Ma, Y-H. (1987). Some properties of linked time systems. *International Journal of General Systems*, vol. 13, pp. 125–134.
Lin, Y., and Ma, Y. (1987a). Remarks on analogy between systems. *International Journal of General Systems*, vol. 13, pp. 135–141.
Lin, Yi and Ma, Y. H. (1990). General feedback systems. *International Journal of General Systems*, vol. 18, no. 2, pp. 143–154.
Lin, Y., and Ma, Y. H. (1993). System – A unified concept. *Cybernetics and Systems: An International Journal*, vol. 24, pp. 375–406
Lin, Y., Ma, Y. and Port, R. (1990). Several epistemological problems related to the concept of systems. *Mathl. Comput. Modeling*, vol. 14, pp. 52–57.
Lin, Y., and Ou Yang, S. C. (2010). *Irregularities and Prediction of Major Disasters*. New York: CRC Press, an imprint of Taylor and Francis.
Liu, Sifeng and Lin, Yi (2006). *Grey Information: Theory and Practical Applications*. London: Springer.
Livshits, M. (2004). *Simplifying Calculus by Using Uniform Estimates*. http://www.mathfoolery.org.
Lorenz, E. N. (1993). *The Essence of Chaos*. Seattle: Washington University Press.
Maclaurin, C. (1742). *Treatise on Fluxions* (reproduced in 1801). Ann Arbor, MI: University of Michigan Library.
Macpherson, C. B. (1962). *The Political Theory of Possessive Individualism: Hobbes to Locke*. Oxford: Oxford University Press.
Mannel, T. (2005). *Effective Field Theories in Flavour Physics*. New York: Springer.
Marchi, P. (1972). The controversy between Leibniz and Bernoulli on the nature of the logarithms of negative numbers. In: *Akten das II Inter. Leibniz-Kongress* (Hanover, 1972), BndII, 1974, pp. 67–75.
Martin, B., and Shaw, G. P. (2008). *Particle Physics, in Manchester Physics Series*. New York, Wiley.
Mathematical Sciences: a unifying and dynamic resource (1985). *Notices of the American Mathematical Society*, vol. 33, pp. 463–479.
McClelland, David (1975) *Power: The Inner Experience*. Somerset, NJ: John Wiley & Sons.
McLaughlin, W. L. (1997). Thompson's lamp is dysfunctional. California Institute of Technology. pp. 1–38.
McMahon, D. (2009). *Calculus without Limits: Self-Study Guide*. Create Space.
Menger, K. (1943). What is dimension? *The American Mathematical Monthly*, vol. 50, pp. 2–7.
Mesarovic, M. and Takahara, Y. (1974). *General Systems Theory: Mathematical Foundations* New York: Academic Press.
Mesarovic, M. and Takahara, Y. (1989). *Abstract Systems Theory*. Berlin: Springer-Verlag.
Mickens, R. E. (1990). *Mathematics and Science*. Singapore: World Scientific.

Mill, John Stuart (2003). On Liberty. Millis, MA: Agora Publications.
Mohd, A. (2005). *Fundamentals of Electrical Machines.* Oxford, England: Alpha Science Intl Ltd.
Moore, A. W. (1990). *The Infinite.* London: Routledge
Nagel, E. (1935). Impossible numbers: A chapter in the history of modern logic. *Stud. in the Hist. of Ideas*, vol. 3, pp. 429–474.
Newton, I. (1704). *Opticks, or A Treatise of the Reflexions, Reflections, Inflexions and Colours of Light. Also Two Treatises of Species and magnitude of Curvilinear Figures.* London: The Royal Society.
Norman, Ron (2000). *Cultivating Imagination in Adult Education* Proceedings of the 41st Annual Adult Education Research.
OuYang, S. C. (1994). *Break-Offs of Moving Fluids and Several Problems of Weather Forecasting.* Chengdu: Press of Chengdu University of Science and Technology.
OuYang, S. C., Chen, Y. G., and Lin, Y. (2009). *Digitization of Information and Prediction.* Beijing: Meteorological Press.
OuYang, S. C., McNeil, D. H., and Lin, Y. (2002). *Entering the Era of Irregularity* (in Chinese). Beijing: Meteorological Press.
OuYang, S. C., Miao, J. H., Wu, Y., Lin, Y., Peng, T. Y., and Xiao, T. G. (2000). Second stir and incompleteness of quantitative analysis. *Kybernetes: The International Journal of Cybernetics, Systems and Management Science*, vol. 29, pp. 53–70.
Peirce, C. S. (1881). On the logic of numbers. *American Journal of Mathematics*, vol. 4, nos. 1–4, pp. 85–95.
Pfaff, Donald W. (2007). *The Neuroscience of Fair Play: Why We (Usually) Follow the Golden Rule.* New York: Dana Press, The Dana Foundation.
Pinel, P. J. (1990). *Biopsychology.* Prentice Hall.
Plato (author) (translated by Allan Bloom) (1991). The Republic Of Plato (2nd edition). Basic Books.
Plato (translated by Christopher Rowe) (2005). Phaedrus. Penguin Classics.
Polansky, Ronald (2007). Aristotle's "De Anima": A Critical Commentary. Cambridge University Press.
Qian, X. S. (1983b). *About Systems Engineering.* Changsha: Hunan Science and Technology Press.
Qiu, M. L. and Zhang, S. C. (1985). *Science of Acupuncture.* Shanghai: Shanghai Press of Science and Technology.
Quine, W. V. (1976). *The Ways of Paradox and Other Essays* (revised and enlarged edition). Cambridge, Massachusetts: Harvard University Press.
Remmert, R. (1991). *Theory of Complex Functions.* Berlin: Springer-Verlag.
Ren, Z. Q. (1996). The resolution of the difficulties in the research of earth science. In: F. H. Wu and X. J. He (eds.), *Earth Science and Development*, Beijing: Earthquake Press, pp. 298–304.
Ren, Z. Q., Lin, Y., and OuYang, S. C. (1998). Conjecture on law of conservation of informational infrastructures. *Kybernetes: The International Journal of Systems and Cybernetics*, vol. 27, pp. 543–52.
Ren, Z. Q. and Nio, T. (1994). Discussions on several problems of atmospheric vertical motion equations. *Plateau Meteorology*, vol. 13, pp. 102–105.
Richard, J. (1905). Les principes des mathématiques et le problème des ensembles. *Revue générale des sciences pure set appliquées*, vol. 16, pp. 541–543.
Robinson, A. (1964). Formalism 64, logic, methodology, and philosophy of science. In: *Proceedings of the 1964 International Congress*, edited by Y. Bar-Hillel, North-Holland and Pubco., pp. 228–246.

Russell, B. (1967). 'Letter to Frege'. In: Jean van Heijenoort (ed.), *From Frege to Godel: A Source Book in Mathematical Logic, 1879–1931*. Cambridge, Massachusetts: Harvard University Press.

Salgado, J. F. (1997). The five factor model of personality and job performance in the European community. *Journal of Applied Psychology*, vol. 82, no. 1, pp. 30–43.

Schedler, Andreas (1999). Conceptualizing accountability in Andreas Schedler, Larry Diamond, Marc F. Plattner. *The Self-Restraining State: Power and Accountability in New Democracies*. London: Lynne Rienner Publishers. pp. 13–28.

Sivin, N. (1990). Science and medicine in Chinese History. In: *Heritage of China: Contemporary Perspectives on Chinese Civilization*, edited by Paul S. Ropp, University of California Press, Berkeley, pp. 164–196.

Skolem, T. (1967). Some remarks on axiomatized set theory. Trans. Stefan Bauer-Mengelberg, in: Jean van Heijenoort (ed.), *From Frege to Godel: A Source Book in Mathematical Logic, 1879–1931*. Cambridge, Massachusetts: Harvard University Press.

Sparks, J. (2004). *Calculus Without Limits: Almost* (3rd edition). Bloomington, Indiana: Author House.

Spencer, Herbert (1841). The Study of Sociology. New York: D. A. Appleton.

Stanford Encyclopedia of Philosophy at http://plato.stanford.edu/entries/stoicism/.

Steen, L. A. (1971). New models of the real-number line. *Scientific American*, vol. 225 (August), pp. 92–99.

Stillwell, J. (2002). *Mathematics and Its History*. London: Springer.

Stoykov, L., and Pacheva, V. (2005). *Public Relationos and Business Communication*. Sofia: Ot Igla Do Konetz.

Stratton, J. A. (2007). Electromagnetic Theory. In: *IEEE Press Series on Electromagnetic Wave Theory*. New York: Wiley-IEEE Press.

Suppes, P. (1993). The transcendental character of determinism. *Midwest Studies in Philosophy*, vol. 18, pp. 242–257.

Tierra, M. (1997). *Biomagnetic and Herbal Theorapy*. Twin Lakes, WI: Lotus Press.

Tinbergen, N. (1951). *The Study of Instinct*. New York: Oxford University Press.

Tucker, Susie (1972). *Enthusiasm: A Study in Semantic Change*. London: Cambridge University Press.

Van Bendegem, J. P. (1994). Ross' paradox is an impossible super-task. *The British Journal for the Philosophy of Science*, vol. 45, no. 2, pp. 74–748.

van Heijenoort, J. (1967). *From Frege to Gödel: A Source Book in Mathematical Logic, 1979–1931*. Cambridge, Massachusetts: Harvard University Press.

Van Wormer, Katherine S., Besthorn, Fred H., and Keefe, Thomas (2007). *Human Behavior and the Social Environment: Macro Level: Groups, Communities, and Organizations*. New York: Oxford University Press.

VandenBos, Gary (2006). *APA Dictionary of Psychology*. Washington, DC: American Psychological Association.

Volkert, K. (1987). Die Geschichte der pathologischen funktionen – Ein Beitrag zur Enstehung der mathematischen methodologie. *Arch. Hist. Ex. Sc.*, vol. 37, pp. 193–232.

Volkert, K. (1989). Zur differentzierbarkeit stetiger funktionen – Ampere's Beweis und seine Folgen. *Arch. Hist. Ex. Sc.*, vol. 40, pp. 37–112.

von Bertalanffy, L. (1924). Einfuhrung in Spengler's werk. Literaturblatt Kolnische Zeitung, May.

Von Bertalanffy, L. (1972). The history and status of general systems theory. In: *Trends in General Systems Theory*, edited by G. Klir. New York: John Wiley.

von Hayek, Friedrich (1991). Freedom and coercion. In: David Miller (ed), *Liberty*, Oxford University Press, New York.

Wageman, Ruth, Nunes, Debra A., Burruss, James A., and Hackman J. Richard (2008). Senior Leadership Teams: What It Takes to Make Them Great. Cambridge, Massachusetts: Harvard Business School Press.

Wagon, S. (1985). *The Banach-Tarski Paradox*. Cambridge University Press.

Wang, L. M., and Xu, S. L. (2003). Analysis on the cohesive stress at half infinite crack tip. *Applied Mathematics and Mechanics* (English Edition), vol. 24, no. 8, pp. 917–927.

Weiss, R. (1995). Medicine's latest miracle. *Health*, Jan/Feb., 1995, pp. 71–78.

Wen, B. Y. (2005). About renovation and logic. *Journal of People's University of China*, no. 1, pp. 59–67.

Wen, B. Y. (2006). The prohibition of self-reference & substitution propositions. *Journal of Anhui University (Edition of Philosophy and Social Sciences)*, vol. 30, no. 5, pp. 13–20.

Weyl, H. (1946). Mathematics and logic. *American Mathematical Monthly*, vol. 53.

Whitrow, G. J. (2001). Laplace, Pierre-Simon, marquis de. *Encyclopaedia Britannica*, Deluxe CDROM edition.

Whyburn, G. T. (1942). What is a curve? *The American Mathematical Monthly*, vol. 49, pp. 493–497.

Wigner, E. P. (1960). The unreasonable effectiveness of mathematics in the natural sciences. *Commun. Pure Appl. Math.*, vol. 13, pp. 1–14.

Wilhalm, R., and Baynes, C. (1967). *The I Ching or Book of Changes* (3rd edition). Princeton, NJ: Princeton University Press.

Wood-Harper, A. T. and Fitzgerald, G. (1982). A taxonomy of current approaches of systems analysis. *The Computer Journal*, vol. 25, pp. 12–16.

Wu, X. M. (1990). *The Pansystems View of the World.* Beijing: People's University of China Press.

Wu, Y. and Lin, Y. (2002). *Beyond Nonstructural Quantitative Analysis: Blown-Ups, Spinning Currents and Modern Science.* River Edge, New Jersey: World Scientific.

Zhang, J. Z. (2010). *Straightforward Calculus.* Beijing: Science Press.

Zeilik, M. (2002). *Astronomy: The Evolving Universe* (9th edition). London: Cambridge University Press.

Zhu, W. J., Lin, Y., Gong, N. S., and Du, G. P. (2008). Modern system of mathematics and a pair of hidden contradictions in its foundation. *Kybernetes: The International Journal of Systems and Cybernetics*, vol. 37, no. ¾, pp. 438–445.

Zhu, W. J. and Xiao, X. A. (1988). Medium axiomatic set theoretic system MS. *Scientia Sinica*, series A, no. 2, pp. 113–123.

Zhu, W. J. and Xiao, X. A. (1991). *Introduction to Set Theory.* Nanjing: Nanjing University Press.

Zhu, W. J. and Xiao, X. A. (1996) *Briefs of Mathematical Foundation.* Nanjing: Press of Nanjing University

Zhu, W. J., and Xiao, X. A. (1996). *Introduction to Mathematics Foundation.* Nanjing: Nanjing University Press

Zhu W. J. and Xiao, Z. A. (1996a). *Essentials of Mathematical Foundation.* Nanjing: Nanjing University Press

Zhu, W. J., Xiao, X. A., Song, F. M., and Gu, H. F. (2002). An outlook on infinity (III): 'every' and 'all'. *Journal of Nanjing University of Aeronautics & Astronautics*, vol. 34, no. 3, pp. 206–210.

Zhu, Y. Z. (1985). *Albert Einstein: The Great Explorer.* Beijing: Beijing People's Press.

Subject index